English Letters *(continued)*

r sample Pearson correlation coefficient   **9.3**

r^2 squared correlation coefficient   **10.8**

r_{pb}^2 squared point biserial correlation   **21.9**

S sample standard deviation (descriptive statistics)   **5.9**

$S_{y|x}$ standard error of prediction   **10.6**

SS sum of squares   **22.8**

s sample standard deviation (inferential statistics)   **18.8**

s_D sample standard deviation of difference scores   **20.12**

$s_{\overline{X}}$ estimated standard error of the mean   **18.9**

$s_{\overline{X}_1 - \overline{X}_2}$ estimated standard error of the difference between two sample means   **19.12**

$s_{\overline{D}}$ estimated standard error of the mean difference scores   **20.12**

s^2 sample variance (inferential statistics)   **19.12**

s_p^2 pooled sample variance   **19.12**

T Wilcoxon **T** test for ranked data   **25.11**

t *t* ratio   **18.4**

U Mann-Whitney **U** test for ranked data   **25.5**

X any unspecified observation or score   **4.3**

$\overline{X}$ sample mean   **4.3**

$\overline{X}_1 - \overline{X}_2$ difference between two sample means   **19.5**

Y a score paired with **X**   **9.2**

Y' predicted score   **10.4**

z $\begin{cases} \text{standard score (descriptive statistics)} & \textbf{6.3} \\ z \text{ ratio (inferential statistics)} & \textbf{14.2} \end{cases}$

z' transformed standard score   **8.4**

STATISTICS

Sixth Edition

Robert S. Witte
San Jose State University

John S. Witte
Case Western Reserve University

Harcourt College Publishers

Fort Worth   Philadelphia   San Diego   New York   Orlando   Austin   San Antonio
Toronto   Montreal   London   Sydney   Tokyo

Publisher	Earl McPeek
Acquisitions Editor	Bradley J. Potthoff
Market Strategist	Katie Matthews
Developmental Editor	Laurie Runion
Project Manager	Barrett Lackey

ISBN 0-15-507157-2
Library of Congress Catalog Card Number: 00-105488

Address for Domestic Orders
Harcourt College Publishers, 6277 Sea Harbor Drive, Orlando, FL 32887-6777
800-782-4479

Address for International Orders
International Customer Service
Harcourt College Publishers, 6277 Sea Harbor Drive, Orlando, FL 32887-6777
407-345-3800
(fax) 407-345-4060
(e-mail) hbintl@harcourtbrace.com

Address for Editorial Correspondence
Harcourt College Publishers, 301 Commerce Street, Suite 3700, Fort Worth, TX 76102

Web Site Address
http//www.harcourtcollege.com

Harcourt Brace College Publishers will provide complimentary supplements or supplement packages to those adopters qualified under our adoption policy. Please contact your sales representative to learn how you qualify. If as an adopter or potential user you receive supplements you do not need, please return them to your sales representative or send them to:
Attn: Returns Department, Troy Warehouse, 465 South Lincoln Drive, Troy, MO 63379.

Printed in the United States of America

0 1 2 3 4 5 6 7 8 9   039   9 8 7 6 5 4 3 2

Harcourt College Publishers

To Doris

Preface

TO THE READER

People often approach statistics with great apprehension. For many, it is a required course to be taken only under the most favorable circumstances, such as during a quarter or semester when the student may carry a light course load; for others, it is as distasteful as a visit to the dentist—to be postponed as long as possible, with the vague hope that the problem might miraculously disappear. Much of this apprehension doubtless rests on the widespread fear of mathematics and mathematically related areas.

This book is written to help you overcome any fear about statistics. All unnecessary quantitative considerations have been eliminated. When not obscured by mathematical treatments better reserved for more advanced books, some of the beauty of statistics, as well as its everyday usefulness, reveals itself to virtually everybody. The chances are excellent that you too will have this experience.

You could go through life quite successfully without ever learning statistics. Having learned some statistics, however, you will be less likely to flinch and change the topic when numbers enter a discussion; you will be more skeptical of conclusions based on loose or erroneous interpretations of sets of numbers; you may even be tempted to initiate a statistical analysis of some problem within your special area of interest.

TO THE INSTRUCTOR

Students who panic at the prospect of any math beyond long division may view the introductory statistics class as cruel and unjust punishment. A half-dozen years of experimentation, first with assorted handouts and then with an extensive set of lecture notes distributed as a second text, convinced us that a text could be written for these students. Representing the culmination of this effort, the present book provides a simple overview of descriptive and inferential statistics for mathematically unsophisticated students in the behavioral sciences, social sciences, health sciences, and education.

PEDAGOGICAL FEATURES

☐ Basic concepts and procedures are explained in plain English, and a special effort has been made to clarify such perennially mystifying topics as the standard deviation, normal curve applications, variance interpretation of the correlation coefficient, hypothesis tests, degrees of freedom, and analysis of variance. For example, the standard deviation is more than a formula; it roughly reflects the average amount by which individual observations deviate from their mean.

☐ Unnecessary math, computational busy work, and subtle technical distinctions are avoided without sacrificing either accuracy or realism. Small batches of data define most computational tasks. Single examples permeate entire chapters, or even several related chapters, serving as handy frames of reference for new concepts and procedures.

☐ Each chapter begins with an outline and ends with a summary, a list of important terms, and numerous exercises.

☐ The two-color format spotlights topic headings and important formulas, keys step-by-step computational instructions to actual computations, and adds an extra dimension to illustrations.

☐ Key statements appear in bold type, and step-by-step summaries of important procedures, such as solving normal curve problems, appear in boxes.

☐ Important terms are defined in the page margins.

☐ Scattered throughout the book are computer outputs for three of the most prevalent programs, Minitab, SPSS, and SAS. As interpretative aids, key ingredients are cross-referenced with explanatory notes. These sections can be either ignored or expanded without disrupting the continuity of the text.

☐ Exercises are introduced within chapters, often section-by-section, in order to minimize the cumulative confusion reported by many students for some chapters (and some students for most chapters). Review exercises terminate each chapter.

☐ Exercises have been selected to appeal to student interests: for example, how standard scores might explain the disappearance of .400 hitters in major league baseball (Exercise 8.7, page 129); probability calculations, based on design flaws, that recreate the chillingly high likelihood of the *Challenger* shuttle catastrophe (Exercise 12.8, page 207); and a *t*-test analysis of global temperatures to evaluate a possible greenhouse effect (Exercise 18.8, page 304).

☐ Appendices supply answers to selected exercises, a practical math review complete with self-diagnostic tests, a glossary of important terms, and a list of formulas.

INSTRUCTIONAL AIDS

Both a student workbook and an instructor's manual accompany the text. Self-paced and self-correcting, the workbook contains problems, discussions, exercises, and tests to be completed by the student. The instructor's manual supplies exercise answers omitted in the text, as well as at least twenty-five multiple-choice items per chapter; and a chapter-by-chapter commentary that reflects the authors' teaching experiences with this material. The multiple-choice items also have been written in ExaMaster, a computerized test bank, and are available on a computer disk upon request.

CHANGES IN THIS EDITION

☐ We maintain a website for this book (http://darwin.cwru.edu/~witte/ statistics) that links to statistically relevant Internet sites, including many student-friendly, interactive (Java Applet) demonstrations, such as the effect of outliers on correlation coefficients and the effect of population shape and sample size on sampling distributions. References to these sites, designated as an "Internet Demonstration" or an "Internet Site", are scattered throughout the book. The Website for the book also supplies explanatory comments to ease the transition to these sites. Students with access to the Internet are encouraged to explore these sites; they are enlightening and entertaining. They can, however, be ignored – whether by choice or necessity—without destroying the continuity of the book.

☐ Seven of the more technically challenging chapters (5, 9, 18, 19, 20, 22, and 23) have been reorganized by relegating most formulas and computations to a DETAILS section at the end of each chapter. The objective is to help first-time readers focus on important statistical topics without being overwhelmed with technical details—as otherwise tends to happen to some of our students. Once an overview has been attained, technical details can be assimilated, as required, at a more leisurely pace. Coincidentally, this new arrangement also should appeal to instructors who wish to de-emphasize or ignore technical details in favor of computer-driven analyses.

☐ The discussion of outliers is extended to paired observations. See Section 9.9.

☐ Exercises, examples, and computer outputs have been updated.

☐ At the urging of a number of instructors, answers are given in the text only for selected exercises, including all exercises embedded within chapters. Selected exercises are marked with asterisks and cross-referenced with answers in Appendix C. Answers for the remaining exercises, numbering slightly less than half of the total exercises, now appear only in the instructor's manual.

☐ A current list of all detected errors now can be accessed by clicking on "Textbook Corrections" in the Website for this book (http://darwin.cwru.edu/~witte/statistics).

USING THE BOOK

The book contains more material than is covered in most one-quarter or one-semester courses. Various chapters can be omitted without interrupting the main development. Typically, during a one semester course we cover the entire book, except for Chapter 22 (Analysis of Variance [One Way]); Chapter 23 (Analysis of Variance [Two Way]); and Chapter 25 (Tests for Ranked Data). An instructor who wishes to emphasize inferential statistics could skim some of the earlier chapters, particularly Chapter 2 (Describing Data with Tables), Chapter 3 (Describing Data with Graphs), and Chapter 8 (More about z Scores), while an instructor who desires a more applied emphasis could omit Chapter 11 (Populations and Samples), Chapter 12 (Probability), and Chapter 16 (Controlling Type I and Type II Errors). Depending on your perspective, the self-contained treatment of levels of measurement in Appendix B can be ignored or assigned at any time during the term.

ACKNOWLEDGMENTS

The authors wish to acknowledge their immediate family: Doris, Steve, Faith, Mike, Sharon, Andrea, Phil, Katie, Keegan, Brittany, Kristen, Scott, Joe, John, Jack, Carson, Sam, and Margaret. The first author also wishes to acknowledge his brothers and sisters: Lila, Henry, the late J. Stuart, A. Gerhart, and Etz; deceased parents: Henry and Emma; and all friends and relatives, past and present, including Arthur, Betty, Bob, Cal, David, Dick, George, Grace, Harold, Helen, Joyce, Kayo, Kit, Mary, Paul, Ralph, Ruth, Shirley, and Suzanne.

Numerous helpful comments were made by those who reviewed the current edition of this book: Ann Barich, Lewis University; Terence Bazzett, State University of New York-Geneseo; Georgjeanna Wilson-Doenges, University of Wisconsin-Green Bay; Margaret C. Dust, Chicago State University; Chwan-Shyang Jih, Lewis University; K. C. Kirasic, University of South Carolina-Columbia; Ron Mulson, Hudson Valley Community College; John C. Nardo, North Georgia College and State University; Robert A. Reeves, Augusta State University; and Stuart A. Vyse, Connecticut College. Thanks also to the reviewers of the previous edition: Bruce G. Rogers, University of Northern Iowa, Kathy Allen Sisk, North Georgia College; and Anna M. Smith, Troy State University.

Excellent editorial support was supplied by Earl McPeek, Publisher; Carol Wada, former Executive Editor; Lisa Hensley, former Associate Acquisitions Editor; Laurie Runion, Developmental Editor; Barrett Lackey, Project Manager; and by Donna King of Progressive Publishing Alternatives.

We are grateful to the literary executor of the late Sir Ronald A. Fisher, F.R.S., to Dr. Frank Yates, F. R. S., and to the Longman Group, Ltd., London, for permission to reprint Table IV from their book Statistical Tables for Biological, Agricultural and Medical Research (6th ed., 1974). Thanks also to Minitab Inc. and SPSS Inc. for supplying us with the most recent versions of Minitab and SPSS.

Contents

Preface   *v*

1   INTRODUCTION   1

   1.1   WHY STUDY STATISTICS?   2

   WHAT IS STATISTICS?   3
   1.2   DESCRIPTIVE STATISTICS   3
   1.3   INFERENTIAL STATISTICS   3
   1.4   IMPORTANCE OF PERSPECTIVE   3

   TWO TYPES OF DATA   4
   1.5   QUANTITATIVE DATA   5
   1.6   QUALITATIVE DATA   5
   1.7   QUANTITATIVE OR QUALITATIVE?   6

   TWO TYPES OF VARIABLES   7
   1.8   GENERAL DEFINITION OF VARIABLES   7
   1.9   INDEPENDENT AND DEPENDENT VARIABLES   7
   1.10   HOW TO USE THIS BOOK   10
   Summary   11
   Important Terms   11
   Review Exercises   11

PART 1 Descriptive Statistics   15
   Organizing and Summarizing Data   15

2   DESCRIBING DATA WITH TABLES   17

   FREQUENCY DISTRIBUTIONS FOR QUANTITATIVE DATA   18
   2.1   UNGROUPED DATA   18
   2.2   GROUPED DATA   19
   2.3   GUIDELINES   19
   2.4   HOW MANY CLASSES?   21
   2.5   GAPS BETWEEN CLASSES   21
   2.6   DOING IT YOURSELF   22
   2.7   OUTLIERS   26

OTHER TYPES OF FREQUENCY DISTRIBUTIONS   27

2.8   RELATIVE FREQUENCY DISTRIBUTIONS   27
2.9   PERCENTS OR PROPORTIONS?   27
2.10   CUMULATIVE FREQUENCY DISTRIBUTIONS   28
2.11   PERCENTILE RANKS   30
2.12   FREQUENCY DISTRIBUTIONS FOR QUALITATIVE DATA   31
2.13   INTERPRETING DISTRIBUTIONS CONSTRUCTED BY
        OTHERS   33
Summary   33
Important Terms   34
Review Exercises   34

3   DESCRIBING DATA WITH GRAPHS   41

GRAPHS FOR QUANTITATIVE DATA   42

3.1   HISTOGRAMS   42
3.2   FREQUENCY POLYGONS   43
3.3   STEM AND LEAF DISPLAYS   46
3.4   TYPICAL SHAPES   49

A GRAPH FOR QUALITATIVE DATA   51

3.5   BAR GRAPHS   51

MISLEADING GRAPHS   52

3.6   SOME TRICKS   52
3.7   DOING IT YOURSELF   53
3.8   COMPUTER OUTPUT   55
Summary   57
Important Terms   58
Review Exercises   58

4   DESCRIBING DATA WITH AVERAGES   61

AVERAGES FOR QUANTITATIVE DATA   62

4.1   MODE   62
4.2   MEDIAN   63
4.3   MEAN   65
4.4   WHICH AVERAGE?   68
4.5   SPECIAL STATUS OF THE MEAN   70
4.6   AVERAGES FOR QUALITATIVE DATA   70
4.7   USING THE WORD *AVERAGE*   71
Summary   72

Important Terms   72
Review Exercises   72

5   DESCRIBING VARIABILITY   75

QUANTITATIVE MEASURES OF VARIABILITY   76
5.1   INTUITIVE APPROACH   75
5.2   RANGE   77
5.3   VARIANCE   77
5.4   WEAKNESS OF VARIANCE   78
5.5   STANDARD DEVIATION: AN INTERPRETATION   79
5.6   STANDARD DEVIATION: SOME GENERALIZATIONS   79
5.7   STANDARD DEVIATION: A MEASURE OF DISTANCE   81
5.8   MEASURES OF VARIABILITY FOR QUALITATIVE DATA   83

DETAILS   83
5.9   DEFINITION FORMULA FOR STANDARD DEVIATION   83
5.10   COMPUTATION FORMULA FOR STANDARD DEVIATION   85
5.11   INTERQUARTILE RANGE (IQR)   87
Summary   89
Important Terms   89
Review Exercises   89

6   NORMAL DISTRIBUTIONS (I): BASICS   93
6.1   THE THEORETICAL NORMAL CURVE   95
6.2   PROPERTIES OF THE NORMAL CURVE   96
6.3   *z* SCORES   97
6.4   STANDARD NORMAL CURVE   98
6.5   STANDARD NORMAL TABLE   99
6.6   EXAMPLE: FBI APPLICANTS   101
6.7   KEY FACTS TO REMEMBER   102
Summary   102
Important Terms   103
Review Exercises   103

7   NORMAL DISTRIBUTIONS (II): APPLICATIONS   105

FINDING PROPORTIONS   106
7.1   EXAMPLE: FINDING PROPORTION *BELOW* A SCORE (TO LEFT OF MEAN)   106
7.2   EXAMPLE: FINDING PROPORTION *BELOW* A SCORE (TO RIGHT OF MEAN)   106

7.3   EXAMPLE: FINDING PROPORTION *BETWEEN* SCORES   108
7.4   EXAMPLE: FINDING PROPORTIONS *BEYOND* PAIRS OF
        SCORES   110

FINDING SCORES   112
7.5   EXAMPLE: FINDING *A* SCORE (TO RIGHT OF MEAN)   113
7.6   EXAMPLE: FINDING *PAIRS* OF SCORES (ON BOTH SIDES OF
        MEAN)   115
Summary   119
Review Exercises   119

8   MORE ABOUT *z* SCORES   123
8.1   *z* SCORES FOR NON-NORMAL DISTRIBUTIONS   124
8.2   STANDARD SCORES   125
8.3   TRANSFORMED STANDARD SCORES   125
8.4   CONVERTING TO TRANSFORMED STANDARD SCORES   126
8.5   PERCENTILE RANKS AGAIN   128
Summary   129
Important Terms   129
Review Exercises   129

9   DESCRIBING RELATIONSHIPS: CORRELATION   133
9.1   AN INTUITIVE APPROACH   134
9.2   SCATTERPLOTS   136
9.3   A CORRELATION COEFFICIENT FOR QUANTITATIVE DATA: *r*   140
9.4   INTERPRETATION OF *r*   141
9.5   INTERPRETATION OF r^2 (SEE SECTION 10.8)   143
9.6   CORRELATION NOT NECESSARILY CAUSE-EFFECT   143

DETAILS   144
9.7   *z* SCORE FORMULA FOR *r*   144
9.8   COMPUTATION FORMULA FOR *r*   148
9.9   OUTLIERS AGAIN   149
9.10   OTHER TYPES OF CORRELATION COEFFICIENTS   151
9.11   COMPUTER OUTPUT   151
Summary   155
Important Terms   156
Review Exercises   156

10   PREDICTION   159

   10.1   TWO ROUGH PREDICTIONS   160
   10.2   A PREDICTION LINE   161
   10.3   LEAST SQUARES PREDICTION LINE   163
   10.4   LEAST SQUARES EQUATION   164
   10.5   GRAPHS OR EQUATIONS?   167
   10.6   STANDARD ERROR OF PREDICTION, $S_{y|x}$ 168
   10.7   ASSUMPTIONS   170
   10.8   INTERPRETATION OF r^2 170
   10.9 r REVISITED   176
   10.10 MORE COMPLEX PREDICTION EQUATIONS   178
   Summary   178
   Important Terms   179
   Review Exercises   179

PART 2 Inferential Statistics   183

Generalizing beyond Data   183

11   POPULATIONS AND SAMPLES   185

   11.1   WHY SAMPLES?   186
   11.2   POPULATIONS   186
   11.3   SAMPLES   187
   11.4   RANDOM SAMPLES   188
   11.5   TABLES OF RANDOM NUMBERS   190
   11.6   SOME COMPLICATIONS   191
   11.7   RANDOM ASSIGNMENT OF SUBJECTS   192
   11.8   AN OVERVIEW: SURVEYS OR EXPERIMENTS?   194
   Summary   194
   Important Terms   195
   Review Exercises   195

12   PROBABILITY   197

   12.1   DEFINITION   198
   12.2   ADDITION RULE   199
   12.3   MULTIPLICATION RULE   200
   12.4   PROBABILITY AND STATISTICS   203
   Summary   205
   Important Terms   206
   Review Exercises   206

13 SAMPLING DISTRIBUTION OF THE MEAN   211
13.1   AN EXAMPLE   212
13.2   CREATING A SAMPLING DISTRIBUTION FROM SCRATCH   213
13.3   SOME IMPORTANT SYMBOLS   215
13.4   MEAN OF ALL SAMPLE MEANS ($\mu_{\overline{X}}$)   217
13.5   STANDARD ERROR OF THE MEAN ($\sigma_{\overline{X}}$)   218
13.6   SHAPE OF THE SAMPLING DISTRIBUTION   220
13.7   WHY THE CENTRAL LIMIT THEOREM WORKS   222
13.8   OTHER SAMPLING DISTRIBUTIONS   223
Summary   223
Important Terms   224
Review Exercises   224

14 INTRODUCTION TO HYPOTHESIS TESTING: THE *z* TEST   227
14.1   TESTING A HYPOTHESIS ABOUT SAT SCORES   228
14.2   *z* TEST FOR A POPULATION MEAN   230
14.3   STEP-BY-STEP PROCEDURE   233
14.4   STATEMENT OF THE RESEARCH PROBLEM   234
14.5   NULL HYPOTHESIS (H_0)   234
14.6   ALTERNATIVE HYPOTHESIS (H_1)   235
14.7   DECISION RULE   236
14.8   CALCULATIONS   237
14.9   DECISION   237
14.10 INTERPRETATION   238
Summary   239
Important Terms   240
Review Exercises   240

15 MORE ABOUT HYPOTHESIS TESTING   243
15.1   HYPOTHESIS TESTING: AN OVERVIEW   244
15.2   STRONG OR WEAK DECISIONS   246
15.3   WHY THE RESEARCH HYPOTHESIS ISN'T TESTED DIRECTLY   246
15.4   ONE-TAILED AND TWO-TAILED TESTS   247
15.5   CHOOSING A LEVEL OF SIGNIFICANCE (α)   252
Summary   253
Important Terms   254
Review Exercises   254

16   CONTROLLING TYPE I AND TYPE II ERRORS   257
     16.1   TESTING A HYPOTHESIS ABOUT VITAMIN C   258
     16.2   FOUR POSSIBLE OUTCOMES   259
     16.3   IF H_0 REALLY IS TRUE   261
     16.4   IF H_0 REALLY IS FALSE BECAUSE OF A *LARGE* EFFECT   263
     16.5   IF H_0 REALLY IS FALSE BECAUSE OF A *SMALL* EFFECT   265
     16.6   INFLUENCE OF SAMPLE SIZE   266
     16.7   SELECTION OF SAMPLE SIZE   268
     16.8   POWER CURVES   269
     Summary   *270*
     Important Terms   *271*
     Review Exercises   *271*

17   ESTIMATION   275
     17.1   ESTIMATING μ FOR SAT SCORES   276
     17.2   POINT ESTIMATE FOR μ 276
     17.3   CONFIDENCE INTERVAL FOR μ 276
     17.4   WHY CONFIDENCE INTERVALS WORK   277
     17.5   CONFIDENCE INTERVAL FOR μ BASED ON z 280
     17.6   INTERPRETATION OF A CONFIDENCE INTERVAL   281
     17.7   LEVEL OF CONFIDENCE   281
     17.8   EFFECT OF SAMPLE SIZE   282
     17.9   HYPOTHESIS TESTS OR CONFIDENCE INTERVALS?   283
     17.10   CONFIDENCE INTERVAL FOR POPULATION PERCENT   284
     17.11   OTHER TYPES OF CONFIDENCE INTERVALS   286
     Summary   *286*
     Important Terms   *286*
     Review Exercises   *286*

18   *t* TEST FOR ONE SAMPLE   289
     18.1   GAS MILEAGE INVESTIGATION   290
     18.2   *t* SAMPLING DISTRIBUTION   291
     18.3   *t* TABLES   292
     18.4   *t* RATIO   293
     18.5   HYPOTHESIS TESTS: A COMMON THEME   295
     18.6   CONFIDENCE INTERVALS FOR μ BASED ON t 295
     18.7   ASSUMPTIONS   297

DETAILS   297

18.8   ESTIMATING THE POPULATION STANDARD DEVIATION   297

18.9   ESTIMATING THE STANDARD ERROR $(s_{\bar{X}})$ 298

18.10   CALCULATION OF t TEST FOR ONE SAMPLE (GAS MILEAGE INVESTIGATION)   299

18.11   DEGREES OF FREEDOM   299

Summary   *302*

Important Terms   *303*

Review Exercises   *303*

19   *t* **TEST FOR TWO INDEPENDENT SAMPLES**   **307**

19.1   BLOOD-DOPING EXPERIMENT   308

19.2   TWO INDEPENDENT SAMPLES   308

19.3   TWO HYPOTHETICAL POPULATIONS   308

19.4   STATISTICAL HYPOTHESES   309

19.5   SAMPLING DISTRIBUTION OF $\bar{X}_1 - \bar{X}_2$ 311

19.6   MEAN OF THE SAMPLING DISTRIBUTION OF $\bar{X}_1 - \bar{X}_2$ 311

19.7   STANDARD ERROR OF THE SAMPLING DISTRIBUTION OF $\bar{X}_1 - \bar{X}_2$ 311

19.8 z TEST   312

19.9 t RATIO   312

19.10   CONFIDENCE INTERVAL FOR $\mu_1 - \mu_2$ 314

19.11   ASSUMPTIONS   316

DETAILS   317

19.12   ESTIMATING THE STANDARD ERROR $(s_{\bar{X}_1 - \bar{X}_2})$ 317

19.13   CALCULATIONS FOR t TEST FOR TWO INDEPENDENT SAMPLES (BLOOD-DOPING EXPERIMENT)   318

Summary   *321*

Important Terms   *322*

Review Exercises   *322*

20   *t* **TEST FOR TWO MATCHED SAMPLES**   **325**

TWO POPULATION MEANS   **326**

20.1   MATCHING PAIRS OF ATHLETES IN THE BLOOD-DOPING EXPERIMENT   326

20.2   TWO MATCHED SAMPLES   326

20.3   DIFFERENCE SCORES (D) 326

20.4   STATISTICAL HYPOTHESES   327

20.5   SAMPLING DISTRIBUTION OF $\bar{D}$ 328

20.6 t RATIO   328

20.7   CONFIDENCE INTERVAL FOR μ_D 329
20.8   TO MATCH OR NOT TO MATCH?   331
20.9   USING THE SAME SUBJECTS IN BOTH GROUPS (REPEATED MEASURES)   332
20.10 ASSUMPTIONS   333
20.11 THREE t TESTS FOR POPULATION MEANS: AN OVERVIEW   333

DETAILS   336
20.12 ESTIMATING THE STANDARD ERROR $(s_{\bar{D}})$ 336
20.13 CALCULATIONS FOR t TEST FOR TWO MATCHED SAMPLES (BLOOD-DOPING EXPERIMENT)   336

POPULATION CORRELATION COEFFICIENT   338
20.14 t TEST FOR THE GREETING CARD EXCHANGE   338
20.15 ASSUMPTIONS   340
20.16 A LIMITATION   341
Summary   341
Important Terms   342
Review Exercises   342

21   BEYOND HYPOTHESIS TESTS: p-VALUES AND EFFECT SIZE   347

p-VALUES   348
21.1   DEFINITION   348
21.2   FINDING APPROXIMATE p-VALUES   349
21.3   READING p-VALUES REPORTED BY OTHERS   350
21.4   MERITS OF LESS STRUCTURED (p-VALUE) APPROACH   350
21.5   LEVEL OF SIGNIFICANCE OR p-VALUE?   351
21.6   A NOTE ON USAGE   352
21.7   COMPUTER OUTPUT   352

EFFECT SIZE   354
21.8   STATISTICALLY SIGNIFICANT RESULTS   354
21.9   SQUARED POINT BISERIAL CORRELATION, r_{pb}^2 355
21.10 SMALL, MEDIUM, OR LARGE EFFECT?   356
21.11 A RECOMMENDATION   357
Summary   358
Important Terms   358
Review Exercises   358

22 ANALYSIS OF VARIANCE (ONE WAY) 361

22.1   TESTING A HYPOTHESIS ABOUT RESPONSIBILITY IN CROWDS 362

22.2   TWO SOURCES OF VARIABILITY 364

22.3 F RATIO 366

22.4 F TEST 366

22.5   ASSUMPTIONS 369

22.6   TWO CAUTIONS 369

DETAILS 369

22.7   VARIANCE ESTIMATES 369

22.8   SUM OF SQUARES (*SS*) 371

22.9   DEGREES OF FREEDOM (*df*) 372

22.10 MEAN SQUARES (*MS*) AND THE F RATIO 374

22.11 F TABLES 376

22.12 NOTES ON USAGE 377

22.13 F TEST IS NONDIRECTIONAL 378

BEYOND THE F TEST 379

22.14 SMALL, MEDIUM, OR LARGE EFFECT? 379

22.15 MULTIPLE COMPARISONS 381

22.16 SCHEFFÉ'S TEST 381

22.17 OTHER MULTIPLE COMPARISON TESTS 383

22.18 COMPUTER OUTPUT 384

Summary   386

Important Terms   387

Review Exercises   387

23 ANALYSIS OF VARIANCE (TWO WAY) 391

23.1   TESTING HYPOTHESES ABOUT REACTIONS OF MALES AND FEMALES IN CROWDS 392

23.2   PRELIMINARY INTERPRETATIONS 392

23.3   THREE F RATIOS 395

23.4   INTERACTION 396

23.5   DESCRIBING INTERACTIONS 400

23.6   ASSUMPTIONS 400

23.7   IMPORTANCE OF EQUAL SAMPLE SIZES 401

23.8   OTHER TYPES OF ANOVA 401

DETAILS 401
23.9   VARIANCE ESTIMATES   401
23.10 SUM OF SQUARES (*SS*)   401
23.11 DEGREES OF FREEDOM (*df*)   404
23.12 MEAN SQUARES (*MS*) AND *F* RATIOS   405
23.13 *F* TABLES   406
23.14 SMALL, MEDIUM, OR LARGE EFFECT?   406
23.15 MULTIPLE COMPARISONS   407
23.16 COMPUTER OUTPUT   407

Summary   409
Important Terms   410
Review Exercises   410

24   CHI-SQUARE (χ^2) TEST FOR QUALITATIVE DATA   415

ONE-WAY χ^2 TEST   416
24.1   SURVEY OF BLOOD TYPES   416
24.2   STATISTICAL HYPOTHESES   416
24.3   OBSERVED AND EXPECTED FREQUENCIES   418
24.4   CALCULATION OF χ^2 419
24.5 χ^2 TABLES AND DEGREES OF FREEDOM   421
24.6 χ^2 TEST   421
24.7 χ^2 TEST IS NONDIRECTIONAL   423

TWO-WAY χ^2 TEST   424
24.8   LOST-LETTER STUDY   424
24.9   STATISTICAL HYPOTHESES   424
24.10 OBSERVED AND EXPECTED FREQUENCIES   425
24.11 χ^2 TABLES AND DEGREES OF FREEDOM   427
24.12 χ^2 TEST   430
24.13 SOME PRECAUTIONS   431
24.14 CHECKING IMPORTANCE   432
24.15 COMPUTER OUTPUT   435

Summary   435
Important Terms   436
Review Exercises   436

25   TESTS FOR RANKED DATA   441
25.1   USE ONLY WHEN APPROPRIATE   442
25.2   A NOTE ON TERMINOLOGY   442

MANN-WHITNEY *U* TEST (TWO INDEPENDENT SAMPLES) 443

25.3   WHY NOT A *t* TEST?   443
25.4   STATISTICAL HYPOTHESES FOR *U*   443
25.5   CALCULATION OF *U*   444
25.6   *U* TABLES   446
25.7   DECISION RULE   447
25.8   DIRECTIONAL TESTS   449

WILCOXON *T* TEST (TWO MATCHED SAMPLES)   449

25.9   WHY NOT A *t* TEST?   449
25.10   STATISTICAL HYPOTHESES FOR *T*   451
25.11   CALCULATION OF *T*   451
25.12   *T* TABLES   452
25.13   DECISION RULE   453

KRUSKAL-WALLIS *H* TEST (THREE OR MORE INDEPENDENT SAMPLES)   454

25.14   WHY NOT AN *F* TEST?   454
25.15   STATISTICAL HYPOTHESES FOR *H*   454
25.16   CALCULATION OF *H*   456
25.17 χ^2 TABLES   456
25.18   DECISION RULE   457
25.19   *H* TEST IS NONDIRECTIONAL   458
25.20   GENERAL COMMENT: TIES   458
Summary   *458*
Important Terms   *459*
Review Exercises   *459*

26   POSTSCRIPT: WHICH TEST?   463

26.1   DESCRIPTIVE OR INFERENTIAL STATISTICS?   464
26.2   HYPOTHESIS TESTS OR CONFIDENCE INTERVALS?   464
26.3   QUANTITATIVE OR QUALITATIVE DATA?   465
26.4   DISTINGUISHING BETWEEN THE TWO TYPES OF DATA   466
26.5   ONE, TWO, OR MORE GROUPS?   467
26.6   CONCLUDING COMMENTS   468
Review Exercises   *468*

APPENDICES   473

A   MATH REVIEW   473
B   LEVELS OF MEASUREMENT   483

C   ANSWERS TO SELECTED EXERCISES   491
D   TABLES   531
E   GLOSSARY   543
F   FORMULAS   553
Index   559

CHAPTER 1

Introduction

1.1   WHY STUDY STATISTICS?

WHAT IS STATISTICS?

1.2   DESCRIPTIVE STATISTICS
1.3   INFERENTIAL STATISTICS
1.4   IMPORTANCE OF PERSPECTIVE

TWO TYPES OF DATA

1.5   QUANTITATIVE DATA
1.6   QUALITATIVE DATA
1.7   QUANTITATIVE OR QUALITATIVE?

TWO TYPES OF VARIABLES

1.8   GENERAL DEFINATION OF VARIABLES
1.9   INDEPENDENT AND DEPENDENT VARIABLES
1.10   HOW TO USE THIS BOOK

Summary

Important Terms

Review Exercises

1.1 WHY STUDY STATISTICS?

There's a good chance that you're taking statistics because it's required, and your feelings about it may be similar to those of the late William Saroyan in regard to death: "Everybody has got to die, but I have always believed an exception would be made in my case."

Let's explore some of the reasons why you should study statistics. If you read news publications or watch TV, you'll encounter many statements with statistical overtones. For instance, recent issues of a daily newspaper carried these items:

- The annual earnings of college graduates exceed, *on the average*, those of high school graduates by $14,000.
- On the basis of existing research, *there is no evidence of a relationship* between the degree of "bonding" (of parents and newborns) and subsequent child development.
- Heavy users of tobacco suffer *significantly more* respiratory ailments than do nonusers.

Having learned some statistics, you will not stumble over the italicized phrases. Nor, as you continue reading, will you hesitate to probe for clarification by asking, "Which average shows a higher lifetime earning?" or "What constitutes a lack of evidence?" or "How many more is significantly more respiratory ailments?"

A statistical background is indispensable in order to understand research reports within your special area of interest. Statistical references punctuate the results sections of most research reports. Often expressed parenthetically and with telegraphic brevity, these references provide statistical support for conclusions in the narrative text:

- Highly anxious students are perceived as less attractive than nonanxious students [$t(48) = 3.21, p < .01$].
- Subjects who engage in daily exercise suffer fewer colds than subjects who don't exercise [$p < .05$].
- Attitudes toward extramarital sex are dependent on socioeconomic status [$\chi^2 (4, n = 185) = 11.49, p < .05$].

Having learned some statistics, you'll be able to decipher the meaning of an assortment of symbols, including t, p, and χ^2, and consequently read these reports more intelligently.

Sometime in the future—possibly sooner than you think—you might want to plan a statistical analysis for your own research project. Having learned some statistics, you'll be able to plan the statistical analysis for modest projects involving straightforward research questions. If your project requires more advanced statistical analysis, you'll know enough to consult someone who is more knowledgeable in statistics. Once you begin to understand basic statistical concepts, you'll discover that, with some guidance, your own efforts often will enable you to use and interpret the more advanced statistical analysis demanded by your research.

WHAT IS STATISTICS?

1.2 DESCRIPTIVE STATISTICS

Statistics has two main subdivisions: descriptive statistics and inferential statistics. The historically older area, **descriptive statistics**, supplies several tools, such as tables, graphs, and averages, for *organizing and summarizing information about a collection of actual observations*. A *table* listing the numbers of different types of crimes reported in your city during the past year; a *graph* showing the daily performance of the Dow-Jones Industrial Average during the past week, and your college grade point *average* are all examples of statistical tools that organize and summarize information about a collection of actual observations. Think of these everyday items as your preliminary exposure to descriptive statistics.

Descriptive statistics
The area of statistics concerned with organizing and summarizing information about a collection of actual observations.

1.3 INFERENTIAL STATISTICS

The other area, **inferential statistics**, is primarily a product of the twentieth century and supplies several tools for *generalizing beyond actual observations*. Tools from inferential statistics permit us to use a relatively small batch of actual observations to check a manufacturer's claim about the performance of an entire population of cars, a researcher's hypothesis that meditators experience fewer headaches than do nonmeditators, and a public official's pronouncement that a majority of citizens oppose gun control. Statistical references in research reports usually employ tools from inferential statistics. Most likely, you will have little or no prior acquaintance with inferential statistics, but with the proper mix of exposure and effort, you will eventually be able to claim the area as your own.

Inferential statistics
The area of statistics concerned with generalizing beyond actual observations.

1.4 IMPORTANCE OF PERSPECTIVE

Strictly speaking, these two areas of statistics overlap. Indeed, depending on your perspective, a given set of observations can exemplify either descriptive or inferential statistics. For instance, the weights reported by fifty-three male statistics students in **Table 1.1** can be viewed either from the perspective of descriptive statistics, because you are concerned about exceeding the load-bearing capacity of an excursion boat (chartered by the fifty-three students to celebrate successfully completing their statistics class!), or from the perspective of inferential statistics, because you wish to draw some conclusions about the weights of *all* male statistics students or *all* male college students. Notwithstanding these complications, the distinction between these two areas will serve as a convenient organizational device for this book, and you will encounter the most essential tools and concepts of descriptive statistics (Part 1), beginning with Chapter 2, and those of inferential statistics (Part 2), beginning with Chapter 11.

Table 1.1 QUANTITATIVE DATA: WEIGHTS (IN POUNDS) OF MALE STATISTICS STUDENTS							
160	168	133	170	150	165	158	165
193	169	245	160	152	190	179	157
226	160	170	180	150	156	190	156
157	163	152	158	225	135	165	135
180	172	160	170	145	185	152	
205	151	220	166	152	159	156	
165	157	190	206	172	175	154	

Exercises will appear at the end of some sections, as well as at the end of each chapter. You are urged to complete section exercises before reading on. Appendix C supplies answers to all exercises marked with asterisks, including all exercises embedded within chapters and selected review exercises at the end of chapters.

Exercise 1.1 Indicate whether each of the following statements typifies *descriptive statistics* (because it describes sets of actual observations) or *inferential statistics* (because it generalizes beyond sets of actual observations). Try to adopt what, in your opinion, represents the most reasonable perspective before deciding whether a given statement typifies descriptive or inferential statistics.

(a) On the average, students in my statistics class are twenty years old.

(b) The population of the world now exceeds six billion (that is, 6,000,000,000 or a million multiplied by 6 thousand).

(c) Four years has been the most frequent term of office served by American presidents.

(d) A recent poll indicates that 74 percent of all college students favor capital punishment.

Answers on page 492

TWO TYPES OF DATA

Data
A collection of observations from a survey or an experiment.

A statistical analysis is performed on data, that is, on a collection of actual observations from a survey or an experiment.

The precise form of a statistical analysis often depends on whether the *data* are numbers or words.

1.5 QUANTITATIVE DATA

Quantitative data

A set of observations where any single observation is a number that represents an amount or a count.

When, among a set of observations, any single observation is a number that represents an amount or a count, the data are **quantitative**. The weights reported by fifty-three male students in Table 1.1 are quantitative data, because any single observation represents an amount of weight. Other examples of quantitative data include sets of observations based on IQ scores (where, for instance, an IQ score of 123 represents an amount of intellectual aptitude), family sizes (where a value of 4 represents a count of people in a particular family), and total numbers of eye contacts during verbal disputes between romantically involved couples (where a value of 13 represents a count of eye contacts between a particular disputing couple during some standard observation period).*

1.6 QUALITATIVE DATA

Qualitative data

A set of observations where any single observation is a word or code that represents a class or category.

When, among a set of observations, any single observation is a word or code that represents a class or category, the data are **qualitative**. Listed in Table 1.2 are the anonymous replies, coded as Y for Yes and N for No, of eighty-three statistics students to the question "Have you ever smoked marijuana?" The resulting data are qualitative since each individual observation is a code that represents a particular class of replies. Other examples of qualitative data include

Table 1.2
QUALITATIVE DATA:
"HAVE YOU EVER SMOKED MARIJUANA?" YES (Y) OR NO (N) REPLIES OF STATISTICS STUDENTS

Y	Y	Y	N	N	Y	Y	Y
Y	Y	Y	N	N	Y	Y	Y
N	Y	N	Y	Y	Y	Y	Y
Y	Y	N	Y	N	Y	N	Y
Y	N	Y	N	N	Y	Y	Y
Y	Y	N	Y	Y	Y	Y	Y
N	N	N	N	Y	N	N	Y
Y	Y	Y	Y	Y	N	Y	N
Y	Y	Y	N	Y	N	Y	Y
N	Y	N	N	Y	Y	Y	Y
Y	Y	N					

*When any single number indicates not an amount or count, but only *relative standing*, such as first, second, or tenth place in a horse race or a class of graduating seniors, the data are *ranked* rather than quantitative. Although ranked data are ignored for the sake of simplicity in the present discussion, special statistical procedures for ranked data are treated in Chapter 25 of this book.

sets of observations based on sexual preferences (where a particular observation represents either heterosexual or homosexual), ethnic backgrounds (where a particular observation represents African American, Asian American, European American, Native American, or Spanish American), and attitudes toward the death penalty (where a particular reply represents favor, oppose, or undecided).

Numerical Codes

Numbers could be assigned to the Yes and No replies in Table 1.2, possibly as a coding device to permit computer processing. For example, 1 might be assigned to Yes and 2 to No, or, as the choice of numbers is arbitrary, 234 might be assigned to Yes and 570 to No. These numbers serve merely as convenient codes for the two classes of responses. It would never be appropriate to claim, because No is 2 and Yes is 1, that No is twice as much of a reply as Yes or even that No is more of a reply than Yes is. Unlike the magic kiss that transformed a frog into a prince, the use of arbitrary numerical codes leaves qualitative data unchanged—it's still qualitative, not quantitative data.

1.7 QUANTITATIVE OR QUALITATIVE?

The identification of data as either quantitative or qualitative often represents an important first step toward a successful statistical analysis. Ordinarily, this shouldn't be too difficult: it's merely a matter of deciding whether any *single* observation from a batch of data represents an amount or count described by a number, or whether it represents a class or category described by a word or code.

Focus on a Single Observation

When distinguishing between quantitative and qualitative data, *always focus on the status of any single observation rather than on that of the entire set of observations*. Otherwise, confusion can occur because, when viewed as a whole, even a set of qualitative observations can be described with numbers. For example, when viewed as a whole, the set of qualitative observations in Table 1.2 consists of 56 Y and 27 N replies. Although these numbers or frequencies provide valuable information—indeed, essential information for many types of statistical procedures—they don't transform qualitative data into quantitative data.

As a matter of fact, entire sections—and, later, entire chapters—of this book are devoted exclusively to statistical procedures for quantitative data. Only a few sections—and, later, all of Chapter 24—are devoted exclusively to statistical procedures for qualitative data.

*Exercise 1.2 Each of the following terms or phrases could describe a single observation. Indicate whether the corresponding sets of observations or data would be *quantitative* or *qualitative*. Recalling the above discussion, attempt to visualize whether any *single* observation is a number or a word. Also attend to keywords, such as *score* or *rating scale*, which typify quantitative data.

(a) ethnic group

(b) age

(c) family size

(d) academic major

(e) political preference

(f) IQ score

(g) net worth (dollars)

(h) favorite sport

(i) gender

(j) temperature

Answers on page 492

TWO TYPES OF VARIABLES

1.8 GENERAL DEFINATION OF VARIABLES

> *Variable*
>
> *A characteristic or property that can take on different values.*

> *Constant*
>
> *A characteristic or property that can take on only one value.*

Another helpful distinction—from the realm of research methods rather than statistics—is based on two types of variables. A **variable** *is a characteristic or property that can take on different values.* Accordingly, the weights in Table 1.1 can be described not only as quantitative data, but also as observations for a quantitative variable, because the various weights take on different numerical values. By the same token, the replies in Table 1.2 can be described as observations for a qualitative variable, because the replies to the marijuana question take on different values of either Yes or No. Given this perspective, any *single* observation in Table 1.1 or 1.2 can be described as a **constant**, because *it takes on only one value.*

1.9 INDEPENDENT AND DEPENDENT VARIABLES

Unlike the simple studies that produced the data in Tables 1.1 and 1.2, most studies raise questions about the nature, if any, of the relationship between two (or even more) variables. For example, a psychologist might wish to determine whether couples who undergo special training in "active listening" tend to have fewer communication breakdowns than do couples who undergo no special training. Type of training qualifies as a variable because it assumes two different values: either special training or no special training. Number of communication breakdowns also qualifies as a variable because it assumes different numerical values for the various couples, once the data have been collected. Let's look more closely at the special properties of each of these two variables.

Independent Variable

The psychologist exposes couples to two different types of training by assigning them (according to impartial rules, much like a coin flip, discussed in

Chapter 11) either to special training in active listening or to no special training. When a variable is manipulated by the investigator, as in the present case, it is an **independent variable**.

The impartial creation of distinct groups, which differ in terms of the independent variable, has a most desirable consequence. Once the data have been collected, any difference between the groups (that survives a statistical analysis, as described in Part 2 of this book) can be interpreted as being *caused* by the independent variable. If, for instance, a difference appears in favor of the active listening group, the psychologist can conclude that training in active listening *causes* fewer communication breakdowns between couples. Having observed this relationship, she can expect that, if new couples were trained in active listening, fewer breakdowns in communication would occur.

Dependent Variable

To test whether training affects communication, the psychologist counts the number of communication breakdowns between each couple, as revealed by inappropriate replies, aggressive comments, verbal interruptions, and so on, while discussing a conflict-provoking topic, such as whether it's acceptable to be intimate with a third person. When a variable is measured, counted, or recorded by the investigator, as in the present case, it is a **dependent variable**.

Unlike the independent variable, the dependent variable isn't manipulated by the investigator. Instead, it represents an outcome: the data produced by the experiment. Accordingly, the values that appear for the dependent variable can't be specified in advance. Although the psychologist suspects that couples with special training will tend to show fewer subsequent communication breakdowns, she has to wait to see precisely how many breakdowns will be observed for each couple. (Incidentally, the data for the simple studies described in Tables 1.1 and 1.2 also qualify as dependent variables.)

Independent or Dependent Variable?

With just a little practice, you should be able to identify these two types of variables. What is being manipulated by the investigator and, therefore, qualifies as the independent variable? What is measured, counted, or recorded by the investigator and, therefore, qualifies as the dependent variable? Once these two variables have been identified, they can be used to describe the problem posed by the study; that is, does the independent variable cause a change in the dependent variable? Such studies are referred to as **experiments**, and when well designed, they yield the most informative and unambiguous conclusions about cause-effect relationships.

Two Dependent Variables

Sometimes it's not feasible or possible to manipulate a particular variable of interest. For instance, even though a sociologist is vitally interested in the possible relationship between poverty and crime, it's not feasible to manipulate poverty as an independent variable by impartially assigning newborns to particular poverty levels for much of their lives and then recording their subsequent criminal activity as the dependent variable. Instead, the sociologist must approach both poverty and crime as dependent variables. If his analysis reveals a relationship between poverty level and crime rate, he can only speculate about the possible basis for this relationship. Poverty might cause crime or vice versa. Or both poverty and crime might be caused by one or some combination of more basic variables, such as inadequate education, racial discrimination, unstable family environment, and so on, whose effects could be neutralized only by the harsh, socially unacceptable experiment alluded to here. Studies with two dependent variables often are referred to as **correlation studies**, and they yield less clear-cut conclusions about possible cause-effect relationships than do experiments.

To detect any relationship between active listening and fewer breakdowns in communication, even our psychologist could have conducted a correlation study rather than an experiment. In this case, she would have made no effort to manipulate active listening skills by assigning couples to special training sessions. Instead, the psychologist might have used a preliminary interview to categorize couples as already possessing either "high" or "low" degrees of active listening skills. Subsequently, our psychologist would have obtained a count of the number of communication breakdowns for each couple during the conflict-resolution session; now both variables would have been dependent variables—being data collected by the psychologist—and the interpretation of any observed relationship would have to be qualified because of the possible effects of other, more basic variables. For example, couples already possessing high degrees of active listening might also tend to be more seriously committed to each other, and this more serious commitment itself might cause both the higher degree of active listening and fewer breakdowns in communication. In this case, any special training in active listening, without regard to the existing degree of commitment, wouldn't reduce the number of breakdowns in communication.

Correlation study

A study with two dependent variables.

---

*Exercise 1.3 For each of the possible relationships described in the following list, indicate whether the corresponding study is an experiment or a correlation study. If it is an experiment, identify the independent variable. An investigator wishes to study the relationship between the following variables:

(a) years of education and annual income

(b) prescribed hours of sleep deprivation and subsequent amount of REM (dream) sleep

(c) self-esteem and body weight

(d) pounds lost among obese males who choose to participate either in a weight-loss program or in a self-esteem enhancement program

(e) estimated study hours and subsequent test score

(f) recidivism among substance abusers assigned to different rehabilitation programs

(g) subsequent grade point averages of college applicants who, as the result of a housing lottery, live either on campus or off campus

Answers on page 492

1.10 HOW TO USE THIS BOOK

This book contains several features that will help your study of statistics. Topic headings and formulas are highlighted with color. Each chapter begins with an outline and ends with a summary and a list of important terms. Use these aids to orientate yourself before reading a new chapter and to facilitate your review of previous chapters. Frequent reviews are desirable, since statistics is cumulative, with earlier topics forming the basis for later topics. For easy reference, important terms are defined in the margins. Exercises appear within chapters, and review exercises appear at the end of each chapter. Don't shy away from the exercises; they will clarify and expand your understanding as well as improve your ability to work with statistics. Also remember that Appendix C supplies answers to all exercises marked with asterisks.

The math review in Appendix A summarizes most of the basic math symbols and operations used throughout this book. If you're anxious about your math background—and almost everyone is—check out Appendix A as soon as possible. Be assured that no special math background is required. If you can add, subtract, multiply, and divide, you can learn (or relearn) the simple math described in Appendix A. If this material looks unfamiliar, it would be a good idea to study Appendix A within the next few weeks and to review this material throughout the term, as necessary. Your instructor also might want you to read Appendix B, which supplements the distinction between quantitative and qualitative data.

A Web site for this book (**http://darwin.cwru.edu/~witte/statistics**) supplies links to interesting statistical sites on the Internet, including many with interactive demonstrations of basic concepts and procedures. References to these sites, designated as an **Internet Demonstration** or an **Internet Site**, are scattered throughout this book. The main Web site also provides explanatory comments to ease the transition to these sites. If you have access to the Internet, do explore these sites; they are enlightening and entertaining. They can, however, be ignored—whether by choice or necessity—without destroying the continuity of the book.

We can't resist ending this chapter with a personal note, as well as a few suggestions based on findings from the learning laboratory. A dear relative lent this book to an elderly neighbor, who not only praised it, saying that he wished he had such a stat text many years ago while he was a student at the University of Pittsburgh, but subsequently died with the book still open next to his bed. Upon being informed of this, the first author's wife commented, "I wonder which chapter killed him." In all good conscience, therefore, we can't recommend this book for casual bedside reading if you are more than eighty-five years old. Otherwise, read it anywhere or anytime. Seriously, not only read assigned material before class, but also reread it as soon as possible after class to maximize the retention of newly learned material. Also in the same vein, end reading sessions with active rehearsal—close the book and attempt to recreate mentally, in an orderly fashion and with little or no peeking at the book, the material that you have just read. With this type of effort, you should find the remaining chapters accessible and statistics to be both understandable and useful.

Summary

There are many reasons for taking a statistics class. Having learned some statistics, you'll be better able to interpret the statistical messages in everyday life, to understand statistical references in research reports, and to plan at least a simple statistical analysis for your own research project. Statistics consists of two main subdivisions: descriptive statistics, which is concerned with organizing and summarizing information for sets of actual observations, and inferential statistics, which is concerned with generalizing beyond sets of actual observations.

It's important to distinguish between quantitative and qualitative data. When, among a set of observations, any single observation is a number that represents an amount or a count, the data are quantitative. When, among a set of observations, any single observation is a word or code that represents a class or category, the data are qualitative.

It's also helpful to distinguish between independent and dependent variables. Independent variables are manipulated by the investigator, while dependent variables are outcomes measured, counted, or recorded by the investigator. Studies with independent and dependent variables are experiments, and, if well-designed, they yield the most clear-cut information about cause-effect relationships. Studies with two dependent variables are correlation studies, and they yield less clear-cut information about cause-effect relationships.

Important Terms

Data	**Constant**
Descriptive statistics	**Independent variable**
Inferential statistics	**Dependent variable**
Quantitative data	**Experiment**
Qualitative data	**Correlation study**
Variable	

REVIEW EXERCISES

1.4   If you have not done so already, familiarize yourself with the various appendices in the back of the book.

(a)   Particularly note the location of Appendix C (Answers to Selected Exercises) and Appendix E (Glossary).

(b)   Browse through Appendix A (Math Review). If this material looks unfamiliar, study Appendix A, using the self-diagnostic tests as guides.

1.5   Indicate whether each of the following statements typifies *descriptive statistics* (because it describes sets of actual observations) or *inferential statistics* (because it generalizes beyond sets of actual observations).

(a)   On the basis of a survey conducted by the Bureau of Labor Statistics, it is estimated that 6.1 percent of the entire workforce was unemployed during the last month.

(b) During a recent semester, the ages of students at my college ranged from 16 to 75 years.

(c) Recent research suggests that an aspirin a day reduces the chance of heart attacks in middle-aged men.

(d) Joe's grade point average has hovered near 3.0 throughout his college career.

(e) There is some evidence that any form of frustration—physical, social, economic, or political—always leads to some form of aggression by the frustrated person.

(f) According to tests conducted by the Environmental Protection Agency, any 2000 Honda Civic should average approximately 37 miles per gallon (highway) and 32 miles per gallon (city).

(g) On the average, Babe Ruth hit 32 home runs during each season of his major league baseball career.

(h) Research on learning suggests that active rehearsal increases the retention of newly read material; therefore, immediately after reading material in this book, you should close the book and try to organize the new material in your mind.

(i) Children with no siblings tend to be more adult oriented than children with one or more siblings.

1.6 Each of the following terms or phrases could describe a single observation. Indicate whether the corresponding sets of observations or data would be *quantitative* or *qualitative*.

(a) height

(b) religious affiliation

(c) math aptitude score

(d) years of education

(e) military rank

(f) favorite TV program

(g) place of birth

(h) diastolic blood pressure

(i) vocational goal

(j) grade point average

(k) daily intake of calories

(l) marital status

(m) highest academic degree

(n) blood type

(o) attitude toward total nuclear disarmament (favor, neutral, oppose)

(p) degree of test anxiety (high, medium, low)

(q) days absent from work

(r) academic letter grade

(s) taxable income

(t) astrological sign

(u) score for psychopathic tendency

(v) hair color

(w) reaction time

(x) mechanical aptitude score

(y) nationality of mother

(z) degree of satisfaction with life (100-point rating scale)

1.7 Indicate whether each of the following studies is an experiment or a correlation study. If it is an experiment, identify the independent variable.

(a) A psychologist uses chimpanzees to test the notion that more crowded living conditions trigger aggressive behavior. Chimps are placed, according to an impartial assignment rule, in cages with either one, several, or many other chimps. Subsequently, during a standard observation period, each chimp is assigned a score on the basis of its aggressive behavior toward a chimplike stuffed doll.

(b) An investigator wishes to test whether, when compared with recognized scientists, recognized artists tend to be born under different astrological signs.

(c) To determine whether there is a relationship between the sexual codes of primitive tribes and their behavior toward neighboring tribes, an anthropologist consults available records, classifying each tribe on the basis of their sexual codes (permissive or repressive) and their behavior toward neighboring tribes (friendly or hostile).

(d) In a study of group problem solving, a investigator assigns college students to groups of two, three, or four students and measures the amount of time required by each group to solve a complex puzzle.

(e) A school psychologist wishes to determine whether reading comprehension scores are related to the number of months of formal education, as reported on school transcripts, for a group of twelve-year-old migrant children.

(f) To determine whether Graduate Record Exam (GRE) scores can be increased by cramming, an investigator assigns college students to either a GRE test-taking workshop or a control (non-test-taking) workshop, then compares GRE scores earned subsequently by the two groups of students.

(g) A social scientist wishes to determine whether there is a relationship between the attractiveness scores (on a 100-point scale) assigned to college students by a panel of peers and their scores on a paper-and-pencil test of anxiety.

(h) A political scientist wishes to determine whether males and females differ with respect to their attitudes toward defense spending by the federal government. She asks each person if he or she thinks the current level of defense spending should be increased, remain the same, or be decreased.

PART 1

Descriptive Statistics
Organizing and Summarizing Data

2   Describing Data with Tables
3   Describing Data with Graphs
4   Describing Data with Averages
5   Describing Variability
6   Normal Distributions (I): Basics
7   Normal Distributions (II): Applications
8   More About z Scores
9   Describing Relationships: Correlation
10 Prediction

CHAPTER
2

Describing Data with Tables

FREQUENCY DISTRIBUTIONS FOR QUANTITATIVE DATA

2.1   UNGROUPED DATA
2.2   GROUPED DATA
2.3   GUIDELINES
2.4   HOW MANY CLASSES?
2.5   GAPS BETWEEN CLASSES
2.6   DOING IT YOURSELF
2.7   OUTLIERS

OTHER TYPES OF FREQUENCY DISTRIBUTIONS

2.8   RELATIVE FREQUENCY DISTRIBUTIONS
2.9   PERCENTS OR PROPORTIONS?
2.10 CUMULATIVE FREQUENCY DISTRIBUTIONS
2.11 PERCENTILE RANKS
2.12 FREQUENCY DISTRIBUTIONS FOR QUALITATIVE DATA
2.13 INTERPRETING DISTRIBUTIONS CONSTRUCTED BY OTHERS

Summary

Important Terms

Review Exercises

**Table 2.1
FREQUENCY
DISTRIBUTION
(UNGROUPED DATA)**

WEIGHT	FREQUENCY
245	1
244	0
243	0
242	0
*	
*	
*	
161	0
160	4
159	1
158	2
157	3
*	
*	
*	
136	0
135	2
134	0
133	1
Total	53

······················

Frequency distribution

*A collection of observations produced by sorting observations into classes and showing their frequency
(or number) of occurrences in each class.*

······················

Frequency distribution for ungrouped data

A frequency distribution produced whenever observations are sorted into classes of single values.

Given a batch of data, as in Table 1.1 on page 4, how do you make sense out of it—both for yourself and for others? Is there an important message hidden among all those observations, possibly one that either supports or fails to support one of your ideas? (Or more realistically, is there a difference between two batches of data—for instance, between Graduate Record Exam (GRE) scores of students who do and students who don't attend a test-taking workshop, between the survival rates of coronary bypass patients who do and those patients who don't own a dog, or between the starting salaries of male and female executives?) At this point, especially if you're facing a fresh set of data in which you have a special interest, statistics can be exciting as well as challenging. Your initial responsibility is to describe the data as clearly, completely, and concisely as possible. Statistics supplies some tools—for instance, tables and graphs—and some guidelines. Beyond that, it's just the data and you. As in life, there is not one single right way. Equally valid descriptions of the same data might appear in tables with different formats. Relax. By following just a few guidelines, your reward will be a well-organized and well-summarized set of data.

FREQUENCY DISTRIBUTIONS FOR QUANTITATIVE DATA

2.1 UNGROUPED DATA

Table 2.1 shows one way to organize the weights of the male statistics students listed in Table 1.1. First arrange a column of consecutive numbers, beginning with the lightest weight (133) at the bottom and ending with the heaviest weight (245) at the top. (Because of the extreme length of this column, many intermediate numbers have been omitted in Table 2.1, a procedure never followed in practice.) Then place a tally—that is, a short vertical stroke—next to a number each time its value appears in the original set of data; once this process has been completed, substitute for each tally count (not shown in Table 2.1) a number indicating the frequency (or number) of occurrences of each weight.

> A *frequency distribution* is a collection of observations produced by sorting observations into classes and showing their frequency (or number) of occurrences in each class.

When observations are sorted into classes of *single* values, as in Table 2.1, the result is referred to as a **frequency distribution for ungrouped data**.

Not Always Appropriate

The frequency distribution shown in Table 2.1 is unwieldy (and hence only partially displayed) because of more than one hundred possible values between the largest and smallest observations. Frequency distributions for ungrouped data are much more informative when the number of possible values is smaller, say, less than about twenty. Under these circumstances, they serve as a straightforward method for organizing data. Otherwise, if there are twenty or more possible values, use a frequency distribution for grouped data, as discussed in the next section.

Exercise 2.1 Students in a theater arts appreciation class rated the film *Titanic* on a 10-point scale, ranging from 1 (poor) to 10 (excellent), as follows:

```
3  7  2   7   8
3  1  4  10   3
2  5  3   5   8
9  7  6   3   7
8  9  7   3   6
```

Because the number of possible values is relatively small—only ten—it's appropriate to construct a frequency distribution for ungrouped data. Do this.

Answers on page 492

2.2 GROUPED DATA

Table 2.2 shows another way to organize the weights in Table 1.1 according to their frequencies of occurrence. When observations are sorted into classes of *more than one* value, as in Table 2.2, the result is referred to as a **frequency distribution for grouped data**. Let's look at the general structure of this frequency distribution. Data are grouped into classes with ten possible values each. The bottom class includes the smallest observation (133), and the top class includes the largest observation (245). The distance between bottom and top is occupied by an orderly series of classes, including classes such as 210 to 219 that contain no observations. The frequency column shows the frequency (or number) of observations in each class and, at the bottom, the total number of observations in all classes.

Still referring to Table 2.2, let's shift our focus from structure to content and summarize the more important properties of the weight data, as revealed by the frequency distribution. Although ranging from the 130s to the 240s, the weights peak in the 150s, with a progressively decreasing but relatively heavy concentration in the 160s and 170s. Furthermore, the distribution of weight isn't balanced about its peak, but tilted in the direction of the heavier weights.

Cost of Grouping

Notice one inevitable by-product of grouping data, found in Table 2.2. The identities of individual observations are sacrificed for a more concise description in terms of classes and frequencies of occurrence. Occasionally this can be a liability if you wish more information than is given in the frequency distribution but lack access to the original data.

2.3 GUIDELINES

The box "Guidelines for Frequency Distributions" lists seven guidelines for producing a frequency distribution. The first three specify some of the basic ingredients of any well-constructed frequency distribution, and they should not be violated unless there are extenuating circumstances. The last four guidelines specify optional ingredients that can be modified or ignored, as circumstances

Frequency distribution for grouped data

A frequency distribution produced whenever observations are sorted into classes of more than one value.

Table 2.2 FREQUENCY DISTRIBUTION (GROUPED DATA)

WEIGHT	FREQUENCY
240–249	1
230–239	0
220–229	3
210–219	0
200–209	2
190–199	4
180–189	3
170–179	7
160–169	12
150–159	17
140–149	1
130–139	3
Total	53

warrant. Before reading further, satisfy yourself that the frequency distribution in Table 2.2 actually complies with each of the seven guidelines.

GUIDELINES FOR FREQUENCY DISTRIBUTIONS

Basic

1. Each observation should be included in one, and only one, class.

Example: 130–139, 140–149, 150–159, etc. It would be incorrect to use 130–140, 140–150, 150–160, etc., in which, because the boundaries of classes overlap, an observation of 140 (or 150) could be assigned to either of two classes.

2. List all classes, even those with zero frequencies.

Example: Listed in Table 2.2 is the class 210–219 and its frequency of zero. It would be incorrect not to list this class because of its zero frequency in this particular set of data.

3. All classes (with both upper and lower boundaries) should be equal in width.

Example: 130–139, 140–149, 150–159, etc. It would be incorrect to use 130–139, 140–159, etc., in which the second class is twice the width of the first class.

Optional

4. All classes should have both an upper boundary and a lower boundary.

Example: 240–249. Less preferred would be 240–above, in which no maximum value can be assigned to observations in this class. (Nevertheless, this type of open-ended class is employed as a space-saving device when many different tables must be listed, as in the *Statistical Abstract of the U.S.* Open-ended classes appear in several exercises at the end of this chapter.)

5. Select the width of classes from convenient numbers, such as 1, 2, 3, . . . 10, particularly 5 and 10 or multiples of 5 and 10.

Example: 130–139, 140–149, 150–159, in which the class width is 10, a convenient number. Less preferred would be 130–142, 143–155, 156–168, etc., in which the class width is 13, an inconvenient number.

6. The lower boundary of each class should be a multiple of the class width.

Example: 130–139, 140–149, 150–159, in which the lower boundaries of 130, 140, 150 are multiples of 10, the class width. Less preferred would be 135–144, 145–154, 155–164, etc., in which the lower boundaries of 135, 145, and 155 are not multiples of 10, the class width.

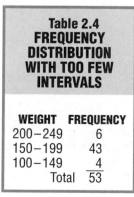

Table 2.3
FREQUENCY DISTRIBUTION WITH TOO MANY INTERVALS

WEIGHT	FREQUENCY
245–249	1
240–244	0
235–239	0
230–234	0
225–229	2
220–224	1
215–219	0
210–214	0
205–209	2
200–204	0
195–199	0
190–194	4
185–189	1
180–184	2
175–179	2
170–174	5
165–169	7
160–164	5
155–159	9
150–154	8
145–149	1
140–144	0
135–139	2
130–134	1
Total	53

Table 2.4
FREQUENCY DISTRIBUTION WITH TOO FEW INTERVALS

WEIGHT	FREQUENCY
200–249	6
150–199	43
100–149	4
Total	53

7. In general, aim for a total of approximately ten classes.

Example: The distribution in Table 2.2 uses twelve classes. Less preferred would be the distributions in Tables 2.3 and 2.4. The distribution in Table 2.3 has too many classes (twenty-four), whereas the distribution in Table 2.4 has too few classes (three).

2.4 HOW MANY CLASSES?

The seventh guideline requires a few more comments. The use of too many classes—as in **Table 2.3**, in which the weight data are grouped into twenty-four classes, each of width five—tends to defeat the major purpose of a frequency distribution, namely, to provide a reasonably concise description of data. Alternatively, the use of too few classes—as in **Table 2.4**, in which the weight data are grouped into three classes, each of width fifty—can mask important data patterns such as the high density of weights in the 150s and, to a lesser extent, in the 160s and 170s.

When There Are Many Observations

But there is nothing sacred about ten, the recommended number of classes. When describing large batches of data, you might aim for considerably more than ten classes in order to portray some of the more fine-grained data patterns that otherwise could vanish. For example, add two zeros after each frequency in Table 2.3 to create a new weight distribution for 5300 students from the original distribution for 53 students. Given this weight distribution (admittedly unlikely because of the abrupt shifts in frequency between adjacent classes), it is important to retain the twenty-four classes in Table 2.3 in order to portray, for instance, the curious presence of several peak frequencies, including the one at 165 to 169 pounds.

When There Are Few Observations

However, when describing small batches of data, you might choose to aim for fewer than ten classes in order to spotlight data regularities that otherwise could be blurred. For example, if the weight distribution had been based on only about twelve students, then it would be desirable to use the three classes in Table 2.4 in order to detect any tendency for frequencies to peak, possibly at 150 to 199 pounds. It's best, therefore, to think of ten, the recommended number of classes, as a rough rule of thumb to be applied with discretion rather than with slavish devotion.

2.5 GAPS BETWEEN CLASSES

In well-constructed frequency tables, the gaps between classes, such as between 139 and 140 in Table 2.2, show clearly that each observation has been assigned to one, and only one, class. The size of the gap should always equal

Unit of measurement

The smallest possible difference between scores.

one **unit of measurement**; that is, it should always equal the smallest possible difference between scores within a particular set of data. Because the gap is never bigger than one unit of measurement, no observation can fall into the gap. In the present case, in which the weights are reported to the nearest pound, one pound is the unit of measurement and, quite appropriately, the gap between the classes equals one pound. These gaps would not be appropriate if the weights had been reported to the nearest tenth of a pound. In this case, one-tenth of a pound is the unit of measurement, and therefore the gap should equal one-tenth of a pound. The smallest class would be 130.0 to 139.9 (not 130 to 139), and the next class would be 140.0 to 149.9 (not 140 to 149), and so on. These new boundaries guarantee that any observation, such as 139.6, will be assigned to one, and only one, class.

More Apparent Than Real

Gaps between classes do not signify any disruption in the essentially continuous nature of the data. It would be erroneous to conclude that because of the gap between 139 and 140 in the frequency distribution for Table 2.2, nobody can weigh between 139 and 140 pounds. As a matter of fact, a man who reports his weight as 139 pounds actually could weigh anywhere between 138.5 and 139.5 pounds, and a man who reports his weight as 140 pounds actually could weigh anywhere between 139.5 and 140.5 pounds. Thus, the gaps between classes are more apparent than real.

Class Interval Width

Class interval width

The distance between the two tabled boundaries, after each boundary has been expanded by one-half of one unit of measurement.

Gaps cannot be ignored when you are determining the width of any class interval. **Class interval width** *equals the distance between the two tabled boundaries, after each boundary has been expanded by one-half of one unit of measurement*—that is, after one-half of one unit of measurement has been subtracted from the lower tabled boundary and added to the upper tabled boundary. For example, the width of the interval 140–149 in Table 2.2 equals 10, the distance between 139.5 (140 minus one-half of the unit of measurement of 1) and 149.5 (149 plus one-half of the unit of measurement of 1).

If weights had been reported to the nearest tenth of a pound, the width of the interval 140.0–149.9 would still equal 10, the distance between 139.95 (140.0 minus one-half of the unit of measurement of .1) and 149.95 (149.9 plus one-half of one unit of measurement of .1).

2.6. DOING IT YOURSELF

Now that you know the properties of well-constructed frequency distributions, study the step-by-step procedure listed in the box "Constructing Frequency Distributions", which shows precisely how the distribution in Table 2.2 was constructed from the weight data in Table 1.1. You'll want to refer to this procedure when, as with some of the exercises in this chapter, you must construct a frequency distribution for grouped data.

CONSTRUCTING FREQUENCY DISTRIBUTIONS

1. Find the data range, that is, the difference between the largest and smallest observation. Inspection of Table 1.1 reveals a data range of 245 − 133 = 112.

2. Find the class width required to span the data range by dividing the range by the desired number of classes (ordinarily ten). In the present example,

$$Class\ width = \frac{data\ range}{desired\ number\ of\ classes} = \frac{112}{10} = 11.2$$

3. Round off to the nearest convenient width (such as 1, 2, 3, . . . 10, particularly 5 or 10 or multiples of 5 or 10). In the present example, the nearest convenient width is 10.

4. Determine where the lowest class should begin. (Ordinarily this number should be a multiple of the class width.) In the present example, the smallest observation is 133, and the lowest class should begin at 130, because 130 is a multiple of 10, the class width.

5. Determine where the lowest class should end by adding the class width to the lower boundary and then subtracting one unit of measurement (defined in Section 2.5). In the present example, add 10 to 130, and then subtract 1, the unit of measurement, to obtain 139—the number at which the lowest class should end.

6. Working upward, list as many equivalent classes (usually a total of about ten) as are required to include the largest observation. In the present example, list 130–139, 140–149, . . . , 240–249, in which the last class includes 245, the largest observation.

7. Indicate with a tally the class in which each observation falls. For example, the first observation in Table 1.1, 160, produces a tally next to 160–169; the next observation, 193, produces a tally next to 190–199; and so on.

8. Replace the tally count for each class with a frequency and show the total of all frequencies. (Tally marks aren't usually shown in the final frequency distribution.)

9. Supply headings for both columns and a title for the table.

EXAMPLE: CONSTRUCTING FREQUENCY DISTRIBUTION FOR GROUPED DATA

The grade point averages for a group of college students are as follows:

3.67	2.50	3.50	2.80	2.83	3.25	2.90	2.34	3.59
3.78	2.75	2.67	2.65	3.10	2.76	2.10	3.20	3.00
3.00	1.90	2.90	2.58	3.37	2.86	2.66	2.67	3.08

Solution

*The largest observation minus the smallest observation equals the data range:

$$3.78 - 1.90 = 1.88$$

*Divide the data range by ten, the desired number of classes, to find the class width:

$$\frac{1.88}{10} = .188$$

*Round off to the nearest convenient class width (usually a multiple of 5 or 10):

$$.20$$

*The lowest class should begin at 1.80 (because it is a multiple of .20, the class width, and it includes the smallest observation, 1.90)
*The lowest class should end at 1.99 (obtained by adding .20 to 1.80, then subtracting .01, the unit of measurement)
*Work upward with equivalent classes to the largest observation:

3.60–3.79 (because 3.78 is the largest observation)

*Use tallies to sort observations into classes (not shown) and then replace the tally count for each class with a frequency and show the total of all frequencies.
*Supply headings for both columns and a title for the table.

GRADE POINT AVERAGES FOR COLLEGE STUDENTS

GPA	FREQUENCY
3.60–3.79	2
3.40–3.59	2
3.20–3.39	3
3.00–3.19	4
2.80–2.99	5
2.60–2.79	6
2.40–2.59	2
2.20–2.39	1
2.00–2.19	1
1.80–1.99	1
Total	27

*Exercise 2.2 The IQ scores for a group of thirty-five high school dropouts are as follows:

91	85	84	79	80
87	96	75	86	104
95	71	105	90	77
123	80	100	93	108
98	69	99	95	90
110	109	94	100	103
112	90	90	98	89

Using the steps described in the text, construct a frequency distribution for grouped data.

*Exercise 2.3 What are some possible poor features of the following frequency distribution?

ESTIMATED WEEKLY TV VIEWING TIME (HRS) FOR 250 SIXTH GRADERS

VIEWING TIME	FREQUENCY
35–above	2
30–34	5
25–30	29
20–22	60
15–19	60
10–14	34
5–9	31
0–4	29
Total	250

Answers on pages 492–493

2.7 OUTLIERS

When working with data, be prepared to deal occasionally with the appearance of one or more *very extreme* observations, or **outliers**. A GPA of 0.06, an IQ of 170, summer wages of $32,000—each requires special attention because of its potential impact on a summary of the data.

Check for Accuracy

Whenever you encounter an observation with an outrageously extreme value, such as a GPA of 0.06, attempt to verify its accuracy. For instance, was a respectable GPA of 3.06 recorded erroneously as 0.06? If the outlier survives an accuracy check—or if its accuracy can't be checked because the original observation is irretrievable—it should be treated as a legitimate observation.

Might Exclude from Summaries

Even if an outlier qualifies as a legitimate observation, however, you might choose to segregate it (but not to suppress it!) from any summary of the data. For example, you might relegate it to a footnote instead of using excessively wide classes in order to include it in a frequency distribution. Or you might use various numerical summaries, such as the median and interquartile range (discussed in Chapters 4 and 5), that ignore all extreme observations, including outliers.

Might Enhance Understanding

Insofar as a valid outlier can be viewed as the product of special circumstances, it might enhance your understanding of the data. For example, you might increase your understanding of why crime rates differ among communities by studying the special circumstances that produce a community with an extremely low crime rate, or you could investigate why learning rates differ among third graders by studying a third grader who learns very rapidly.

*Exercise 2.4 Identify any outliers in each of the following sets of data collected from nine college students. First, inspect the nine summer incomes for outliers, then repeat this process for the nine observations for age, for family size, and for GPA.

SUMMER INCOME	AGE	FAMILY SIZE	GPA
$6,450	20	2	2.30
$4,820	19	4	4.00
$5,650	61	3	3.56
$1,720	32	6	2.89
$600	19	18	2.15
$0	22	2	3.01
$3,482	23	6	3.09
$25,700	27	3	3.50
$8,548	21	4	3.20

Answers on page 493

OTHER TYPES OF FREQUENCY DISTRIBUTIONS

2.8 RELATIVE FREQUENCY DISTRIBUTIONS

An important variation of the frequency distribution is the relative frequency distribution.

***Relative frequency distributions* show the frequency of each class as a part or fraction of the total frequency for the entire distribution.**

This type of distribution allows us to focus on the relative concentration of observations among different classes within the same distribution. In the case of the weight data in Table 2.2, it permits us to see that the 160s account for about one-fourth (12/53 = .23, or 23%) of all observations. This type of distribution also helps us compare two or more distributions based on different total numbers of observations. For instance, as in Exercise 2.14, you might wish to compare the distribution of ages for 500 residents of a small town with that for the more than 250 million residents of the United States. The conversion from absolute to relative frequencies allows us to compare directly the two distributions without having to worry about the radically different total numbers of observations.

Constructing Relative Frequency Distributions

To convert a frequency distribution into a relative frequency distribution, divide the frequency for each class by the total frequency for the entire distribution. **Table 2.5** illustrates a relative frequency distribution based on the weight

---

Relative frequency distribution

A frequency distribution showing the frequency of each class as a part or fraction of the total frequency for the entire distribution.

Table 2.5
RELATIVE FREQUENCY DISTRIBUTIONS

WEIGHT	FREQUENCY	RELATIVE FREQUENCY	
		PROPORTION	PERCENT
240–249	1	.02	2
230–239	0	.00	0
220–229	3	.06	6
210–219	0	.00	0
200–209	2	.04	4
190–199	4	.08	8
180–189	3	.06	6
170–179	7	.13	13
160–169	12	.23	23
150–159	17	.32	32
140–149	1	.02	2
130–139	3	.06	6
Total	53	1.02*	102*

The sums do not equal 1.00 and 100 percent because of rounding-off errors.

distribution of Table 2.2. The conversion to proportions is straightforward. For instance, to obtain the proportion of .06 for the class 130 to 139, simply divide the frequency of 3 for that class by the total frequency of 53. Repeat this process until a proportion has been calculated for each class.

2.9 PERCENTS OR PROPORTIONS?

Some people prefer to deal with percents rather than proportions because percents usually lack decimal points. A proportion always varies between 0 and 1, whereas a percent always varies between 0 percent and 100 percent. To convert the relative frequencies in Table 2.5 from proportions to percents, multiply each proportion by 100; that is, move the decimal point two places to the right. For example, multiply .06 (the proportion for the class 130 to 139) by 100 to obtain 6 percent. In practice, a relative frequency distribution consists of either proportions or percents, but not both, as shown in Table 2.5 only for illustrative purposes.

*Exercise 2.5 Graduate Record Exam (GRE) scores for a group of applicants are distributed as follows:

GRE	FREQUENCY
725–749	1
700–724	3
675–699	14
650–674	30
625–649	34
600–624	42
575–599	30
550–574	27
525–549	13
500–524	4
475–499	2
Total	200

Convert to a relative frequency distribution, first using proportions, then percents. When calculating proportions, round numbers to two digits to the right of the decimal point, using the rounding procedure specified in Section 7 of Appendix A.

Answers on page 493

2.10 CUMULATIVE FREQUENCY DISTRIBUTIONS

Cumulative frequency distributions show the total number of observations in each class and in all lower-ranked classes.

This type of distribution can be used effectively with sets of observations—such as test scores for intellectual or academic aptitude—when relative standing within the distribution assumes primary importance. Under these circumstances, cumulative frequencies are often converted, in turn, to cumulative proportions or percents. Cumulative percents are often referred to as percentile ranks and are discussed more thoroughly in the next section.

Cumulative frequency distribution

A frequency distribution showing the total number of observations in each class and all lower-ranked classes.

Constructing Cumulative Frequency Distributions

To convert a frequency distribution into a cumulative frequency distribution, add to the frequency of each class the sum of the frequencies of all classes ranked below it. This gives the cumulative frequency for that class. The most efficient procedure is to begin with the lowest-ranked class in the frequency distribution and work upward, finding the cumulative frequencies in ascending order. In **Table 2.6**, the cumulative frequency for the class 130 to 139 is 3, because there are no classes ranked lower. The cumulative frequency for the class 140 to 149 is 4, because 1 is the frequency for that class, and 3 is the frequency of all lower-ranked classes. The cumulative frequency for the class 150 to 159 is 21, because 17 is the frequency for that class, and 4 is the sum of the frequencies of all lower-ranked classes.

Cumulative Proportions or Percents

As has been suggested, if relative standing within a distribution is particularly important, then cumulative frequencies are converted to cumulative proportions or percents. A glance at Table 2.6 reveals that .75, or 75 percent, of all weights are the same as or lighter than the weights between 170 and 179 pounds. To obtain this cumulative proportion (.75), the cumulative frequency of 40 for the class 170 to 179 should be divided by the total frequency of 53 for the entire distribution. To determine the cumulative percent (75%), simply multiply the cumulative proportion (.75) by 100.

*Exercise 2.6

(a) Convert the distribution of GRE scores shown in Exercise 2.5 to a cumulative frequency distribution.

		Table 2.6		
		CUMULATIVE FREQUENCY DISTRIBUTIONS		
WEIGHT	**FREQUENCY**	**CUMULATIVE FREQUENCY**	**CUMULATIVE PROPORTION**	**CUMULATIVE PERCENT**
240–249	1	53	1.00	100
230–239	0	52	.98	98
220–229	3	52	.98	98
210–219	0	49	.92	92
200–209	2	49	.92	92
190–199	4	47	.89	89
180–189	3	43	.81	81
170–179	7	40	.75	75
160–169	12	33	.62	62
150–159	17	21	.40	40
140–149	1	4	.08	8
130–139	3	3	.06	6
Total	53			

(b) Convert the distribution of GRE scores shown in Exercise 2.5 to a cumulative relative frequency distribution, using proportions, then percents.

Answers on page 494

2.11 PERCENTILE RANKS

When used to describe the relative position of any observation within its parent distribution, cumulative percents are referred to as percentile ranks. *The* **percentile rank of an observation** *indicates the percentage of observations in the entire distribution with similar or smaller values than that observation.* Thus a weight has a percentile rank of 80 if equal or lighter weights constitute 80 percent of the entire distribution.

Exact Percentile Ranks (from Ungrouped Data)

When cumulative percents have been obtained from frequency distributions for ungrouped data, the assignment of *exact* percentile ranks is fairly straightforward. **Table 2.7** shows ungrouped data based on the reports of twenty college students. First obtain cumulative frequencies (by adding to the frequency for each observation the frequencies of all smaller observations), and then calculate the cumulative percents or percentile ranks (by dividing each cumulative frequency by the total frequency of 20 and multiplying by 100). Table 2.7 lists the exact percentile ranks for each different number of romantic affairs. For example, located at the upper end of the distribution, five affairs has a percentile rank of 90—that is, students who report five affairs equal or exceed 90 percent of their classmates (and, therefore, probably qualify for the runner-up award as "most fickle").

Percentile rank of an observation

Percentage of observations in the entire distribution with similar or smaller values than that observation.

Table 2.7
NUMBER OF ROMANTIC AFFAIRS DURING HIGH SCHOOL REPORTED BY TWENTY COLLEGE STUDENTS

NO. OF AFFAIRS	FREQUENCY	CUMULATIVE FREQUENCY	CUMULATIVE PERCENT OR PERCENTILE RANK
8	1	20	100
7	0	19	95
6	1	19	95
5	2	18	90
4	1	16	80
3	2	15	75
2	4	13	65
1	4	9	45
0	5	5	25
Total	20		

To find the exact percentile rank for any observation, first locate that observation in the frequency distribution for ungrouped data, as in Table 2.7, and then interpret the corresponding cumulative percent as a percentile rank.

Approximate Percentile Ranks (from Grouped Data)

The assignment of *exact* percentile ranks requires that cumulative percents be obtained from frequency distributions for ungrouped data. If we have access only to a frequency distribution for grouped data, as in Table 2.6, cumulative percents can be used to assign *approximate* percentile ranks. In Table 2.6, for example, any weight between 170 and 179 pounds could be assigned an approximate percentile rank of 75, because 75 is the cumulative percent for this class.

*Exercise 2.7

(a) Referring to Table 2.7, find the *exact* percentile rank for students who report 1 romantic affair.

(b) Referring to Table 2.6, find the *approximate* percentile rank of any weight between 200 and 209 pounds.

Answers on page 494

2.12 FREQUENCY DISTRIBUTIONS FOR QUALITATIVE DATA

When, among a set of observations, any single observation is a word or code, the data are qualitative. Frequency distributions for qualitative data are easy to construct. Simply determine the frequency with which observations occupy each class, and report these frequencies as shown in **Table 2.8** for the marijuana smoking survey. This frequency distribution reveals that Yes replies are approximately twice as prevalent as No replies. (Remember, just because qualitative data can be summarized numerically with frequencies doesn't change the status of the original observations from qualitative to quantitative.)

Ordered Qualitative Data

Notice that it's totally arbitrary whether Yes is listed above or below No in Table 2.8. When qualitative data can be ordered from least to most, however, that order should be preserved in the frequency table, as illustrated in **Table 2.9**, in which military ranks are listed in descending order from general to warrant officer. (In Appendix B, ordered qualitative data also are described as data with an ordinal level of measurement.)

Relative and Cumulative Distributions for Qualitative Data

Frequency distributions for qualitative variables can always be converted into relative frequency distributions, expressed either as proportions or percents, as illustrated in Table 2.9. Furthermore, if observations can be ordered from least to most, cumulative frequencies (and cumulative relative frequencies) can be used, as illustrated in Table 2.9, and it is appropriate to claim, for

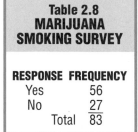

Table 2.8 MARIJUANA SMOKING SURVEY

RESPONSE	FREQUENCY
Yes	56
No	27
Total	83

Table 2.9
RANKS OF OFFICERS IN THE U.S. ARMY (1997)

RANK	FREQUENCY	PROPORTION	CUMULATIVE PERCENT
General	304	.004*	100.0
Colonel	12,778	.163	99.6
Major	13,038	.166	83.3
Captain	23,098	.295	66.7
Lieutenant	17,279	.221	37.2
Warrant Officer	11,816	.151	15.1
Total	78,313		

Source: Defense 97 Almanac.
*To avoid a value of .00 for general, proportions are carried three places to the
right of the decimal point.

example, that a captain has a percentile rank of 67 among officers in the U.S. Army. If observations can't be ordered, as in Table 2.8, a cumulative frequency distribution will be meaningless.

*Exercise 2.8 Motion picture ratings can be ordered from most to least restrictive: NC-17, R, PG-13, PG, and G. The ratings of some films shown recently in San Francisco are as follows:

PG	PG	PG	PG-13	G
G	PG-13	R	PG	PG
R	PG	R	PG	R
NC-17	NC-17	PG	G	PG-13

(a) Construct a frequency distribution.

(b) Convert to relative frequencies, expressed as percents.

(c) Construct a cumulative frequency distribution.

(d) Find the percentile rank for those films with a PG rating.

*Exercise 2.9

(a) Construct a frequency distribution for the blood types of a group of prospective donors.

O	O	A	A	AB
A	O	O	A	O
A	A	A	B	A
A	O	A	O	O
O	O	A	O	A
B	O	A	O	O

(b) Convert to relative frequencies, expressed as proportions and percents.

(c) If appropriate, construct a cumulative frequency distribution. (Remember, it's only appropriate to construct a cumulative frequency distribution if data can be ordered from least to most.)

Answers on page 494

2.13 INTERPRETING DISTRIBUTIONS CONSTRUCTED BY OTHERS

When inspecting a distribution for the first time, train yourself to look at the entire table, not just the distribution. Read the title, column headings, and any footnotes. Where do the data come from? Is a source cited? Next focus on the form of the frequency distribution. Is it well constructed? For grouped quantitative data, does the selection of classes seem to avoid either oversummarizing or undersummarizing the data?

After these preliminaries, inspect the content of the frequency distribution. What is the approximate range? Does it seem reasonable? (Otherwise you might be misinterpreting the distribution or the distribution might contain one or more outliers that require special attention.) As best you can, disregard the inevitable irregularities that accompany a frequency distribution, and focus on its overall appearance or shape. Do the frequencies arrange themselves around a single peak (high point) or several peaks? (More than one peak might signify the presence of several different types of observations—for example, the annual incomes of male and female wage earners—coexisting in the same distribution.) Is the distribution fairly balanced around its peak? (An obviously unbalanced distribution might reflect the presence of a numerical boundary, such as a score of 100 percent correct on an extremely easy exam, beyond which no observation is possible.)

When interpreting distributions, including distributions constructed by someone else, keep an open mind. Follow not only the preceding suggestions but also any questions stimulated by your inspection of the entire table.

Particularly attend to the review exercises at the end of this chapter that ask you to interpret data patterns. With practice, you will develop the important skill of interpreting distributions, whether presented in the form of a table, as in this chapter, or in the form of a graph, as in Chapter 3.

Summary

Frequency distributions organize observations according to their frequencies of occurrence.

Frequency distributions for ungrouped data are produced whenever observations are sorted into classes of single values. This type of frequency distribution is the most informative when there are fewer than about twenty possible values between the largest and smallest observations.

Frequency distributions for grouped data are produced whenever observations are sorted into classes of more than one value. This type of frequency distribution should be constructed, step by step, to comply with several guidelines. Essentially, a well-constructed frequency distribution consists of a string of non-overlapping, equal classes that occupy the entire distance between the largest and smallest observations.

Avoid oversummarizing data (by using too few classes) or undersummarizing data (by using too many classes). In general, aim for about ten classes, but treat this number as a rough rule of thumb to be modified, depending on circumstances.

Very extreme observations or outliers require special attention. Given a valid outlier, you might choose to relegate it to a footnote because of its potential for distortion, or you even might choose to concentrate on it because of its potential for clarification.

When comparing two or more frequency distributions based on appreciably different numbers of observations, it's often helpful to express frequencies as relative frequencies. Either proportions or percents can be used in relative frequency distributions.

When relative standing within the distribution is important, it's helpful to convert frequency distributions into cumulative frequency distributions, particularly those involving cumulative proportions or percents.

When used to describe the relative position of any observation within its parent distribution, cumulative percents are referred to as percentile ranks. The percentile rank of an observation indicates the percentage of observations in the entire distribution with similar or smaller values. The assignment of percentile ranks is exact when cumulative percents are obtained from ungrouped data, but only approximate when cumulative percents are obtained from grouped data.

Frequency distributions for qualitative data are easy to construct. They also can be converted to relative frequency distributions involving either proportions or percents and—if the data can be ordered from least to most—into cumulative frequency distributions involving frequencies, proportions or percents, and percentile ranks.

Important Terms
....

Frequency distribution

Frequency distribution for ungrouped data

Frequency distribution for grouped data

Unit of measurement

Class interval width

Outlier

Relative frequency distribution

Cumulative frequency distribution

Percentile rank of an observation

REVIEW EXERCISES

2.10 A public transit district maintains records of the average number of miles between breakdowns for each of its fleet of fifty secondhand buses. The rather dismal record looks like this:

654	429	1630	1970	1800
789	1708	1825	1723	1701
961	1644	752	880	1788
893	1798	847	576	1903
763	423	743	889	567
421	821	926	593	998
562	756	233	431	741
923	543	378	776	632
759	1789	1754	679	771
677	777	555	660	638

(a) What is the unit of measurement for these data?

(b) Construct a frequency distribution for grouped data.

(c) Describe any interesting data patterns.

(d) Add an outlier with a value of 9750 to the above fifty observations. Given that this is a valid outlier, would it, in your opinion, be desirable to include the outlier (without using an open-ended class) in a frequency distribution for the entire batch of data? *Hint:* Note the effect of the outlier on the frequency distribution for the original fifty observations in Exercise 2.10(b).

2.11 Let's pretend that student volunteers were assigned arbitrarily (according to a coin toss) to be trained either to meditate or to behave as usual. To determine whether meditation training (the independent variable) influenced grade point averages (the dependent variable), GPAs were calculated for each student at the end of the one-year experiment, yielding these results for the two groups:

MEDITATORS			NONMEDITATORS		
3.25	2.25	2.75	3.67	3.79	3.00
3.56	3.33	2.25	2.50	2.75	1.90
3.57	2.45	3.75	3.50	2.67	2.90
2.95	3.30	3.56	2.80	2.65	2.58
3.56	3.78	3.75	2.83	3.10	3.37
3.45	3.00	3.35	3.25	2.76	2.86
3.10	2.75	3.09	2.90	2.10	2.66
2.58	2.95	3.56	2.34	3.20	2.67
3.30	3.43	3.47	3.59	3.00	3.08

(a) What is the unit of measurement for these data?

(b) Construct separate frequency distributions using grouped data. (First construct the frequency distribution for the group having the larger range. Then, to facilitate comparisons, use the same set of classes for the other frequency distribution.)

(c) Do the two groups tend to differ?

2.12   Given the following frequency distribution:

U.S. HOUSEHOLDS BY NUMBER OF PERSONS, 1997	
HOUSEHOLD SIZE (PEOPLE)	**FREQUENCY**
7 or more	1,300,000*
6	2,300,000
5	6,800,000
4	15,400,000
3	17,100,000
2	32,700,000
1	25,400,000
Total	101,000,000

Source: 1998 Statistical Abstract of the U.S.
*Rounded to nearest 100,000.

Note: The top class (7 or more) has no upper boundary. Although less preferred, as discussed previously, this type of open-ended class is employed as a space-saving device when, as in the *Statistical Abstract of the U.S.*, many different tables must be listed. The tables in Exercises 2.13 and 2.14 also contain open-ended classes.

(a)   Convert to relative frequencies, expressed as percents. (When computing relative frequencies, you can ignore the five zeros to the right.)

(b)   Convert to a cumulative frequency distribution.

(c)   Convert to a cumulative relative frequency distribution, expressed as percents.

(d)   Find the percentile rank for a household of four people.

2.13   As has been noted, relative frequency distributions are particularly helpful when you're comparing two frequency distributions based on appreciably

U.S. HOUSEHOLDS BY NUMBER OF PERSONS 1950 AND 1997 (IN MILLIONS)		
	FREQUENCY	
HOUSEHOLD SIZE (PEOPLE)	**1950**	**1997**
7 or more	2.1	1.3
6	2.2	2.3
5	4.4	6.8
4	7.7	15.4
3	9.8	17.1
2	12.5	32.7
1	4.7	25.4
Total	43.4	101.0

Source: 1998 Statistical Abstract of the U.S.

different numbers of observations, as in this exercise. The distribution for 1997 is based on 101.0 million households, whereas that for 1950 is based on only 43.4 million households. To facilitate further comparisons, both frequency distributions share a common set of classes.

(a) In the frequency distribution for 1950, how many households contain exactly three people?

(b) Convert each of these distributions to relative frequency distributions (percents). If you've already done Exercise 2.12, you will have to do this only for the 1950 frequencies.

(c) Compare the two relative frequency distributions, noting any conspicuous differences between 1950 and 1997.

*2.14 Are there any conspicuous differences between the two distributions in the following table (one reflecting the ages of all residents of a small town and the other reflecting the ages of all U.S. residents)?

AGE	SMALL TOWN FREQUENCY	U.S. POPULATION 1997 RELATIVE FREQUENCY (%)
TWO AGE DISTRIBUTIONS		
65–above	105	13
60–64	53	4
55–59	45	4
50–54	40	6
45–49	44	7
40–44	38	8
35–39	31	9
30–34	27	8
25–29	25	7
20–24	20	7
15–19	20	7
10–14	19	7
5–9	17	7
0–4	16	7
Total	500	101%

Source: 1998 Statistical Abstract of the U.S.

(a) As an aid in making the desired comparison, convert the age distribution for the small town to a relative frequency distribution.

(b) Describe any conspicuous differences between the two distributions.

Answers on page 495

2.15 The following table shows two different distributions of bachelor's degrees earned in 1995 for selected fields of study. More specifically, the single list of different fields of study is paired, in turn, with the frequencies for all male graduates and for all female graduates.

BACHELOR'S DEGREES EARNED IN 1995 BY FIELD OF STUDY AND GENDER (IN THOUSANDS)		
MAJOR FIELD OF STUDY	**MALES**	**FEMALES**
Life Sciences	26.7	29.3
Business	121.8	112.5
Education	25.7	80.4
Engineering	66.0	12.2
Fine Arts	19.8	28.9
Health Sciences	14.5	65.4
Letters	17.8	34.1
Psychology	19.5	52.5
Social Sciences	68.2	60.0
Total	380.0	475.3

Source: 1998 Statistical Abstract of the U.S.

(a) How many female psychology majors graduated in 1995?

(b) Because the total numbers of male and female graduates are fairly different—380,000 and 475,300—it's helpful to convert first to relative frequencies before making comparisons. Then, inspect these relative frequencies and note the most important differences between male and female graduates.

(c) Would it be meaningful to cumulate the frequencies in either of the previous two frequency distributions?

*2.16 Slightly more complex than any previous table in Chapter 2, the following table shows both a frequency distribution and a relative frequency distribution (of race and Hispanic origin) for the U.S. population in 1980 and also for the U.S. population in 1997. Furthermore, it also shows a comparable frequency distribution for the *change* between the 1980 and 1997 populations and a comparable relative frequency distribution for the percent *change* between the 1980 and 1997 populations.

RACE/HISPANIC ORIGIN OF U.S. POPULATION (IN MILLIONS)						
	1997		**1980**		**1980–97**	
ORIGIN	**FREQUENCY**	**PERCENT**	**FREQUENCY**	**PERCENT**	**CHANGE**	**% CHANGE**
African American	32.3	12%	26.7	12%	5.6	21%
Asian American*	11.4	4%	5.2	2%	6.2	119%
Hispanic	29.3	11%	14.6	6%	14.7	101%
White	194.6	73%	180.1	79%	14.5	8%
Total	267.6	100%	226.6	100%	41.0	18%

Source: 1998 Statistical Abstracts of the U.S.
*Mostly Asians, but also other races, such as American Indians and Eskimos.

Note: The last column expresses the 1980–97 change as a percent of the 1980 population for that row.

(a) Which group changed the most in terms of actual number of people?

(b) Relative to its size in 1980, which group increased most?

(c) Relative to its size in 1980, which group increased less rapidly than the general population?

(d) What is the most striking trend in these data?

Answers on page 495

CHAPTER 3

Describing Data with Graphs

GRAPHS FOR QUANTITATIVE DATA

3.1   HISTOGRAMS
3.2   FREQUENCY POLYGONS
3.3   STEM AND LEAF DISPLAYS
3.4   TYPICAL SHAPES

A GRAPH FOR QUALITATIVE DATA

3.5   BAR GRAPHS

MISLEADING GRAPHS

3.6   SOME TRICKS
3.7   DOING IT YOURSELF
3.8   COMPUTER OUTPUT

Summary

Important Terms

Review Exercises

Data can be described clearly, concisely, and completely with the aid of a well-constructed frequency distribution. And data often can be described even more vividly, particularly when you're attempting to communicate with a general audience, by converting frequency distributions into graphs. This chapter explores some of the most common types of graphs for quantitative and qualitative data.

GRAPHS FOR QUANTITATIVE DATA

3.1 HISTOGRAMS

Histogram

A bar-type graph for quantitative data. No gaps between adjacent bars.

The weight distribution described in Table 2.2 on page 19 appears as a **histogram** in **Figure 3.1**. A casual glance at this histogram confirms previous conclusions: a dense concentration of weights among the 150s, 160s, and 170s, with a spread in the direction of the heavier weights.

Let's pinpoint some of the more important features of histograms. Equal units along the horizontal axis (the X-axis, or abscissa) reflect the various class intervals of the frequency distribution. Equal units along the vertical axis (the Y-axis, or ordinate) reflect increases in frequency. (The units along the vertical

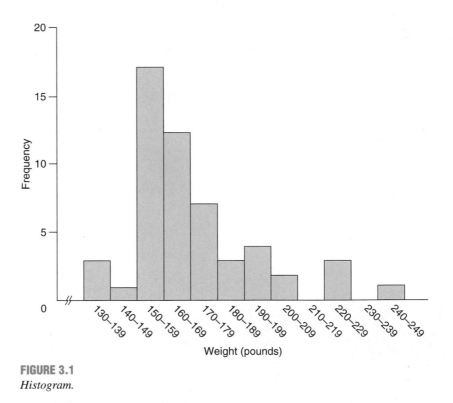

FIGURE 3.1
Histogram.

axis do not have to be the same size as those along the horizontal axis.) The intersection of the two axes defines the origin at which both weight and frequency equal 0. Numbers always increase from left to right along the horizontal axis and from bottom to top along the vertical axis. It's considered good practice to use wiggly lines, such as those along the horizontal axis in Figure 3.1, to spotlight breaks in scale between the origin of 0 and the smallest class interval of 130 to 139. The body of the histogram consists of a series of bars whose heights reflect the frequencies for the various class intervals. Notice that adjacent bars in histograms have common boundaries. The introduction of gaps between adjacent bars would suggest an artificial disruption in the data.

The extensive set of numbers along the horizontal scale of Figure 3.1 can be replaced with a few convenient numbers, as in panel A of **Figure 3.2**. This concession helps avoid excessive cluttering along the horizontal axis.

3.2 FREQUENCY POLYGONS

Frequency polygon

A line graph for quantitative data.

An important variation on a histogram is the **frequency polygon**, or line graph. *Frequency polygons may be constructed directly from frequency distributions.* However, let's follow the step-by-step transformation of a histogram into a frequency polygon, as described in panels A, B, C, and D of Figure 3.2.

A Panel A shows the histogram for the weight distribution.

B Place dots at the midpoints of each bar top (or, in the absence of bar tops, at midpoints on the horizontal axis) and connect them with straight lines.* Notice that the resulting line tends to smooth out the frequency pattern by cutting protruding corners and filling in recessed corners, although at the expense of a completely accurate portrait of the original frequency distribution.

C Anchor the frequency polygon to the horizontal axis. First extend the upper tail to the midpoint of the first unoccupied interval (250–259) on the upper flank of the histogram. Then extend the lower tail to the midpoint of the first unoccupied interval (120–129) on the lower flank of the histogram. Notice the smoothing effect here, too. In addition, all of the area under the frequency polygon is enclosed completely.

D Finally, erase all of the histogram bars, leaving only the frequency polygon.

Comparing Two Frequency Distributions

Frequency polygons are particularly useful when two or more frequency distributions (or relative frequency distributions) are to be included in the same graph. **Figure 3.3** depicts two distributions of grade point averages (GPA), one for a group of meditators and the other for a group of nonmeditators, as described in Exercise 2.11 on page 35. There is considerable variability in both distributions. The most conspicuous difference is the tendency for the GPAs of the meditators to peak about one point higher than do the GPAs of nonmeditators. Should we emphasize this difference or should we ignore it? The techniques for inferential statistics (described in Part 2 of this book) will help us decide.

---

*To find the midpoint of any interval, such as 160–169, simply add the two tabled boundaries (160+169=329) and divide this sum by two (329/2=164.5).

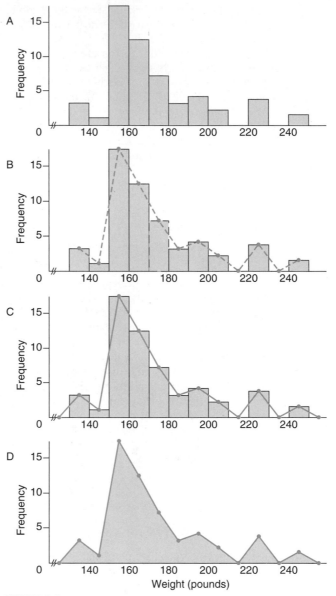

FIGURE 3.2
Transition from histogram to frequency polygon.

*Exercise 3.1 The frequency distribution on page 45 shows the annual incomes in dollars for a group of college graduates.

(a) Construct a histogram. (Notice that the necessary string of class intervals is already given in the following frequency distribution. For step-by-step instructions, refer to the box "Constructing Graphs" on page 55.)

INCOME	FREQUENCY
130,000 – 139,999	1
120,000 – 129,999	0
110,000 – 119,999	1
100,000 – 109,999	3
90,000 – 99,999	1
80,000 – 89,999	5
70,000 – 79,999	7
60,000 – 69,999	10
50,000 – 59,999	14
40,000 – 49,999	23
30,000 – 39,999	17
20,000 – 29,999	10
10,000 – 19,999	8
0 – 9,999	3
Total	103

(b) Construct a frequency polygon.

(c) Is the shape of this distribution balanced or lopsided?

Answers on pages 495–496

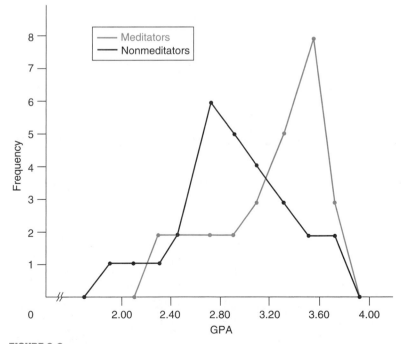

FIGURE 3.3
Two frequency polygons.

3.3 STEM AND LEAF DISPLAYS

Stem and leaf display

A device for sorting quantitative data on the basis of leading and trailing digits.

Still another technique for summarizing quantitative data is a **stem and leaf display**. Relatively recent in origin, it represents a cross between a frequency table and a histogram (and, therefore, could have been introduced in the previous chapter). Stem and leaf displays are ideal for summarizing distributions, such as that for the weight data, without destroying the identities of raw scores (individual observations).

Constructing a Display

The leftmost panel of **Table 3.1** recreates the weights of fifty-three male statistics students listed in Table 1.1. To construct the stem and leaf display for these data, first note that when counting by tens, the weights range from the 130s to the 240s. Arrange a column of numbers, the stems, beginning with 13 (representing the 130s) and ending with 24 (representing the 240s). Draw a vertical line to separate the stems, which represent multiples of ten, from the space to be occupied by the leaves, which represent multiples of one.

Next, enter each raw score into the stem and leaf display. Thus, as suggested by the color coding in Table 3.1, the first raw score of 160 reappears as a leaf of 0 on a stem of 16. By the same token, the next raw score of 193 reappears as a leaf of 3 on a stem of 19, whereas the third raw score of 226 reappears as a leaf of 6 on a stem of 22, and so on, until each raw score reappears as a leaf on its appropriate stem.

Table 3.1
CONSTRUCTING STEM AND LEAF DISPLAY FROM WEIGHTS OF MALE STATISTICS STUDENTS

RAW SCORES				STEM AND LEAF DISPLAY	
160	165	135	175		
193	168	245	165	13	3 5 5
226	169	170	185	14	5
152	160	156	154	15	2 7 1 7 8 0 2 0 2 6 9 8 2 6 4 7 6
180	170	160	179	16	0 3 5 8 9 0 0 0 6 5 5 5
205	150	225	165	17	2 0 0 0 2 5 9
163	152	190	206	18	0 0 5
157	160	159	165	19	3 0 0 0
151	190	172	157	20	5 6
157	150	190	156	21	
220	133	166	135	22	6 0 5
145	180	158		23	
158	152	152		24	5
172	170	156			

Interpretation

Notice that the weight data have been sorted by the stems. All weights in the 130s are listed together; all of those in the 140s are listed together, and so on. A glance at the stem and leaf display in Table 3.1 shows essentially the same pattern of weights as depicted by the frequency distribution in Table 2.2 and the histogram in Figure 3.1.

Resemblance to Histograms

A simple maneuver demonstrates the resemblance between a stem and leaf display and a histogram. As shown in **Figure 3.4**, if the entire stem and leaf display for weights is rotated counterclockwise one-quarter of a full turn, the silhouette of the display is the same as the histogram for the weight data. (Notice that this simple maneuver only works if, as in the present display, stems are listed from smallest at top to largest at bottom—one reason why the customary ranking for most tables in this book has been reversed for stem and leaf displays.)

Selection of Stems

Stem values are not limited to units of ten. Depending on the data, you might identify the stem with one or more leading digits that culminates in some variation on a stem value of 10, such as 1, 100, 1000, or even .1, .01, .001, and so on. For instance, an annual income of 23,784 dollars could be displayed as a stem of 23 (thousands) and a leaf of 784. (Leaves, such as 784, consisting of two or more digits are separated by commas.) A Scholastic Assessment Test (SAT) test score of 689 could be displayed as a stem of 6 (hundreds) and a leaf of 89. A grade point average (GPA) of 3.25 could be displayed as a stem of 3 (ones) and a leaf of 25, or if you wanted more than a few stems, 3.25 could be displayed as having stem of 32 (one-tenths) and a leaf of 5.

Stem and leaf displays represent statistical "bargains." Just a few minutes of work produce a description of data that is both clear and complete. Although rarely appearing in official reports and publications, they have won favor among many practitioners, and they often serve as the first step toward a summary based on more traditional devices, such as frequency tables and histograms.

*Exercise 3.2 Construct a stem and leaf display for the following IQ scores obtained from a group of four-year-old children.

120	98	118	117	99	111
126	85	88	124	104	113
108	141	123	137	78	96
102	132	109	106	143	

Answers on page 496

ROTATED STEM AND LEAF DISPLAY

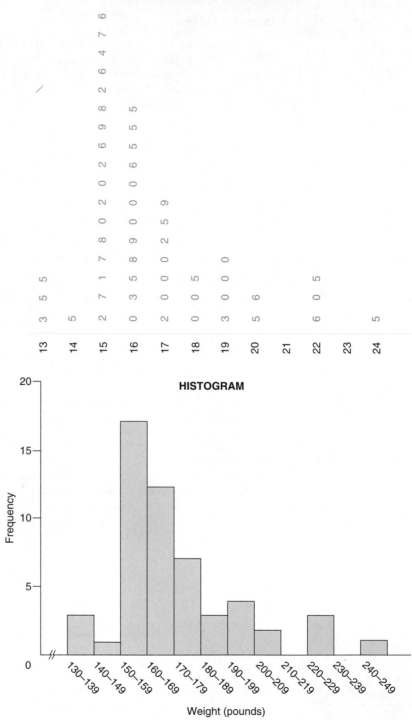

HISTOGRAM

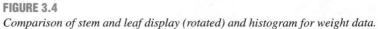

FIGURE 3.4

Comparison of stem and leaf display (rotated) and histogram for weight data.

3.4 TYPICAL SHAPES

Whether expressed as a histogram, a frequency polygon, or a stem and leaf display, an important characteristic of a frequency distribution is its shape. **Figure 3.5** shows some of the more typical shapes for smoothed frequency polygons (that ignore the inevitable irregularities of real data).

Normal

Any distribution that approximates the normal shape in panel A of Figure 3.5 can be analyzed, as we'll see in Chapters 6 and 7, with the aid of the well-documented normal curve. The familiar bell-shaped silhouette of the normal curve can be superimposed on many frequency distributions, including those for uninterrupted gestation periods of human fetuses, scores on standardized tests, and even the popping times of individual kernels in a batch of popcorn.

Bimodal

Any distribution that approximates the bimodal shape in panel B of Figure 3.5 might, as suggested in the preceding chapter, reflect the coexistence of two different types of observations in the same distribution. For instance, the distribution of the ages of residents in a neighborhood consisting largely of new

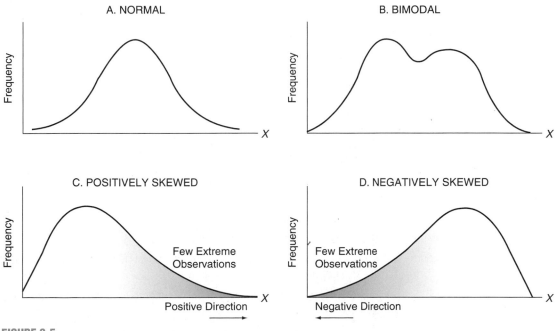

FIGURE 3.5
Typical shapes.

parents and their infants has a bimodal shape. That distributions with two peak frequencies should be described as bimodal will make more sense after you've read Section 4.1.

Positively Skewed

The two remaining shapes in Figure 3.5 are lopsided. *Lopsided distributions caused by a few extreme observations with relatively large values, as in panel C of Figure 3.5, are* **positively skewed distributions,** because the few extreme observations are in the positive direction (to the right of the majority of observations). The distribution of income among U.S. families has a pronounced positive skew, with most family incomes under $100,000 and relatively few family incomes spanning a wide range of values above $100,000. The distribution of weight in Figures 3.1, 3.2, and 3.4 also is positively skewed.

Negatively Skewed

Lopsided distributions caused by a few extreme observations with relatively small values, as in panel D of Figure 3.5, are **negatively skewed distributions**, because the few extreme observations are in the negative direction (to the left of the majority of observations). The distribution of age at retirement among U.S. job holders has a pronounced negative skew, with most retirement ages at sixty years or older and relatively few retirement ages spanning the wide range of ages younger than sixty.

Positively or Negatively Skewed?

Some people have difficulty with this terminology, probably because an entire distribution is labeled on the basis of the relative location, in the positive or negative direction, of a few extreme observations, rather than on the basis of the location of the majority of observations. To make this distinction, always force yourself to focus on the relative locations of the few extreme observations. If you get confused, use panels C and D of Figure 3.5 as guides, noting which silhouette in these two panels best approximates the shape of the distribution in question.

*Exercise 3.3 Describe the probable shape—normal, bimodal, positively skewed, or negatively skewed—for each of the following distributions:

(a) Female beauty contestants' scores on a masculinity test, with a higher score indicating a greater degree of masculinity.

(b) Scores on a standardized IQ test for a group of people selected from the general population.

(c) Test scores for a group of high school students on a very difficult college-level exam.

(d) Reading achievement scores for a fourth grade class consisting of about equal numbers of regular students and learning-impaired students.

Positively skewed distribution

A distribution that includes a few extreme observations with relatively large values in the positive direction.

Negatively skewed distribution

A distribution that includes a few extreme observations with relatively small values in the negative direction.

(e) Scores of students at the Eastman School of Music on a test of music aptitude (designed for use with the general population).
Answers on page 496

A GRAPH FOR QUALITATIVE DATA

3.5 BAR GRAPHS

Bar graph

A bar-type graph for qualitative data. Gaps between adjacent bars.

The distribution in Table 2.8, based on replies to the question "Have you ever smoked marijuana?" appears as a **bar graph** in **Figure 3.6.** A glance at this graph confirms that Yes replies occur approximately twice as often as No replies.

As with histograms, equal segments along the horizontal axis are allocated to the different words or classes that appear in the frequency distribution for qualitative data. Likewise, equal segments along the vertical axis reflect in-creases in frequency. The body of the bar graph consists of a series of bars whose heights reflect the frequencies for the various words or classes.

A person's answer to the question "Have you ever smoked marijuana?" is either Yes or No, not some impossible intermediate value, such as 40 percent Yes and 60 percent No. Gaps are placed between adjacent bars of bar graphs to emphasize the essentially discontinuous nature of qualitative data. That's why we rarely use histograms and frequency polygons to depict qualitative data.

*Exercise 3.4 Referring to the box on page 55 for step-by-step instructions, construct a bar graph for the data shown in the table on the next page.

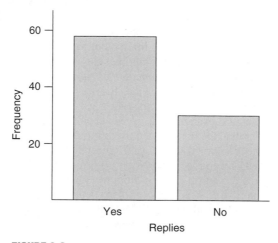

FIGURE 3.6
Bar graph.

RACE/HISPANIC ORIGIN OF U.S. POPULATION, 1997 (IN MILLIONS)	
ORIGIN	**FREQUENCY**
African American	32.3
Asian American	11.4
Hispanic	29.3
White	194.6
Total	267.6

Source: 1998 Statistical Abstracts of the U.S.

Answer on page 496

MISLEADING GRAPHS

3.6 SOME TRICKS

Graphs can be constructed in an unscrupulous manner to support a particular point of view. Indeed, this type of statistical fraud gives credibility to a variety of popular quotations, including "Numbers don't lie, but statisticians do" and "There are three kinds of lies—lies, damned lies, and statistics."

For example, to imply that comparatively many students responded Yes to the marijuana-smoking question, an unscrupulous person might resort to the various tricks shown in **Figure 3.7**. First, the width of the Yes bar is more than three times that for the No bar, thus violating the custom that bars be equal in width. Second, the lower end of the frequency scale is omitted, thus violating the custom that the entire scale be reproduced, beginning with zero. (Otherwise a broken scale should be highlighted by wiggly lines, as in Figures 3.1, 3.2, and 3.3.) Third, the height of the vertical axis is several times the width of the horizontal axis, thus violating the custom, heretofore unmentioned, that the vertical axis be *approximately* as tall as the horizontal axis is wide. Beware of graphs in which, because the vertical axis is many times larger than the horizontal axis, as in Figure 3.7, frequency differences are exaggerated, or in which, because the vertical axis is many times smaller than the horizontal axis, frequency differences are suppressed.

The combined effect of Figure 3.7 is to imply that virtually all of the students responded Yes. Notice the radically different impressions created by Figures 3.6 and 3.7, even though both are based on exactly the same data. To heighten your sensitivity to this type of distortion, and to other types of statistical frauds, read the highly entertaining book by Darrell Huff and Irving Geis, *How to Lie with Statistics* (New York: Norton, 1954).

*Exercise 3.5 Criticize the graphs that appear at the top of page 54 (ignore the inadequate labeling of both axes).

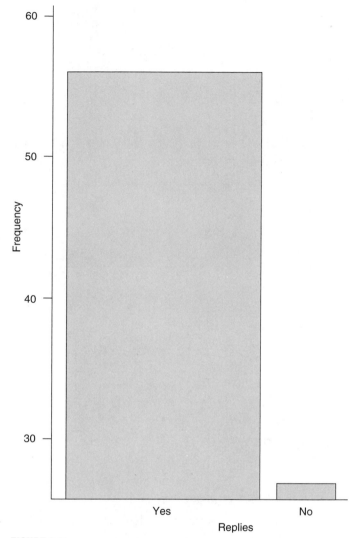

FIGURE 3.7
Distorted bar graph.

3.7 DOING IT YOURSELF

When you are constructing a graph, attempt to depict the data as clearly, concisely, and completely as possible. The blatant distortion shown in Figure 3.7 can be easily avoided by complying with the several customs described in the preceding section and by following the step-by-step procedure listed in the box on page 55. Otherwise, equally valid graphs of the same data might appear in

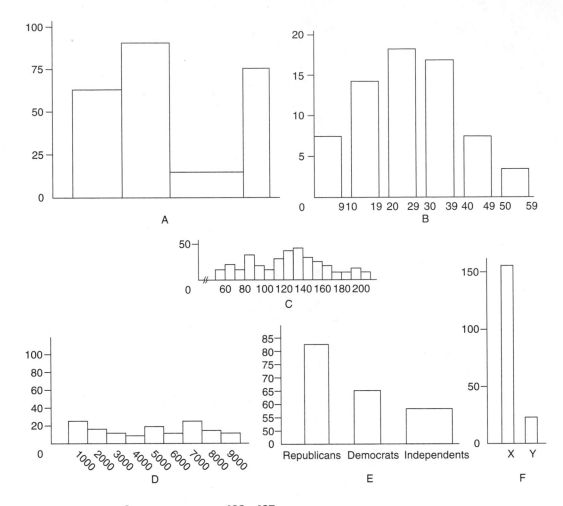

Answers on pages 496–497

different formats. It's often a matter of personal preference whether, for instance, a histogram or a frequency polygon should be used with quantitative data. Once again, as with the construction of frequency distributions, relax. With just a little practice, such as that provided by the exercises in this chapter, you should become a competent constructor of graphs.

INTERNET DEMONSTRATION

Go to the Web site for this book **(http://darwin.cwru.edu/ ~witte/statistics)** and click on Histogram to see the effect of varying bar width on the shape of a batch of data.

CONSTRUCTING GRAPHS

1. Decide on the appropriate type of graph, recalling that histograms and frequency polygons are appropriate for quantitative data, whereas bar graphs are appropriate for qualitative data.

2. Within the available space, use a ruler to *draw the horizontal axis, then the vertical axis,* remembering that the vertical axis should be about as tall as the horizontal axis is wide.

3. Identify the string of class intervals that eventually will be superimposed on the horizontal axis. For qualitative data or ungrouped quantitative data, this is easy—merely use the classes suggested by the data. For grouped quantitative data, proceed as if you were creating a set of class intervals for a frequency distribution. (See the box in Chapter 2 on page 20.)

4. Superimpose the string of class intervals (with gaps for bar graphs) along the entire length of the horizontal axis. For histograms and frequency polygons, be prepared for some trial and error—use a pencil! Don't use a string of empty class intervals to bridge a sizable gap between the origin of 0 and the smallest class interval. Instead, use wiggly lines to signal a break in scale, then begin with the smallest class interval. Also, don't clutter the horizontal axis with excessive numbers—just use a few convenient numbers.

5. Along the entire length of the vertical axis, superimpose a progression of convenient numbers, beginning at the bottom with 0 and ending at the top with a number as large as or slightly larger than the maximum observed frequency. If there's a considerable gap between the origin of 0 and the smallest observed frequency, use wiggly lines to signal a break in scale.

6. Using the scaled axes, *construct bars (or dots and lines) to reflect the frequency of observations within each class interval.* For frequency polygons, dots should be located above the midpoints of class intervals, and both tails of the graph should be anchored to the horizontal axis, as described in Section 3.2.

7. Supply labels for both axes and a title (or even an explanatory sentence) for the graph.

3.8 COMPUTER OUTPUT

For simplicity, most analyses in this book are performed by hand on small batches of data. When analyses are based on large batches of data, as often happens in practice, it's much more efficient to use a computer. Although this book won't attempt to describe how to enter instructions and data into a computer, it will describe how to interpret the most relevant portions of computer outputs. Once you've learned to ignore irrelevant details and references to more advanced statistical procedures, you'll find that statistical results produced by computers aren't any more difficult to interpret than those produced by hand.

Three of the most widely used statistical programs, Minitab, SPSS (Statistical Package for the Social Sciences), and SAS (Statistical Analysis System), are used to generate computer outputs in this book. As interpretive aids, key results are cross-referenced with explanatory comments at the bottom of the printout. Since these outputs are based on data already analyzed by hand, computer-produced results can be compared with familiar results. For example, the computer-produced stem and leaf display in **Table 3.2** can be compared with the manually produced stem and leaf display in Figure 3.1.

INTERNET SITES:

Go to the Web site for this book **(http://darwin.cwru.edu/ ~witte/statistics)** and click on **Minitab, SPSS,** or **SAS** to obtain more information about these statistical packages, as well as demonstration software.

*Exercise 3.6 The following Minitab output shows a histogram for the grade point averages (GPAs) of students who are meditators, as described in Exercise 2.11 on page 35.

(a) What is the frequency count for the class interval with a midpoint of 3.30?

(b) What is the shape of this histogram?

Answers on page 497

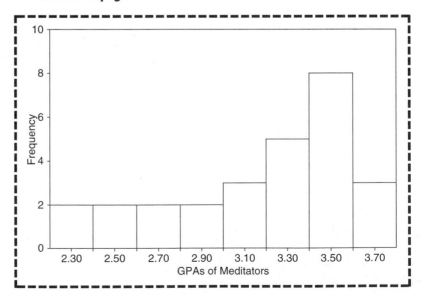

Table 3.2
MINTAB OUTPUT: STEM AND LEAF DISPLAY FOR WEIGHTS OF FEMALE STATISTICS STUDENTS

CHARACTER STEM-AND-LEAF DISPLAY

1 Steam-and-leaf of Weight   N = 53
Leaf Unit = 10

2		
3	13	355
4	14	5
21	15	00122224666777889
(12)	16	000035555689
20	17	0002259
13	18	005
10	19	0003
6	20	56
4	21	
4	22	056
1	23	
1	24	5

Comments

1 Note that the silhouette of this stem and leaf display is identical to that shown in Table 3.1.

2 Although this particular column requires more effort to understand than its value merits, it is cross-referenced for the sake of completeness. (And you might choose not to read any further) Parentheses enclose a number indicating how many leaves (12) lie on the stem line that contains the middle observation(s). Remaining numbers in this column indicate how many leaves lie on that particular line and all lines farther away from the middle. For example, the line for a stem of 10 has a value of 4 beacuse one leaf lies on that stem line and three leaves lie farther away from the middle.

Summary
. . . .

Frequency distributions can be converted to graphs.

If the data are quantitative, histograms, frequency polygons, or stem and leaf displays are often used. Histograms may be transformed into frequency polygons first by constructing a line that skims the peaks of the bars and then by erasing the bars. Frequency polygons are particularly useful when two or more frequency distributions are to be included in the same graph. Stem and leaf displays often serve as the first step toward a summarization based on more traditional devices.

Shape is an important characteristic of a histogram or a frequency polygon. Smoothed frequency polygons were used to describe four of the more typical shapes: normal, bimodal, positively skewed, and negatively skewed.

If the data are qualitative, bar graphs are often used. Bar graphs resemble histograms except that gaps separate adjacent bars in bar graphs.

When interpreting graphs, beware of various unscrupulous techniques, such as using bizarre combinations of axes, either to exaggerate or to suppress a particular data pattern. When you are constructing graphs, refer to the box "Constructing Graphs."

Important Terms

Histogram	**Positively skewed distribution**
Frequency polygon	**Negatively skewed distribution**
Stem and leaf display	**Bar graph**

REVIEW EXERCISES

3.7

(a) Construct a histogram, a frequency polygon, and a stem and leaf display for the data in Exercise 2.2 on page 25.

(b) Describe the shape of this distribution.

3.8

(a) Construct a histogram for the frequency distribution in Exercise 2.12 on page 00. Note: When segmenting the horizontal axis, treat the open-ended interval (7 or more) the same as any other class interval. (This tactic causes some distortion at the upper end of the histogram, because one class interval is doing the work of several. Nothing is free, including the convenience of open-ended intervals.)

(b) Describe the shape of this distribution.

3.9

(a) Construct a frequency polygon and a stem and leaf display for the data in Exercise 2.10 on page 34.

(b) Describe the shape of the distribution.

3.10

(a) Construct a frequency polygon for the relative frequency distribution obtained in Exercise 2.5 on page 27.

(b) Describe the shape of this distribution.

3.11 Using just one graph, construct frequency polygons for the two relative frequency distributions obtained in Exercise 2.14 on page 37. Note: When segmenting the horizontal axis, treat the open-ended intervals (65–above) the same as any other class interval. See the parenthetical statement in Exercise 3.8.

3.12 Using just one graph, construct bar graphs for the distribution of field of study for all male graduates in 1995 and that for all female graduates in 1995, as shown in Exercise 2.15 on page 37. *Hint:* Alternate unshaded and shaded bars for males and females, respectively.

CHAPTER
4

Describing Data with Averages

AVERAGES FOR QUANTITATIVE DATA

4.1   **MODE**
4.2   **MEDIAN**
4.3   **MEAN**
4.4   **WHICH AVERAGE?**
4.5   **SPECIAL STATUS OF THE MEAN**
4.6   **AVERAGES FOR QUALITATIVE DATA**
4.7   **USING THE WORD *AVERAGE***

Summary

Important Terms

Review Exercises

You might give up a bad habit such as smoking because, *on the average,* heavy smokers have a shorter life expectancy than nonsmokers. You might buy a particular car because, *on the average,* that type of car gives better gas mileage. You might strengthen your resolve to graduate from college upon hearing that, *on the average,* the lifetime earnings of college graduates exceed those of the general population. Or, in a moment of weakness, you even might consider altering your hair color because acquaintances with other hair colors, *on the average,* seem to have more fun. Averages occur regularly in our everyday life, and they are important tools in statistics. A well-chosen average consists of a single number (or word) about which the data are, in some sense, centered. Actually, even for a given set of data, there can be several different types of averages or, as they're sometimes called, **measures of central tendency.** This chapter describes both how to calculate and how to interpret three commonly employed measures of central tendency.

AVERAGES FOR QUANTITATIVE DATA

4.1 MODE

The *mode* reflects the value of the most frequently occurring observation.

Table 4.1 shows the number of years served by the last twenty U.S. presidents, beginning with Benjamin Harrison (four years) and ending with Bill Clinton (eight years). Four years is the modal term, because the greatest number of presidents, seven, served this term. Note that the mode equals the value of the most frequently occurring observation, four, *not* the frequency of that observation, seven.

It's easy to assign a value to the mode. If the data are organized, a glance will often be enough. But if the data are not organized, some counting may be required. The mode is readily understood as the value of the most prevalent observation.

More Than One Mode

Distributions can have more than one mode (or no mode at all!). *Distributions with two obvious peaks, even though they're not exactly the same height, are referred to as* **bimodal.** Distributions with more than two peaks are referred to as multimodal. The presence of more than one mode, particularly in large sets of data, might reflect important differences among subsets of data. For instance, the distribution of weights for both male and female statistics students would most likely be bimodal, reflecting the combination of two separate weight distributions—one for males and the other for females. Notice that even the distribution of presidential terms in **Figure 4.1** tends to be bimodal, with a major peak at four years and a minor peak at eight years, reflecting the two normal terms of office.

*Exercise 4.1 Determine the mode for the following distribution of retirement ages: 60, 63, 45, 63, 65, 70, 55, 63, 60, 65, 63.

Measures of central tendency

General term for the various averages.

Mode

The value of the most frequent observation.

Bimodal

Describes any distribution with two obvious peaks.

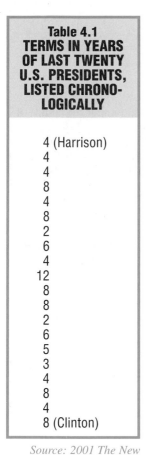

Table 4.1
TERMS IN YEARS OF LAST TWENTY U.S. PRESIDENTS, LISTED CHRONOLOGICALLY

4 (Harrison)
4
4
8
4
8
2
6
4
12
8
8
2
6
5
3
4
8
4
8 (Clinton)

Source: 2001 The New York Times Almanac.

Median

The middle value when observations are ordered.

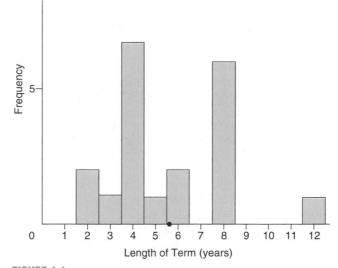

FIGURE 4.1
Distribution of presidential terms.

*Exercise 4.2 The owner of a new car conducts a series of six gas mileage tests and obtains the following results, expressed in miles per gallon: 20.3, 22.7, 21.4, 20.6, 21.4, 20.9. Find the mode for these data.

Answers on page 497

4.2 MEDIAN

The *median* reflects the middle value when observations are ordered from least to most, or vice versa.

The median splits a set of ordered observations into two equal parts, the upper and lower halves. In other words, the median has a percentile rank of 50, because observations with equal or smaller values constitute 50 percent of the entire distribution.＊

Finding the Median

Table 4.2 shows how to find the median for two different sets of observations. The numbers in shaded squares cross-reference instructions in the top panel with examples in the bottom panel. Study Table 4.2 before reading on.

To find the median, *observations always must be ordered from least to most, or vice versa.* This task is straightforward with small sets of data but becomes

---

＊Strictly speaking, the median always has a percentile rank of *exactly* 50 only insofar as interpolation procedures, not discussed in this book, identify the value of the median with a single point along the numerical scale for the data.

Table 4.2
FINDING THE MEDIAN

A. INSTRUCTIONS
  1 Order observations from least to most.
  2 Find the middle position by adding one to the total number (or count) of observations and dividing by two.
  3 *If the middle position is a whole number,* as in the left-hand panel below, use this number to *count* into the set of ordered observations.
  4 The value of the median equals the value of the observation located at the middle position.
  5 *If the middle position is not a whole number,* as in the right-hand panel below, use the two nearest whole numbers to *count* into the set of ordered observations.
  6 The value of the median equals the value midway between those of the two middlemost observations; to find the midway value, add the two given values and divide by two.

B. EXAMPLES

Set of five observations:	Set of six observations:
2, 8, 2, 7, 6	3, 8, 9, 3, 1, 8
1 2, 2, 6, 7, 8	1 1, 3, 3, 8, 8, 9
2 $\dfrac{5+1}{2} = 3$	2 $\dfrac{6+1}{2} = 3.5$
2, 2, 6, 7, 8	
3 1 2 3	
4 median = 6	
	1, 3, 3, 8, 8, 9
	5 1 2 3 4
	6 median = $\dfrac{3+8}{2} = 5.5$

increasingly cumbersome with larger sets of data, particularly when you must start from scratch with unorganized data.

When the total number of observations is odd, as in the lower left-hand panel of Table 4.2, there is a single middle-ranked observation, and the value of the median equals the value of this observation. When the total number of observations is even, as in the lower right-hand panel of Table 4.2, the value of the median is set equal to a value midway between the values of the two middlemost observations. In either case, the value of the median always reflects the value of middle-ranked observations, not the position of these observations among the set of ordered observations.

The median term can, of course, be found for the last twenty presidents. First, the terms must be ordered from longest (Franklin Roosevelt) to shortest

Table 4.3
TERMS IN YEARS OF LAST TWENTY U.S. PRESIDENTS

ARRANGED BY LENGTH	DEVIATION FROM MEAN	SUM OF DEVIATIONS
12	6.40	
8	2.40	
8	2.40	
8	2.40	
8	2.40	21.6
8	2.40	
8	2.40	
6	0.40	
6	0.40	
(mean = 5.60)		0
5	−0.60	
4	−1.60	
4	−1.60	
4	−1.60	
4	−1.60	
4	−1.60	−21.6
4	−1.60	
4	−1.60	
3	−2.60	
2	−3.60	
2	−3.60	

(Harding and Kennedy), as shown in the left-hand column of **Table 4.3.** Then, following the instructions in Table 4.2, you can verify that the median term for the last twenty presidents equals 4.5 years, because 4.5 is the value midway between the values (4 and 5) of the two middlemost (tenth- and eleventh-ranked) observations in Table 4.3.

*Exercise 4.3 Find the median for the following distribution of retirement ages: 60, 63, 45, 63, 65, 70, 55, 63, 60, 65, 63.

*Exercise 4.4 Find the median for the following gas mileage tests: 20.3, 22.7, 21.4, 20.6, 21.4, 20.9.

Answers on page 497

4.3 MEAN

The mean is the most common average, one you have doubtless calculated many times. Among the three averages, it also is the most useful.

The mean is found by adding all observations and then dividing by the number of observations.

That is,

$$Mean = \frac{sum\ of\ all\ observations}{number\ of\ observations}$$

To find the mean term for the last twenty presidents, add all twenty terms in Table 4.1 $(4 + \ldots + 4 + 8)$ to obtain a sum of 112 years, and then divide this sum by 20, the number of presidents, to obtain a mean of 5.60 years.

Note that observations need not be ordered from least to most before calculating the mean. Even when large sets of unorganized data are involved, the calculation of the mean is usually straightforward, particularly with the aid of a calculator.

Sample or Population?

Statisticians distinguish between two types of means—the population mean and the sample mean—depending on whether the data are viewed as a **population**, that is, *a complete set of observations*, or as a **sample**, that is, *a subset of observations from a population*. For example, if the terms of the last twenty U.S. presidents are viewed as a population, then 5.60 years qualifies as a population mean. On the other hand, if the terms of the last twenty U.S. presidents are viewed as a sample from the terms of *all* U.S. presidents, then 5.60 years qualifies as a sample mean. Not only is the present distinction entirely a matter of perspective, but it also produces exactly the same numerical value of 5.60 for both means. Nevertheless, this distinction will be introduced here because of its importance in later chapters. *Until then, unless noted otherwise, you can assume that we will be dealing with the sample mean.*

Formula for Sample Mean

It's usually more efficient to substitute symbols for words in statistical formulas, including the word formula just stated for the sample mean. When symbols are used, $\overline{X}$, designates the sample mean, and the formula becomes and

SAMPLE MEAN	
$$\overline{X} = \frac{\Sigma X}{n}$$	(4.1)

reads: "X-bar equals the sum of the variable X divided by the sample size n." (Note that the uppercase Greek letter sigma [Σ] is read as *the sum of,* not as *sigma*. To avoid confusion, read only the lowercase Greek letter sigma [σ] as "sigma," which, as described in Chapter 5, has an entirely different meaning in statistics.)

In Formula 4.1, the variable X can be replaced, in turn, by each of the twenty presidential terms in Table 4.1, beginning with 4 and ending with 8. The symbol Σ, the uppercase Greek letter sigma, specifies that all observations represented by the variable X should be added $(4 + \ldots + 4 + 8)$ to find the sum

---

Population

A complete set of observations.

Sample

A subset of observations from a population.

Sample mean ($\overline{X}$)

The balance point for a sample, found by dividing the total value of all observations in the sample by the sample size.

Sample size (n)

The total number of observations in the sample.

of 112. (Notice that this sum contains the values of *all* observations *including duplications*.) Then this sum should be divided by *n*, the sample size—20 in the present example—to obtain the mean presidential term of 5.60 years.

Formula for Population Mean

The formula for the population mean differs from that for the sample mean only because of a change in symbols. Now, the population mean is represented by μ, the Greek letter for m (pronounced "mu"),

POPULATION MEAN

$$\mu = \frac{\Sigma X}{N} \qquad (4.2)$$

Population mean (μ)

The balance point for a population, found by dividing the total value of all observations in the population by the population size.

Population size (N)

The total number of observations in the population.

where the uppercase letter *N* refers to the **population size.** Otherwise, the calculations are exactly the same as those for the sample mean.

Mean as Balance Point

The mean serves as the balance point for a frequency distribution.

Imagine that the histogram in Figure 4.1, showing the terms of the last twenty presidents, has been constructed out of some rigid material such as wood. Furthermore, imagine that you wish to lift the histogram by using only one finger placed under its base and without disturbing the horizontal balance of the histogram. To accomplish this, your finger should coincide with 5.60, the value of the mean, shown as a dot in Figure 4.1. If your finger were to the right of this point, the entire histogram would seesaw down to the left; if your finger were to the left of this point, the histogram would seesaw down to the right.

The mean serves as the balance point for any distribution because of a special property: *The sum of all observations, expressed as positive and negative deviations from the mean, always equals zero*. In the right-hand column of Table 4.3 on page 65, each original value reappears as a deviation from the mean, obtained by taking each original value (including duplications) one at a time and subtracting the mean. Original values above the mean reappear as positive deviations (for example, 12 reappears as a positive deviation of 6.40 from the mean, because 12 − 5.60 = 6.40), whereas original values below the mean reappear as negative deviations (for example, 2 reappears as a negative deviation of −3.60 from the mean, because 2 − 5.60 = −3.60). As suggested in Table 4.3, when the total of all positive deviations, 21.6 is combined with the total of all negative deviations, −21.6, the resulting sum equals zero.

In its role as balance point, the mean describes the single point of equilibrium at which, once all observations have been expressed as deviations from the mean, those above the mean counterbalance those below the mean. You can appreciate, therefore, why a change in the value of a single observation produces a change in the value of the mean for the entire distribution. In other words, the mean is affected by the values of all observations, not just by those that are middle ranked, as with the median, or by those that occur most frequently, as with the mode.

*Exercise 4.5 Find the mean for the following distribution of retirement ages: 60, 63, 45, 63, 65, 70, 55, 63, 60, 65, 63.

*Exercise 4.6 Find the mean for the following gas mileage tests: 20.3, 22.7, 21.4, 20.6, 21.4, 20.9.

    Answers on page 497

4.4 WHICH AVERAGE?

If Distribution Is Not Skewed

When a distribution of observations is not too skewed, the values of the mode, the median, and the mean are similar, and any of them may be used to describe the central tendency of the distribution. This is the case in Figure 4.1, in which the mode, median, and mean equal 4, 4.5, and 5.60 years, respectively.

If Distribution Is Skewed

When, however, extreme observations cause a distribution to be skewed, as for the infant death rates for selected countries listed in **Table 4.4,** the values of the three averages can differ appreciably. The modal infant death rate equals 6 (because the largest number of countries, 7, has this rate). The median infant death rate equals 8 (because the United States, with a death rate of 8, occupies the middle-ranked, or eleventh, position among the 21 ranked countries). Finally, the mean infant death rate equals 29.29 (from the sum of all rates, 615, divided by the number of countries, 21).

The mode and the median are not very sensitive to extreme observations. The mode emphasizes the most frequently occurring observation, and the median emphasizes the middle-ranked observation. On the other hand, the mean is very sensitive to extreme observations. Any extreme observation, such as the high infant death rate of 158 listed for Afghanistan in Table 4.4, contributes directly to the calculation of the mean and, with arithmetic inevitability, sways the value of the mean in its direction.

Interpreting Differences between Mean and Median

Ideally, when a distribution is skewed, both the mean and the median should be reported. Appreciable differences between the values of the mean and median signal the presence of a skewed distribution. If the mean exceeds the median, as it does slightly for the presidential terms and much more so for the infant death rates, the underlying distribution will be positively skewed because of one or more observations with extremely large values: for the previous distributions, Franklin Roosevelt's twelve-year presidential term and a number of countries, including Afghanistan, with extremely high infant death rates. On the other hand, if the median exceeds the mean, the underlying distribution will be negatively skewed because of one or more observations with extremely small values. **Figure 4.2** summarizes the relationship between the various averages and the two types of skewed distributions (shown as smoothed curves).

Table 4.4 INFANT DEATH RATES FOR SELECTED COUNTRIES	
COUNTRY	**INFANT DEATH RATE***
Afghanistan	158
Cambodia	108
India	68
Turkey	48
Brazil	44
China	34
Mexico	33
Syria	32
Poland	14
Hungary	11
United States	8
Spain	7
Australia	6
Canada	6
Denmark	6
France	6
Germany	6
Netherlands	6
United Kingdom	6
Japan	4
Sweden	4

Source: *1997 World Development Indicators.* **Rates are per 1000 live births.*

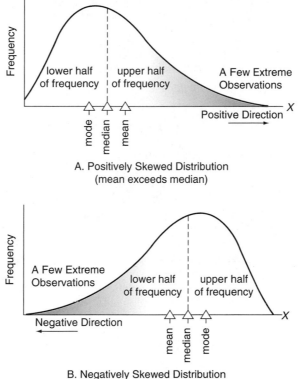

FIGURE 4.2
Mode, median, and mean in positively and negatively skewed distributions.

INTERNET DEMONSTRATION

Go to the Web site for this book **(http://darwin.cwru.edu/~witte/statistics)** and click on **Mean and Median** to see how these two averages vary with changes in histograms specified by you.

---

*Exercise 4.7 Indicate whether the mean exceeds the median, or vice versa, for each of the following skewed distributions.

(a) A negatively skewed distribution of test scores on an easy test, with most students scoring high, but with a few students scoring low.

(b) A positively skewed distribution of ages of college students, with most students in their late teens or early twenties, but with some students in their fifties or sixties.

(c) A positively skewed distribution of loose change carried by classmates, with most carrying less than one dollar, but with some carrying as much as three or four dollars' worth of loose change.

(d) A negatively skewed distribution of crowd sizes in attendance at a popular movie theater, with most audiences at or near capacity.

Answers on page 497

4.5 SPECIAL STATUS OF THE MEAN

As has been seen, the mean sometimes fails to typify a set of observations and, therefore, should be used in conjunction with another average, such as the median. In the long run, however, *the mean serves as the single most preferred average for quantitative data.* There are many reasons for this. Essentially, because the mean describes the balance point of any distribution, it reappears—out of mathematical necessity—throughout the remainder of this book. Sometimes, as in each of the next four chapters, it serves as a key component in a more complex statistical measure. At other times, as in some of the later chapters, it emerges as a well-documented measure to use when generalizing from samples to populations. Indeed, after this chapter, the mean will be used almost to the exclusion of the median and mode.

4.6 AVERAGES FOR QUALITATIVE DATA

Mode Always Appropriate

So far, we've been talking about quantitative data for which, in principle, all three averages can be used. But when the data are qualitative, your choice among averages is restricted. *The mode always can be used with qualitative data.* For instance, Yes qualifies as the modal response for the marijuana-smoking question, because Yes occurred more frequently than No. By the same token, it would be appropriate to report that type O is the modal blood type of prospective donors (Exercise 2.9 on page 32) and that White is the modal race of Americans (Exercise 3.4 on page 5).

Median Sometimes Appropriate

The median can be used whenever it's possible to order qualitative data from least to most, or vice versa. It's easiest to determine the median class for ordered qualitative data by using relative frequencies, as in **Table 4.5.** (Otherwise, first convert regular frequencies to relative frequencies.) Cumulate the relative frequencies, working up from the bottom of the distribution, until the cumulative proportion or percent first equals or exceeds .50, or 50 percent. Because the corresponding class includes the median and, roughly speaking, splits the distribution into an upper and a lower half, it is designated as the median class. For instance, the qualitative data in Table 4.5 can be ordered from lowly warrant officer to exalted general. Starting at the bottom of Table 4.5 and cumulating upward (in color), we have a cumulative percent of 37.2 for the class of lieutenant and 66.7 percent for the class of captain. Accordingly, because it includes a cumulative percent of 50, captain is the median rank of officers in the U.S. Army.

Table 4.5
FINDING THE MEDIAN FOR ORDERED QUALITATIVE DATA: RANKS OF OFFICERS IN THE U.S. ARMY (1997)

RANK	%	CUMULATIVE %
General	0.4	
Colonel	16.3	
Major	16.6	
Captain	29.5	+37.2 = 66.7
Lieutenant	22.1	+15.1 = 37.2
Warrant Officer	15.1	15.1
	100.0	

Source: *Defense 97 Almanac.*

One caution when you are finding the median for ordered qualitative data: Avoid a common error that identifies the median simply with the middle or two middlemost classes, such as "between captain and major," without regard to the cumulative relative frequencies and the location of the 50th percentile. In other words, don't treat the various classes as though they have the same frequencies when they actually have different frequencies.

Inappropriate Averages

It would not be appropriate to report a median for unordered qualitative data, such as the blood types of prospective donors or the ancestries of Americans. Nor would it be appropriate to report a mean for *any* qualitative data, such as the ranks of officers in the U.S. Army. After all, words can't be added and then divided as required by the formula for the mean.

*Exercise 4.8 College students were surveyed to determine where they would most like to spend their spring vacation: Fort Lauderdale (FL), Palm Springs (PS), Mexico (M), Lake Havasu (LH), or Other (O). The results were as follows:

FL	FL	M	LH	FL
M	PS	LH	FL	O
O	PS	M	FL	LH
FL	M	FL	O	FL

Find the mode and, if possible, the median.

Answer on page 497

4.7 USING THE WORD *AVERAGE*

Strictly speaking, an *average* can refer to the mode, median, or mean—or even to some more exotic average, such as the geometric mean or the harmonic mean. Conventional usage prescribes that *average* usually signifies *mean*, and this connotation is often reinforced by the context. For instance, *grade point average* is virtually synonymous with *mean grade point*. To our knowledge, not even the most enterprising grade-point-impoverished student has attempted to satisfy graduation requirements by exchanging a more favorable modal or median grade point for the customary mean grade point. Unless context and usage make it clear, however, it is a good policy to specify the particular average with which you are dealing, even if it entails a short explanation. When dealing with controversial topics, it is always wise to insist that the exact nature of the average be specified.

Summary

The mode equals the value of the most frequently occurring observation.

The median equals the value of the middle-ranked observation (or observations), and because it splits frequencies into upper and lower halves, it has a percentile rank of 50.

The value of the mean, whether defined for a sample or a population, is found by adding all observations and dividing them by the size of the sample or population. The mean is the preferred average for quantitative data, mainly because it describes the balance point of any distribution.

When frequency distributions are not skewed, the values of all three averages tend to be similar and equally representative of the central tendencies within the distributions. When frequency distributions are skewed, the values of the three averages differ appreciably, with the mean being particularly sensitive to extreme observations. Ideally, in this case, both the mean and the median should be reported.

The mode can be used with qualitative data. If it's possible to order qualitative data from least to most, the median also can be used.

Conventional usage prescribes that *average* usually signifies *mean*, but when dealing with controversial topics, it's wise to insist that the exact nature of the average be specified.

Important Terms

Measures of central tendency	**Sample**
Mode	**Sample mean ($\overline{X}$)**
Bimodal	**Sample size (n)**
Median	**Population mean (μ)**
Population	**Population size (N)**

REVIEW EXERCISES

Note on Computational Accuracy

Whenever necessary, round numbers to two digits to the right of the decimal point, using the rounding procedure described in Section 7 of Appendix A.

*4.9 To the question "During your lifetime, how often have you changed your permanent residence?" a group of eighteen college students replied as follows: 1, 3, 4, 1, 0, 2, 5, 8, 0, 2, 3, 4, 7, 11, 0, 2, 3, 4. Find the mode, median, and mean.

Answers on page 497

4.10   During their first swim through a water maze, fifteen laboratory rats made the following number of errors (blind alleyway entrances): 2, 17, 5, 3, 28, 7, 5, 8, 5, 6, 2, 12, 10, 4, 3.

(a)   Find the mode, median, and mean for these data.

(b)   Without constructing a frequency distribution or graph, would you characterize the shape of this distribution as balanced, positively skewed, or negatively skewed?

4.11   In some racing events, downhill skiers receive the average of their times for three trials. Would you prefer the average time to be the mean or the median if usually you have

(a)   one very poor time and two average times?

(b)   one very good time and two average times?

(c)   two good times and one average time?

(d)   three different times, spaced at about equal intervals?

*4.12   During the 1998 strike by Northwest Airline's pilots, management claimed that the salaries of pilots averaged $133,000 per year, whereas the pilots' union claimed that pilots averaged only $120,000 per year. Given the focus of the present chapter, what could be the cause of this discrepancy?

Answer on page 497

4.13   The mean serves as the balance point for any distribution because the sum of all observations, expressed as positive and negative distances from the mean, always equals zero.

(a)   Show that the mean possesses this property for the following set of scores: 3, 6, 2, 0, 4.

(b)   Satisfy yourself that the mean identifies the only point that possesses this property. More specifically, select some other number, preferably a whole number (for convenience), and then find the sum of all of the above scores, expressed as positive or negative distances from the newly selected number. This sum should not equal zero.

4.14   Given that, as mentioned previously, White is the modal race of Americans, would it be possible to find the median for this distribution listed in Exercise 3.4 on page 5?

4.15   Find the mode and, if possible, the median for the film ratings listed in Exercise 2.8 on page 32.

4.16   Do the distributions for males and females in Exercise 2.15 on page 00 differ with respect to their modes?

4.17   Specify the single average—either the mode, median, or mean—described by the following statements:

(a)   Never can be used with qualitative data.

(b)   Sometimes can be used with qualitative data.

(c)   Always can be used with qualitative data.

CHAPTER 5

Describing Variability

QUANTITATIVE MEASURES OF VARIABILITY

5.1   INTUITIVE APPROACH
5.2   RANGE
5.3   VARIANCE
5.4   WEAKNESS OF VARIANCE
5.5   STANDARD DEVIATION: AN INTERPRETATION
5.6   STANDARD DEVIATION: SOME GENERALIZATIONS
5.7   STANDARD DEVIATION: A MEASURE OF DISTANCE
5.8   MEASURES OF VARIABILITY FOR QUALITATIVE DATA

DETAILS

5.9   DEFINITION FORMULA FOR STANDARD DEVIATION
5.10   COMPUTATION FORMULA FOR STANDARD DEVIATION
5.11   INTERQUARTILE RANGE (IQR)

Summary

Important Terms

Review Exercises

If you lived in a nightmarish world where, for instance, everybody was cloned from the same person and all expressions of individuality were suppressed—in other words, if you lived in a world of little or no *variability*—the relevance of statistics would be diminished (and a few of us probably would be looking for work). When describing groups of people, you could ignore the techniques in previous chapters, as well as the measures of variability to be discussed in this chapter, and simply observe one person's weight or IQ score or attitude toward reincarnation as being representative of everybody else's. In a very real sense then, statistics flourishes because we live in a world of variability—no two people are identical, and a few are really far out.

Ordinarily, when attempting to summarize a batch of data, we specify not only a measure of central tendency, such as the mean, but also a **measure of variability,** that is, a measure of the amount of variation or differences among observations in a distribution. This chapter describes several measures of variability, including the range, interquartile range, variance, and particularly the standard deviation.

Measures of variability

General term for various measures of the amount of variation or differences among observations in a distribution.

QUANTITATIVE MEASURES OF VARIABILITY

5.1 INTUITIVE APPROACH

You probably already possess an intuitive feel for gross differences in variability. In **Figure 5.1,** each of the three frequency distributions consists of seven observations with the same mean (10) but different variabilities. (Ignore the numbers in the shaded boxes; their significance will be explained later.) Before reading on, rank the three distributions from least to most variable. Your intuition was correct if you concluded that distribution A has the smallest variability, distribution B has intermediate variability, and distribution C has the greatest variability.

If this conclusion isn't obvious, look at each of the three distributions, one at a time, and note any differences among the values of individual observa-

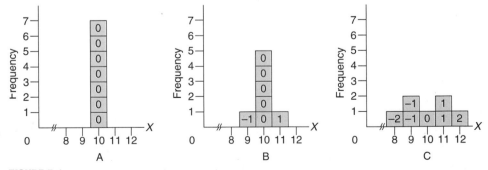

FIGURE 5.1

Three distributions with the same mean (10) but different amounts of variability. Numbers in the shaded boxes indicate distances from the mean.

tions. For the distribution with the smallest variability (A), all seven observations have the same value (10). For the distribution with intermediate variability (B), the values of observations vary slightly (one 9 and one 11), and for that with greatest variability (C), they vary even more (one 8, two 9s, two 11s, and one 12).

5.2 RANGE

Range

The difference between the largest and smallest observations.

More exact measures of variability not only enhance communication but also are essential for subsequent quantitative work in statistics. One such exact measure is the range. *The* **range** *is the difference between the largest and smallest observations.* In Figure 5.1, the least variable distribution (A) has a range of 0 (from 10 to 10); the moderately variable distribution (B) has a range of 2 (from 11 to 9); and the most variable distribution (C) has a range of 4 (from 12 to 8), in agreement with the intuitive judgments about differences in variability. The range is a handy measure of variability; it can be readily calculated and understood.

Shortcomings

The range has several shortcomings. First, its value depends only on two observations—the largest and the smallest—and thus it fails to use the information provided by the remaining observations. Furthermore, the value of the range tends to increase with increases in the total number of observations. For instance, the range of adult heights might be 6 or 8 inches for a distribution of half a dozen people, whereas it might be 14 or 16 inches for a distribution of six dozen people. Larger sets of observations are more likely to include very short or very tall people who, of course, inflate the value of the range. Instead of being a relatively "pure" measure of variability, the size of the range is determined, in part, by how many observations are included in the distribution.

5.3 VARIANCE

Although both the range and its most important spinoff, the interquartile range (discussed later in Section 5.11), serve as valid measures of variability, neither ranks among the statistician's preferred measures of variability. Those roles are reserved for the variance and *particularly for its square root, the standard deviation,* because, as will become apparent in later chapters, these measures serve as key components in other more complex statistical measures. Accordingly, the variance and standard deviation occupy the same exalted position among measures of variability as does the mean among measures of central tendency.

Following the computational procedures described in later sections, we could calculate the value of the variance for each of the three distributions in Figure 5.1. Its value equals 0.00 for the least variable distribution (A), 0.29 for the moderately variable distribution (B), and 1.71 for the most variable distribution (C), in agreement with our intuitive judgments about the relative variability of these three distributions.

Reconstructing the Variance

To understand the variance better, let's reconstruct it bit by bit. Although a measure of variability, the variance also qualifies as a type of mean, that is, as the balance point for some distribution. To qualify as a type of mean, the values of all of the observations must be added and then divided by the total number of observations. In the case of the variance, original observations are reexpressed as distances or deviations from their mean, $\overline{X}$. For each of the three distributions in Figure 5.1 on page 76, the face values of the seven original observations (shown as numbers along the X axis) have been reexpressed as deviations from their mean of 10 (shown as numbers in the shaded boxes). For example, in distribution C, one observation coincides with the mean of 10, four observations (two 9s and two 11s) deviate one unit from the mean, and two observations (one 8 and one 12) deviate two units from the mean, yielding a set of seven deviations about the mean: one 0, two −1s, two 1s, one −2, and one 2.

Mean of the Deviations Not a Useful Measure

No useful measure of variability can be produced by calculating the mean of these seven deviations, because, as you'll recall, the sum of all deviations from their mean always equals zero. In effect, negative deviations counterbalance positive deviations, yielding a sum of zero, regardless of the amount of variability in the distribution.[1]

Mean of the Squared Deviations

Before calculating the variance (a type of mean), negative signs must be eliminated from deviation scores. Squaring each deviation—that is, multiplying each deviation by itself—generates a set of squared deviations, all of which are positive. (Remember, the product of any two numbers with similar signs is always positive.) Now it's merely a matter of adding the consistently positive values of all squared deviations and then dividing by the total number of squared deviations to produce *the mean of all squared deviations from the mean, also known as the* **variance.**

· ·

Variance

The mean of all squared deviations from the mean.

· ·

5.4 WEAKNESS OF VARIANCE

In the case of the weights of male statistics students, discussed first in Chapter 1, it's useful to know that the mean for the distribution of weights equals 169.51 pounds, but it is confusing to know that the variance for the same distribution equals 533.83 *squared pounds* (because of the squared deviations). What, you might reasonably ask, are squared pounds?

···

[1] A measure of variability, known as the *mean absolute deviation* (or *m.a.d.*), can be salvaged by summing all *absolute* deviations from the mean, that is, by ignoring negative signs. This measure of variability is not preferred because, in the long run, the simple act of ignoring negative signs has undesirable mathematical and statistical repercussions.

Emergence of Standard Deviation

To rid ourselves of these mind-boggling units of measurement, simply take the square root of the variance.[2] This produces a new measure, known as the standard deviation, that describes variability in terms of the original units of measurement. For example, the standard deviation for the distribution of weights equals the square root of 533.83 squared pounds, that is, 23.10 pounds.

The variance often assumes a special role in more advanced statistical work, including that described in Chapters 10, 22, and 23 of this book. Otherwise, because of its unintelligible units of measurement, the variance serves mainly as a stepping-stone, only a square root away from a more preferred measure of variability, the standard deviation, *the square root of the mean of all squared deviations from the mean.*

5.5 STANDARD DEVIATION: AN INTERPRETATION

Standard deviation

A rough *measure of the average amount by which observations deviate on either side of their mean.*

You might find it helpful to think of the *standard deviation* as a *rough* measure of the average amount by which observations deviate on either side of their mean.

For distribution C in Figure 5.1 the square root of the variance of 1.71 yields a standard deviation of 1.31. Given this perspective, a standard deviation of 1.31 is a rough measure of the average amount by which the seven observations in distribution C (8, 9, 9, 10, 11, 11, 12) deviate on either side of their mean of 10. In other words, the standard deviation of 1.31 is a rough measure of the average amount (actually, 1.14) for the seven deviation scores in distribution C, namely, one 0, four 1s, and two 2s.

Actually Exceeds Average Deviation

Strictly speaking, the standard deviation usually exceeds by 10 to 20 percent the average deviation (or more accurately, the mean absolute deviation mentioned in the footnote on page 78). Nevertheless, it's reasonable to describe the standard deviation as the average amount by which observations deviate on either side of their mean—as long as you remember that an approximation is involved.

5.6 STANDARD DEVIATION: SOME GENERALIZATIONS

Majority within One Standard Deviation

A slightly different perspective makes the standard deviation even more accessible.

---

[2]The square root of a number is the number that when multiplied by itself yields the original number. For example, the square root of 16 is 4, because 4 times 4 equals 16. To extract the square root of any number, consult a calculator with a square root key (usually denoted by the symbol $\sqrt{}$).

> **For most frequency distributions, a majority (often as many as 68 percent) of all observations are within one standard deviation on either side of the mean.**

This generalization applies to all of the distributions in Figure 5.1. For instance, among the seven deviations in distribution C, a majority of five deviate within one standard deviation (1.31) on either side of the mean; that is, they deviate less than 1.31 either above or below the mean.

Essentially the same pattern describes a wide variety of frequency distributions including the two shown in Figure 5.2, where the capital letter S represents the standard deviation. As suggested in the top panel of Figure 5.2, if the distribution of IQ scores for a class of fourth graders has a mean ($\overline{X}$) of 105 and a standard deviation (S) of 15, a majority of the class scores should be within 15 points on either side of the mean, that is, between 90 and 120. By the same token, as suggested in the bottom panel of Figure 5.2, if the distribution of weekly study times, estimated to the nearest hour by a group of college

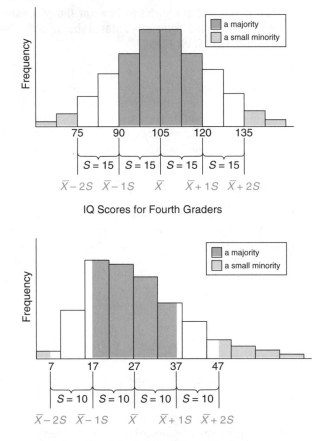

IQ Scores for Fourth Graders

Study Times (hours) for College Students

FIGURE 5.2

Some generalizations that apply to most frequency distributions.

students, has a mean ($\overline{X}$) of 27 hours and a standard deviation (S) of 10 hours, a majority of these study times should be within one standard deviation on either side of the mean, that is, between 17 and 37 hours.

Small Minority Deviate Outside Two Standard Deviations

The standard deviation also can be used in a generalization about the extremities or tails of frequency distributions:

> **For most frequency distributions, a small minority (often as small as 5 percent) of all observations deviate more than two standard deviations on either side of the mean.**

This generalization describes each of the distributions in Figure 5.1. For instance, among the seven deviations in distribution C, none deviates more than two standard deviations ($2 \times 1.31 = 2.62$) on either side of the mean. As suggested in Figure 5.2, relatively few fourth graders have IQ scores that deviate more than two standard deviations ($2 \times 15 = 30$) on either side of the mean of 105, that is, IQ scores less than 75 (from $105 - 30$) or more than 135 (from $105 + 30$). Likewise, relatively few college students estimate their weekly study times to be more than two standard deviations ($2 \times 10 = 20$) on either side of the mean of 27, that is, less than 7 hours (from $27 - 20$) or more than 47 hours (from $27 + 20$).

Generalizations Are for All Distributions

These two generalizations are independent of the particular shape of the distribution. In Figure 5.2, they apply to both the balanced distribution of IQ scores and the positively skewed distribution of study times. As a matter of fact, the balanced distribution of IQ scores approximates an important theoretical distribution, the normal distribution, which is discussed in Chapters 6 and 7. As will be seen, much more precise generalizations are possible for normal distributions.

*Exercise 5.1 Employees of Corporation A earn annual salaries described by a mean of $70,000 and a standard deviation of $5,000.

(a) The majority of all salaries fall between what two values?

(b) A small minority of all salaries are less than what value?

(c) A small minority of all salaries are more than what value?

(d) Answer (a), (b), and (c) for Corporation B's employees, who earn annual salaries described by a mean of $70,000 and a standard deviation of $2,000.

Answers on page 497

5.7 STANDARD DEVIATION: A MEASURE OF DISTANCE

There's an important difference between the standard deviation and its indispensable co-measure, the mean. *The mean is a measure of position, but the standard deviation is a measure of distance (on either side of the mean of the*

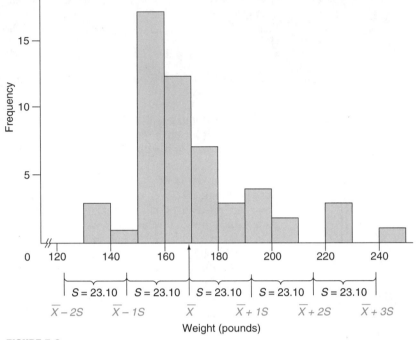

FIGURE 5.3
Weight distribution with mean and standard deviation.

distribution). **Figure 5.3** describes the weight distribution for male statistics students, originally shown in Figure 3.1. Note that the mean $(\overline{X})$ of 169.51 pounds has a particular position or location along the horizontal axis: It's located at the point, and only at the point, corresponding to 169.51 pounds. On the other hand, the standard deviation (S) of 23.10 pounds for the same distribution has no particular location along the horizontal axis. Using the standard deviation as a measure of distance on either side of the mean, we could describe one person's weight as two standard deviations above the mean, $\overline{X} + 2S$, another person's weight as two-thirds of one standard deviation below the mean, $\overline{X} - \frac{2}{3}S$, and so on.

Value of Standard Deviation Can't Be Negative

Standard deviation distances always originate from the mean and are expressed as positive or negative deviations above or below the mean. Note, however, that although the actual value of the standard deviation can be zero or a positive number, it can't ever be a negative number (because any negative deviation disappears when squared). When a negative sign appears next to the standard deviation, as in the expression $\overline{X} - \frac{1}{2}S$, the negative sign indicates that one half of a standard deviation unit (always positive) must be subtracted from the mean to identify a weight located one half of a standard deviation *below* the mean weight. More specifically, the expression $\overline{X} - \frac{1}{2}S$ translates into a weight of 158 pounds (from $169.51 - \frac{1}{2}(23.10) = 169.51 - 11.55 = 157.96$).

5.8 MEASURES OF VARIABILITY FOR QUALITATIVE DATA

Measures of variability are virtually nonexistent for qualitative data. It's probably adequate to note merely whether observations are evenly divided among the various classes (maximum variability), concentrated mostly in one class (minimum variability), or unevenly divided among the various classes (intermediate variability). For example, if the ethnic composition of the residents of a city is about evenly divided among several groups, the variability with respect to ethnic groups will be maximum; in other words, there is considerable heterogeneity. (An inspection of the 1997 data for the Population Estimates Program of the U.S. Census Bureau, available on the Internet at **http://www.census.gov/population/www/,** reveals that the greatest ethnic variability occurs in large urban counties, such as Bronx County, New York, and San Francisco County, California.) At the other extreme, if almost all of the residents are concentrated in a single ethnic group, the variability will be minimum; there is little heterogeneity. (According to the above source, virtually no ethnic variability occurs in sparsely populated rural counties, such as Hooker County, Nebraska, and King County, Texas, with an almost exclusively white population.) If, as is true of many U.S. cities and counties, the ethnic composition falls between these two extremes because of an uneven division among several large ethnic groups, the variability will be intermediate.

Ordered Qualitative Data

If qualitative data can be ordered, then it is appropriate to describe variability by identifying extreme observations. For instance, the active membership of an officer's club might include no one with a rank below first lieutenant or above brigadier general.

DETAILS

5.9 DEFINITION FORMULA FOR STANDARD DEVIATION

There are several formulas for the standard deviation: This section describes the more easily understood definition formula, whereas the next section describes the more efficient computation formula.

Sample or Population?

As with the mean, statisticians distinguish between sample and population standard deviations, depending on whether the data, such as in **Table 5.1,** are viewed as a sample (that is, a subset from a population) or as a population (that is, a complete set). When you are concerned only with descriptive statistics, as in Part 1 of this book, the formulas for sample and population standard deviations differ only because of a change in symbols. Once again, this distinction will be introduced here because of its importance in later chapters.

Table 5.1
CALCULATION OF THE VARIANCE AND STANDARD DEVIATION (DEFINITION FORMULA)

A. COMPUTATIONAL SEQUENCE

Assign a value to n **1** representing the number of X scores.

Add all X scores **2**, and obtain the mean of these scores **3**.

Express each X score, one at a time, as a deviation from the mean, that is, subtract the mean from each X score **4**.

Square each deviation score **5**, one at a time, and then add all squared deviation scores **6**.

Substitute numbers into the formula **7**, and solve for S^2 (the variance).

Extract the square root of S^2 **8**, to obtain S (the standard deviation).

B. DATA AND COMPUTATIONS

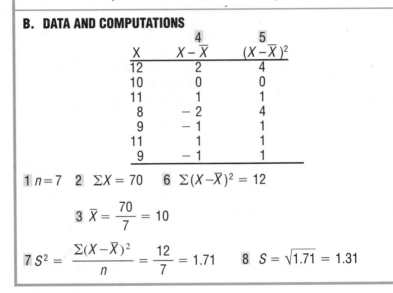

	4	**5**
X	$X - \overline{X}$	$(X - \overline{X})^2$
12	2	4
10	0	0
11	1	1
8	-2	4
9	-1	1
11	1	1
9	-1	1

1 $n = 7$ **2** $\Sigma X = 70$ **6** $\Sigma(X - \overline{X})^2 = 12$

$$\textbf{3} \ \ \overline{X} = \frac{70}{7} = 10$$

$$\textbf{7} \ \ S^2 = \frac{\Sigma(X - \overline{X})^2}{n} = \frac{12}{7} = 1.71 \qquad \textbf{8} \ \ S = \sqrt{1.71} = 1.31$$

The definition formula for the sample standard deviation, symbolized as S, takes the form and may be read as "S equals the square root of the sum of the

---

Sample standard deviation (S)

A rough measure of the average amount by which observations in the sample deviate on either side of the sample mean.

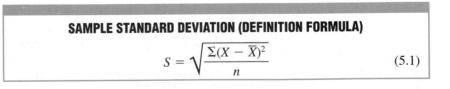

SAMPLE STANDARD DEVIATION (DEFINITION FORMULA)

$$S = \sqrt{\frac{\Sigma(X - \overline{X})^2}{n}} \qquad (5.1)$$

squared deviations from X-bar, divided by n." With the aid of Formula 5.1, which contains only familiar symbols, you can reconstruct the standard deviation step by step. Briefly, original observations are expressed as deviations from their sample mean, $X - \overline{X}$; each deviation is squared $(X - \overline{X})^2$; all squared deviations are added, $\Sigma(X - \overline{X})^2$; and this sum is divided by the sample size, $\Sigma(X - \overline{X})^2/n$, to produce the sample variance, the mean of all squared deviations in the sample. Finally, the square root of the sample variance yields the sample standard deviation, a rough measure of the average amount by which observations deviate on either side of the sample mean.

Table 5.1 illustrates how to calculate the sample standard deviation for the seven original observations (one 8, two 9s, one 10, two 11s, and one 12) of distribution C in Figure 5.1. Notice that these observations needn't be ranked in any special order when calculating the standard deviation, just as when calculating the mean.

Definition Formula for Population Standard Deviation (σ)

The formula for the population standard deviation differs from that for the sample standard deviation only because of a change in symbols. Now, the population variance is represented by the lowercase Greek letter σ, (pronounced "sigma"),

---

Population standard deviation (σ)

A rough measure of the average amount by which observations in the population deviate on either side of the population mean.

POPULATION STANDARD DEVIATION (DEFINITION FORMULA)

$$\sigma = \sqrt{\frac{\Sigma(X - \mu)^2}{N}} \tag{5.2}$$

where μ represents the population mean and N represents the population size. Otherwise, as has been noted, the calculations are exactly the same as for the sample standard deviation.

5.10 COMPUTATION FORMULA FOR STANDARD DEVIATION

The definition formula is cumbersome when the mean equals some complex number, such as 169.51, or when the number of observations is large. In these cases, which occur often, use the more efficient computation formula, defined as follows:

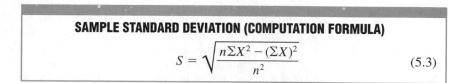

SAMPLE STANDARD DEVIATION (COMPUTATION FORMULA)

$$S = \sqrt{\frac{n\Sigma X^2 - (\Sigma X)^2}{n^2}} \tag{5.3}$$

Table 5.2
CALCULATION OF THE VARIANCE
AND STANDARD DEVIATION (COMPUTATION FORMULA)

A. COMPUTATIONAL SEQUENCE
Assign a value to n **1** representing the number of X scores.
Sum all X scores **2**.
Square each X score **3**, one at a time, and then add all squared X scores **4**.
Substitute numbers into the formula **5**, and solve for S^2 (the variance).
Extract the square root of S^2 **6**, to obtain S (the standard deviation).

B. DATA AND COMPUTATIONS

X	**3** X^2
12	144
10	100
11	121
8	64
9	81
11	121
9	81

1 $n = 7$ **2** $\Sigma X = 70$ **4** $\Sigma X^2 = 712$

$$\textbf{5}\quad S^2 = \frac{(n)(\Sigma X^2) - (\Sigma X)^2}{(n)^2} = \frac{(7)(712) - (70)^2}{(7)^2}$$

$$= \frac{4984 - 4900}{49} = \frac{84}{49} = 1.71$$

6 $\quad S = \sqrt{1.71} = 1.31$

Note that no mean appears in the computation formula. **Table 5.2** illustrates the use of this formula to calculate the standard deviation for the same seven observations as in Table 5.1. It's wise to adopt the computational format shown in the bottom panel of Table 5.2. (No computation formula is shown for the population standard deviation; its formula is the same as that for the sample standard deviation, once the N for population size replaces the n for sample size.)

Not unexpectedly, because of their mathematical equivalence, the definition and computation formulas yield the same standard deviation of 1.31 for distribution C. (Hereafter, in any cases where the two formulas produce *slightly* different results, these differences probably are due to the larger rounding error in computations based on the definition formula.) The tremendous efficiency of the computation formula becomes more apparent when dealing with large sets of observations, as in Exercise 5.12 at the end of this chapter.

Some Computational Advice

A common confusion involves $(\Sigma X)^2$, obtained by the first adding all scores and then squaring the total, and ΣX^2, obtained by first squaring each score and then adding all squared scores. It's a helpful computational check to remember that neither the square of the summed scores, $(\Sigma X)^2$, nor the sum of the squared scores, ΣX^2, can be negative numbers. Nor can the expression $n\Sigma X^2 - (\Sigma X)^2$ be a negative number. As an almost foolproof method for avoiding computational errors, calculate everything twice, and proceed only when the computational results agree.

Sample Standard Deviation for Inferential Statistics

As a matter of fact, Formulas 5.1 and 5.3 for the sample standard deviation are only appropriate in descriptive statistics, when the standard deviation is used merely to *describe* the variability of a set of actual observations. Later, in inferential statistics, when the sample standard deviation is used to *estimate* the unknown value of the population standard deviation, these formulas must be changed (by entering $n - 1$ for n in the denominator of Formula 5.1 and by entering $n(n - 1)$ for n^2 in the denominator of Formula 5.3) to more accurately represent the true variability in the population. More about this in Sections 18.8 and 18.11.

*Exercise 5.2 Using the definition formula, calculate the standard deviation, *S,* for the following distribution of four numbers: 1, 3, 4, 4.

*Exercise 5.3 Using the computation formula, find the standard deviation for each of the following distributions:

(a) 1, 3, 7, 2, 0, 4, 7, 3　　　**(b)** 10, 8, 5, 0, 1, 1, 7, 9, 2
　　　Answers on page 497

5.11 INTERQUARTILE RANGE (IQR)

Interquartile range (IQR)
The range for the middle 50 percent of all observations.

The most important spin-off of the range, the **interquartile range (IQR),** *is simply the range for the middle 50 percent of the observations.* More specifically, the interquartile range equals the distance between the third quartile (or 75th percentile) and the first quartile (or 25th percentile) of a frequency distribution. In other words, it equals the difference between the largest and smallest scores that remain after the highest quarter (or top 25 percent) and the lowest quarter (or bottom 25 percent) have been trimmed from the original set of observations. Insofar as most distributions have tails in their extremities, the interquartile range tends to be less than half the size of the range.

The calculation of the interquartile range is relatively straightforward, as you can verify by studying **Table 5.3.** This table shows that the IQR equals 2 for distribution C (8, 9, 9, 10, 11, 11, 12) in Figure 5.1 on page 76.

Not Sensitive to Extreme Observations

A key property of the interquartile range is its resistance to the distorting effect of extreme observations. For example, if the smallest observation (8) in distribution C were replaced by a much smaller observation (for instance, 1)

Table 5.3
CALCULATION OF THE INTERQUARTILE RANGE

A. INSTRUCTIONS

1 Order observations from least to most.
2 To determine how far to penetrate the set of ordered observations, beginning at either end, add 1 to the total number of observations and divide by 4. If necessary, round the result to the nearest whole number.
3 Beginning with the largest observation, count the requisite number of steps into the ordered observation to find the location of the third quartile.
4 The third quartile equals the value of the observation at this location.
5 Beginning with the smallest observation, again count the requisite number of steps into the ordered observations to find the location of the first quartile.
6 The first quartile equals the value of the observation at this location.
7 The interquartile range equals the third quartile minus the first quartile.

B. EXAMPLE

1 8, 9, 9, 10, 11, 11, 12
2 $(7 + 1)/4 = 2$
3 8, 9, 9, 10, 11, 11, 12
          ↑
        2 1
4 third quartile = 11
5 8, 9, 9, 10, 11, 11, 12
     ↑
   1 2
6 first quartile = 9
7 interquartile range $= 11 - 9 = 2$

---

the value of the IQR would remain the same (2), although the value of the range would be almost tripled (11). Thus, if you're concerned about possible distortions caused by extreme observations, including outliers, use the interquartile range as the measure of variability, along with the median (or second quartile) as the measure of central tendency.

*Exercise 5.4 Determine the values of the range and the interquartile range for the following sets of data:

(a) Retirement ages: 60, 63, 45, 63, 65, 70, 55, 63, 60, 65, 63.

(b) Residence changes: 1, 3, 4, 1, 0, 2, 5, 8, 0, 2, 3, 4, 7, 11, 0, 2, 3, 4.
Answers on page 498

Summary

The simplest measure of variability, the range, is readily calculated and understood, but it has two shortcomings. Among measures of variability, the variance and particularly the standard deviation occupy the same exalted position as does the mean among measures of central tendency.

Although the variance is a measure of variability, it also is a type of mean: the mean of all squared deviations about their mean. To avoid mind-boggling squared units of measurement, take the square root of the variance to obtain the standard deviation.

Some people find it helpful to view the standard deviation as a rough measure of the average or mean amount by which observations deviate on either side of their mean.

For most frequency distributions, a majority of all observations are within one standard deviation of the mean, and a small minority (often as small as 5 percent) of all observations deviate more than two standard deviations on either side of their mean.

Unlike the mean, which is a measure of position, the standard deviation is a measure of distance.

Measures of variability are virtually nonexistent for qualitative data.

This book describes two formulas for the sample standard deviation and one for the population standard deviation. The computation formula (5.3) presents the sample standard deviation in its most convenient computational form.

The interquartile range is resistant to the distorting effects of extreme observations.

Important Terms

Measures of variability

Range

Variance

Standard deviation

Sample standard deviation (S)

Population standard deviation (σ)

Interquartile range (IQR)

REVIEW EXERCISES

5.5   Assume that the distribution of IQ scores for all college students has a mean of 120, with a standard deviation of 15. These two bits of information imply which of the following?

  (a) All students have IQs of either 105 or 135 because everybody in the distribution is either one standard deviation above or below the mean. True or false?

  (b) All students score between 105 and 135 because everybody is *within* one standard deviation on either side of the mean. True or false?

  (c) On the average, students deviate approximately fifteen points on either side of the mean. True or false?

(d) Some students deviate more than one standard deviation above or below the mean. True or false?

(e) All students deviate more than one standard deviation above or below the mean. True or false?

(f) Scott's IQ score of 150 deviates two standard deviations above the mean. True or false?

*5.6 For each of the following pairs of distributions, first decide whether both distributions should have standard deviations that are about the same or fairly different. Then, if they are different, indicate which distribution should have the larger standard deviation. *Hint:* The distribution with the more dissimilar set of scores or individuals should produce the larger standard deviation *regardless of whether, on the average, scores or individuals in one distribution differ from those in the other distribution.*

(a) Scholastic Assessment Test scores for all graduating high school seniors (a_1) or all college freshmen (a_2).

(b) Ages of patients in a community hospital (b_1) or a children's hospital (b_2).

(c) Motor skill reaction times of major league baseball players (c_1) or college students (c_2).

(d) Grade point averages of students at some university as revealed by a representative sample (d_1) or a census of the entire student body (d_2).

(e) Anxiety scores (on a scale from 0 to 50) of a representative sample of college students taken from the senior class (e_1) or those who plan to attend an anxiety-reduction clinic (e_2).

(f) Annual incomes of recent college graduates (f_1) or of twenty-year alumni (f_2).

Answers on page 498

5.7 When not interrupted artificially, the duration of human pregnancies can be described, we'll assume, by a mean of nine months (270 days) and a standard deviation of one-half month (15 days).

(a) Between what two times, in days, will a majority of babies arrive?

(b) A small minority of all babies will arrive sooner than _____?

(c) A small minority of all babies will arrive later than _____?

(d) In a paternity suit, the suspected father claims that because he was overseas during the entire ten months prior to the baby's birth, he couldn't possibly be the father. Any comment?

5.8 Add 10 to each of the observations in Exercise 5.2 (1, 3, 4, 4) to produce a new distribution (11, 13, 14, 14). Would you expect the value of *S* to be the same for both the original and present distributions? Explain your answer and then calculate *S* for the present distribution.

5.9 Add 10 to only the smallest observation in Exercise 5.2 (1, 3, 4, 4) to produce still another distribution (11, 3, 4, 4). Would you expect the value of *S* to be the same for both this distribution and the original distribution in Exercise 5.2? Explain your answer and then calculate S for the present distribution.

*5.10 **(a)** While in office, a former governor of California proposed that all state employees receive the same pay raise of $70 per month. What effect, if any, would this raise have on the mean and the standard deviation for the distribution of monthly wages in existence before the proposed raise? *Hint:* Imagine the effect of adding $70 to the monthly wages of each state employee on the mean and on the standard deviation (or another, more easily visualized measure of variability, such as the range).

(b) Other California officials suggested that all state employees receive a pay raise of 5 percent. What effect, if any, would this raise have on the mean and the standard deviation for the distribution of monthly wages in existence before the proposed raise? *Hint:* Imagine the effect of multiplying the monthly wages of each state employee by 5 percent on the mean and on the standard deviation or the range.

Answers on page 498

5.11 Verify that the value of S is the same for the set of scores (12, 6, 3, 4, 5), regardless of whether the definition or the computation formula is used.

5.12 **(a)** Using the computation formula, verify that the standard deviation, S, equals 23.10 pounds for the distribution of weights in Table 1.1.

(b) Verify that a majority of all observations fall within one standard deviation of the mean (169.51) and that a small minority of all observations deviate more than two standard deviations from the mean.

5.13 In what sense is the variance

(a) a type of mean?

(b) not a good indicator of variability?

(c) a stepping-stone to the standard deviation?

5.14 Specify an important difference between the standard deviation and the mean.

5.15 As nontechnically as possible, briefly describe what the standard deviation measures.

5.16 Why can't the value of the standard deviation ever be negative?

5.17 Referring to Exercise 2.15 on page 38, would you describe the distribution for all male graduates as having maximum, intermediate, or minimum variability?

CHAPTER 6

Normal Distributions (I): Basics

6.1   THE THEORETICAL NORMAL CURVE
6.2   PROPERTIES OF THE NORMAL CURVE
6.3   *z* SCORES
6.4   STANDARD NORMAL CURVE
6.5   STANDARD NORMAL TABLE
6.6   EXAMPLE: FBI APPLICANTS
6.7   KEY FACTS TO REMEMBER

Summary

Important Terms

Review Exercises

In the movie *The President's Analyst,* the director of the Federal Bureau of Investigation, rather short himself, encourages the recruitment of similarly short FBI agents. If, in fact, FBI agents are to be selected only from applicants who are no taller than exactly 65 inches, what proportion of all of the original applicants will be eligible? This question is difficult to answer without more information.

One source of additional information is the relative frequency distribution of heights for the 3091 men shown in **Figure 6.1.** To find the proportion of men who are a particular height, merely note the value of the vertical scale that corresponds to the top of any bar in the histogram. For example, .10 of these men, that is, one-tenth of 3091, or about 309 men, stand 69 inches tall.

When expressed as proportions, any conclusion based on the 3091 men can be generalized to other comparable sets of men, even sets containing an unspecified number. For instance, if the distribution in Figure 6.1 is viewed as representative of all men who apply for FBI jobs, we can estimate that .10 of all applicants will stand 69 inches tall. Or, given the director's preference for shorter agents, we can use the same distribution to estimate the proportion of applicants who will be eligible. To obtain the estimated proportion of eligible applicants (.165) from Figure 6.1, add the values associated with the shaded bars. (Only half of the bar at 65 inches is shaded to adjust for the fact that any height between 64.5 and 65.5 inches is reported as 65 inches, whereas eligible applicants must be shorter than *exactly* 65 inches, that is, 65.0 inches.)

The distribution in Figure 6.1 has an obvious limitation—it's based on a group of just 3091 men that, at most, only resembles the distributions for other groups of men, including the group of FBI applicants. Any generalization will, therefore, contain inaccuracies due to chance irregularities in the original distribution. (Part 2 of this book deals with how to estimate the effects of chance and, thereby, to improve the accuracy of our generalizations.)

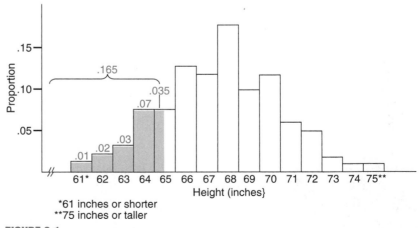

*61 inches or shorter
**75 inches or taller

FIGURE 6.1
*Relative frequency distribution for heights of 3091 men. (*Source: *National Center for Health Statistics, Series 11, No. 14.)*

6.1 THE THEORETICAL NORMAL CURVE

More accurate generalizations usually can be obtained from distributions based on larger numbers of men. A distribution based on 30,910 men usually is more accurate than one based on 3091, and a distribution based on 3,091,000 usually is even more accurate. But it is prohibitively expensive in both time and money to survey even 30,910 people. Fortunately, it's a fact that the distribution of heights for all American men—not just 3091 or even 3,091,000—approximates the normal curve, a well-documented theoretical curve.

In **Figure 6.2,** the idealized normal curve has been superimposed on the original distribution for 3091 men. Irregularities in the original distribution, most likely due to chance, are ignored by the smooth normal curve. Accordingly, any generalizations based on the smoothed normal curve will tend to be more accurate than those based on the original distribution.

Interpreting the Shaded Area

The total area under the normal curve in Figure 6.2 can be identified with all FBI applicants. Viewed relative to the total area, the shaded area represents the proportion of applicants that will be eligible because they're shorter than exactly 65 inches. This new, more accurate proportion will differ from that obtained from the original histogram (.165) because of discrepancies between the two distributions.

Finding a Proportion for the Shaded Area

To find this new proportion, we can't rely on the vertical scale in Figure 6.2, as it describes as proportions the areas in the rectangular bars of histograms, not the areas in the various curved sectors of the normal curve. Instead, in Section 6.5 we'll learn how to use a special table to find the proportion represented by any area under the normal curve, including that represented by the shaded area in Figure 6.2.

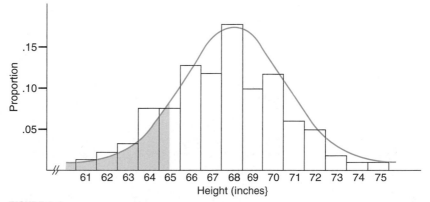

FIGURE 6.2
Normal curve superimposed on distribution of heights.

6.2 PROPERTIES OF THE NORMAL CURVE

Let's note several important properties of the normal curve:

■ Obtained from a mathematical equation, the **normal curve** is a theoretical curve noted for its symmetrical bell-shaped form, as revealed in Figure 6.2.

■ Being symmetrical, the lower half of the normal curve is simply the mirror image of the upper half of the curve.

■ Being bell shaped, the normal curve peaks above a point midway along the horizontal spread and then tapers off gradually in either direction from the peak (without actually touching the horizontal axis, since in theory, the tails of a normal curve extend infinitely far).

■ The values of the mean, median (or 50th percentile), and the mode are always identical for the normal curve.

Importance of Mean and Standard Deviation

When using the normal curve, two bits of information are indispensable: values for the mean and for the standard deviation. For example, before the normal curve can be used to answer the question about eligible FBI applicants, it must be established that for the original distribution of 3091 men, the mean height equals 68 inches and the standard deviation equals 3 inches.

Different Normal Curves

Having established that the normal curve for this example has a mean of 68 inches and a standard deviation of 3 inches, we can't arbitrarily change these values, as an arbitrary change in the value of either the mean or the standard deviation (or both) produces a new normal curve that no longer describes the original distribution of heights. Nevertheless, as a theoretical exercise, it's instructive to note the various types of normal curves that are produced by an arbitrary change in the value of either the mean (μ) or the standard deviation (σ).[1] For example, changing the mean height from 68 to 78 inches produces a new normal curve which, as shown in the top panel of **Figure 6.3,** is displaced 10 inches to the right of the original curve. Dramatically new normal curves are produced by changing the value of the standard deviation. As shown in the bottom panel of Figure 6.3, changing the standard deviation from 3 inches to 1.5 inches produces a more peaked normal curve, whereas changing the standard deviation from 3 to 6 inches produces a shallower normal curve.

Obvious differences in appearance among normal curves are less important than you might suspect. Because of their common mathematical origin, every normal curve can be interpreted in exactly the same way once the distance from the mean is expressed in standard deviation units. For example, .68, or 68 percent, of the total area under a normal curve—any normal curve—is within one standard deviation above and below the mean, and only .05, or 5 percent,

....

[1]Because the normal curve is an idealized curve that is presumed to describe complete sets of observations or populations, the symbols μ and σ, representing the mean and standard deviation of the population, respectively, will be used throughout Chapters 6 and 7.

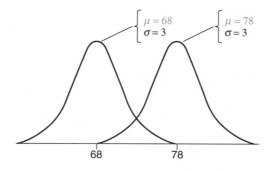

A. Different Means, **Same Standard Deviation**

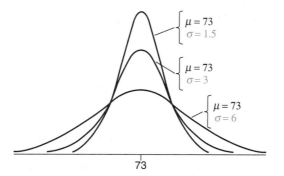

B. Same Mean, Different Standard Deviations

FIGURE 6.3
Different normal curves.

of the total area is more than two standard deviations above and below the mean. And this is only the tip of the iceberg. Once any distance from the mean has been expressed in standard deviation units, we'll be able to consult the special table, described in Section 6.5, to determine the corresponding proportion of area under the normal curve.

6.3 *z* SCORES

z score

A score that indicates how many standard deviations an observation is above or below the mean of the distribution.

A *z* score indicates how many standard deviations an observation is above or below the mean of the distribution.

To obtain a *z* score, express any original score, such as the maximum height of 65 inches for FBI applicants, as a distance from the mean (by subtracting the mean), and split this distance into standard deviation units (by dividing by the standard deviation). In effect, the resulting *z* score always indicates the position of the original score relative to its mean and standard deviation. Regardless of whether the original units of measurement are inches, miles, pounds, or dollars, a *z* score of 2.00 always signifies that the original score is exactly two standard deviations above its mean. Similarly, a *z* score of -1.27 signifies that the original score is exactly 1.27 standard deviations below its mean. A *z* score of 0 signifies that the original score coincides with the mean.

Converting to z Scores

To convert any original observation into a z score, use the following formula:

---

z SCORE

$$z = \frac{X - \mu}{\sigma}$$

(6.1)

---

where X is the original score, and μ and σ are the mean and the standard deviation, respectively, for the normal distribution of original scores. To answer the question about eligible FBI applicants, replace X with 65 (the maximum permissible height), μ with 68 (the mean height), and σ with 3 (the standard deviation of heights), and solve for z as follows:

$$z = \frac{65 - 68}{3} = \frac{-3}{3} = -1.00$$

This informs us that the cutoff height is exactly one standard deviation below the mean. Armed with a value of z, we can refer to the table for the normal curve. First, however, we shall make a few comments about the standard normal curve.

*Exercise 6.1 Express each of the following scores as a z score:

(a) Margaret's IQ of 135, given a mean of 100 and a standard deviation of 15
(b) a verbal score of 470 on the SAT, given a mean of 500 and a standard deviation of 100
(c) a daily production of 2100 units, given a mean of 2180 units and a standard deviation of 50 units
(d) Sam's height of 68 inches, given a mean of 68 and a standard deviation of 3
(e) a meter-reading error of −3 degrees, given a mean of 0 degrees and a standard deviation of 2 degrees

Answers on page 498

6.4 STANDARD NORMAL CURVE

....

Standard normal curve

The one tabled normal curve
with a mean of 0 and a
standard deviation of 1.

If the original distribution approximates a normal curve, then the shift to z scores always will produce a new distribution that approximates the **standard normal curve.** This is the one normal curve for which a table is actually available. It's a mathematical fact—not proven in this book—that the standard normal curve always has a mean of 0 and a standard deviation of 1.

> **Although there are an infinite number of different normal curves, each with its own mean and standard deviation, there is only one *standard* normal curve, with its mean of 0 and standard deviation of 1.**

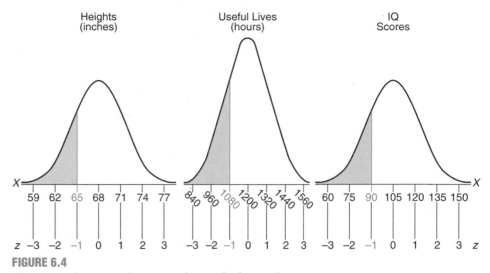

FIGURE 6.4

Converting three normal curves to the standard normal curve.

Figure 6.4 illustrates the emergence of the standard normal curve from three different normal curves: that for the men's heights, with a mean of 68 inches and a standard deviation of 3 inches; that for the useful lives of 100-watt electric lightbulbs, with a mean of 1200 hours and a standard deviation of 120 hours; and that for the IQ scores of fourth graders, with a mean of 105 points and a standard deviation of 15 points.

Converting all original observations into *z* scores leaves the normal shape intact, but not the units of measurement. Sixty-five inches, 1080 hours, and 90 IQ points all reappear as a *z* score of −1.00. Verify this by using the *z* score formula. Showing no traces of the original units of measurement, this *z* score contains the one crucial bit of information common to the three original observations: all are located one standard deviation below the mean. Accordingly, to find the proportion for the shaded areas in Figure 6.4 (that is, the proportion of applicants who stand less than exactly 65 inches, or lightbulbs that burn for fewer than 1080 hours, or fourth graders whose IQ scores are less than 90), we can use the same *z* score of −1.00 when referring to the table for the standard normal curve, the one table for all normal curves.

6.5 STANDARD NORMAL TABLE

Essentially the standard normal table consists of columns of *z* scores coordinated with columns of proportions. In a typical problem, access to the table is gained through a *z* score, such as −1.00, and the answer is read as a proportion or relative frequency, such as the proportion of eligible FBI applicants.

Using the Top Legend of the Table

Let's look at the table of the standard normal curve as presented briefly in **Table 6.1** and more completely in Table A in Appendix D on page 532. The columns are arranged in sets of three, designated as A, B, and C in the legend at the top of each page. When using the top legend, all entries refer to the upper

Table 6.1
PROPORTIONS (OF AREAS) UNDER STANDARD NORMAL CURVE
FOR VALUES OF z (FROM TABLE A OF APPENDIX D)

A z	B	C	A z	B	C	A z	B	C
0.00	0000	5000	0.40	1554	3446	0.80	2881	2119
0.01	0040	4960	0.41	1591	3409	0.81	2910	2090
•	•	•	•	•	•	•	•	•
						•	•	•
						•	•	•
						•	•	•
•	•	•	•	•	•	•	•	•
						0.99	3389	1611
						1.00	3413 →	1587
•	•	•	•	•	•	1.01	3438	1562
						•	•	•
						•	•	•
•	•	•	•	•	•	•	•	•
0.38	1480	3520	0.78	2823	2711	1.18	3810	1190
0.39	1517	3483	0.79	2852	2148	1.19	3830	1170

−z A′	B′	C′	−z A′	B′	C′	−z A′	B′	C′

half of the standard normal curve. The entries in column A are z scores, beginning with 0.00 and ending (on the full-length table) with 4.00. Given a z score of zero or more, columns B and C indicate how the z score splits the area in the upper half of the normal curve. As suggested by the shading in the top legend of the table, column B indicates the proportion of area between the mean and the z score, and column C indicates the proportion of area beyond the z score, in the upper tail of the standard normal curve.

Using the Bottom Legend of the Table

Because of the symmetry of the normal curve, the entries in Table 6.1 and Table A of Appendix D also can refer to the lower half of the normal curve. Now the columns are designated as A′, B′, and C′ in the legend at the bottom of each page. When using the bottom legend, all entries refer to the lower half of the standard normal curve.

Imagine that the nonzero entries in column A′ are negative z scores, beginning with -0.01 and ending with -4.00. Given a negative z score, columns B′ and C′ indicate how that z score splits the lower half of the normal curve. As suggested by the shading in the bottom legend of the table, column B′ indicates the proportion of area between the mean and the negative z score, and column C′ indicates the proportion of area beyond the negative z score, in the lower tail of the standard normal curve.

*Exercise 6.2 Using Table A in Appendix D, find the proportion of the total area identified with the following statements:

(a) above a z score of 1.80

(b) between the mean and a z score of -0.43

(c) below a z score of -3.00

(d) between the mean and a z score of 1.65

(e) above a z score of 0.60

(f) below a z score of -2.65

(g) between z scores of 0 and -1.96

   Answers on page 498

6.6 EXAMPLE: FBI APPLICANTS

Now, let's use a step-by-step procedure, adopted throughout much of Chapter 7, to answer the question at the beginning of Chapter 6: What proportion of all FBI applicants will be shorter than exactly 65 inches, given that the distribution of heights approximates a normal curve with a mean of 68 inches and a standard deviation of 3 inches?

1. **Sketch a normal curve and shade in the target area,** as in the top panel of **Figure 6.5** on page 102. The target area represents the proportion of men shorter than 65 inches.

2. **Plan your solution according to the normal table.** Decide precisely how you will find the value of the target area. In the present case, the answer will be obtained from column C′, because the target area coincides with the type of area identified with column C′, that is, the area in the lower tail beyond a negative z.

3. **Convert X to z.** Express 65 as a z score that, you'll recall, equals -1.00.

4. **Find the target area.** Refer to the standard normal table, using the bottom legend, as the z score is negative. The arrows in Table 6.1 show how to read the table. Look up column A′ to 1.00 (representing a z score of -1.00), and note the corresponding proportion of .1587 or .16 in column C′. This is the answer, as suggested in the bottom panel of Figure 6.5. It can be concluded that only .16 of all of the FBI applicants will be shorter than 65 inches.

*Exercise 6.3 Assume that Graduate Record Exam (GRE) scores approximate a normal curve with a mean of 500 and a standard deviation of 100.

(a) Sketch a normal curve and shade in the target area described by each of the following statements:

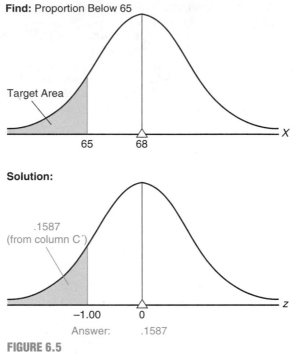

Find: Proportion Below 65

Target Area

65   68

Solution:

.1587
(from column C′)

−1.00   0
Answer: .1587

FIGURE 6.5
Finding proportions.

(a₁) less than 400

(a₂) more than 650

(b) Plan a solution (in terms of columns B, C, B′, or C′ of the standard normal table on page 532) for each of the preceding target areas.

(c) Convert to z scores and find the proportion that corresponds to each of the preceding target areas.

Answers on page 498

6.7 KEY FACTS TO REMEMBER

When using the standard normal table, it's important to remember that for any z score, the corresponding proportions in columns B and C (or columns B′ and C′) always sum to .5000. By the same token, the total area under the normal curve always equals 1.0000, the sum of the proportions in the lower and upper halves, that is, .5000 + .5000. Finally, although a z score can be either positive or negative, the proportions of area under the curve are always positive or zero but *never* negative (because an area can't be negative). **Figure 6.6** summarizes how to interpret the normal curve table in this book.

Summary

Many observed frequency distributions approximate the well-documented normal curve, an important theoretical curve noted for its symmetrical bell-

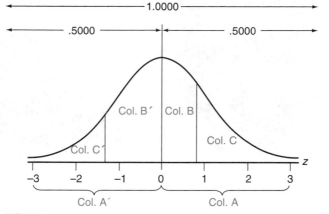

FIGURE 6.6
Interpretation of Table A, Appendix D.

shaped form. The normal curve can be used as the basis for calculating the answers to a wide variety of questions.

Although there are an infinite number of different normal curves, each with its own mean and standard deviation, there's only one standard normal curve, with its mean of 0 and its standard deviation of 1. Only the standard normal curve is actually tabled. To gain access to the standard normal table (Table A in Appendix D on page 532), original scores must be expressed as the number of standard deviations that each deviates above or below its mean—that is, as z scores.

Important Terms
• • • • • • • • • • • • • • • • • • • •

Normal curve   **Standard normal curve**

z score

REVIEW EXERCISES

6.4 Fill in the blank spaces.

To identify a particular normal curve, you must know the __(a)__ and __(b)__ for that distribution. To convert a particular normal curve to the standard normal curve, you must convert original observations into __(c)__ scores. A z score indicates how many __(d)__ an observation is __(e)__ or __(f)__ the mean of the distribution. Although there are an infinite number of different normal curves, there is __(g)__ standard normal curve. The standard normal curve has a __(h)__ of zero and a __(i)__ of one.

The total area under the standard normal curve equals __(j)__. When using the standard normal table, it's important to remember that for any z score, the corresponding proportions in columns B and C (or columns B' and C') always sum to __(k)__. Furthermore, the proportion in column B (or B') always specifies the proportion of area between the __(l)__ and the z score, whereas the proportion in column C (or C') always specifies the proportion of area __(m)__ the z score. Although any z score can be either positive or negative, the proportions of area, specified in columns B and C (or columns B' and C'), are never __(n)__.

Normal Distributions (II): Applications

FINDING PROPORTIONS

7.1   EXAMPLE: FINDING PROPORTIONS *BELOW* A SCORE (TO LEFT OF MEAN)

7.2   EXAMPLE: FINDING PROPORTIONS *BELOW* A SCORE (TO RIGHT OF MEAN)

7.3   EXAMPLE: FINDING PROPORTION *BETWEEN* TWO SCORES

7.4   EXAMPLE: FINDING PROPORTIONS *BEYOND* PAIRS OF SCORES

FINDING SCORES

7.5   EXAMPLE: FINDING *A* SCORE (TO RIGHT OF MEAN)

7.6   EXAMPLE: FINDING *PAIRS* OF SCORES (ON BOTH SIDES OF MEAN)

Summary

Review Exercises

Many observed frequency distributions in the real world approximate a normal curve, including scores on IQ tests, slight measurement errors made by a succession of people who attempt to measure the same distance, and even the heights of cornstalks and the useful lives of 100-watt electric lightbulbs. (More advanced statistics books supply techniques for testing whether an observed frequency distribution adequately approximates a normal curve.) Thanks to the standard normal table, we can answer a wide variety of questions about *any* normal distribution with a known mean and standard deviation. In the long run, this proves to be both more accurate and more efficient than dealing directly with each observed frequency distribution.

Two Types of Problems

This chapter provides representative examples of two main types of normal curve problems. In the first type of problem, we must find a *proportion* (or area under the normal curve). In the second type of problem, we must find a *score* (or value along the numerically scaled base of the normal curve). Initially, the two types of problems will be treated separately, but in Review Exercises 7.6 and 7.7, solutions require that you decide first whether a proportion or a score is to be found.

Solve Problems Logically

Don't rush through these examples, memorizing solutions to particular problems or, even worse, looking for some magic formula. Do concentrate on the logic of the solution, using rough graphs of normal curves as an aid to visualizing the solution. Then, with just a little practice, you'll view the wide variety of normal curve problems not as a bewildering assortment but as many slight variations on two distinctive types.

FINDING PROPORTIONS

7.1 EXAMPLE: FINDING PROPORTION *BELOW* A SCORE (TO LEFT OF MEAN)

(See example in Section 6.6 on page 101.)

7.2 EXAMPLE: FINDING PROPORTION *BELOW* A SCORE (TO RIGHT OF MEAN)

What's the relative standing of a student with a score of 640 on the Graduate Record Exam (GRE), given that these scores approximate a normal curve with a mean of 500 and a standard deviation of 100?

1. **Sketch a normal curve and shade in the target area,** as in the top panel of **Figure 7.1.**
2. **Plan your solution according to the normal table.** The target area represents a combination of two areas, as suggested in the bottom panel of

Find: Proportion Below 640

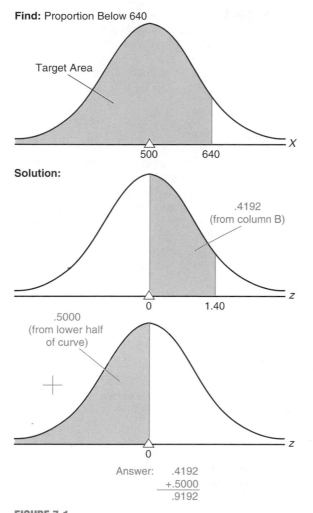

Solution:

Answer: .4192
+.5000
.9192

FIGURE 7.1
Finding proportions.

Figure 7.1. The right area coincides with that type of area identified with column B, and the left area represents the entire lower half of the normal curve and, therefore, has a value of .5000.

3. **Convert from *X* to *z* by expressing 640 as**

$$z = \frac{X - \mu}{\sigma} = \frac{640 - 500}{100} = \frac{140}{100} = 1.40$$

4. **Find the target area.** Refer to Table A in Appendix D, using the top legend, as the *z* score is positive. Look down column A to a *z* score of 1.40, and note the corresponding proportion of .4192 in column B. To this proportion add .5000. The resulting sum, .9192 or .92, represents the proportion of GRE scores lower than 640. In other words, a student with a GRE score of 640 has done better than 92 percent of all other students.

Comment: Different Strategy, but Same Answer

There is always more than one way to solve a normal curve problem. Using a less direct approach, we could have found the desired proportion by first finding the proportion of GRE scores higher than 640 (represented by the unshaded area in Figure 7.1) and then subtracting this proportion from that which represents all GRE scores (1.0000). Again look down column A to a *z* score of 1.40, but now note the corresponding proportion of .0808 in column C. In agreement with the previous answer, the proportion of GRE scores lower than 640 is found to be .9192 or .92, because

$$1.0000 - .0808 = .9192 = .92$$

Important Reminder about Interpreting Areas

When read from left to right, the numerical scale along the base of the normal curve always increases in value. Accordingly, the area to the left of a given score represents the proportion of smaller or lower scores, and the area to the right of a given score represents the proportion of larger or higher scores.

7.3 EXAMPLE: FINDING PROPORTION *BETWEEN* TWO SCORES

Let's assume that, when not interrupted artificially, the gestation periods for human fetuses approximate a normal curve with a mean of 270 days (9 months) and a standard deviation of 15 days. What proportion of gestation periods will be between 245 and 255 days?

1. **Sketch a normal curve and shade in the target area,** as in the top panel of **Figure 7.2.** Satisfy yourself that, in fact, the shaded area represents just those gestation periods between 245 and 255 days.
2. **Plan your solution according to the normal table.** Because the value of the target area can't be read directly from Table A, this type of problem requires more effort to solve. As suggested in the bottom panel of Figure 7.2, the basic idea is to identify the target area with the difference between two overlapping areas whose values can be read from column C′ of Table A. The larger area (less than 255 days) contains two sectors: the target area (between 245 and 255 days) and a remainder (less than 245 days). The smaller area contains only the remainder (less than 245 days).

Find: Proportion Between 245 and 255

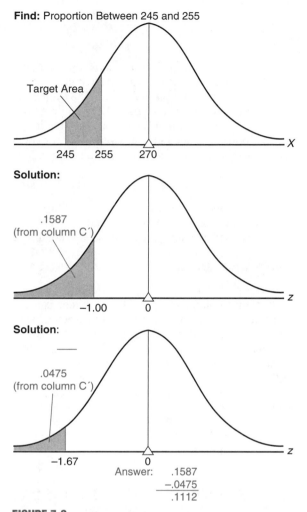

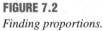

FIGURE 7.2
Finding proportions.

Subtracting the smaller area (less than 245 days) from the larger area (less than 255 days), therefore, eliminates the common remainder (less than 245 days), leaving only the target area (between 245 and 255 days).

3. **Convert** *X* **to** *z* by expressing 255 as

$$z = \frac{255 - 270}{15} = \frac{-15}{15} = -1.00$$

and by expressing 245 as

$$z = \frac{245 - 270}{15} = \frac{-25}{15} = -1.67$$

4. **Find the target area.** Look up column A′ to a negative z score of -1.00 (remember, you must imagine the negative sign), and note the corresponding proportion of .1587 in column C′. Likewise, look up column A′ to a z score of -1.67, and note the corresponding proportion of .0475 in column C′. Subtract the smaller proportion from the larger proportion to obtain the answer, .1112, or .11. Thus, only .11, or 11 percent, of all gestation periods will be between 245 and 255 days.

Warning: Enter Table Only with Single z Score

When solving problems with two z scores, as above, resist the temptation to subtract one z score from another and to enter Table A with this difference. Table A is designed only for individual z scores, not differences between z scores.

Comment: Different Strategy, but Same Answer

The problem can be solved in another way, using entries from column B′ rather than column C′. Now the basic idea is to identify the target area with the difference between two overlapping sectors whose values can be read from column B′. The larger area (between 245 and the mean of 270) contains two sectors: the target area (between 245 and 255) and the remainder (between 255 and the mean of 270). The smaller area contains only the remainder (between 255 and the mean of 270). Once again, subtracting the smaller area (between 255 and 270) from the larger area (between 245 and 270) eliminates the common remainder (between 255 and 270), leaving only the target area (between 245 and 255 days). Before reading on, visualize this alternative solution as a graph of the normal curve, and verify that the answer still equals .1112, even though column B′ is used.

....

7.4 EXAMPLE: FINDING PROPORTIONS *BEYOND* PAIRS OF SCORES

School district officials believe that their students' intellectual aptitudes approximate a normal distribution with a mean of 105 and a standard deviation of 15. Assuming that their belief is correct, what proportion of student IQs should be more than 30 points either above or below the mean?

1. **Sketch a normal curve and shade in the two target areas,** as in the top panel of **Figure 7.3.**
2. **Plan your solution according to the normal table.** Because each of the target areas can be read directly from Table A, the solution to this type of problem is fairly straightforward. The target area in the tail to the right can be obtained from column C, and that in the tail to the left can be obtained from column C′, as suggested in the bottom panel of Figure 7.3.

Find: Proportion Beyond 30 Points from Mean

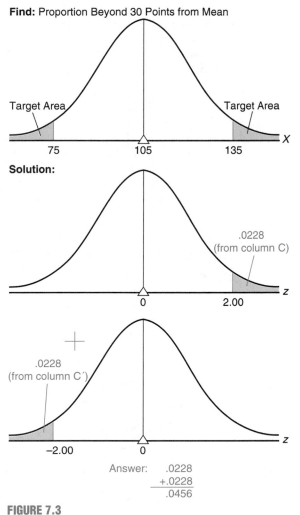

Solution:

.0228
(from column C)

.0228
(from column C′)

−2.00   0

Answer: .0228
          +.0228
           .0456

FIGURE 7.3
Finding proportions.

3. **Convert X to z** by expressing IQ scores of 135 and 75 as

$$z = \frac{135 - 105}{15} = \frac{30}{15} = 2.00$$

$$z = \frac{75 - 105}{15} = \frac{-30}{15} = -2.00$$

4. **Find the target area.** In Table A, locate a z score of 2.00 in column A, and note the corresponding proportion of .0228 in column C. Because of

the symmetry of the normal curve, you needn't enter the table again to find the proportion below a z score of -2.00. Instead, merely double the above proportion of .0228 to obtain .0456 or .05, which represents the proportion of students with IQs more than 30 points either above or below the mean.

Comment: Read Problems Carefully

Attend closely to the wording of problems. As we've just seen, "*more* than 30 points either above or below the mean" translates into two target areas, one in each tail of the normal curve. Slight changes in wording create entirely different target areas. For example, "*within* 30 points either above or below the mean" translates into two entirely new target areas, each sharing a boundary at the mean, but one area extending 30 points above the mean, and the other area extending 30 points below the mean.

*Exercise 7.1 Assume that verbal SAT scores approximate a normal curve with a mean of 500 and a standard deviation of 100.

  (a) Sketch a normal curve and shade in the target area(s) described by each of the following statements:

(a_1) more than 570

(a_2) less than 515

(a_3) between 520 and 540

(a_4) either less than 470 or more than 570

(a_5) between 470 and 520

(a_6) more than 50 points above the mean

(a_7) more than 100 points either above or below the mean

(a_8) within 50 points either above or below the mean

(a_9) more than 520 but less than 560

  (b) Plan a solution (in terms of columns B, C, B′, and C′) for each of the above target areas.

  (c) Convert to z scores and find the value of the proportion that corresponds to each of the preceding target areas.
  Answers on pages 498–499

FINISHING SCORES

So far, we've concentrated on normal curve problems for which Table A on page 532 must be consulted to find the unknown proportion (of area) associated with some score or pair of scores. For instance, given a GRE score of 640, we found that the proportion of scores smaller than 640 equals .92. Now we'll concentrate on the opposite type of normal curve problem for which Table A must be consulted to find the unknown score or scores associated with some area. For instance, given that a GRE score must be in the upper 25 percent of the distribution (in order for an applicant to be considered for admission to graduate school), we must find the smallest value of a qualifying GRE score. Essentially, this type of problem requires that we completely reverse our usage of Table A: Now we'll enter columns B, C, B′, or C′ and read out the scores listed in columns A or A′.

7.5 EXAMPLE: FINDING *A* SCORE (TO RIGHT OF MEAN)

Exam scores for a large biology class approximate a normal curve with a mean of 230 and a standard deviation of 50. Furthermore, students are graded "on a curve," with the upper 20 percent being awarded grades of A. What's the lowest score on the exam to earn a grade of A?

1. **Sketch a normal curve and, on the correct side of the mean, draw a line representing the target score,** as in **Figure 7.4.**

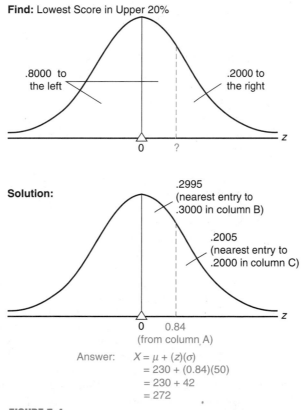

Find: Lowest Score in Upper 20%

.8000 to the left

.2000 to the right

Solution:

.2995 (nearest entry to .3000 in column B)

.2005 (nearest entry to .2000 in column C)

0   0.84 (from column A)

Answer: $X = \mu + (z)(\sigma)$
$= 230 + (0.84)(50)$
$= 230 + 42$
$= 272$

FIGURE 7.4
Finding scores.

This is often the most difficult step, and it involves semantics rather than statistics. It's often helpful to visualize the target score as splitting the total area into two sectors—one to the left of (below) the target score and one to the right of (above) the target score. For example, in the present case, the target score is the point along the base of the curve that splits the total area into 80 percent, or .8000 to the left, and 20 percent, or .2000 to the right. Serving as a frame of reference, the mean of a normal curve always splits the total area into two equal halves—.5000 to the left and .5000 to the right. Because more than .5000—that is, .8000—of the total area is to the left of the target score, this score must be on the upper or right side of the mean. If less than .5000 of the total area had been to the left of the target score, this score would have been placed on the lower or left side of the mean.

2. **Plan your solution according to the normal table.** In problems of this type, you must plan how to find the z score for the target score. Because the target score is on the right side of the mean, concentrate on the area in the upper half of the normal curve, as described in columns B and C. The bottom panel of Figure 7.4 indicates that either column B or C can be used to locate a z score in column A. It's crucial, however, to search for the single value (.3000) that is valid for column B or the single value (.2000) that's valid for column C. Note that we look in column B for .3000, not for .8000. Table A is not designed for sectors such as the lower .8000 that span the mean of the normal curve.

3. **Find z.** Refer to Table A. Scan column C to find .2000. If this value doesn't appear in column C, as typically will be the case, approximate the desired value (and the correct score) by locating the entry in column C nearest to .2000. If adjacent entries are equally close to the value being sought, use either entry—it's your choice! As shown in the lower panel of Figure 7.4, the entry in column C closest to .2000 is .2005, and the corresponding z score in column A equals 0.84. Verify this by checking Table A. Also note that exactly the same z score of 0.84 would have been identified if column B had been searched to find the entry (.2995) nearest to .3000. The z score of 0.84 represents the point that separates the upper 20 percent of the area from the rest of the area under the normal curve.

4. **Convert z to the target score.** Finally, convert the z score of 0.84 into an exam score, given a distribution with a mean of 230 and a standard deviation of 50. You'll recall that a z score indicates how many standard deviations the original score is above or below its mean. In the present case, the target score must be located .84 of a standard deviation above its mean. The distance of the target score above its mean equals 42 (from .84 × 50), which, when added to the mean of 230, yields a value of 272. Therefore, 272 is the smallest score on the exam to qualify for a grade of A.

When converting z scores to original scores, you'll probably find it more efficient to use the following equation (derived from the z score equation on page 98):

CONVERTING *z* SCORE TO ORIGINAL SCORE

$$X = \mu + (z)(\sigma) \tag{7.1}$$

---

in which X is the target score, expressed in original units of measurement; μ and σ are the mean and the standard deviation for the normal curve; and z is the standard score read from column A or A′ of Table A on page 532. When appropriate numerical substitutions are made, as shown in the bottom of Figure 7.4, 272 is found to be the answer, in agreement with our earlier conclusion.

Comment: Place Target Score on Correct Side of Mean

When finding scores, it's crucial that the target score be placed on the correct side of the mean. This placement dictates how the normal table will be read — whether down from the top legend, with entries in column A interpreted as positive z scores, or up from the bottom legend, with entries in column A′ interpreted as negative z scores. In the previous problem, the incorrect placement of the target score on the left side of the mean would have led to a z score of -0.84, rather than 0.84, and an erroneous answer of 188 (from $230 - 42$), rather than the correct answer of 272 (from $230 + 42$).

To make correct placements, you must properly interpret the specifications for the target score. Expand potentially confusing one-sided specifications, such as the "upper 20 percent, or upper .2000," into "left .8000 and right .2000." Having identified the left and right areas of the target score, which sum to 1.000, you can compare the specification of the target score with those of the mean. Remember that the mean of a normal curve always splits the total area into .5000 to the left and .5000 to the right. Accordingly, if the area to the left of the target score is more than .5000, the target score should be placed on the upper or right side of the mean. Otherwise, if the area to the left of the target score is less than .5000, the target score should be placed on the lower or left side of the mean. In the previous example, the target score splits the total area into .8000 to the left and .2000 to the right. Because the area to the left is more than .5000, the target score should be placed on the upper or right side of the mean, as it was in Figure 7.4.

7.6 EXAMPLE: FINDING *PAIRS* OF SCORES (ON BOTH SIDES OF MEAN)

Assume that the annual rainfall in the San Francisco area approximates a normal curve with a mean of 22 inches and a standard deviation of 4 inches. What are the rainfalls for the more atypical years, defined as the driest 2.5 percent of all years and the wettest 2.5 percent of all years?

1. **Sketch a normal curve. On either side of the mean, draw two lines representing the two target scores, as in Figure 7.5.** The smaller target

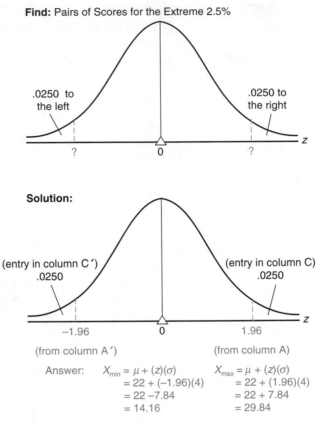

FIGURE 7.5

Finding scores.

score splits the total area into .0250 to the left and .9750 to the right, and the larger target score does the exact opposite.

2. **Plan your solution according to the normal table.** Because the smaller target score is located on the lower or left side of the mean, we'll concentrate on the area in the lower half of the normal curve, as described in columns B′ and C′. The target z score can be found by scanning either column B′ for .4750 or column C′ for .0250. After finding the smaller target score, we'll capitalize on the symmetrical properties of normal curves to find the value of the larger target score.

3. **Find z.** Referring to Table A, let's scan column B′ for .4750, or the entry nearest to .4750. In this case, .4750 appears in column B′, and the corresponding z score in column A′ equals −1.96. The same z score of −1.96 would have been obtained if column C′ had been searched for a value of .0250.

4. **Convert z to the target score.** When the appropriate numbers are substituted in Formula 7.1, as shown in the bottom panel of Figure 7.5, the smaller target score equals 14.16 inches, the amount of annual rainfall that separates the driest 2.5 percent of all years from all of the other years.

The location of the larger target score is the mirror image of that for the smaller target score. Therefore, we needn't even consult Table A to establish that its z score equals 1.96—that is, the same value as that for the smaller target score, but without the negative sign. When 1.96 is converted to inches of rainfall, as shown in the bottom of Figure 7.5, the larger target equals 29.84 inches, the amount of annual rainfall that separates the wettest 2.5 percent of all years from all other years.

Comment: Common and Rare Events

In the preceding problem, we drew attention to the atypical or rare years by concluding that 2.5 percent of the driest years registered less than 14.16 inches of rainfall, whereas 2.5 percent of the wettest years registered more than 29.84 inches. Had we wished, we also could have drawn attention to the typical or common years by concluding that 95 percent of the most moderate years registered between 14.16 and 29.84 inches of rainfall. In other words, the interval from 14.16 to 29.84 inches includes the "middle" 95 percent of annual rainfalls. The middle 95 percent straddles the line perpendicular to the mean or 50th percentile, with half or 47.5 percent above this line and the other half or 47.5 percent below this line.

In many statistical applications, we'll wish to distinguish between *common* and *rare* events, that is, between common years of moderate rainfall and rare years of relatively scarce or frequent rainfall, between IQs that are more or less average and those that are exceptional, and between people whose height is intermediate and those who are relatively short or tall. This distinction often reflects the statistician's preference for identifying common events with the middle 95 percent of the total area under the normal curve and rare events with the extreme 2.5 percent in each tail. In other words, you'll encounter many variations based on the preceding split—all of which involve z scores of -1.96 and $+1.96$ (or ± 1.96). Accordingly, in subsequent sections you'll often find yourself working with z scores of ± 1.96.

You now have the necessary information to solve most normal curve problems, but there is no substitute for actually working a variety of normal curve problems, such as those offered at the end of this chapter. For your convenience, a complete set of guidelines appears in **Figure 7.6.** Spend a few moments studying it now, and then refer to it as often as necessary.

*Exercise 7.2 The burning times of electric lightbulbs approximate a normal curve with a mean of 1200 hours and a standard deviation of 120 hours. If a large number of new lights are installed at the same time (possibly along a newly opened freeway), at what time will

(a) 1 percent fail? (*Hint:* This splits the total area into .0100 to the left and .9900 to the right.)

(b) 5 percent fail?

(c) 10 percent fail?

(d) 50 percent fail?

(e) 95 percent fail?

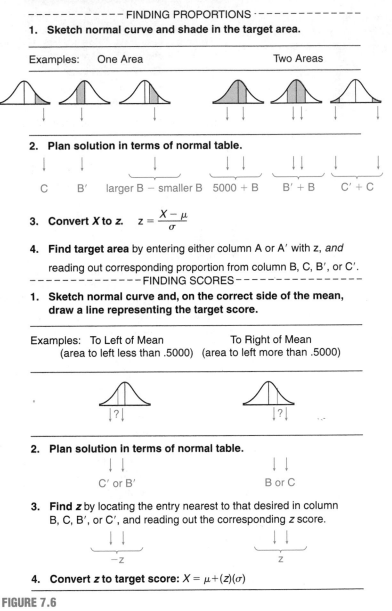

DOING NORMAL CURVE PROBLEMS

Read the problem carefully to determine whether a proportion or a score is to be found.

– – – – – – – – – – – FINDING PROPORTIONS · – – – – – – – – – – –

1. **Sketch normal curve and shade in the target area.**

Examples: One Area   Two Areas

2. **Plan solution in terms of normal table.**

    C   B′   larger B − smaller B   5000 + B   B′ + B   C′ + C

3. **Convert X to z.** $z = \dfrac{X - \mu}{\sigma}$

4. **Find target area** by entering either column A or A′ with z, *and* reading out corresponding proportion from column B, C, B′, or C′.

– – – – – – – – – – – – FINDING SCORES – – – – – – – – – – – – –

1. **Sketch normal curve and, on the correct side of the mean, draw a line representing the target score.**

Examples: To Left of Mean   To Right of Mean
    (area to left less than .5000)   (area to left more than .5000)

2. **Plan solution in terms of normal table.**

        C′ or B′   B or C

3. **Find z** by locating the entry nearest to that desired in column B, C, B′, or C′, and reading out the corresponding z score.

            −z   z

4. **Convert z to target score:** $X = \mu + (z)(\sigma)$

FIGURE 7.6

Guidelines for normal curve problems.

(f) If a new inspection procedure eliminates the weakest 8 percent of all lights before being marketed, the manufacturer can safely offer customers a money-back guarantee on all lights that fail before _____ hours of burning time.

Answers on page 499

Summary
. . . .

There are two general types of normal curve problems: (1) those that require you to find the unknown proportion (of area) associated with some score or pair of scores and (2) those that require you to find the unknown score or scores associated with some area. Answers to the former type of problem usually require you to convert original scores into z scores (Formula 6.1), and answers to the latter type of problem usually require you to translate a z score back into an original score (Formula 7.1).

REVIEW EXERCISES

Finding Proportions

7.3 Scores on the Wechsler Intelligence Scale approximate a normal curve with a mean of 100 and a standard deviation of 15.
What proportion of IQ scores are

(a) above 125?

(b) below 82?

(c) within 9 points either above or below the mean?

(d) more than 40 points either above or below the mean?

7.4 Suppose the burning times of electric lightbulbs approximate a normal curve with a mean of 1200 hours and a standard deviation of 120 hours. What proportion of lights burn for

(a) less than 960 hours?

(b) more than 1500 hours?

(c) within 50 hours either above or below the mean?

(d) between 1300 and 1400 hours?

Finding Scores

7.5 Scores on the Wechsler IQ test approximate a normal curve with a mean of 100 and a standard deviation of 15. What IQ score is identified with

(a) the upper 2 percent, that is, 2 percent to the right (and 98 percent to the left)?

(b) the lower 10 percent?

(c) the upper 60 percent?

(d) the middle 95 percent? [Remember, the middle 95 percent straddles the line perpendicular to the mean (or the 50th percentile) with half of 95 percent or 47.5 percent above this line and the remaining 47.5 percent below this line.]

(e) the middle 99 percent?

(f) the lower 95 percent?

Finding Proportions and Scores

Important note: When doing Exercises 7.6 and 7.7, remember to decide first whether a proportion or a score is to be found.

*7.6   An investigator polls a representative sample of common-cold sufferers, asking them to estimate the number of hours of physical discomfort caused by their most recent colds. Their estimates approximate a normal curve with a mean of 83 hours and a standard deviation of 20 hours.

(a) What is the estimated number of hours for the shortest-suffering 5 percent?

(b) What proportion of sufferers estimate that their colds lasted for longer than 48 hours?

(c) What proportion suffered for fewer than 61 hours?

(d) What is the estimated number of hours for the extreme 1 percent either above or below the mean?

(e) What proportion suffered for between one and three days?

(f) What is the estimated number of hours for the shortest-suffering 10 percent?

(g) What is the estimated number of hours for the middle-suffering 95 percent? (See the comment about "middle 95 percent" in Exercise 7.5 [d].)

(h) What proportion suffered for between two and four days?

(i) A medical researcher wishes to concentrate on the 20 percent who suffered the most. She will work only with those who estimate that they suffered for more than _____ hours.

(j) Another researcher wishes to compare those who suffered least with those who suffered most. If each group is to consist of only the extreme 3 percent, the mild group will consist of those who suffered for fewer than _____ hours, and the severe group will consist of those who suffered for more than _____ hours.

(k) Another survey found that people with colds who took daily doses of vitamin C suffered, on the average, for 61 hours. What proportion of the preceding representative sample suffered for longer than 61 hours?

Answers on page 499

7.7   Admission to a state university depends partially on the applicant's high school GPA. Assume that the applicants' GPAs approximate a normal curve with a mean of 3.20 and a standard deviation of 0.30.

(a) If applicants with GPAs of 3.50 or above are automatically admitted, what proportion of applicants will be in this category?

(b) If applicants with GPAs of 2.50 or below are automatically denied admission, what proportion of applicants will be in this category?

(c) A special honors program is open to all applicants with GPAs of 3.75 or better. What proportion of applicants is eligible?

(d) If the special honors program is limited to students whose GPAs rank in the upper 10 percent, what GPA will be required for admission to this program?S

CHAPTER 8

More about *z* Scores

8.1   *z* SCORES FOR NON-NORMAL DISTRIBUTIONS
8.2   STANDARD SCORES
8.3   TRANSFORMED STANDARD SCORES
8.4   CONVERTING TO TRANSFORMED STANDARD SCORES
8.5   PERCENTILE RANKS AGAIN

Summary

Important Terms

Review Exercises

8.1 *z* SCORES FOR NON-NORMAL DISTRIBUTIONS

In Chapters 6 and 7, *z* scores have been used only in the context of the normal curve, and throughout the remainder of this book, the reappearance of *z* scores will usually prompt an interpretation based on the standard normal table. It would be erroneous to conclude, however, that *z* scores are limited to distributions that approximate a normal curve. Non-normal distributions also can be transformed into sets of *z* scores. *In this case, the standard normal table can't be consulted,* because the shape of the distribution of *z* scores is the same as that for the original non-normal distribution. For instance, if the original distribution is positively skewed, the distribution of *z* scores also will be positively skewed. *Regardless of the shape of the distribution, the shift to z scores always produces a distribution with a mean of 0 and a standard deviation of 1.*

Interpreting Test Scores

Under most circumstances, *z* scores provide efficient descriptions of relative performance on one or more tests (or even, as in Exercise 8.7, help explain the absence of .400 batting averages among major league hitters during the last half century). Without additional information, it's meaningless to know that Sharon earned a raw score of 159 on a math test, but it's very informative to know that she earned a *z* score of 1.80. The latter score suggests that she did relatively well on the math test, being almost two standard deviation units above the mean. More precise interpretations of this score could be made, of course, if it is known that the distribution of test scores approximates a normal curve.

The use of *z* scores can help you identify a person's relative strengths and weaknesses on several different tests. For instance, **Table 8.1** shows Sharon's scores on college achievement tests in three different subjects. The evaluation of her test performance is greatly facilitated by converting her raw scores into the *z* scores listed in the final column of Table 8.1. Now a glance at the *z* scores suggests that although she did relatively well on the math test, her performance on the English test was only slightly above average, as indicated by a *z* score of 0.50, and her performance on the psychology test was slightly below average, as indicated by a *z* score of −0.67.

Importance of Reference Group

Remember that *z* scores reflect performance relative to some group, rather than relative to an absolute standard. A meaningful interpretation of *z* scores requires, therefore, that the nature of the reference group be specified. In the

Table 8.1
SHARON'S ACHIEVEMENT TEST SCORES

SUBJECT	RAW SCORE	MEAN	STANDARD DEVIATION	z SCORE
Math	159	141	10	1.80
English	83	75	16	0.50
Psych	23	27	6	−0.67

present example, it is important to know whether Sharon's scores were relative to those of the other students at her college or to those of students at a wide variety of colleges, as well as to any other special characteristics of the reference group.

Exercise 8.1 Convert each of the following test scores to z scores:

	TEST SCORE	MEAN	STANDARD DEVIATION
(a)	53	50	9
(b)	38	40	10
(c)	45	30	20
(d)	28	20	20

Exercise 8.2 **(a)** Referring to the previous problem, which one test score would you prefer?

(b) Still referring to the previous problem, if, in fact, you had earned a score of 64 on some test, which of the four distributions (a, b, c, or d) would have permitted the most favorable interpretation of this score?

Answers on page 500

8.2 STANDARD SCORES

Whenever scores are expressed relative to a known mean and standard deviation, they are referred to as **standard scores.** Thus z scores qualify as standard scores because they are expressed relative to a mean of 0 and a standard deviation of 1.

Standard score

Any score expressed relative to a known mean and a known standard deviation.

Transformed standard score

A standard score that, unlike a z score, usually lacks negative signs and decimal points.

8.3 TRANSFORMED STANDARD SCORES

Being by far the most important standard score, z scores are often viewed as synonymous with standard scores. *For convenience, particularly when reporting test results, z scores can be transformed to other types of standard scores that lack negative signs and decimal points.* These transformations change neither the shape of the original distribution nor the relative standing of any test score within the distribution. A test score located one standard deviation below the mean might be reported not as a z score of -1.00 but as a T score of 40 in a distribution of T scores with a mean of 50 and a standard deviation of 10. The important point to realize is that although reported as a score of 40, this T score accurately reflects the relative location of the original z score of -1.00: A T score of 40 is located at a distance of one standard deviation (of size 10) below the mean (of size 50). **Figure 8.1** shows the values of some of the more common types of transformed standard scores relative to the various portions of the area under the normal curve. Also illustrated in Figure 8.1 are percentile ranks, to be discussed in Section 8.5.

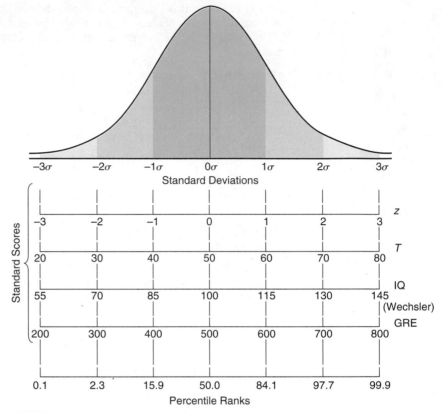

FIGURE 8.1

Common transformed scores (and percentile ranks) associated with normal curves.

8.4 CONVERTING TO TRANSFORMED STANDARD SCORES

To convert any original standard score, z, into a transformed standard score, z', having a distribution with any desired mean and standard deviation, use the following formula:

TRANSFORMED STANDARD SCORE

$$z' = \text{desired mean} + (z)(\text{desired standard deviation}) \qquad (8.1)$$

---

where z' (called "z prime") is the transformed standard score, and z is the original standard score.

For instance, if you wish to convert a z score of -1.50 into a new distribution of z' scores for which the desired mean equals 500 and the desired standard deviation equals 100, substitute these numbers into the preceding formula to obtain

$$z' = 500 + (-1.50)(100)$$
$$= 500 - 150$$
$$= 350$$

Again, notice that the transformed standard score accurately reflects the relative location of the original standard score of -1.50: The transformed score of 350 is located at a distance of 1.5 standard deviation units (each of size 100) below the mean (of size 500). The change from a z score of -1.50 to a z' score of 350 eliminates negative signs and decimal points without distorting the relative location of the original score, expressed as a distance from the mean in standard deviation units.

Substitute Pairs of Convenient Numbers

You could substitute any mean or any standard deviation in Formula 8.1 to generate a new distribution of transformed scores. Traditionally, substitutions have been limited mainly to the pairs of convenient numbers shown in Figure 8.1: a mean of 50 and a standard deviation of 10 (T scores), a mean of 100 and a standard deviation of 15 (IQ scores), and a mean of 500 and a standard deviation of 100 (GRE scores). The substitution of other arbitrary pairs of numbers serves no purpose; indeed, because of their peculiarity, they might make the new distribution, even though it lacks the negative signs and decimal points common to z scores, slightly less comprehensible to people who have been exposed to the traditional pairs of numbers.

*Exercise 8.3 Assume that each of the following raw scores originates from a distribution with the specified mean and standard deviation. After converting each raw score into a z score, transform each z score into a series of new standard scores with means and standard deviations of 50 and 10, 100 and 15, and 500 and 100, respectively. (In practice, you would transform a particular z into only one new standard score.)

	RAW SCORE	MEAN	STANDARD DEVIATION
(a)	24	20	5
(b)	37	42	3
(c)	346	310	20
(d)	1263	1400	74

Answers on page 500

8.5 PERCENTILE RANKS AGAIN

Sometimes original scores are interpreted as **percentile ranks.** As defined in Section 2.11, *the percentile rank of a score indicates the percentage of scores in the entire distribution with similar or smaller values.* For example, a score will have a percentile rank of 80 if similar or smaller scores constitute 80 percent of the entire distribution. The median score of a distribution has a percentile rank of 50 because similar or smaller scores constitute 50 percent of the entire distribution. Even standard scores can be identified with a percentile rank. *If the parent distribution is normal,* then a z score of 2, a T score of 70, a Wechsler IQ score of 130, and a GRE score of 700 will share a percentile rank of 97.7, as can be verified by referring to Figure 8.1. Whether used in conjunction with original scores or standard scores, percentile ranks are ideally suited for exchanges, as between parents and teachers or between patients and doctors, that often require the interpretation of test scores or lab findings with a minimum of numerical explanation.

Limitations of Percentile Ranks

The limitations of percentile ranks become apparent when detailed comparisons are required between scores based on normal distributions. This problem can be appreciated by referring to the bottom scale of Figure 8.1, which shows the percentile ranks for various portions of the area under the normal curve. Notice that the scale of percentile ranks lacks the orderly increase in values apparent in each of the other scales for standard scores. Instead, in the vicinity of the mean, a distance of one standard deviation unit separates the percentile ranks of 50.0 and 84.1, and in the upper tail of the normal curve, a distance of one standard deviation unit separates the percentile ranks of 97.7 and 99.9. Clearly, differences between percentile ranks must be interpreted cautiously — partly on the basis of their location along the scale of percentile ranks. Because the same problem doesn't occur for standard scores, with their equal spacing along the horizontal scale, those who regularly deal with test scores often prefer standard scores to percentile ranks.

*Exercise 8.4 Assume that all of the following transformed standard scores originate from distributions that approximate normal curves.

(a) What's the percentile rank of a GRE score of 640? *Hint:* This is a normal curve problem where you must find an unknown proportion. First refer to Figure 8.1 to find the mean and standard deviation for GRE scores; next convert 640 to a z score; and then, with the aid of a sketch of the target area and the normal curve table, find the proportion of the *entire* normal curve to the left of this z score.

(b) Given that a score on the Wechsler IQ scale has a percentile rank of 93, what's the value of the IQ score? *Hint:* This is a normal curve problem where you must find an unknown score. First locate in a sketch the unknown z score on the correct side of the mean; next, with the aid of the normal curve table find the z score that has the *tabular equivalent* of 93 percent of the area below it; then refer to Figure 8.1 to find the mean and standard deviation for IQ scores and convert the obtained z score to an IQ score.

(c) What's the percentile rank of a *T* score of 35?

(d) Given a percentile rank of 60 for some GRE score, what's the value of that score?

Answers on page 500

Summary
....

Even when distributions fail to approximate normal curves, z scores can provide efficient descriptions of relative performance on one or more tests.

When reporting test results, z scores are often transformed into other types of standard scores that lack negative signs and decimal points. These conversions change neither the shape of the original distribution nor the relative standing of any test score within the original distribution.

The percentile rank of a score indicates the percentage of scores in the entire distribution with similar or smaller values.

Important Terms
....

Standard score   **Percentile rank**
Transformed standard score

REVIEW EXERCISES

8.5   It has been claimed that the distribution of *z* scores always has a mean of 0 and a standard deviation of 1. To convince yourself of this—short of a mathematical proof—transform each of the following raw scores into *z* scores, and then verify that the mean of the *z* scores does, in fact, equal 0 and that the standard deviation of the *z* scores does, in fact, equal 1. The raw scores are 5, 7, 8, 9, 11. *Hint:* First find the mean and standard deviation of the raw scores. Then convert each raw score to a *z* score. Finally, calculate the mean and standard deviation of the *z* scores. If, after doing this exercise, you suspect that these five numbers might be special, use any set of arbitrary numbers to verify that a set of *z* scores always has a mean of 0 and a standard deviation of 1. (Your answers might deviate slightly from 0 or 1 because of rounding errors.)

8.6   When describing test results, someone objects to the conversion of raw scores into standard scores, claiming that this constitutes an arbitrary change in the value of the test score. How might you respond to this objection?

*8.7 Measures of relative standing, such as standard scores, might help explain the absence of .400 batting averages among major league baseball hitters during the last half-century. (For nonbaseball fans, .400 indicates simply that 40 percent of a player's official "at bats" are hits.) Essentially, the mean batting average for all players has declined fairly steadily, year by year, from a preponderance of .290s during the early 1900s to the

.260s during the late 1900s; by the same token, the standard deviation also has declined from roughly .035 during the early 1900s to about .030 during the late 1900s.∗(Some fans have speculated that this decline has been caused by a finer balance between offense and defense as athletes have become more talented.)

(a)   The last player to achieve the magic .400, Ted Williams, batted .406 during the 1941 season when the mean for regularly playing American League hitters was .278 and the standard deviation was .035. What would Williams's .406 be equivalent to during the 1990 season when the mean for regularly playing American League hitters was .260 and the standard deviation was .027? In other words, when adjusted relative to the distribution of 1990 hitters, would Ted Williams still be batting over .400?

Hint: First express Williams's batting average as a *z* score relative to the 1941 season; then, using the information given for the 1990 season, convert Williams's *z* score back to his equivalent batting average for the 1990 season. Notice that this final conversion differs slightly from that described in Section 8.4 for transformed standard scores. Now, instead of any desired mean and any desired standard deviation, you must substitute the observed mean (.260) and the observed standard deviation (.027) for 1990 hitters in Formula 8.1 to obtain Williams's equivalent batting average for the 1990 season.

Comment about decimal points: If you're disturbed by decimal points, treat .406 as 406, .035 as 35, and so on, in your calculations. Do place a decimal point to the left of all numbers appearing in your final answer.

(b)   George Brett batted .390 during the 1980 season when the mean for regularly playing American League hitters was .270 and the standard deviation was .032. What would Brett's equivalent batting average have been during the 1941 season (when Ted Williams batted .406)?

(c)   Rogers Hornsby hit .424, the highest batting average of the twentieth century, during the 1924 season when the mean for regularly playing National League hitters was .290 and the standard deviation was .035. What would Hornsby's .424 be equivalent to during the 1990 season when the mean for regularly playing National League hitters was .269 and the standard deviation was .028?

Answers on page 500

8.8   You have probably taken the Scholastic Assessment Test, or SAT. Prior to 1941, scores on the verbal portion of the SAT were recalibrated each year by converting to transformed standard scores with a mean of 500 and a standard deviation of 100. After 1941, in order to make meaningful comparisons from year to year, the scores of each subsequent group were based on the scale created by the 1941 group. During recent decades, SAT scores have declined steadily, and during the 1990s, the average verbal score dropped almost 80 points, to the 420s. (Numerous

∗Special thanks to Pete Palmer of Lexington, Mass., who graciously supplied the means and standard deviations for this exercise. Some comments by Stephen Jay Gould in *Against All Odds,* the excellent instructional TV course on statistics, inspired this exercise.

explanations of this drop have been suggested, including the adverse effect of extensive TV viewing on academic achievement, as well as the much greater heterogeneity of recent college-bound groups when compared to the group of 10,000 students in 1941, who were mostly well-educated, affluent white men from the Northeast.)

(a) Using Formula 8.1, reconstruct the expression that transformed verbal scores (or more accurately, their *z*-score equivalents) for each person in the 1941 group into a set of transformed standard scores with a mean of 500 and a standard deviation of 100.

(b) If, contrary to fact, the scores of each subsequent group had been expressed as transformed standard scores relative to *only their own group,* as with the group in 1941, what would have been the mean and standard deviation of these transformed standard scores for each subsequent group?

(c) Beginning with 1995, SAT scores have been "recentered" by replacing the 1941 scale with one based on the 1990 group. Essentially, this caused 80 points to be added to the verbal score for each person in the 1995 group. Given that the mean verbal score for the 1995 group would have been 420 on the original 1941 scale, what would be the mean verbal score for this group on the new 1990 scale? *Hint:* Refer to Exercise 5.10, page 91, which examines the effect of adding a constant to the mean.

Answers on page 500

CHAPTER

9

Describing Relationships: Correlation

9.1   AN INTUITIVE APPROACH

9.2   SCATTERPLOTS

9.3   A CORRELATION COEFFICIENT FOR QUANTITATIVE DATA: r

9.4   INTERPRETATION OF r

9.5   INTERPRETATION OF r^2 (SEE SECTION 10.8)

9.6   CORRELATION NOT NECESSARILY CAUSE-EFFECT

DETAILS

9.7 z SCORE FORMULA FOR r

9.8   COMPUTATION FORMULA FOR r

9.9   OUTLIERS AGAIN

9.10   OTHER TYPES OF CORRELATION COEFFICIENTS

9.11   COMPUTER OUTPUT

        Summary

        Important Terms

        Review Exercises

Table 9.1 GREETING CARDS GIVEN AND RECEIVED BY FIVE FRIENDS		
	NUMBER OF CARDS	
FRIEND	**GIVEN**	**RE- CEIVED**
Andrea	5	10
Mike	7	12
Doris	13	14
Steve	9	18
John	1	6

The familiar saying "You get what you give" has many ramifications, including, for instance, the exchange of holiday greeting cards. An investigator suspects that a relationship exists between the number of greeting cards *given* and the number of greeting cards *received* by individuals. Prior to a full-fledged survey, he obtains estimates for the most recent holiday season from five friends, as shown in **Table 9.1.** (The data in Table 9.1 represent a very simple correlation study with two dependent variables, as defined in Section 1.9, because the number of cards given and received are outcomes not under the investigator's control.)

9.1 AN INTUITIVE APPROACH

You don't need elaborate statistical tests to determine whether two variables are related when, as in the present example, the analysis involves only a few pairs of observations. If the suspected relationship does exist between cards given and cards received, then an inspection of the data might reveal, as one possibility, that "big givers" tend to be "big receivers" and that "small givers" tend to be "small receivers." In other words, there is a tendency for pairs of observations to occupy similar relative positions in their respective distributions.

Positive Relationship

Trends among pairs of observations can be detected most easily by constructing a list of paired observations in which the observations along one variable are arranged from largest to smallest. In panel A of **Table 9.2,** the five pairs of observations are rearranged from the largest number of cards given (13) to the smallest number given (1). This table reveals a pronounced tendency for pairs of observations to occupy similar *relative* positions in their respective distributions. For example, John gave relatively few cards (1) and received relatively few cards (6), whereas Doris gave relatively many cards (13) and received relatively many cards (14). We can conclude, therefore, that the two variables are related. Furthermore, this relationship implies that "You get what you give." *Insofar as relatively low values are paired with relatively low values, and relatively high values are paired with relatively high values, the relationship is* **positive.**

In panels B and C of Table 9.2, each of the five friends continues to give the same number of cards as in panel A, but new pairs are created to illustrate two other possibilities—a negative relationship and no relationship. (In practice, of course, the pairs are fixed by the data and can't be changed.)

Negative Relationship

Notice the pattern among the pairs in panel B. Now there is a pronounced tendency for pairs of observations to occupy dissimilar and opposite relative positions in their respective distributions. For example, although John gave relatively few cards (1), he received relatively many (18). From this pattern, we can conclude that the two variables are related. Furthermore, this relationship implies that "You get the opposite of what you give." *Insofar as relatively low values are paired with relatively high values, and relatively high values are paired with relatively low values, the relationship is* **negative.**

Positive relationship

Occurs insofar as pairs of observations tend to occupy similar relative positions (high with high and low with low) in their respective distributions.

Negative relationship

Occurs insofar as pairs of observations tend to occupy dissimilar relative positions (high with low and vice versa) in their respective distributions.

Table 9.2
THREE TYPES OF RELATIONSHIPS

A. POSITIVE RELATIONSHIP

FRIEND	NUMBER OF CARDS	
	GIVEN	RE-CEIVED
Doris	13	14
Steve	9	18
Mike	7	12
Andrea	5	10
John	1	6

B. NEGATIVE RELATIONSHIP

FRIEND	GIVEN	RE-CEIVED
Doris	13	6
Steve	9	10
Mike	7	14
Andrea	5	12
John	1	18

C. LITTLE OR NO RELATIONSHIP

FRIEND	GIVEN	RE-CEIVED
Doris	13	10
Steve	9	18
Mike	7	12
Andrea	5	6
John	1	14

No Relationship

No regularity is apparent among the pairs of observations in panel C. For instance, although both Andrea and John gave relatively few cards (5 and 1), Andrea received relatively few cards (6), and John received relatively many cards (14). Given this lack of regularity, we can conclude that little, if any, relationship exists between the two variables and "What you get has no bearing on what you give."

Review

Whether we're concerned about the relationship between cards given and cards received, years of heavy smoking and life expectancy, educational level and annual income, or scores on a screening test and subsequent ratings as a police officer,

two variables are *positively* related if pairs of observations tend to occupy similar relative positions *(high with high and low with low)* in their respective distributions, and they are *negatively* related if pairs of observations tend to occupy dissimilar relative positions *(high with low and vice versa)* in their respective distributions.

Preview

The remainder of this chapter deals with how best to describe and interpret a relationship between pairs of variables. The intuitive method of searching for regularity among pairs of observations is cumbersome and inexact when the analysis involves more than a few pairs of observations. Hence, although this technique has much appeal, it must be abandoned in favor of several other more efficient and exact statistical techniques, namely, a special graph known as a *scatterplot* and a measure known as a *correlation coefficient*.

It will become apparent in the next chapter that once a relationship has been identified, it can be used for predictive purposes. Having established that years of heavy smoking is negatively related to length of life (because heavier smokers tend to have shorter lives), we can use this relationship to predict the life expectancy of someone who has smoked heavily for the past ten years. This type of prediction serves a variety of purposes, ranging from the calculation of life insurance premiums to its use as educational material in an antismoking workshop.

*Exercise 9.1 Indicate whether the following statements suggest a positive or negative relationship:

(a) High school students with lower IQs have lower GPAs.

(b) Increasing rates of unemployment accompany decreasing rates of inflation.

(c) More densely populated areas have higher crime rates.

(d) Schoolchildren who often watch TV perform more poorly on academic achievement tests than those who seldom watch.

(e) Heavier automobiles are more costly to run.

(f) Heavier automobiles yield poorer gas mileage.

(g) Better-educated people have higher incomes.

(h) More anxious people are willing to spend more time performing a simple repetitive task than less anxious people are.

Answers on page 500

9.2 SCATTERPLOTS

A **scatterplot** *is a graph containing a cluster of dots that represents all pairs of observations.* With a little training, you can use any dot cluster as a preview of a fully measured relationship.

Construction

To construct a scatterplot, as in **Figure 9.1,** scale each of the two variables along the horizontal (X) and vertical (Y) axes, and use each pair of observations to locate a dot within the scatterplot. For example, the pair of numbers for Mike, 7 and 12, define points along the X- and Y-axes, respectively. Using these points to anchor lines perpendicular (at right angles) to each axis, locate Mike's dot where the two lines intersect. Repeat this process, with imaginary lines, for each of the four remaining pairs of numbers to create the scatterplot of Figure 9.1.

Our simple example of greeting cards has shown the basic idea of correlation and the construction of a scatterplot. We shall now examine more complex sets of data in order to learn how to interpret scatterplots.

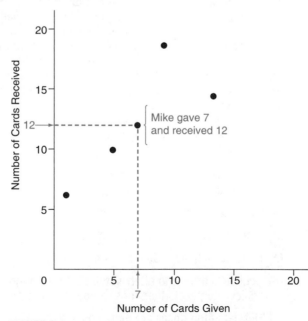

FIGURE 9.1
Scatterplot for greeting card exchange.

Positive, Negative, or No Relationship?

The first step is to note the tilt or slope, if any, of a dot cluster. *A dot cluster that has a slope from the lower left to the upper right,* as in panel A of **Figure 9.2,** *reflects a positive relationship.* Small values of one variable are paired with small values of the other variable, and large values are paired with large values. In panel A, short people tend to be light, and tall people tend to be heavy.

On the other hand, *a dot cluster that has a slope from the upper left to the lower right,* as in panel B of Figure 9.2, *reflects a negative relationship.* Small values of one variable tend to be paired with large values of the other variable,

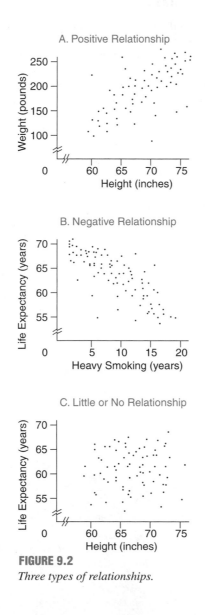

FIGURE 9.2

Three types of relationships.

and vice versa. In panel B, people who have smoked heavily for a few years or not at all tend to live longer, and people who have smoked heavily for many years tend to live shorter lives.

Finally, *a dot cluster that lacks any apparent slope,* as in panel C of Figure 9.2, *reflects little or no relationship.* Small values of one variable are just as likely to be paired with small, medium, or large values of the other variable. In panel C, notice that the dots are strewn about in an irregular shotgun fashion, suggesting that there is little or no relationship between the height of young adults and their life expectancies.

Strong or Weak Relationship?

Having established that a relationship is either positive or negative, next note how closely the dot cluster approximates a straight line. *The more closely the dot cluster approximates a straight line, the stronger (the more regular) the relationship will be.* **Figure 9.3** shows a series of scatterplots, each representing a different positive relationship between IQ scores for pairs of people whose backgrounds reflect different degrees of overlap, ranging from minimum overlap, as with foster parents and foster children, to maximum overlap, as with identical twins. (Ignore the parenthetical expressions involving *r,* to be discussed later.) Notice that the dot cluster more closely approximates a straight line for people with greater degrees of genetic overlap—for parents and children in panel B of Figure 9.3 and even more so for identical twins in panel C.

Perfect Relationship

A dot cluster that equals (rather than merely approximates) a straight line reflects a perfect relationship between two variables. In practice, perfect relationships are most unlikely.

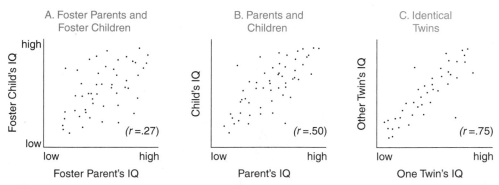

FIGURE 9.3

Three positive relationships. (Scatterplots simulated from a 50-year literature survey.
Source: *Erlenmeyer-Kimling, L. and L. F. Jarvik. "Genetics and Intelligence: A Review." Science, 142, 1477–1479.)*

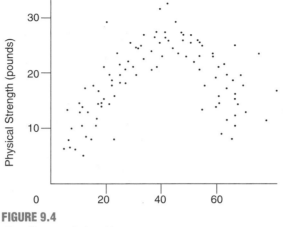

FIGURE 9.4
Curvilinear relationship.

Curvilinear Relationship

. . . .

Linear relationship

A relationship that can be described best with a straight line.

. . . .

Curvilinear relationship

A relationship that can be described best with a curved line.

The previous discussion assumes that if a dot cluster has a discernible slope, it approximates a *straight* line and, therefore, reflects a **linear relationship.** But this is not always the case.

Sometimes a dot cluster approximates a *bent* or *curved* line, as in **Figure 9.4,** and therefore reflects a **curvilinear relationship.** Descriptions of these relationships are more complex than those for linear relationships. For instance, we see in Figure 9.4 that physical strength, as measured by the force of a person's handgrip, is less for youngsters, more for adults, and then less again for older people. Otherwise, the scatterplot can be interpreted as before, that is, the more closely the dot cluster approximates a curved line, the stronger the curvilinear relationship will be.

Take another look at the scatterplot in Figure 9.1 for the greeting card data. Although the small number of dots in Figure 9.1 hinders any interpretation, the dot cluster appears to approximate a straight line, stretching from the lower left to the upper right. This suggests a positive relationship between greeting cards given and received, in agreement with the earlier intuitive analysis for these data.

*Exercise 9.2 Students A, B, C, D, E, F, G, H, and I took the Scholastic Assessment Test (SAT), and their verbal and math scores are shown in the scatterplot on the next page.

(a) Which student(s) scored about the same on both tests?

(b) Which student(s) scored higher on the verbal test than on the math test?

(c) Which student(s) scored above 600 on both tests?

(d) Which student(s) will be eligible for an honors program that requires minimum scores of 700 (verbal) and 500 (math)?

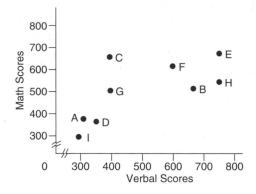

(e) Generally, students who perform well on the verbal test also perform well on the math test. True or false?

(f) There is a negative relationship between verbal and math scores. True or false?

Answers on page 500

9.3 A CORRELATION COEFFICIENT FOR QUANTITATIVE DATA: *r*

A *correlation coefficient* is a number between −1 and 1 that describes the relationship between variables.

Correlation coefficient

A number between −1 and 1 that describes the relationship between pairs of variables.

The next few sections concentrate on the type of correlation coefficient, designated as *r*, that *describes the linear relationship between pairs of variables for quantitative data.* Many other types of correlation coefficients have been introduced to handle specific types of data, including qualitative data, and a few of these will be described briefly in Section 9.10. (Detailed descriptions of various types of correlation coefficients, including the one that best describes the occasional curvilinear relationship, can be found in more advanced statistics books.)

Key Properties of *r*

Pearson correlation coefficient (r)

A number between −1 and 1 that describes the linear relationship between pairs of quantitative variables.

Named in honor of the British scientist Karl Pearson, the **Pearson correlation coefficient, *r*,** can equal any value between +1.00 and −1.00. Furthermore,

1. *the sign of r indicates the type of linear relationship, whether positive or negative,* and

2. *the value of r without regard to sign, indicates the strength of the linear relationship.*

Sign of *r*

A number with a plus sign (or no sign) indicates a positive relationship, and a number with a minus sign indicates a negative relationship. For example, an *r* with a plus sign describes the positive relationship between height and weight shown in panel A of Figure 9.2, and an *r* with a minus sign describes the negative relationship between heavy smoking and life expectancy shown in panel B.

Numerical Value of *r*

The more closely a value of *r* approaches either -1.00 or $+1.00$, the stronger (more regular) the relationship. Conversely, the more closely the value of *r* approaches 0, the weaker (less regular) the relationship. For example, an *r* of $-.90$ indicates a stronger relationship than an *r* of $-.70$ does, and an *r* of $-.70$ indicates a stronger relationship than an *r* of .50 does. (Remember, if no sign appears, it's understood to be plus.) In Figure 9.3, notice that the values of *r* shift from .75 to .27 as the analysis for pairs of IQ scores shifts from a relatively strong relationship for identical twins to a relatively weak relationship for foster parents and foster children.

From a slightly different perspective, the value of *r* is a measure of how well a straight line (representing the linear relationship) describes the cluster of dots in the scatterplot. In other words, the value of *r* is a measure of how well a straight line fits the data. Again referring to Figure 9.3, notice that an imaginary straight line describes the dot cluster or data less well as the values of *r* shift from .75 to .27.

INTERNET DEMONSTRATION

Go to the Web site for this book **(http://darwin.cwru.edu/~witte/statistics)** and click on **Guessing Correlations** to practice matching values of correlation coefficients with various scatterplots.

9.4 INTERPRETATION OF *r*

Located along a scale from -1.00 to $+1.00$, the value of *r* supplies information about the direction of a linear relationship—whether positive or negative—and, generally, information about the relative strength of a linear relationship—whether relatively weak (and a poor describer of the data), because *r* is in the vicinity of 0, or relatively strong (and a good describer of the data), because *r* deviates from 0 in the direction of either $+1.00$ or -1.00.

As discussed in Section 20.13, a valid interpretation of the significance of *r* depends not merely on the value of *r*, but also on the actual number of pairs of observations used to calculate *r*. On the assumption that reasonably large numbers of pairs of observations are involved (preferably in the hundreds and certainly many more than the five pairs of observations in the greeting card example), an *r* of .50 or more, in either the positive or the negative direction, would represent a *very strong* relationship in most areas of behavioral and educational research. But there are exceptions. An *r* of at least .80 or more would be expected when correlation coefficients measure "test reliability," as determined, for example, from pairs of IQ scores for people who take the same IQ test twice, or two forms of the same IQ test, (to establish that any person's two scores tend to be similar and, therefore, that the test scores are reproducible or "reliable").

Caution

Be careful when interpreting the actual numerical value of *r*. An *r* of .70 for height and weight doesn't signify that the strength of this relationship equals either .70, or 70 percent of the strength of a perfect relationship. In other words, *the value of r can't be interpreted as a proportion or percent of some perfect relationship.*

Verbal Descriptions

When interpreting a brand new *r*, you'll find it helpful to translate the numerical value of *r* into a verbal description of the relationship. An *r* of .70 for the height and weight of college students could be translated into "taller students tend to weigh more" (or some other equally valid statement, such as "lighter students tend to be shorter"); an *r* of −.42 for time spent taking an exam and subsequent exam score could be translated into "students who take less time tend to make higher scores"; and an *r* in the neighborhood of 0 for shoe size and IQ could be translated into "little, if any, relationship exists between shoe size and IQ."

If you have trouble verbalizing the value of *r*, refer to the original scatterplot or, if necessary, visualize a rough scatterplot corresponding to the value of *r*. Use any detectable dot cluster to think your way through the relationship. Does the dot cluster have a slope from the lower left to the upper right—that is, does low go with low and high go with high? Or does the dot cluster have a slope from the upper left to the lower right—that is, does low go with high, and vice versa? It's crucial that you translate abstractions such as "low goes with low, and high goes with high" into concrete terms such as "shorter students tend to weigh less, and taller students tend to weigh more."

*Exercise 9.3 Supply a verbal description for each of the following correlations. (If necessary, visualize a rough scatterplot for *r*, using the scatterplots in Figure 9.3 as a frame of reference.)

(a) an *r* of −.84 between total mileage and automobile resale value

(b) an *r* of −.35 between the number of days absent from school and performance on a math achievement test

(c) an *r* of −.05 between height and IQ

(d) an *r* of .89 between gross annual income and the total dollar value of claimed tax deductions

(e) an *r* of .03 between anxiety level and college GPA

(f) an r of .56 between age of schoolchildren and reading comprehension

(g) an r of .21 between length of pregnancy and weight of infant at birth
　　Answers on page 501

9.5 INTERPRETATION OF r^2 (SEE SECTION 10.8)

Still another interpretation, favored by many statisticians, requires that the correlation coefficient be squared. Because an understanding of this interpretation presupposes a familiarity with prediction, a discussion of r^2 will be postponed until Section 10.8. Once mastered, you'll find this interpretation very useful.

9.6 CORRELATION NOT NECESSARILY CAUSE-EFFECT

Given a correlation between the prevalence of poverty and crime in U.S. cities, you can *speculate* that poverty causes crime—that is, poverty produces crime with the same degree of inevitability as the flip of a switch illuminates a room. According to this view, any widespread reduction in poverty should cause a corresponding decrease in crime. As suggested in Chapter 1, you also can *speculate* that a common cause, such as inadequate education, overpopulation, racial discrimination, and so on, or some combination of these factors, produces both poverty and crime. According to this view, a widespread reduction in poverty shouldn't have any effect on crime. Which speculation is correct? Unfortunately, this issue can't be resolved merely on the basis of an observed correlation.

> **A correlation coefficient, regardless of size, never provides information about whether an observed relationship reflects a simple cause-effect relationship or some more complex state of affairs.**

In the past, the interpretation of the correlation between cigarette smoking and lung cancer has been vigorously disputed. American Cancer Society representatives interpret the correlation as a causal relationship: Smoking produces lung cancer. On the other hand, tobacco industry representatives interpret the correlation as, at most, an indication that both the desire to smoke cigarettes and lung cancer are caused by some more basic but as yet unidentified factor or factors, such as the body metabolism or the personality of some people. According to this line of reasoning, people with a high body metabolism might be more prone to smoke and, quite independent of their smoking, more vulnerable to lung cancer. In other words, smoking correlates with lung cancer because both are effects of some common cause or causes.

Role of Experimentation

Sometimes experimentation can resolve this kind of controversy. In the present case, laboratory animals were trained to inhale different amounts of tobacco tars and then were sacrificed. Autopsies revealed that the observed incidence of lung cancer (the dependent variable) varied directly with the amount of inhaled tobacco tars (the independent variable), even though possible "contaminating" factors, such as different body metabolisms or personalities, had been neutralized either through experimental control or random assignment of the subjects to different test conditions. As has been noted in Chapter 1, experimental confirmation of a correlation provides strong evidence in favor of a cause-effect interpretation of the observed relationship; indeed, in the smoking-cancer controversy, cumulative experimental findings overwhelmingly support the conclusion that smoking causes lung cancer.

*Exercise 9.4 Speculate on whether each of the following correlations reflects a simple cause-effect relationship or a more complex state of affairs. (*Hint:* A cause-effect relationship implies that, if all else remains the same, any change in the causal variable always should produce a predictable change in the other variable.)

(a) caloric intake and body weight

(b) height and weight

(c) price of Jamaican rum and salaries of Presbyterian ministers

(d) automobile weight and gas mileage

(e) SAT math score and score on a calculus test

(f) poverty and crime

Answers on page 501

DETAILS

9.7 z SCORE FORMULA FOR r

The simplest formula for *r* reads

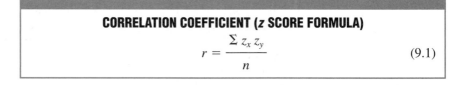

CORRELATION COEFFICIENT (z SCORE FORMULA)

$$r = \frac{\sum z_x z_y}{n} \qquad (9.1)$$

in which z_x and z_y are the *z* score equivalents for each pair of original observations, *X* and *Y*, and *n* refers to the number of pairs of observations. The term $\sum z_x z_y$ is found by first multiplying each pair of z_x and z_y scores, then adding the products for all pairs. Because Formula 9.1 directs us to add the products of all pairs of *z* scores and then to divide by the total number of pairs, *r* qualifies as a type of mean—the mean of the products of paired *z* scores or, simply, the mean of the products.

Calculating r (z Score Formula)

In actual practice, you should *never* use the *z* score formula to calculate the value of *r* because of several complications, including the extra effort of converting original observations to *z* scores. (A much more efficient formula for calculating *r* is given in the next section.) Nevertheless, in order to clarify some important properties of *r,* **Table 9.3** illustrates how the *z* score formula could be used to calculate the value of *r* for the original greeting card data in panel A of Table 9.2.

As in previous computational tables, the numbers in the shaded squares are used to cross-reference the top and bottom panels in Table 9.3. Notice the series of steps that produce each pair of *z* scores. For example, Doris's z_x score of 1.50 is obtained by first subtracting $\overline{X}$, the mean of 7, from her *X* score of 13, then dividing the resulting deviation of 6 by S_x, the standard deviation of 4. Her z_y of 0.50 is obtained in the same fashion by first subtracting $\overline{Y}$, the mean of 12, from her *Y* score of 14, then dividing the resulting deviation of 2 by S_y, the standard deviation of 4. A complete inspection of Table 9.3—well worth several moments of your time—reveals that *r* equals .80, a strong positive correlation, as you might have anticipated from our previous discussion of these data.

Table 9.3
CALCULATION OF *r*: *z* SCORE FORMULA

A. COMPUTATIONAL SEQUENCE

Assign a value to *n* **1**, representing the number of pairs of scores.
Obtain a mean **2** and standard deviation **3** for the *X* scores.
Obtain a mean **4** and standard deviation **5** for the *Y* scores.
Determine z_x scores **6**, one at a time, by subtracting $\overline{X}$ from each *X* and dividing the difference by S_x.
Determine z_y scores **7**, one at a time, by subtracting $\overline{Y}$ from each *Y* and dividing the difference by S_y.
Find the product of each pair of *z* scores **8**, one at a time, then add all of these products **9**.
Substitute numbers into the formula **10** and solve for *r*.

B. DATA AND COMPUTATIONS

FRIEND	GIVEN, *X*	CARDS RECEIVED, *Y*	6 z_x	7 z_y	8 $z_x z_y$
Doris	13	14	1.50	0.50	0.75
Steve	9	18	0.50	1.50	0.75
Mike	7	12	0.00	0.00	0.00
Andrea	5	10	−0.50	−0.50	0.25
John	1	6	−1.50	−1.50	2.25

1 $n = 5$ $\Sigma X = 35$ $\Sigma Y = 60$ **9** $\Sigma z_x z_y = 4.00$

$$\textbf{2}\ \overline{X} = \frac{35}{5} = 7 \qquad\qquad \textbf{4}\ \overline{Y} = \frac{60}{5} = 12$$

$$\textbf{3}\ S_x = 4^* \qquad\qquad \textbf{5}\ S_y = 4^*$$

$$\textbf{10}\ r = \frac{\Sigma z_x z_y}{n} = \frac{4.00}{5} = 80$$

* Computations not shown. Verify, if you wish, using Formula 5.1.

Let's look more closely at how three different types of correlations—a strong positive correlation, a strong negative correlation, and a zero correlation—are processed by Formula 9.1.

Strong Positive Correlation (*z* Score Formula)

If there is a strong positive correlation, such as that illustrated in Table 9.3, the pairs of *X* and *Y* scores will tend to have similar relative locations (high with high and low with low) in their respective distributions, and therefore, the pairs of *z* scores will tend to have the same sign. As can be seen in Table 9.3, positive *z* scores are paired with positive *z* scores, and negative *z* scores are paired with negative *z* scores. Consequently, and this is a crucial point, the products of paired *z* scores, $z_x z_y$, are positive because multiplication involves pairs of numbers with like signs, either both plus or both minus. As a result, the numerator term in Formula 9.1, $\Sigma z_x z_y$, becomes a relatively large positive number, which, when divided by *n*, the number of pairs, yields a positive *r*—in the present case, an *r* of .80.

Perfect Positive Correlation (*z* Score Formula)

If pairs of *z* scores have both the same magnitude and the same sign (for instance, one pair might be 2.34 and 2.34, and another might be -1.25 and -1.25), *r* will equal a value of 1.00, indicating a perfect positive correlation. Under these circumstances, the relationship would display total regularity, with pairs of *z* scores occupying *exactly* the same relative locations in their respective distributions. Returning to Table 9.3, you might verify that *r* would have been equal to 1.00 if the top two entries in the z_x column had been reversed, causing each pair of *z* scores to have both the same magnitude and the same sign.

Strong Negative Correlation (*z* Score Formula)

If there is a strong negative correlation, such as that between cards given and received in panel B of Table 9.2, pairs of *X* and *Y* scores will tend to have relative locations that are reversed (high with low and vice versa) in their respective distributions, and therefore, pairs of *z* scores will tend to have unlike signs. Positive *z* scores are paired with negative scores, and vice versa. Consequently, the products of paired *z* scores, $z_x z_y$, are negative because multiplication involves pairs of numbers with unlike signs, one plus and the other minus. As a result, the numerator term in Formula 9.1, $\Sigma z_x z_y$, is a relatively large negative term, which, when divided by *n*, yields a negative *r*.

Perfect Negative Correlation (*z* Score Formula)

If pairs of *z* scores have the same magnitude but unlike signs (for instance, one pair might be 1.75 and -1.75, and another pair might be .53 and $-.53$), *r* will equal a value of -1.00, indicating a perfect negative correlation. Under these circumstances, the relationship would also display perfect regularity, with pairs of *z* scores occupying relative locations that are *exactly reversed* in their respective distributions.

Nearly Zero Correlation (z Score Formula)

If there is a zero or nearly zero correlation, such as that between cards given and received in panel C of Table 9.2, no consistent pattern will describe the relative locations of pairs of X and Y scores in their respective distributions. Therefore, a positive z score is equally likely to be paired with either a positive or a negative z score, and vice versa. Consequently, about half of all products of paired z scores, $z_x z_y$, are positive because multiplication involves numbers with like signs, and about half of all products of paired z scores, $z_x z_y$, are negative because multiplication involves numbers with unlike signs. Since positive and negative products tend to cancel each other, the numerator term in Formula 9.1, $\Sigma z_x z_y$, tends toward a small positive or negative number that, when divided by n, yields a value of r in the vicinity of 0.

Review

To summarize, an understanding of correlation, as measured by r, can be gained from the z score formula (9.1). The pattern among pairs of z scores can be used to anticipate the value of r. If pairs of z scores are similar in both magnitude and sign, the value of r will tend toward 1.00, indicating a strong positive correlation. But if pairs of z scores are similar in magnitude but opposite in sign, the value of r will tend toward -1.00, indicating a strong negative correlation. As the pattern among pairs of z scores becomes less apparent, the value of r tends toward 0, indicating a weak or nonexistent correlation.

r Is Independent of Units of Measurement

The z score formula also pinpoints another important property of r—its independence of the original units of measurement. As a matter of fact, the same value of r describes the correlation between height and weight for a group of adults, regardless of whether height is measured in inches or centimeters or whether weight is measured in pounds or grams. In effect, the value of r depends only on the pattern among pairs of z scores, which in turn show no traces of the units of measurement for the original X and Y observations. If you think about it, this is the same as saying that

a positive value of r reflects a tendency for pairs of observations to occupy *similar* relative locations (high with high and low with low) in their respective distributions, while a negative value of r reflects a tendency for pairs of observations to occupy *dissimilar* relative locations (high with low and vice versa) in their respective distributions.

*Exercise 9.5 Pretend that there's a perfect positive $(+1.00)$ relationship between height and weight for adults. (Actually, it's in the vicinity of .70 for college students.) In this case, if John stands 2 standard deviations above the mean height, John's weight will be __(a)__ standard deviation units __(b)__ the mean. (*Hint:* Remember the discussion in Section 9.7 about the z score formula for r and a perfect positive relationship.)

If Kristen stands $1\frac{1}{2}$ standard deviation units below the mean height, Kristen's weight will be __(c)__ standard deviation units __(d)__ the mean.

If Carson is $\frac{1}{3}$ of a standard deviation above the mean weight, Carson's height will be __(e)__ of a standard deviation __(f)__ the mean.

*Exercise 9.6 Repeat Exercise 9.5, assuming a perfect negative (-1.00) relationship between height and weight.

Answers on page 501

....

9.8 COMPUTATION FORMULA FOR *r*

Except to provide a more intuitive picture of correlation, as in the previous section, *the z score formula should not be used to calculate the value of r*. It would be most laborious first to convert each *X* and *Y* score into its *z* score equivalents, using means and standard deviations that typically aren't whole numbers (and therefore produce an appreciable rounding error), and then to calculate *r* using the *z* score formula. It's more efficient and usually more accurate to calculate a value for *r* by using the computation formula, which reads:

CORRELATION COEFFICIENT (COMPUTATION FORMULA)

$$r = \frac{n\Sigma XY - (\Sigma X)(\Sigma Y)}{[\sqrt{n\Sigma X^2 - (\Sigma X)^2}][\sqrt{n\Sigma Y^2 - (\Sigma Y)^2}]} \qquad (9.2)$$

All terms in this expression have been encountered previously, except for ΣXY, which is found by first multiplying each pair of *X* and *Y* scores and then adding the products for all pairs.

Table 9.4 illustrates the calculation of *r* for the original greeting card data by using the computation formula. This formula yields the same value of .80 for *r* as did the *z* score formula (because the means and standard deviations used in the *z* score formula happen to be exact whole numbers—a most unlikely situation in practice).

*Exercise 9.7 Couples who attend a clinic for first pregnancies are asked to estimate (independently of each other) the ideal number of children. Given that *X* and *Y* represent the estimates of married females and males, respectively, the results are as follows:

COUPLE	X	Y
A	1	2
B	3	4
C	2	3
D	3	2
E	1	0
F	2	3

Calculate a value for *r*, using the computation formula (9.2).

Answer on page 501

Table 9.4
CALCULATION OF r: COMPUTATION FORMULA

A. COMPUTATIONAL SEQUENCE
Assign a value to n **1**, representing the number of pairs of scores.
Sum all scores for X **2** and for Y **3**.
Find the product of each pair of X and Y scores **4**, one at a time, then add all of these products **5**.
Square each X score **6**, one at a time, then add all squared X scores **7**.
Square each Y score **8**, one at a time, then add all squared Y scores **9**.
Substitute numbers into the formula **10** and solve for r.

B. DATA AND COMPUTATIONS

	CARDS		**4**	**6**	**8**
FRIEND	**GIVEN, X**	**RECEIVED, Y**	**XY**	**X^2**	**Y^2**
Doris	13	14	182	169	196
Steve	9	18	162	81	324
Mike	7	12	84	49	144
Andrea	5	10	50	25	100
John	1	6	6	1	36

1 $n = 5$ **2** $\Sigma X = 35$ **3** $\Sigma Y = 60$ **5** $\Sigma XY = 484$ **7** $\Sigma X^2 = 325$ **9** $\Sigma Y^2 = 800$

$$\mathbf{10}\ r = \frac{(n)(\Sigma XY) - (\Sigma X)(\Sigma Y)}{[\sqrt{n\Sigma X^2 - (\Sigma X)^2}][\sqrt{n\Sigma Y^2 - (\Sigma Y)^2}]}$$

$$= \frac{(5)(484) - (35)(60)}{[\sqrt{(5)(325) - (35)^2}][\sqrt{(5)(800) - (60)^2}]}$$

$$= \frac{2420 - 2100}{[\sqrt{1625 - 1225})][\sqrt{4000 - 3600}]} = \frac{320}{[\sqrt{400}][\sqrt{400}]}$$

$$= \frac{320}{[20][20]} = \frac{320}{400} = 0.80$$

9.9 OUTLIERS AGAIN

In Section 2.7, *outliers* are defined as very extreme observations that require special attention because of their potential impact on a summary of data. This is also true when outliers appear among sets of paired observations. Although quantitative techniques can be used to detect these outliers, we simply focus on dots in scatterplots that deviate conspicuously from the main dot cluster.[1]

Greeting Card Study Revisited

Figure 9.5 shows the effect of each of two possible outliers, substituted one at a time for Doris's dot, (13, 14), on the original value of *r* (.80) for the greeting card data. Although both outliers A and B deviate conspicuously from the dot cluster, they have radically different effects on the value of *r*. Outlier A (33, 34) contributes to a new value of .98 for *r* that merely reaffirms the original positive relationship between cards given and received. On the other hand, outlier B (13, 4) causes a dramatically new value of .04 for *r* that entirely neutralizes the original positive relationship.

Neither of the values for outlier B, taken singularly, are extreme. Rather, it is their unusual combination—13 cards given and only 4 received—that yields a radically different value of .04 for *r*, indicating that the new dot cluster is not remotely approximated by a straight line.

Dealing with Outliers

Serious practitioners will, of course, be using many more than five pairs of observations, and therefore the effect of outliers on the value of *r* will tend not to be as dramatic as the preceding example. Nevertheless, outliers can have a considerable impact on the value of *r* and, therefore, pose problems of interpretation. Unless there is some reason for discarding an outlier (because of a failed accuracy check or because, for example, you establish that the 13 cards sent by the friend who received only 4 cards were substitutes for a customary holiday monetary gift), the most defensible strategy is to report the values of *r* both with and without any outliers.

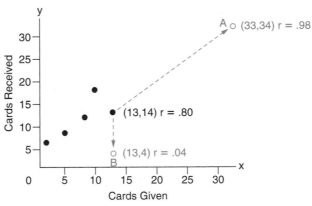

FIGURE 9.5

Effect of each of two outliers on the value of r.

[1]For more information about the quantitative detection of outliers among sets of paired observations, see D. C. Howell, *Statistical Methods for Psychology*, 4th ed., (Belmont, CA: Duxbury, 1997), chapter 15.

INTERNET DEMONSTRATION
Go to the Web site for this book **(http://darwin.cwru.edu/~witte/statistics)** and click on **Outliers** to see the effect of outliers on the correlation coefficient.

9.10 OTHER TYPES OF CORRELATION COEFFICIENTS

There are many other types of correlation coefficients, but we'll discuss only some that are direct descendants of the Pearson correlation coefficient. Although designed originally for use with quantitative data, the Pearson r has been extended, sometimes under the guise of new names and customized versions of Formula 9.2, to other kinds of situations. For example, to describe the correlation between *ranks* assigned independently by two judges to a set of science projects, simply substitute the numerical ranks into Formula 9.2, then solve for a value of Pearson r (also designated as Spearman's correlation coefficient for ranked data, or Spearman's rho). To describe the correlation between quantitative data (annual income) and *qualitative data with only two categories* (male and female), assign arbitrary numerical codes, such as 0 and 1, or 1 and 2, to the two qualitative categories, then solve Formula 9.2 for a value of Pearson r (also designated as a point biserial correlation coefficient). Or to describe the relationship between *two ordered qualitative variables,* such as attitude toward legal abortion (favor, neutral, or opposed) and educational level (high school only, some college, college graduate), assign any *ordered* numerical codes to the categories for both qualitative variables, then solve Formula 9.2 for a value of Pearson r (also designated as Cramer's phi coefficient).

Most computer outputs would simply report each of the above correlations as a Pearson r. Given the widespread use of computers, these more specialized names for the Pearson r probably will survive, if at all, as artifacts of an earlier age, when most calculations were manual and some computational relief was obtained by customizing Formula 9.2 for situations involving ranks and qualitative data.

9.11 COMPUTER OUTPUT

Correlation matrix

Table showing correlations for all possible pairs of variables.

Before reading on, study **Table 9.5,** particularly the lower half of the output and comments 2, 3, and 4. When every possible pairing of variables is reported, as in lower half of the output, a **correlation matrix** is produced. The value of .8000 occurs twice in the matrix, because the correlation is the same whether the relationship is described as that between cards given and cards received or vice versa. The value of 1.000, which also occurs twice, reflects the trivial fact that any variable correlates perfectly with itself.

Table 9.5
SPSS OUTPUT: SCATTERPLOT AND CORRELATION
FOR GREETING CARD DATA

GRAPH

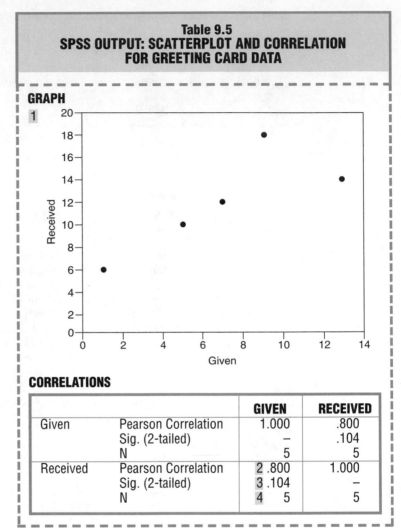

CORRELATIONS

		GIVEN	**RECEIVED**
Given	Pearson Correlation	1.000	.800
	Sig. (2-tailed)	–	.104
	N		5	5
Received	Pearson Correlation	2 .800	1.000
	Sig. (2-tailed)	3 .104	–
	N	4 5	5

Comments:

1 Scatterplot for greeting card data (using slightly different scales than in Figure 9.1).

2 The correlation for cards given and cards received equals .800, in agreement with the calculations in Tables 9.3 and 9.4.

3 The value of Sig. helps us interpret the statistical significance of a correlation by evaluating the observed value of r relative to the actual number of pairs of observations used to calculate r. Discussed later in Section 21.1, Sig.- values are referred to as p-values in this book. At this point, perhaps the easiest way to view a Sig.-value is as follows: The smaller the value of Sig. (on a scale from 0 to 1), the more likely that you would observe a correlation with the same sign, either positive or negative, if the study were repeated with new observations. Investigators often focus on only those correlations with Sig.-values smaller than .05.

4 Number of cases or paired observations.

Reading a Larger Correlation Matrix

Because correlation matrices can be expanded to incorporate any number of variables, they are useful devices for showing correlations between all possible pairs of variables when, in fact, many variables are being studied. For example, in **Table 9.6,** four variables generate a correlation matrix with four times four or sixteen correlation coefficients. The four perfect (but trivial) correlations of 1.00, produced by pairing each variable with itself, split the remainder of the matrix into two triangular sections, each containing six nontrivial correlations. Because the correlations in these two sectors are mirror images, you can attend to just the values of the six correlations in one sector in order to evaluate all relevant correlations among the four original variables.

Interpreting a Larger Correlation Matrix

Three of the six color-coded correlations in Table 9.6 involve gender. [As suggested in the previous section, gender qualifies for a correlation analysis once arbitrary numerical codes have been assigned to male (0) and female (1).] Looking across the bottom row, gender is positively correlated with age (.0813); with college GPA (.2069); and with high school GPA (.2981). Looking across the next row, high school GPA is negatively correlated with age (−.0376); and positively correlated with college GPA (.2521). Lastly, college GPA is positively correlated with age (.2228).

Table 9.6
SPSS OUTPUT: CORRELATION MATRIX FOR FOUR VARIABLES (BASED ON 336 STATISTICS STUDENTS)

CORRELATIONS

		AGE	COLLEGE GPA	HIGH SCHOOL GPA	GENDER
AGE	Pearson Correlation	1.000	.2228	−.0376	.0813
	Sig. (2-tailed)	−	.000	.511	.138
	N	335	333	307	335
COLLEGE GPA	Pearson Correlation	.2228	1.000	.2521	.2069
	Sig. (2-tailed)	.000	−	.000	.000
	N	333	334	306	334
HIGH SCHOOL GPA	Pearson Correlation	−.0376	.2521	1.000	.2981
	Sig. (2-tailed)	.511	.000	−	.000
	N	307	306	307	307
GENDER	Pearson Correlation	.0813	.2069	.2981	1.000
	Sig. (2-tailed)	.138	.000	.000	−
	N	335	334	307	336

As suggested in comment 3 at the bottom of Table 9.5, values of Sig. help us judge the statistical significance of the various correlations. A smaller value of Sig. implies that if the study were repeated, the same positive or negative sign of the corresponding correlation probably would reappear, even though calculations are based on an entirely new group of similarly selected students. Therefore, we can conclude that the four correlations with Sig.-values close to zero (.000) probably would reappear as positive relationships. In other words, for new groups of similarly selected students, females would tend to have higher high school and college GPAs, and students with higher college GPAs would tend to have higher high school GPAs and also tend to be older. Because of the larger Sig.-value of .138 for the correlation between gender and age, we can't be as confident that a new group would show the observed tendency of female students to be older than male students. Because of the even larger Sig.-value of .511 for the negative correlation between age and high school GPA, this correlation would be just as likely to reappear as either a positive or negative relationship. In other words, the small negative correlation between age and high school GPA shouldn't be taken seriously.

Finally, the numbers in the last row of each cell in Table 9.6 show the total number of cases actually used to calculate the corresponding correlation. Excluded from these totals are those cases where students failed to supply the requested information.

*Exercise 9.8 The following SAS output shows the various correlations between variables (VAR) X, Y, and Z for the paired observations in Exercise 9.9 at the end of the chapter. Although the form of this output differs slightly from that shown in the SPSS output of Table 9.5, you still should be able to answer the following questions:

(a) What is the correlation between Y and Z?

SAS OUTPUT:
CORRELATIONS BETWEEN VARIABLES
X, Y, AND Z

THE SAS SYSTEM
12:46 FRIDAY, DECEMBER 12, 1999
CORRELATION ANALYSIS

Pearson Correlation Coefficients / Prob > |R| under Ho: Rho = 0 / N = 9

	VARX	VARY	VARZ
VARX	1.00000 0.0000	−0.91413 0.0006	−0.81459 0.0075
VARY	−0.91413 0.0006	1.00000 0.0000	0.77854 0.0134
VARZ	−0.81459 0.0075	0.77854 0.0134	1.00000 0.0000

(b) Using only correlations, reproduce the entire correlation matrix.

(c) In your reconstruction of the correlation matrix, draw a single line through all of the trivial correlations of a variable with itself.

(d) In your reconstruction of the correlation matrix, draw a triangle around the remaining subset of nontrivial correlations.

(e) Indicate whether each of the nontrivial correlations, for example, the one between X and Y, reflects a positive or negative relationship.

(f) Returning to the original output, specify how many pairs of observations contribute to each correlation.

(g) As suggested in comment 2 at the bottom of Table 9.5, investigators usually attend only to those correlations with smaller values of Sig. Assume that if the Sig.-value for a correlation is less than .010, the sign of that correlation probably would reappear in a new study consisting of similarly selected observations. Identify those correlations whose signs probably would reappear in a new study. (In the correlation matrix from SAS, each correlation coefficient has its Sig.-value listed beneath it. For example, the correlation coefficient of -0.91413 between X and Y has a Sig.-value of 0.0006.)

Answers on page 501

Summary

The presence of regularity among pairs of X and Y scores indicates that the two variables are related, and the absence of any regularity suggests that the two variables are, at most, only slightly related. When the regularity consists of relatively low X scores being paired with relatively low Y scores and relatively high X scores being paired with relatively high Y scores, the relationship is positive. When it consists of relatively low X scores being paired with relatively high Y scores and vice versa, the relationship is negative.

A scatterplot is a graph containing a cluster of dots that represents all pairs of observations. A dot cluster that has a slope from the lower left to the upper right reflects a positive relationship, and a dot cluster that has a slope from the upper left to the lower right reflects a negative relationship. A dot cluster that lacks any apparent slope reflects little or no relationship.

In a positive or negative relationship, the more closely the dot cluster approximates a straight line, the stronger the relationship will be.

When the dot cluster approximates a straight line, the relationship is linear; when it approximates a bent line, the relationship is curvilinear.

Located on a scale from -1.00 to $+1.00$, the value of r indicates the direction of a linear relationship and, generally, the relative strength of a linear relationship, that is, the adequacy with which a linear relationship describes the data. Plus and minus signs indicate positive and negative relationships, respectively. Values of r in the general vicinity of either -1.00 or $+1.00$ indicate a relatively strong relationship, and values of r in the neighborhood of 0 indicate a relatively weak relationship.

Although the value of r can be used to formulate a verbal description of the relationship, it doesn't indicate a proportion or percent of a perfect relationship.

The presence of a correlation, by itself, doesn't resolve the issue of whether it reflects a simple cause-effect relationship or a more complex state of affairs.

The Pearson correlation coefficient, r, describes best the linear relationship between pairs of variables for quantitative data. An understanding of correlation, as described by r, can be gained from the z score formula (9.1). In practice, it's both more efficient and more accurate to calculate r by using the computation formula (9.2).

Outliers can have a considerable impact on the value of r and, therefore, pose problems of interpretation.

Important Terms
. . . .

Positive relationship	**Curvilinear relationship**
Negative relationship	**Correlation coefficient**
Scatterplot	**Pearson correlation coefficient (*r*)**
Linear relationship	**Correlation matrix**

REVIEW EXERCISES

9.9 (a) Ignoring the observations under Z (which were referred to earlier in Exercise 9.8), estimate whether the following pairs of observations for X and Y reflect a positive relationship, a negative relationship, or no relationship. (*Hint:* Note any tendency for pairs of X and Y observations to occupy similar or dissimilar relative locations.)

X	Y	Z
64	66	31
40	79	36
30	98	43
71	65	15
55	76	38
31	83	40
61	68	17
42	80	41
57	72	33

   (b) Construct a scatterplot for X and Y. Verify that the scatterplot doesn't describe a pronounced curvilinear trend.

   (c) Calculate r for X and Y, using the computation formula (9.2).

9.10 Verify that a strong negative relationship is depicted in panel B of Table 9.2 by using the numerical observations to calculate a value of r (Formula 9.2).

9.11 Verify that little or no relationship exists between the observations in panel C of Table 9.2 by using the numerical observations to calculate a value of r (Formula 9.2).

*9.12 On the basis of an extensive survey, the California Department of Education reported an r of $-.32$ for the relationship between the amount of time spent watching TV and the achievement test scores of schoolchildren. Each of the following statements represents a possible interpretation of this finding. Indicate whether each is true or false.

(a) *Every* child who watches lots of TV will perform poorly on the achievement tests.

(b) Extensive TV viewing causes a decline in test scores.

(c) Children who watch little TV will tend to perform well on the tests.

(d) Children who perform well on the tests will tend to watch little TV.

(e) If, over a long period of time, a child's TV-viewing time is reduced by one half, we can expect a substantial improvement in test scores.

(f) TV viewing couldn't possibly cause a decline in test scores.

Answers on pages 501–502

9.13 Assume that an r of .80 describes the relationship between daily food intake, measured in ounces, and body weight, measured in pounds, for a group of adults. Would a shift in the units of measurement from ounces to grams and from pounds to kilograms change the value of r? Justify your answer.

9.14 An extensive correlation study indicates that a longer life is experienced by people who follow the seven "golden rules" of behavior, including moderate drinking, no smoking, regular meals, some exercise, and eight hours of sleep each night. Can we conclude, therefore, that this type of behavior *causes* a longer life?

*9.15 When discussing the positive correlation of .2981 between gender and high school GPA, shown in Table 9.6, it was concluded that females tend to have higher high school GPAs than do males, given the assignment of codes of 0 and 1 to males and females, respectively.

(a) Would the same positive correlation of .2981 have been obtained if the assignment of codes had been reversed, with females being coded as 0 and males coded as 1? Explain your answer.

(b) Given the new coding of females (0) and males (1), would the results still permit you to conclude that females tend to have higher high school GPAs than do males?

(c) Would the same positive correlation of .2981 have been obtained if, instead of the original coding of males as 0 and females as 1, males were coded as 10 and females as 20? Explain your answer.

Answers on page 502

CHAPTER 10

Prediction

10.1 **TWO ROUGH PREDICTIONS**

10.2 **A PREDICTION LINE**

10.3 **LEAST SQUARES PREDICTION LINE**

10.4 **LEAST SQUARES EQUATION**

10.5 **GRAPHS OR EQUATIONS?**

10.6 **STANDARD ERROR OF PREDICTION, $S_{y|x}$**

10.7 **ASSUMPTIONS**

10.8 **INTERPRETATION OF r^2**

10.9 **r REVISITED**

10.10 **MORE COMPLEX PREDICTION EQUATIONS**

Summary

Important Terms

Review Exercises

A correlation analysis of the exchange of greeting cards by five friends for the most recent holiday season suggests a strong positive relationship between cards given and cards received. When informed of these results, another friend, Emma, who greatly enjoys receiving greeting cards, asks you to predict how many cards she will receive during the next holiday season, assuming that she plans to send 11 cards.

10.1 TWO ROUGH PREDICTIONS

Predict "Relatively Large Number"

You could offer Emma a very rough prediction by recalling that cards given and received tend to occupy *similar* relative locations in their respective distributions. Looking at these distributions, Emma can expect to receive a *relatively large* number of cards, as she plans to send a *relatively large* number of cards.

Predict "between 14 and 18 Cards"

To obtain a slightly more precise prediction for Emma, refer to the scatterplot for the original five friends shown in **Figure 10.1.** Notice that Emma's plan to send 11 cards locates her along the *X*-axis between Steve's 9 and Doris's 13. Using the dots for Steve and Doris as guides, construct two strings

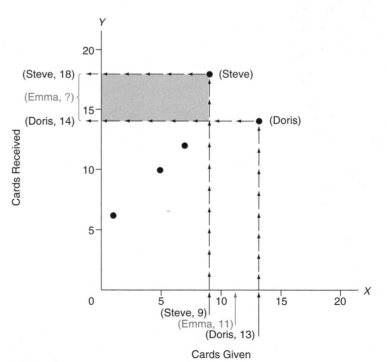

FIGURE 10.1
A rough prediction for Emma (using dots for Steve and Doris).

of arrows, one beginning at 9 and ending at 18 for Steve and the other beginning at 13 and ending at 14 for Doris. (The direction of the arrows reflects our attempt to predict cards received [Y] from cards given [X]. Although not required, it's customary to predict from X to Y.) Focusing on the interval along the Y-axis between the two strings of arrows, you could predict that Emma's return should be between 14 and 18 cards, the numbers received by Doris and Steve.

The latter prediction might satisfy Emma, but it wouldn't win any statistical awards. Although each of the five dots in Figure 10.1 supplies valuable information about the exchange of greeting cards, our prediction for Emma is based only on the two dots for Steve and Doris.

10.2 A PREDICTION LINE

All five dots contribute to the more precise prediction, illustrated in **Figure 10.2,** that Emma will receive 15.20 cards. Let's look more closely at the single straight line designated as the prediction line in Figure 10.2, which guides the string of arrows, beginning at 11, toward the predicted value of 15.20. It's a straight line rather than a curved line because of the linear relationship between cards given and cards received. As will become apparent, it can be used over and over to predict cards received. Regardless of whether Emma decides to give 5 or 15 or 25 cards, it will guide a new string of arrows, beginning at 5 or 15 or 25, toward a new predicted value along the Y axis.

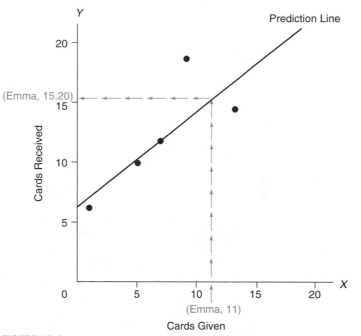

FIGURE 10.2

Prediction of 15.20 for Emma (using prediction line).

Placement of Line

For the time being, forget about any prediction for Emma, and concentrate on how the five dots dictate the placement of the prediction line. If all five dots had defined a single straight line, placement of the prediction line would have been simple; merely let it pass through all dots. When the dots fail to define a single straight line, as in the scatterplot for the five friends, placement of the prediction line represents a compromise. It passes through the main cluster, possibly touching some dots but missing others.

Predictive Errors

Figure 10.3 illustrates the predictive errors that would have occurred if the prediction line had been used to predict the number of cards received by the five friends. Solid dots reflect the *actual* number of cards received, and open dots, always located along the prediction line, reflect the *predicted* number of cards received. (To avoid clutter in Figure 10.3, the strings of arrows have been omitted. But you might find it helpful to imagine a string of arrows, ending along the *Y*-axis, for each dot, whether solid or open.) The largest predictive error, shown as a broken vertical line, occurs for Steve, who gave 9 cards. Although he actually received 18 cards, he should have received slightly fewer than 14 cards, according to the prediction line. The smallest predictive error—none whatsoever—occurs for Mike, who gave 7 cards. He actually received the 12 cards that he should have received, according to the prediction line.

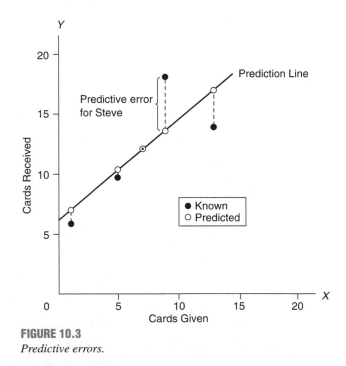

FIGURE 10.3
Predictive errors.

Total Predictive Error

We engage in the seemingly silly activity of "predicting" what is known already for the five friends not as an end in itself, but to check the adequacy of our predictive effort, as well as to estimate the error that will occur when predicting what isn't known for Emma. The smaller the total for all predictive errors in Figure 10.3, the more favorable the prognosis for the predictive effort will be. Clearly, it's desirable for the prediction line to be placed in a position that *minimizes* the total predictive error, that is, that minimizes the total of the vertical discrepancies between the solid and open dots shown in Figure 10.3.

*Exercise 10.1 To check your understanding of the first part of this chapter, make predictions using the accompanying graph.

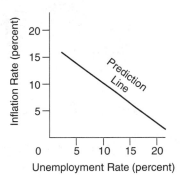

(a) Predict the approximate rate of inflation, given an unemployment rate of 5 percent.

(b) Predict the approximate rate of inflation, given an unemployment rate of 15 percent.

Answers on page 502

10.3 LEAST SQUARES PREDICTION LINE

To avoid the arithmetic standoff of zero always produced by adding positive and negative predictive errors (associated with errors above and below the prediction line, respectively), *the placement of the prediction line minimizes* not the total predictive error but *the total squared predictive error*, that is, the total for all squared predictive errors. When located in this fashion, the prediction line is often referred to as the *least squares regression line* or, in this book, as the *least squares prediction line*. Although more difficult to visualize, this approach is consistent with the original aim—to minimize the total predictive error or some version of the total predictive error, thereby providing a more favorable prognosis for the predictive effort.

Need a Mathematical Solution

Without the aid of mathematics, the search for a least squares prediction line would be a frustrating job. Scatterplots would be proving grounds cluttered with tentative prediction lines, discarded because of their excessively large totals for squared discrepancies. Even the most time-consuming effort, conducted with the precision of a draftsperson and the patience of a Zen master, would culminate in only a close approximation to the least squares prediction line.

INTERNET DEMONSTRATION

Go to the Web site for this book **(http://darwin.cwru.edu/~ witte/statistics)** and click on **Prediction** to try fitting by eye the least squares prediction line to a cluster of dots.

10.4 LEAST SQUARES EQUATION

Happily, we have recourse to an equation that pinpoints the exact least squares prediction line for any scatterplot. Most generally, this equation reads

---

LEAST SQUARES PREDICTION EQUATION
$$Y' = bX + a \qquad (10.1)$$

---

where Y' represents the predicted value (the predicted number of cards that will be received by any new friend, such as Emma); X represents the known value (the known number of cards given by any new friend, such as Emma); and b and a represent numbers calculated from the original correlation analysis, as described in the following section.[1]

Finding Values of *b* and *a*

To obtain a working prediction equation, solve each of the following expressions, first for b and then for a, using data from the original correlation analysis. The expression for b reads

---

[1]You might recognize that the least squares equation defines a straight line with a slope of b and a Y-intercept of a.

SOLVING FOR *b*

$$b = \frac{S_y}{S_x} r \qquad (10.2)$$

---

where S_y represents the standard deviation for all observations along Y (the cards received by the five friends); S_x represents the standard deviation for all observations along X (the cards given by the five friends); and r represents the correlation between X and Y (cards given and received by the five friends).

The expression for a reads

SOLVING FOR *a*

$$a = \overline{Y} - b\overline{X} \qquad (10.3)$$

---

where $\overline{Y}$ and $\overline{X}$ refer to the means for all observations along Y and X, respectively, and b is defined by the preceding expression.

The values of all terms in the expressions for b and a can be obtained from the original correlation analysis either directly, as with the value of r, or indirectly, as with the values of the remaining terms: S_x, S_y, $\overline{Y}$, and $\overline{X}$. **Table 10.1** illustrates the computational sequence that culminates in a least squares prediction equation for the greeting card example, namely,

$$Y' = .80\,(X) + 6.40$$

where .80 and 6.40 represent the values computed for b and a, respectively.

Key Property

Least squares prediction equation

The equation that minimizes the total of all squared prediction errors for known Y scores in the original correlation analysis.

Once numbers have been assigned to b and a, as just described, the **least squares prediction equation** emerges as a working equation with a most desirable property: It automatically *minimizes the total of all squared predictive errors for known Y scores in the original correlation analysis.*

Solving for *Y'*

In its present form, the prediction equation can be used to predict the number of cards that Emma will receive, assuming that she plans to give 11 cards. Simply substitute 11 for X and solve for the value of Y' as follows:

$$\begin{aligned} Y' &= .80(11) + 6.40 \\ &= 10.80 + 6.40 \\ &= 15.20 \end{aligned}$$

Table 10.1
DETERMINING THE LEAST SQUARES PREDICTION EQUATION

A. COMPUTATIONAL SEQUENCE
Determine values of S_x, S_y, and r [1] by referring to the original correlation analysis in Table 9.3.
Substitute numbers into the formula [2] and solve for b.
Assign values to $\overline{X}$ and $\overline{Y}$ [3] and by referring to the original correlation analysis in Table 9.3.
Substitute numbers into the formula [4] and solve for a.
Substitute numbers for b and a in the least squares prediction equation [5].

B. COMPUTATIONS
[1] $S_x = 4^*$
$\quad S_y = 4^*$
$\quad r = .80$

[2] $b = \left(\dfrac{S_y}{S_x}\right)(r) = \left(\dfrac{4}{4}\right)(.80) = .80$

[3] $\overline{X} = 7^{**}$
$\quad \overline{Y} = 12^{**}$
[4] $a = \overline{Y} - (b)(\overline{X}) = 12 - (.80)(7) = 12 - 5.60 = 6.40$
[5] $Y' = (b)(X) + a$
$\quad\quad = (.80)(X) + 6.40$

*Computations not shown. Verify, if you wish, using Formula 5.1.
**Computations not shown. Verify, if you wish, using Formula 4.1.

Table 10.2
PREDICTED CARD RETURNS (Y') FOR DIFFERENT CARD INVESTMENTS (X)

X	Y'
0	6.40
4	9.60
8	12.80
10	14.40
12	16.00
20	22.40
30	30.40

Notice that the predicted card return for Emma, 15.20, qualifies as a genuine prediction, that is, a forecast of an unknown event based on information about some other event. This prediction appeared earlier in Figure 10.2.

Our working prediction equation provides an inexhaustible supply of predictions for the card exchange. Each prediction emerges simply by substituting some value for X and solving the equation for Y', as just described. Table 10.2 lists the predicted card returns for a number of different card investments. Verify that you can obtain a few of the Y' values shown in Table 10.2 from the prediction equation.

Notice that, even when no cards are given ($X = 0$), we predict 6.40 cards, the value of a, will be received. Also notice that for each additional card given, we predict an additional .80 cards received, the value of b. The fact that b has a value less than 1.00 indicates that increments in predicted cards received lag a bit (.80 or 80 percent) behind increments in cards given. (If the value of b had been negative, because of an underlying negative correlation, then increments in cards given would have triggered decrements, not increments, in predicted cards received.)

A Limitation

Emma might survey these predicted card returns before committing herself to a particular card investment. However, this strategy could backfire because there's no evidence of a simple *cause-effect* relationship between cards given and cards received. The desired effect might be completely missing if, for instance, Emma expands her usual card distribution to include casual acquaintances and even strangers, as well as her friends and relatives.

*Exercise 10.2 Assume that an r of .30 describes the relationship between educational level (highest grade completed) and estimated number of hours spent reading each week. More specifically,

EDUCATIONAL LEVEL (X)	WEEKLY READING TIME (Y)
$\overline{X} = 13$	$\overline{Y} = 8$
$S_x = 2$	$S_y = 4$
$r = .30$	

(a) Determine the least squares equation for predicting weekly reading time from educational level.

(b) Faith's education level is 15. What is her predicted reading time?

(c) Keegan's educational level is 11. What is his predicted reading time?

Answers on page 502

···

10.5 GRAPHS OR EQUATIONS?

Encouraged by Figures 10.2 and 10.3, you might be tempted to generate predictions from graphs rather than equations. But unless constructed skillfully, graphs yield less accurate predictions than equations do. In the long run, it's more accurate and easier to generate predictions from equations.

Constructing Graphs of Prediction Equations

Occasionally, after having solved the prediction equation, you might wish to construct, for descriptive rather than predictive purposes, a graph showing the prediction line. Merely identify any two points along the prediction line, as described in the following sentences, and then, because two points define a straight line, connect these points to produce the prediction line. To identify two points along the prediction line, substitute any two values for X, one at a time, into the prediction equation and solve for Y'. Use the resulting two pairs of X and Y' values, for instance, $X = 0$ and $Y' = 6.40$, and $X = 10$ and $Y' = 14.40$ from Table 10.2, to identify the two points that, when connected, produce the prediction line. (Verify this by using these two pairs of values to reconstruct the prediction line shown in Figure 10.2.)

10.6 STANDARD ERROR OF PREDICTION, $S_{y|x}$

Having predicted that Emma's investment of 11 cards will yield a return of 15.20 cards, we would, nevertheless, be surprised if she actually received 15.20 (or 15) cards. It's more likely that because of the imperfect relationship between cards given and cards received, Emma's return will be some number other than 15. Although designed to minimize predictive error, the least squares equation does not eliminate it. Thus, our next task is to estimate the amount of error associated with our predictions. The smaller the estimated error is, the better the prognosis will be for the predictive effort.

Finding the Standard Error of Prediction

The estimate of error for new predictions is based on known data: the failures to predict the number of cards received by the original five friends, as depicted by the discrepancies between solid and open dots in Figure 10.3. Although we can estimate the overall predictive error by dealing directly with these discrepancies, as we will in Section 10.9, it's more efficient to invoke the following formula:

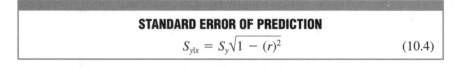

STANDARD ERROR OF PREDICTION
$$S_{y|x} = S_y\sqrt{1 - (r)^2} \qquad (10.4)$$

where $S_{y|x}$ denotes the estimated predictive error (of Y given values of X); S_y represents the standard deviation for all observations along Y (cards received by the five friends); and r represents the correlation between X and Y (cards given and received). The estimated predictive error, $S_{y|x}$, also is referred to as the standard error of prediction (or the standard error of the estimate), and its symbol, $S_{y|x}$, is read as "S sub y given x."

Key Property

Although not obvious from Formula 10.4, the standard error of prediction represents a special kind of standard deviation that reflects the magnitude of predictive error.

Standard error of prediction
($S_{y|x}$)

A rough measure of the average amount of predictive error.

You might find it helpful to think of the *standard error of prediction, $S_{y|x}$,* as a rough measure of the average amount of predictive error—that is, as a rough measure of the average amount by which known *Y* values deviate from their predicted *Y′* values.[2]

[2]Strictly speaking, the standard error of prediction exceeds the average predictive error by 10 to 20 percent. Nevertheless, it's reasonable to describe the standard error in this fashion—as long as you remember that, as with the corresponding definition for the standard deviation in Chapter 5, an approximation is involved.

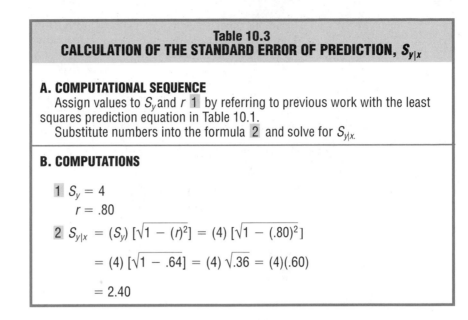

Table 10.3
CALCULATION OF THE STANDARD ERROR OF PREDICTION, $S_{y|x}$

A. COMPUTATIONAL SEQUENCE
Assign values to S_y and r **1** by referring to previous work with the least squares prediction equation in Table 10.1.
Substitute numbers into the formula **2** and solve for $S_{y|x}$.

B. COMPUTATIONS

1 $S_y = 4$
$r = .80$

2 $S_{y|x} = (S_y) [\sqrt{1 - (r)^2}] = (4) [\sqrt{1 - (.80)^2}]$

$= (4) [\sqrt{1 - .64}] = (4) \sqrt{.36} = (4)(.60)$

$= 2.40$

The value of 2.40 for $S_{y|x}$, as calculated in **Table 10.3,** represents the standard deviation for the discrepancies between known and predicted card returns originally shown in Figure 10.3. In its role as an estimate of predictive error, the value of $S_{y|x}$ can be attached to any new prediction. Thus, a concise prediction statement may read "The predicted card return for Emma equals 15.20 ± 2.40," in which the latter term serves as a rough estimate of the average amount of predictive error, that is, the average amount by which 15.20 will either overestimate or underestimate Emma's true card return.

Importance of *r*

To appreciate the importance of the correlation coefficient in any predictive effort, let's substitute a few extreme values for *r* in Formula 10.4 and note the resulting effect on predictive error. Substituting a value of 1 for *r* in Formula 10.4, we obtain

$$S_{y|x} = S_y\sqrt{1 - (1)^2} = S_y\sqrt{1 - 1} = S_y\sqrt{0} = S_y(0) = 0$$

As expected, when predictions are based on perfect relationships, there is no predictive error. At the other extreme, substituting a value of 0 for *r* in Formula 10.4, we obtain

$$S_{y|x} = S_y\sqrt{1 - (0)^2} = S_y\sqrt{1 - 0} = S_y\sqrt{1} = S_y(1) = S_y$$

Again, as expected, when predictions are based on a nonexistent relationship, there is no reduction in predictive error; it simply equals the standard deviation for all observations along *Y*.

Clearly, the prognosis for a predictive effort is most favorable when predictions are based on strong relationships, as reflected by a sizable positive or negative value of *r*. The prognosis is most dismal—and a predictive effort shouldn't even be attempted—when predictions must be based on a weak or nonexistent relationship, as reflected by a value of *r* in the vicinity of 0.

*Exercise 10.3 (a) Calculate the standard error of prediction for the data in Exercise 10.2 on page 167.

(b) Supply a rough interpretation of the standard error of prediction.
  Answers on page 502

10.7 ASSUMPTIONS
Linearity

Use of the prediction equation requires that the underlying relationship be linear. You need to worry about violating this assumption only when the scatterplot for the original correlation analysis reveals an obviously bent or curvilinear dot cluster, such as illustrated in Figure 9.4 on page 139. In the unlikely event that a dot cluster describes a pronounced curvilinear trend, consult more advanced statistics books for appropriate procedures.

Homoscedasticity

Use of the standard error of prediction, $S_{y|x}$, presumes that except for chance, the dots in the original scatterplot will be dispersed equally about all segments of the prediction line or about some straight line that roughly denotes the main trend of the dot cluster. You need to worry about violating this assumption, officially known by its tongue-twisting designation as the assumption of *homoscedasticity* (pronounced "ho-mo-skee-das-ti-ci-ty"), only when the scatterplot reveals a dramatically different type of dot cluster, such as that shown in **Figure 10.4.** At the very least, the standard error of prediction for the data in Figure 10.4 should be used cautiously, because its value overestimates the variability of dots about the lower half of the prediction line and underestimates the variability of dots about the upper half of the prediction line.

10.8 INTERPRETATION OF r^2

Alluded to in Section 9.5, the squared correlation coefficient, r^2, provides us with not only a key interpretation of the correlation coefficient, but also a measure of predictive accuracy that supplements the standard error of prediction, $S_{y|x}$. Understanding r^2 requires that we return to the problem of predicting the number of greeting cards received by the five friends. (Remember, we engage in the seemingly silly activity of predicting that which we already know not as an end in itself, but to check the adequacy of our predictive effort.)

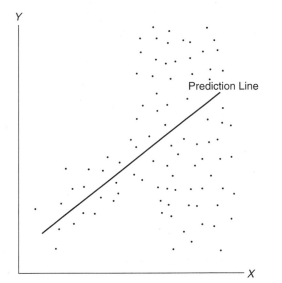

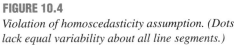

FIGURE 10.4
Violation of homoscedasticity assumption. (Dots lack equal variability about all line segments.)

Repetitive Prediction of the Mean

For the sake of the present argument, pretend that we know the Y scores (cards received), but not the corresponding X scores (cards given), for each of the five friends. Lacking information about the relationship between X and Y scores, we could not construct a prediction equation and use it to generate a customized prediction, Y', for each friend. We could, nonetheless, mount a primitive predictive effort by always predicting the mean, $\overline{Y}$, for each of the five friends' Y scores. (Under the present circumstances, statisticians recommend repetitive predictions of the mean for a variety of reasons, including the fact that, although the predictive error for any individual might be quite large, at least the *sum* of all of the resulting five predictive errors [deviations of Y scores about $\overline{Y}$] always equals zero, as you may recall from Section 4.3.) Most important for our purposes, using the repetitive prediction of $\overline{Y}$ for each of the Y scores of all five friends supplies us with a *frame of reference against which to evaluate our customary predictive effort* with a least squares equation based on the correlation between cards given (X) and cards received (Y). Any predictive effort that capitalizes on an existing correlation between X and Y should be able to generate more accurate predictions of Y than a primitive effort based only on the repetitive prediction of $\overline{Y}$. Indeed, as we shall see, r^2 precisely measures the proportionate increase in predictive accuracy when predictions based on the least squares equation replace those based only on $\overline{Y}$.

Predictive Errors for John

Let's look more closely at the errors associated with the repetitive prediction of $\overline{Y}$ and also with the customary least squares prediction equation, first for a single friend, John, then for all five friends. The top panel of **Figure 10.5** shows

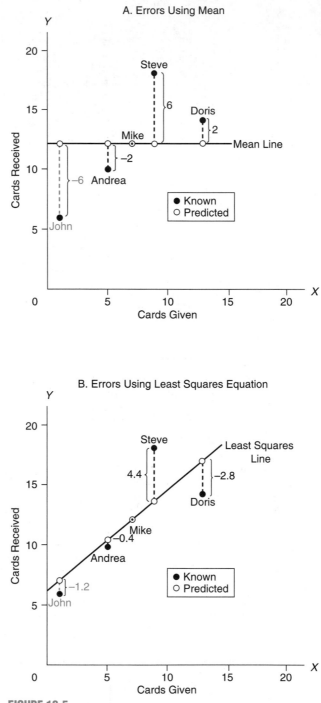

FIGURE 10.5
Predictive errors for five friends.

in color the error for John when the mean for all five friends, $\overline{Y}$, of 12 is used to predict his Y score of 6. Shown as a broken vertical line, the error of -6 for John (from $Y - \overline{Y} = 6 - 12 = -6$) indicates that $\overline{Y}$ overestimates John's Y score by 6 cards. The bottom panel of Figure 10.5 shows *in color* a smaller error of -1.20 for John when a Y' value of 7.20 is used to predict the same Y score of 6. This Y' value of 7.20 is obtained from the least squares equation, encountered previously in Table 10.1, page 166, for predicting cards received from cards given, after the number of cards given by John, 1, has been substituted for X, that is,

$$Y' = .80(X) + 6.40$$
$$= .80(1) + 6.40$$
$$= 7.20$$

Predictive Errors for All Five Friends

The top panel of Figure 10.5 shows the errors for all five friends when the mean for all five friends, $\overline{Y}$, of 12 (shown as the mean line) always is used to predict each of their five Y scores, whereas the bottom panel shows the corresponding errors for all five friends when a series of Y' values, obtained from the least squares equation (shown as the least squares line), is used to predict each of their five Y scores. Positive and negative errors indicate that Y scores are either above or below their corresponding predicted scores, respectively. Overall, as expected, errors are smaller when customized predictions of Y' from the least squares equation can be used (because X scores are known) than when only the repetitive prediction of $\overline{Y}$ can be used (because X scores are ignored). As with most statistical phenomena, there are exceptions: the predictive error for Doris is slightly larger when the least squares equation is used.

Variance of Errors

To more precisely evaluate the accuracy of our two predictive efforts, we need some measure of the collective errors produced by each effort. Such a measure is the variance, the mean of the squared deviations (or squared errors), introduced in Section 5.3. The smaller the variance of errors, the greater the accuracy of the predictive effort. As you may recall, the variance of any set of deviations, now called *errors,* can be calculated by first squaring each error (to eliminate negative errors); then summing all squared errors; and finally dividing by the number of errors (or predictions).

The variance of errors for the repetitive prediction of the mean can be designated as S_y^2, because each Y score is expressed as a deviation from $\overline{Y}$ and the variance of these deviations is simply the variance for any set of observations described in Table 5.1, page 184. Using the errors for the five friends shown in the top panel of Figure 10.5, we obtain

$$S_y^2 = \frac{[(-6)^2 + (-2)^2 + 0^2 + 6^2 + 2^2}{5} = \frac{80}{5} = 16$$

(which equals the square of $S_y = 4$ shown in Table 10.1, page 166).

The variance of errors for the customized predictions from the least squares equation can be designated as $S_{y|x}^2$, because each Y score is expressed as a deviation from its corresponding Y' and the variance of these deviations is simply the square of $S_{y|x}$, the standard error of prediction for the least squares equation, discussed in Section 10.6. Using the errors for the five friends shown in the bottom panel of Figure 10.5, we obtain

$$S_{y|x}^2 = \frac{[(-1.2)^2 + (-0.4)^2 + 0^2 + (4.4)^2 + (-2.8)^2}{5} = \frac{28.5}{5} = 5.76$$

(which equals the square of $S_{y|x} = 2.40$ shown in Table 10.3, page 169).

Proportion of *Predicted* Variance

Clearly, the error variance of 5.76 for the least square predictions is much smaller than the error variance of 16 for the repetitive prediction of $\overline{Y}$, confirming the greater accuracy of the least squares predictions apparent in Figure 10.5. To obtain a measure of the actual *gain in accuracy* due to the least squares predictions, simply subtract $S_{y|x}^2$ from S_y^2, that is,

$$S_y^2 - S_{y|x}^2 = 16 - 5.76 = 10.24$$

It's more meaningful to express this difference, 10.24, as a gain in accuracy *relative* to the original error variance for the repetitive prediction of $\overline{Y}$, by simply dividing the above difference by S_y^2, that is,

$$\frac{S_y^2 - S_{y|x}^2}{S_y^2} = \frac{16 - 5.76}{16} = \frac{10.24}{16} = .64$$

This result, .64, or 64 percent, represents the proportion or percent gain in predictive accuracy when the repetitive prediction of $\overline{Y}$ is replaced by a series of customized Y' predictions based on the least squares equation. In other words, .64 or 64 percent represents the proportion or percent of the total variance of the Y variable, S_y^2, that is predictable from its relationship with the X variable.

To the delight of statisticians, when squared, the value of the correlation coefficient equals this proportion of predictable variance. Recalling that an r of .80 was obtained for the correlation between cards given and cards received by the five friends, we can verify that $r^2 = (.80)(.80) = .64$, which, of course, also is the proportion of predictable variance. Given this perspective,

---

· ·

Squared correlation coefficient (r^2)

The proportion of the total variance in one variable that is predictable from its relationship with the other variable.

---

the *square of the correlation coefficient*, r^2, *always* indicates the proportion of total variance in one variable that is predictable from its relationship with the other variable.

Accordingly, r^2 provides us with a straightforward measure of the worth of our least squares predictive effort.[3]

r^2 Doesn't Describe Individual Scores

Don't attempt to apply the variance interpretation of r^2 to individual scores. For instance, the fact that 64 percent of the variance in cards received by the five friends (Y) is predictable from their cards given (X) doesn't signify, therefore, that 64 percent of the five friends' Y scores can be predicted perfectly. As can be seen in the bottom panel of Figure 10.5, only one of the Y scores for the five friends, that for Mike, was predicted perfectly by the least squares equation, and even this perfect prediction isn't guaranteed just because r^2 equals .64. To the contrary, the 64 percent must be interpreted as applying to the variance for the *entire* set of Y scores, that is, to the variance of errors produced by the repetitive prediction of $\overline{Y}$. The variance of *all* Y scores can be reduced by 64 percent when each of the five Y scores is expressed, in turn, not as a squared deviation from the mean of all observed scores, but as a squared deviation from its corresponding predicted score Y'. Thus, the 64 percent represents a reduction in the variance for the five Y scores when the single mean for all observed scores is replaced by a succession of more specialized means or predicted scores, given the least squares equation and various values of X.

*Exercise 10.4 Assume, as in Exercise 10.2, that an r of .30 describes the relationship between educational level and estimated hours spent reading each week.

(a) According to the variance interpretation of r^2, what percent of the variance in weekly reading time is predictable from its relationship with educational level?

(b) What percent of variance in weekly reading time isn't predictable from this relationship?

....

[3]In practice, to actually calculate the value of r, never use the previous procedure, designed only as an aid to understanding the variance interpretation of r^2. Instead, always use the vastly more efficient computational Formula 9.2 on page 148.

(c) Someone claims that 9 percent of *each* person's estimated reading time is predictable from the relationship. What's wrong with this claim?

Answers on page 502

INTERNET DEMONSTRATION

Go to the Web site for this book **(http://darwin.cwru.edu/~witte/statistics)** and click on **Explained Variance** to see the effect of varying scatterplots on explained and unexplained variance and the numerical value of the correlation coefficient.

····································

10.9 r REVISITED

Let's return to the correlation coefficient, r, now using the variance interpretation, r^2. Recalling that an r of approximately .70 describes the correlation between height and weight for college students, we can claim that, since $r^2 = (.70)(.70) = .49$, fully .49 or 49 percent of the variance in heights of college students is predictable from differences in weight, or vice versa. In other words, the total variance in height (that is, the error variance if the mean height were used to predict each college student's height) can be divided into two portions: .49 or 49 percent that is predictable from differences in weight and the remainder, .51 or 51 percent that isn't predictable from differences in weight (because of the imperfect relationship between height and weight).

As another example, imagine that a child psychologist finds a correlation of .20 between the scores of sixth graders on a mental health test and their weaning ages as infants. Because $(.20)(.20) = .04$, only 4 percent of the variance of mental health scores is predictable from differences in weaning ages, whereas a huge 96 percent of the variance of mental health scores isn't predictable from differences in weaning ages.

Direct Measure of Strength

The value of r^2 supplies us with a direct measure of the strength of a relationship. For example, an r^2 value of .50 can be interpreted either as one-half as strong as a perfect relationship, because it signifies that one-half of the total variance is predictable from the current relationship, or as twice as strong as an r^2 value of .25, because it signifies that twice as much variance is predictable when the current relationship is compared with the weaker relationship.

This perspective also reinforces the warning in Chapter 9 against interpreting the face value of r as a proportion of a perfect relationship. From the perspective of r^2, an r of .50 is not one-half, but the square of one-half or only one-fourth, as strong as a perfect relationship of 1.00, because $(.50)(.50) = .25$. That same r of .50 is not five times, but the square of five times or twenty-five times, as strong as an r of .10, because $(.10)(.10) = .01$.

Small Values of r^2

Don't expect to routinely encounter large values of r^2 in behavioral and educational research. In these areas, where measures of complex phenomena, such as intellectual aptitude, psychopathic tendency, or self-esteem, fail to correlate highly with any single variable, values of r^2 larger than about .25 are most unlikely. (And, therefore, as suggested in Section 10.10, many investigators resort to advanced statistical tools, such as correlation coefficients and prediction equations for more than two variables.)

Under these circumstances, even values of r^2 close to zero might merit our attention. For instance, if just 4 percent — or even as little as 1 or 2 percent — of the variance of mental health scores of sixth graders actually could be predicted from a single variable, such as differences in weaning age, most investigators probably would view this as an important finding, worthy of additional investigation.

r^2 Doesn't Ensure Cause-Effect

The question of cause-effect, raised in Section 9.6, can't be resolved merely by squaring the correlation coefficient to obtain a value of r^2. If, as we imagined earlier, the correlation between mental health scores of sixth graders and their weaning ages as infants equals .20, we can't claim, therefore, that $(.20)(20) = .04$, or 4 percent, of the total variance in mental health scores is *caused* by the differences in weaning ages. Instead, it's possible that this correlation reflects some more basic factor or factors, such as, for example, a tendency for more economically secure, less stressed mothers both to create a family environment that perpetuates good mental health and, coincidentally, to nurse their infants longer. Certainly, in the absence of additional evidence, it would be foolhardy to encourage mothers, regardless of their circumstances, to postpone weaning because of its projected effect on mental health scores.

Although we have consistently referred to r^2 as indicating the proportion or percent of *predictable* variance, you also might encounter references to r^2 as indicating the proportion or percent of *explained* variance. In this context, "explained" signifies only predictability, *not* causality. Thus, you could assert that .04, or 4 percent, of the variance in mental health scores is "explained" by differences in weaning age, insomuch as .04, or 4 percent, is predictable from — or statistically attributable to — differences in weaning age.

*Exercise 10.5 As indicated in Figure 9.3, page 138, the correlation between the IQ scores of parents and children is .50, and that between the IQ scores of foster parents and foster children is .27.

(a) Does this signify, therefore, that the relationship between foster parents and foster children is about one-half as strong as the relationship between parents and children?

(b) Use r^2 to compare the strengths of these two correlations.
   Answers on page 502

10.10 MORE COMPLEX PREDICTION EQUATIONS

Any serious predictive effort usually culminates in a more complex equation that contains not just one but several "predictor" or X variables. For instance, a serious effort to predict college GPA might culminate in the following equation:

$$Y' = .410(X_1) + .005(X_2) + .001(X_3) + 1.03$$

where Y' represents predicted college GPA, and X_1, X_2, and X_3 refer to high school GPA, IQ score, and verbal SAT score, respectively. By capitalizing on the combined predictive power of several predictor variables, these complex prediction equations (or *multiple regression* equations, as they often are called) supply more accurate predictions than could be obtained from a simple prediction equation.

Common Features

Although more difficult to visualize, complex prediction equations possess many features in common with their simple counterparts. For instance, the complex equation still qualifies as a least squares equation, because it minimizes the sum of the squared predictive errors. By the same token, it is accompanied by a standard error of prediction that roughly measures the average amount of predictive error. Be assured, therefore, that the present chapter will serve as a good point of departure if, sometime in the future, you must deal with complex prediction or multiple regression equations.

Summary

If a linear relationship exists between two variables, then one variable can be predicted from the other by using the least squares prediction equation, as described in Formulas 10.1, 10.2, and 10.3. (See Table 10.1.)

The least squares equation minimizes a variation on the total predictive error, that is, the total of all squared predictive errors that would have occurred if the equation had been used to predict known Y scores from the original correlation analysis.

An estimate of predictive error can be obtained from Formula 10.4. Known as the standard error of prediction, this estimate is a special kind of standard deviation that roughly reflects the average amount of predictive error. The prognosis for a predictive effort, as reflected by the value of the standard error of prediction, depends mainly on the size of the correlation coefficient. The larger the size of the correlation coefficient, in either the positive or negative direction, the smaller the standard error of prediction, and the more favorable the prognosis for a predictive effort.

Several assumptions underlie any predictive effort. The prediction equation assumes a linear relationship between variables, and the standard error of prediction assumes homoscedasticity—approximately equal dispersion of data points about all segments of the prediction line.

The square of the correlation coefficient, r^2, indicates the proportion of the total variance in one variable that is predictable from its relationship with the other variable, as embodied in the least squares equation. The value of r^2 also supplies us with a direct measure of the strength of a relationship.

Serious predictive efforts usually involve complex prediction equations composed of more than one predictor or X variable. These complex equations share many common features with the simple prediction equations discussed in this chapter.

Important Terms

Least squares prediction equation　　**Squared correlation coefficient (r^2)**
Standard error of prediction ($S_{y|x}$)

REVIEW EXERCISES

10.6　Assume that an r of $-.80$ describes the strong negative relationship between years of heavy smoking (X) and life expectancy (Y). Assume, furthermore, that the distributions of heavy smoking and life expectancy each have the following means and standard deviations:

$$\overline{X} = 5 \quad \overline{Y} = 60$$
$$S_x = 3 \quad S_y = 6$$

(a) Determine the least squares equation for predicting life expectancy from years of heavy smoking.

(b) Determine the standard error of prediction, $S_{y|x}$.

(c) Supply a rough interpretation of $S_{y|x}$.

(d) Predict the life expectancy for Sara, who has smoked heavily for eight years.

(e) Predict the life expectancy for Katie, who has never smoked heavily.

10.7   Each of the following pairs represents the number of licensed drivers (X) and the number of cars (Y) for houses in my neighborhood:

DRIVERS (X)	CARS (Y)
5	4
5	3
2	2
2	2
3	2
1	1
2	2

(a) Construct a scatterplot to verify a lack of pronounced curvilinearity.

(b) Determine the least squares equation for these data. (Remember, you'll first have to calculate r, S_x, and S_y.)

(c) Determine the standard error of prediction, $S_{y|x}$.

(d) Predict the number of cars for each of two new families with two and five drivers, respectively.

10.8   In Exercise 10.7, a least squares equation was used to predict the number of cars (Y) from the number of licensed drivers (X). Would a different equation be required to predict the number of licensed drivers from the number of cars? First speculate about your answer and then actually calculate the least squares equation for predicting the number of licensed drivers (Y) from the number of cars (X).

10.9   Construct a graph of the equation for predicting weekly reading time (Y') from years of education (X) found in Exercise 10.2; that is, construct a graph of the equation $Y' = .60(X) + .20$. *Hint:* Find two separate pairs of X and Y' values that can be used to identify the prediction line.

10.10 (a) At a large bank, length of service is the best single predictor of employees' salaries. Can we conclude, therefore, that there is a cause-effect relationship between length of service and salary?

(b) Assuming that r equals .60 for the correlation between length of service and employees' salaries, give the variance interpretation for r^2, the squared correlation coefficient.

10.11   Pretend that women always marry men who are exactly 5 inches taller than they are.

(a) Given that Brittany is 69 inches tall, how tall would we predict her husband to be?

(b) Construct a graph to verify that, in fact, the relationship between the heights of husbands (Y) and wives (X) is linear.

(c) What would be the value of r for the heights of married couples?

(d) What would be the value of the standard error of prediction, given that you plan to predict the heights of husbands from those of their wives?

(e) What proportion of the variance of husbands' heights would be predictable, given a least squares equation for predicting husbands' heights from the heights of their wives?

PART 2

Inferential Statistics
Generalizing beyond Data

11   **Populations and Samples**

12   **Probability**

13   **Sampling Distribution of the Mean**

14   **Introduction to Hypothesis Testing: The *z* Test**

15   **More about Hypothesis Testing**

16   **Controlling Type I and Type II Errors**

17   **Estimation**

18   *t* **Test for One Sample**

19   *t* **Test for Two Independent Samples**

20   *t* **Test for Two Matched Samples**

21   **Beyond Hypothesis Tests: *p*-Values and Effect Size**

22   **Analysis of Variance (One Way)**

23   **Analysis of Variance (Two Way)**

24   **Chi Square (χ^2) Test for Qualitative Data**

25   **Tests for Ranked Data**

26   **Postscript: Which Test?**

CHAPTER 11

Populations and Samples

11.1 WHY SAMPLES?

11.2 POPULATIONS

11.3 SAMPLES

11.4 RANDOM SAMPLES

11.5 TABLES OF RANDOM NUMBERS

11.6 SOME COMPLICATIONS

11.7 RANDOM ASSIGNMENT OF SUBJECTS

11.8 AN OVERVIEW: SURVEYS OR EXPERIMENTS?

Summary

Important Terms

Review Exercises

In everyday life, we regularly generalize from limited sets of observations (samples) to broader sets of observations (populations). One sip indicates that the batch of soup is too salty; dipping a toe in the swimming pool reassures us before taking the first plunge; a test drive triggers suspicions that the used car is not what it was advertised to be; and a casual encounter with a stranger stimulates fantasies about a deeper relationship. In inferential statistics, we are supplied with more sophisticated tools for generalizing from samples to populations.

Generalizations can, of course, backfire if the sample misrepresents the population. For example, perhaps the sip contains unstirred salt; the toe dips too near the warm-water outlet; the test drive merely reflects an untuned engine; and the casual encounter was really very contrived. In inferential statistics, we can specify and control the chances that generalizations will backfire.

11.1 WHY SAMPLES?

Faced with the possibility of erroneous generalizations, you might prefer to bypass the uncertainties of inferential statistics by surveying an entire population. This is often done if the size of the population is fairly small. For instance, you calculate your GPA from all of your course grades, not just a sample. If the size of the population is large, however, complete surveys are often prohibitively expensive and sometimes impossible. Under these circumstances, you might have to use samples and risk the possibility of erroneous generalizations. For instance, you might have to use a sample to estimate the mean study time of all students at a large university.

11.2 POPULATIONS

Population

Any complete set of observations (or potential observations).

Any complete set of observations (or potential observations) may be characterized as a **population.** Accurate descriptions of populations specify the nature of the observations to be taken. For example, a population might be described as "attitudes toward religion of currently enrolled students at Rutgers University" or as "SAT scores of currently enrolled students at Rutgers University."

Real Populations

Populations may be real or hypothetical. A *real* population is one in which all potential observations are accessible at the time of sampling. Examples of real populations include the two described in the previous paragraph, as well as the ages of all visitors to Disneyland on a given day, the ethnic backgrounds of all current employees of the U.S. Postal Service, and presidential preferences of all currently registered voters in the United States. Incidentally, federal law requires that a complete survey be taken every ten years of the real population of all U.S. households—at a considerable expense, involving many thousands of data collectors—as a means of revising election districts for the House of Representatives. (An estimated undercount of four million people in the 1990 census has revived a suggestion, long endorsed by statisticians, that the entire U.S. population could be estimated more accurately if a highly trained group of data collectors focused only on an impartially selected sample of households. In the 2000 census, the Census Bureau proposed to supplement the traditional head count with a sampling plan designed to yield a more accurate estimate of the total population.)

INTERNET SITE

Go to the Web site for this book **(http://darwin.cwru.edu/~witte/statistics)** and click on the **U.S. Census Bureau** to view its Web site, including links to its many reports and to digital clocks that show the growth rates of current population estimates for the United States and the world.

Hypothetical Populations

Pollsters, such as the Gallup Organization, deal with real populations. But research workers usually deal with hypothetical populations. A *hypothetical* population is one in which all potential observations are not accessible at the time of sampling. In most experiments, subjects are selected from very small real populations: the lab rats housed in the local animal colony or student volunteers from general psychology classes. Experimental subjects are viewed, nevertheless, as a sample from a much larger hypothetical population, loosely described as "the scores of all similar animal subjects (or student volunteers) who, either now or *in the future,* could conceivably undergo the present experiment." Experimental findings are generalized to this much larger population that contains many potential observations that, because of their hypothetical status in future experiments, couldn't possibly be included in the present experiment.

According to the rules of inferential statistics, generalizations should be made only to real populations that have in fact been sampled. Generalizations to hypothetical populations should be viewed, therefore, as provisional conclusions based on the wisdom of the researcher rather than on any logical or statistical necessity. In effect, it's an open question—resolved only through additional experimentation—whether a given experimental finding merits the generality assigned to it by the researcher.

11.3 SAMPLES

Sample

Any subset of observations from a population.

Any subset of observations from a population may be characterized as a **sample.** In typical applications of inferential statistics, sample size is small relative to population size. Less than 1 percent of all U.S. households are included in the monthly survey of the Bureau of Labor Statistics to estimate the current rate of unemployment. Although, at most, only about 4000 voters have been sampled in recent presidential election polls by Gallup, predictions have been amazingly accurate since 1952—missing the actual percent of votes for the winning candidate by an average of only 2 percent, according to the Internet Web site for the Gallup Organization.

Optimal Sample Size

There is no simple rule of thumb for determining the best or optimal sample size for any particular situation. Optimal sample size depends on the answers to several questions, including "What is the estimated variability among observations?" and "What is an acceptable amount of probable error?" Once these questions have been answered, with the aid of guidelines such as those discussed in Chapter 16, specific procedures can be followed to determine the optimal sample size for any situation.

*Exercise 11.1 For each of the following pairs, indicate with a Yes or No whether the relationship between the first and second expressions could describe that between a sample and its population, respectively.

(a) spoonful of soup; cupful of soup

(b) students in the last row; students in class

(c) students in class; students in college

(d) a student in the last row; students in the last row

(e) citizens of Wyoming; citizens of New York

(f) every tenth California income tax return; every California income tax return

(g) twenty lab rats in an experiment; all lab rats, similar to those used, that could undergo the same experiment

(h) all U.S. presidents; all registered Republicans

(i) two tosses of a coin; all possible tosses of a coin

*Exercise 11.2 List all expressions in Exercise 11.1 that involve a hypothetical population.

Answers on pages 502–503

11.4 RANDOM SAMPLES

In order to use techniques from inferential statistics, the analysis should be based on random samples.

....

Random sample

*A sample produced when all
potential observations in the
population have equal chances
of being selected.*

A sample is *random* if, at each stage of sampling, the selection process guarantees that all remaining observations in the population have equal chances of being included in the sample.

It's important to note that randomness describes the *selection process,* that is, the conditions under which the sample is taken, and not the particular pattern of observations in the sample.

Having established that a sample is random, you still can't predict anything about the particular pattern of observations in that sample. The observations in the sample should be representative of those in the population, but there is no guarantee that they actually will be.

Casual or Haphazard, Not Random

A casual or haphazard sample doesn't qualify as a random sample. Not every student at Rutgers University has an equal chance of being sampled if, for instance, a pollster casually selects only students who enter the main library. Obviously excluded from this sample are all those students who never enter the main library. Even the final selection of students from among those who do enter the main library might reflect the pollster's various biases, such as an unconscious preference for well-dressed, attractive students who are walking alone.

"Fishbowl" Method

How, then, do you select a random sample? Several techniques have been used with varying degrees of success. Probably the best-known technique is the "fishbowl" method. This method requires that all potential observations be represented in some fashion on slips of paper. For instance, the names of all students at some college might be written on separate slips of paper. All of the slips of paper are then deposited in a bowl and stirred. Some person draws a slip from the bowl, and the name on that slip designates the first student to be included in the sample. Drawings continue until the desired sample size has been reached.

1970 Draft Lottery

The adequacy of the fishbowl method depends on the thoroughness of the stirring. A truly thorough stirring of slips of paper is more difficult to achieve than might be suspected, as evidenced by the draft lottery of 1970. Essentially, in this lottery, the population consisted of the 366 days of the year, representing the birthdays of all draft-eligible young men. All of these days (rather than a sample) were to be selected from the fishbowl, with the order of selection from the fishbowl establishing the order of induction of draft-eligible males into military service.

If the lottery were random, as required by law, then all 366 days should have had an equal chance of being drawn first; all remaining 365 days should have had an equal chance of being drawn second, and so forth. Although anything can happen by chance, the results of this lottery looked suspiciously nonrandom. There was a noticeable tendency for the days of some months—notably those months during the latter part of the year—to be drawn early and, therefore, to be assigned a high draft-priority number. Subsequent reports indicate that, after the chronologically ordered birthday capsules had been deposited into a box, the box was shaken only several times and then the capsules were poured into the selection bowl.[1] To correct this situation, succeeding draft lotteries involved elaborate efforts to guarantee randomization, particularly in the original placement of capsules in the bowl.

*Exercise 11.3 Indicate whether each of the following statements is true or false. A random selection of ten playing cards from a deck of fifty-two implies that

(a) the random sample of ten cards accurately represents the important features of the whole deck.

(b) each card in the deck has an equal chance of being selected.

(c) it is impossible to get ten cards from the same suit (for example, ten hearts).

(d) any outcome, however unlikely, is possible.

Answers on page 503

11.5 TABLES OF RANDOM NUMBERS

A more adequate method for generating a truly random sample involves tables of random numbers—a method used to randomize the initial placement of capsules in post-1970 selective service lotteries. These tables are generated by a computer designed to equalize as much as possible the occurrence of any one of the ten digits: 0, 1, 2, . . . , 8, 9. For convenience, many random number tables are spaced in columns of five-digit numbers. Table G in Appendix D shows a specimen page of random numbers from a book devoted almost entirely to random digits.

How Many Digits?

The size of the real population determines whether you deal with numbers having one, two, three, or more digits. The only requirement is that you have at least as many different numbers as you have potential observations within the population. For example, if you were attempting to take a random sample from a real population consisting of 679 students at some college, you could use the 1000 three-digit numbers ranging from 000 to 999. In this case, you could identify each of the potential observations, as represented by a particular student's name, with a single number. For instance, if a student directory were available, the first person, Alice Aakins, might be assigned the three-digit number 001, and so on through to the last person in the directory, Zachary Ziegler, who might be assigned 679.

-----

[1] For more details about the 1970 draft lottery, as well as many other interesting articles about everyday statistical applications, see M. Hollander & F. Proschan, *The Statistical Exorcist* (New York: Dekker, 1984).

Using Tables

Enter the random number table (Table G in Appendix D) at some arbitrarily determined place. Ordinarily this should be determined haphazardly. Open a book of random numbers to any page and begin with the number closest to a blind pencil stab. For illustrative purposes, however, let's use the upper left-hand corner of the specimen page as our entry point. (Ignore the column of numbers that identifies the various rows.) Read in a consistent direction, for instance, from left to right. Then as each row is used up, shift down to the start of the next row and repeat the entire process. As a given number between 001 and 679 is encountered, the person identified with that number is included in the random sample.

Because the first number on the specimen page in Table G is 100 (disregard the fourth and fifth digits in each five-digit number), the person identified with that number is included in the sample. The next three-digit number, 325, identifies the second person. Ignore the next number, 765, because none of the numbers between 680 and 999 is identified with any names in the student directory. Also ignore repeat appearances of any number between 001 and 679. The next three-digit number, 135, identifies the third person. Continue this process until the specified sample size has been realized.

Efficient Use of Tables

The inefficiency of the previous procedure becomes apparent when a random sample must be obtained from a large population, such as that defined by a city telephone directory. It would be most laborious to assign a different number to each name in the directory prior to consulting the table of random numbers. Instead, most investigators refer directly to the random number table, using each random number as a guide to a particular name in the directory. For example, a six-digit random number, such as 239421, identifies the name on page 239 (the first three digits) and line 421 (the last three digits). This process is repeated for a series of six-digit random numbers until the required number of names has been sampled.

*Exercise 11.4 Describe how you would use the tables of random numbers to take

(a) a random sample of size 5 from your statistics class.

(b) a random sample of size 40 from a large city telephone directory consisting of 3041 pages, with 480 lines per page.

Answers on page 503

11.6 SOME COMPLICATIONS

No Population Directory

Lacking the convenience of an existing population directory, investigators resort to various embellishments of the previous procedure. For instance, the Gallup Organization makes a separate presidential survey in each of the four geographical areas of the United States: Northeast, South, Midwest, and West.

Within each of these areas, a series of random selections culminates in the identification of particular election precincts: small geographical districts with a single polling place. Once household directories have been obtained for each of these precincts, households are randomly selected, and specific household members are interviewed.

Many pollsters now use *random digit dialing* in an effort to give each residential telephone number in the United States an equal chance of being called for an interview. Essentially, the first six digits of a ten-digit phone number, including the area code, are randomly selected from tens of thousands of telephone exchanges, whereas the final four digits are taken directly from random numbers. This technique ensures that even unlisted telephone numbers will be sampled and called.

Hypothetical Populations

As has been noted, the researcher, unlike the pollster, usually deals with hypothetical populations. Unfortunately, it's impossible to take random samples from hypothetical populations. All potential observations can't have an equal chance of being included in the sample if, in fact, some observations aren't accessible at the time of sampling. It's a common practice, nonetheless, for researchers to treat samples from hypothetical populations *as if* they were random samples and to analyze sample results with techniques from inferential statistics.

11.7 RANDOM ASSIGNMENT OF SUBJECTS

Typically, experiments consist of an independent variable with at least two conditions: an experimental condition and a control condition. Even though subjects in experiments can't be selected randomly from a hypothetical population, they should be assigned randomly, that is, with equal likelihood, to these conditions. This procedure has a number of desirable consequences, including the production of several groups of subjects who are similar at the outset of the experiment. For instance, to determine whether vitamin C improves academic performance, volunteer subjects should be assigned randomly either to the vitamin C (or experimental group), whose members receive a daily dose of vitamin C, or to the fake vitamin C (or control group), whose members receive only fake vitamin C.

One purpose of the random assignment of subjects is to ensure that, except for random differences, groups of subjects are similar with respect to any uncontrolled variables,

such as academic preparation, motivation, and IQ. At the conclusion of such a well-designed experiment, therefore, any observed differences in academic performance between these two groups, not attributable to random differences, would provide the most clear-cut evidence of a cause-effect relationship between the independent variable (vitamin C) and the dependent variable (academic performance), as suggested in Chapter 1.

How to Assign Subjects

The random assignment of subjects can be accomplished in several ways. For instance, as each new subject arrives to participate in the experiment, a flip of a coin can decide whether that subject should be assigned to the vitamin C group (if "heads" turns up) or the fake vitamin C group (if "tails" turns up). An even better procedure, because it eliminates any biases of a human coin tosser, relies on tables of random numbers. Once the tables have been entered at some arbitrary point, they can be consulted, much like a string of coin tosses, to determine whether each new subject should be assigned to the vitamin C group (if, for instance, the random number is odd) or to the fake vitamin C group (if the random number is even).

Creating Equal Groups

For a variety of reasons, it's highly desirable that equal numbers of subjects be assigned to the experimental and control groups. Given this restriction, the random assignment should involve pairs of subjects. If the table of random numbers assigns the first volunteer to the vitamin C group, the second volunteer will be assigned *automatically* to the fake vitamin C group. If the random numbers assign the third volunteer to the fake vitamin C group, the fourth volunteer will be assigned *automatically* to the vitamin C group, and so forth. This procedure guarantees that at any stage of the random assignment, equal numbers of subjects will be assigned to the two groups.

More Than Two Groups

Many experiments consist of more than two groups. For example, if three groups are involved, then random assignment should involve trios of subjects. After the first subject is randomly assigned to any one of the three groups, the second subject should be randomly assigned to one of the two remaining groups, and then the third subject will be assigned *automatically* to the one remaining group. As in the case for two groups, this procedure guarantees that at any stage of the random assignment, equal numbers of subjects will be assigned to the three groups.

More Extensive Sets of Random Numbers

Incidentally, the page of random numbers in Table G serves only as a specimen. For serious applications, refer to a more extensive collection of random numbers, such as that in the book by the Rand Corporation, described on page 532 of Appendix D. Or if you have access to a computer, you might refer to the list of random numbers that can be generated, almost effortlessly, by computers.

*Exercise 11.5 Assume that twelve subjects arrive, one at a time, to participate in an experiment. Use random numbers to assign the twelve subjects in equal numbers to group A and group B. In other words, random numbers should be used to identify the first subject as either A or B, the second subject as either A or B, and so forth, until all twelve subjects have been identified. Furthermore, there should be six subjects identified with A and six with B.

(a) Formulate an acceptable rule for single-digit random numbers. Incorporate into this rule a procedure that will ensure equal numbers of subjects in the two groups. *Check your answer in Appendix C before proceeding.*

(b) Reading from left to right in the top row of the random number page (Table G, Appendix D), use the random digits of each random number in conjunction with your assignment rule to determine whether the first subject is A or B, and so forth. List the assignment for each subject.

*Exercise 11.6 (a) Sometimes experiments involve more than two groups. Describe a rule, similar to those listed in Exercise 11.5, that could be used to assign subjects randomly to *four* groups.

(b) It was suggested earlier that random assignment should involve pairs of subjects if equal numbers of subjects are to be assigned to two groups. How might this rule be modified if equal numbers of subjects are to be assigned to all four groups?

Answers on pages 503–504

11.8 AN OVERVIEW: SURVEYS OR EXPERIMENTS?

When using random numbers, it's important to have a general perspective. Are you engaged in a *survey* (because subjects are being sampled from a real population) or in an *experiment* (because subjects are being assigned to various groups)? In the case of surveys, the object is to obtain a random sample from some real population. Short-circuit unnecessary clerical work as much as possible, but use random numbers in a fashion that complies with the notion of *random sampling—that all subjects in the population have an equal opportunity of being sampled.* In the case of experiments, one objective is to obtain various groups of comparable subjects at the outset of the experiment. Introduce any restrictions required to generate equal group sizes (for example, the restriction that every other subject be assigned to the smaller group), but use random numbers in a fashion that complies with the notion of *random assignment—that all subjects have an equal opportunity of being assigned to each of the various groups.*

Summary

Any set of potential observations may be characterized as a population. Any subset of observations constitutes a sample.

Populations are either real or hypothetical, depending on whether all observations are accessible at the time of sampling.

The valid application of techniques from inferential statistics requires that the samples be random. A sample is random if at each stage of sampling, the selection process guarantees that all remaining observations in the population have equal chances of being included in the sample.

Tables of random numbers provide the most effective method both for taking random samples in surveys and for randomly assigning subjects to various conditions in experiments. Some type of randomization always should occur during the early stages of any investigation, whether a survey or an experiment.

Important Terms
....

Population   **Random sample**
Sample

REVIEW EXERCISES

11.7 Television networks sometimes solicit feedback volunteered by viewers about a televised event. For example, following a widely watched televised debate between Jimmy Carter and Ronald Reagan in the 1980 presidential election campaign, ABC conducted a telephone poll to determine the "winner." Callers were given two phone numbers, one for Carter and the other for Reagan, to register their opinions automatically. Some 477,815 callers voluntarily dialed the Reagan number, but only 243,554 voluntarily dialed the Carter number.

(a) Comment on whether this was a random sample.

(b) How might this poll have been improved?

11.8 As subjects arrive to participate in an experiment, tables of random numbers are used to make random assignments to either group A or group B. Indicate with a Yes or No whether each of the following rules would work:

(a) Assign the subject to group A if the random number is *even* and to group B if the random number is *odd*.

(b) Assign the subject to group A if the first digit of the random number is between **0 and 4** and to group B if the first digit is between **5 and 9.**

(c) Assign the subject to group A if the first two digits of the random number are between **00 and 40** and to group B if the first two digits are between **41 and 99.**

(d) Assign the subject to group A if the first three digits of the random number are between **000 and 499** and to group B if the first three digits are between **500 and 999.**

11.9 What is accomplished by randomly assigning subjects to various groups at the outset of an experiment?

CHAPTER 12

Probability

12.1   **DEFINITION**

12.2   **ADDITION RULE**

12.3   **MULTIPLICATION RULE**

12.4   **PROBABILITY AND STATISTICS**

Summary

Important Terms

Review Exercises

According to Bruce Hoadley, a statistician at Bell Communications Research, the catastrophic failure of the *Challenger* shuttle in 1986, which took the lives of seven astronauts, wasn't that unlikely. Indeed, Hoadley speculated that, prior to this tragedy, the chances of catastrophic failure of each of a succession of shuttles were actually about one in nine, an intolerably high risk even for professional astronauts. "One in nine" represents the *probability* of a catastrophic failure, given certain assumptions. Exercise 12.8 on page 207 shows how "one in nine" was arrived at and also suggests why this space mission was even attempted.

Probability considerations permeate our life: the *probability* that it will rain this weekend (20 percent or one in five, according to my morning newspaper), that a projected family of two children will consist of a boy and a girl (one half, on the assumption that boys and girls are equally likely), or that you'll win a state lottery (one in many millions).

As you will see, probability considerations also permeate much of inferential statistics and, therefore, much of the remainder of this book. In this chapter, we'll define probability for a simple outcome, then we'll discuss two rules for finding probabilities of more complex outcomes, such as those just mentioned.

12.1 DEFINITION

***Probability* refers to the proportion or fraction of times that a particular outcome is likely to occur.**

Probability

The proportion or fraction of times that a particular outcome is likely to occur.

The probability of an outcome can be determined in several ways. We can *speculate* that if a coin is truly fair, heads and tails should be equally likely to occur whenever the coin is tossed, and therefore, the probability of heads should equal .50, or $\frac{1}{2}$. By the same token, ignoring the slight differences in the lengths of the twelve months of the year, we can *speculate* that if a couple's wedding is equally likely to occur in each of the twelve months, then the probability of a June wedding should be .08 or $\frac{1}{12}$.

On the other hand, we might actually *observe* the outcomes of a long string of coin tosses and conclude, on the basis of these observations, that the probability of heads equals approximately .50, or $\frac{1}{2}$. Or we might collect extensive data on wedding months and *observe* that the probability of a June wedding actually is not only much higher than the speculated .08 or $\frac{1}{12}$, but higher than that for any other month. In this case, assuming that the observed probability is well substantiated, we would use it rather than the erroneous speculative probability.

Probability Distribution of Heights

Sometimes we'll use probabilities that are based on a mixture of observation and speculation, as in **Table 12.1.** This table shows a probability distribution of heights for all American men (derived from the observed distribution of heights for 3091 men by superimposing—and this is the speculative component—the idealized normal curve, originally shown in Figure 6.2). These probabilities indicate the proportion of men in the population who attain a particular height. They also indicate the likelihood of observing a particular height when a single man is randomly selected from the population. For exam-

Table 12.1 PROBABILITY DISTRIBUTION FOR HEIGHTS OF AMERICAN MEN

HEIGHT (INCHES)	RELATIVE FREQUENCY
≥ 75	.02
74	.02
73	.03
72	.05
71	.08
70	.11
69	.12
68	.14
67	.12
66	.11
65	.08
64	.05
63	.03
62	.02
≤ 61	.02
Total	1.00

Source: *See Figure 6.2.*

ple, the probability is .14 that a randomly selected man will stand 68 inches tall. Each of the probabilities in Table 12.1 can vary in value between zero (impossible) and one (certain). Furthermore, an entire set of probabilities always sums to one.

Probabilities of Complex Outcomes

Often you can find the probabilities of more complex outcomes by using two rules—the addition and multiplication rules—for combining the probabilities of various simple outcomes. Each rule will be introduced, in turn, to solve a problem based on the probabilities from Table 12.1.

12.2 ADDITION RULE

What's the probability that a randomly selected man will be at least 72 inches tall? That's the same as asking "What's the probability that a man will stand 72 inches tall *or* taller?" To answer this type of question, which involves a cluster of simple outcomes connected by the word *or,* merely add the respective probabilities. The probability that a man will stand 72 or more inches tall equals the sum of the probabilities (in Table 12.1) that a man will stand 72 or 73 or 74 or 75 inches or more, that is,

$$.05 + .03 + .02 + .02 = .12$$

Mutually Exclusive Outcomes

The addition of probabilities, as just stated, works only when none of the outcomes can occur together. This is true in the present case because, for instance, a single man can't stand both 72 and 73 inches tall. By the same token, a single person's blood type can't be both O and B (or any other type); nor can a single person's birth month be both January and February (or any other month). Whenever outcomes can't occur together—that is, more technically, whenever **outcomes** are **mutually exclusive**—the probability that any one of these several outcomes will occur is given by the addition rule. In other words, whenever you must find the probability for two or more sets of mutually exclusive outcomes connected by the word *or,* use the addition rule.

The **addition rule** *tells us to add together the separate probabilities of several mutually exclusive outcomes in order to find the probability that any one of these outcomes will occur.* Stated generally, the addition rule reads

Mutually exclusive outcomes

Outcomes that cannot occur together.

Addition rule

Add together the separate probabilities of several mutually exclusive outcomes to find the probability that any one of these outcomes will occur.

ADDITION RULE FOR MUTUALLY EXCLUSIVE OUTCOMES
$$\text{Pr}(A \text{ or } B) = \text{Pr}(A) + \text{Pr}(B) \qquad (12.1)$$

---

where Pr( ) refers to the probability of the outcome in parentheses.

Nonmutually Exclusive Outcomes

When outcomes aren't mutually exclusive, because they can occur together, the addition rule must be adjusted for the overlap between outcomes. For example, let's assume that drivers who cause fatal auto accidents are under the influence of alcohol with probability .40, under the influence of illegal drugs with probability .20, and under the influence of both alcohol and drugs with probability .12. To determine the probability that a driver who causes a fatal accident is *either* drunk *or* drugged, add the first two probabilities (.40 + .20 = .60), but then subtract the third probability (.60 − .12 = .48), because drivers who are both drunk and drugged were counted twice—once because they are drunk and once because they are drugged.

Ordinarily, in this book you'll be able to use the addition rule for mutually exclusive outcomes. Before doing so, however, always satisfy yourself that the various outcomes are, in fact, mutually exclusive. Otherwise, the preceding addition rule yields an inflated answer that must be reduced by subtracting the overlap between nonmutually exclusive outcomes.

*Exercise 12.1 Assuming that people are equally likely to be born during any one of the twelve months, what's the probability of Jack being born during

(a) June?

(b) any month other than June?

(c) either May or June?

Answers on page 504

12.3 MULTIPLICATION RULE

Given a probability of .12 that a randomly selected man will be at least 72 inches tall, what's the probability that two randomly selected men will be at least 72 inches tall? That's the same as asking "What's the probability that the first man will stand at least 72 inches tall *and* that the second man will stand at least 72 inches tall?"

To answer this type of question, which involves clusters of simple outcomes connected by the word *and*, merely multiply the respective probabilities. The probability that both men will stand at least 72 inches tall equals the product of the probabilities (in Table 12.1) that the first man will stand at least 72 inches tall *and* that the second man will stand at least 72 inches tall, that is,

$$(.12)(.12) = .0144$$

Notice that the probability of two simple outcomes occurring together (.0144) is smaller than the probability of either simple outcome occurring alone (.12). If you think about it, this should make sense; the combined occurrence of two simple outcomes, with probabilities other than either zero or one, always is less likely than the solitary occurrence of just one of the two simple outcomes.

Independent Outcomes

The multiplication of probabilities, as above, works only because the occurrence of one outcome has no effect on the probability of the other outcome. This is true in the present case because, when randomly selecting from the population of American men, the initial appearance of a man at least 72 inches tall has no effect, practically speaking, on the probability that the next man also will be at least 72 inches tall. By the same token, the birth of a girl in a family has no effect on the probability (approximately .50) that the next family addition also will be a girl, and the winning lottery number for this week has no effect on the probability that a particular lottery number will be a winner for the next week. Whenever one outcome has no effect on the other—that is, more technically, whenever **outcomes** are **independent**—the probability of the combined or joint occurrence of both outcomes is given by the multiplication rule.

Whenever you must find the probability for two or more sets of independent outcomes connected by the word *and,* use the multiplication rule. The **multiplication rule** *tells us to multiply together the separate probabilities of several independent outcomes in order to find the probability that these outcomes will occur together.* Stated generally, for the independent outcomes A and B, the multiplication rule reads

---

Independent outcomes

The occurrence of one outcome has no effect on the probability that the other outcome will occur.

Multiplication rule

Multiply together the separate probabilities of several independent outcomes to find the probability that these outcomes will occur together.

> **MULTIPLICATION RULE FOR INDEPENDENT OUTCOMES**
> $$\Pr(A \text{ and } B) = [\Pr(A)] \, [\Pr(B)] \tag{12.2}$$

*Exercise 12.2 Still assuming that people are equally likely to be born during any of the twelve months, and also assuming (possibly over the objections of astrological fans) that the birthdays of married couples are independent, what's the probability of

(a) the husband being born during January and the wife being born during February?

(b) both husband and wife being born during December?

(c) both husband and wife being born during spring (April or May)? (*Hint:* First find the probability of just one person being born during April or May.)

Answers on page 504

Dependent Outcomes

When the occurrence of the one outcome affects the probability of the other outcome, these outcomes are dependent. Although the heights of randomly selected pairs of men are independent, the heights of brothers are dependent—knowing, for instance, that one person is relatively tall increases the probability that his brother also will be relatively tall. Among drivers who cause fatal accidents, alcohol and illegal drugs are dependent insofar as, for instance, knowing that a driver is drunk changes (increases) the probability that the driver also is drugged.

Conditional Probabilities

Before multiplying to obtain the probability that two dependent outcomes occur together, the probability of the second outcome must be adjusted to reflect its dependency on the prior occurrence of the first outcome. This new probability is the **conditional probability** of the second outcome, given the first outcome. Examples of conditional probabilities are the probability that you'll earn a grade of A in a course, given that you've already gotten an A on the midterm, or the probability that you'll graduate from college, given that you've already completed the first two years. Notice that, in both examples, these conditional probabilities are different—they happen to be larger—than the regular or unconditional probabilities of your earning a grade of A (without knowing your grade on the midterm) or of graduating from college (without knowing whether you've completed the first two years). Incidentally, a conditional probability also can be smaller than its corresponding unconditional probability—such is the case for the conditional probability that you'll earn a grade of A in a course, given that you've already gotten (alas) a D on the midterm.

If, as suggested earlier, being drunk and drugged are dependent outcomes among drivers who cause fatal accidents, then it would be incorrect simply to use the multiplication rule for two independent outcomes. More specifically, it would be incorrect simply to multiply the observed unconditional probability of being drunk (.40) and the observed unconditional probability of being drugged (.20), and to conclude that $(.40)(.20) = .08$ is the probability of drivers being both drunk *and* drugged. Instead, first you must determine a conditional probability, such as the conditional probability of being drugged, given that the driver is known to be drunk. In other words, among the set of drivers who cause fatal accidents, you must go beyond knowing merely the proportion of drugged drivers (the unconditional probability of being drugged) to find the proportion of drugged drivers *among the subset of drunk drivers* (the conditional probability of being drugged, given that the driver is drunk). For example, extensive data for fatal accidents might reveal that, although 20 percent of *all* drivers are drugged (for an observed unconditional probability of .20), 30 out of every 100 *drunk* drivers are also drugged (for an observed conditional probability of .30). Therefore, it would be correct to multiply the observed unconditional probability of being drunk (.40) and the observed conditional probability of being drugged, given that the driver is drunk (.30), and to conclude that $(.40)(.30) = .12$, not .08, is the probability of drivers being both drunk *and* drugged.

Ordinarily, in this book you'll be able to use the multiplication rule for independent outcomes (including when it appears, slightly disguised, in Chapter 24 as a key ingredient in the important statistical test known as chi-square). Before using this rule to calculate the probabilities of complex events, however, satisfy yourself—mustering any information at your disposal, whether speculative or observational—that the various outcomes lack any obvious dependency. That is, satisfy yourself that, just as the outcome of the last coin toss has no obvious effect on the outcome of the next toss, the prior occurrence of one event has no obvious effect on the occurrence of the other event. Otherwise, proceed only if one outcome can be expressed, *most likely on the basis of some data collection,* as a conditional probability of the other.[1]

*Exercise 12.3   Assume, just for the sake of this exercise, that there is a tendency for married couples to have been born during the same month. Furthermore, we wish to calculate the probability of a husband and wife both being born during December.

(a) It would be inappropriate to use the multiplication rule for independent outcomes. (True or false?)

(b) According to the information above, the *conditional* probability of a person being born in December, given that his or her spouse was, in fact, born in December, *is smaller than, equals, or is larger than* [choose one] the *unconditional* probability of a person being born in December.

(c) The probability of a married couple both being born during December *is smaller than, equals, or is larger than* $(\frac{1}{12})(\frac{1}{12}) = \frac{1}{144}$.

(d) With only the above information, it would be possible to calculate the actual probability of a married couple both being born during December. (True or false?)

Answers on page 504

12.4 PROBABILITY AND STATISTICS

As was mentioned at the beginning of this chapter, probability considerations permeate the remainder of this book. From one perspective (that assumes all outcomes are random), statisticians help us decide whether an observed outcome, such as a correlation, is large enough to qualify as a rare outcome. With the aid of some theoretical curves, such as the normal curve (which now serves as a model for random outcomes), statisticians assign a probability to the observed outcome. Depending on the size of this probability, the observed outcome is viewed as either common or rare.

Common Outcomes

Generally speaking, a common outcome suggests that nothing special may be happening in the underlying population. For instance, it suggests that the observed correlation coefficient—whatever its value—could have originated

----

[1] Don't confuse independent and dependent *outcomes* with independent and dependent *variables*. Independent and dependent *outcomes* refer to whether the occurrence of one outcome affects the probability that the other outcome will occur and dictate the precise form of the multiplication rule. On the other hand, as described in Chapter 1, independent and dependent *variables* refer to the manipulated and measured variables in experiments. Usually, the context—whether calculating the probabilities of complex outcomes or describing the essential features of an experiment—will make the meanings of these terms clear.

from a population where the true correlation coefficient equals zero and, there-fore, that any comparable study will just as likely produce, depending on the luck of the draw, either a positive or a negative correlation coefficient. In other words, the observed correlation should not be taken seriously because, in the language of statisticians, it lacks "statistical significance."

Rare Outcomes

On the other hand, a rare outcome suggests that something special could be happening in the underlying population. For instance, it suggests that the ob-served correlation coefficient probably originates from a population where the true correlation coefficient equals some nonzero value and, therefore, any com-parable study will most likely produce a correlation coefficient with the same sign and, depending on the luck of the draw, a value in the neighborhood of the original observed correlation coefficient. In other words, the observed correla-tion should be taken seriously because it has "statistical significance." More about this in later chapters.

Common or Rare?

As an aid to determining whether random outcomes are common or rare, statisticians interpret different *proportions of area under theoretical curves,* such as the normal curve shown in **Figure 12.1,** *as probabilities of random out-comes.* For instance, the standard normal table indicates that .9500 is the proportion of total area between z scores of -1.96 and $+1.96$. (Verify this proportion by referring to Table A in Appendix D and, if necessary, to the latter

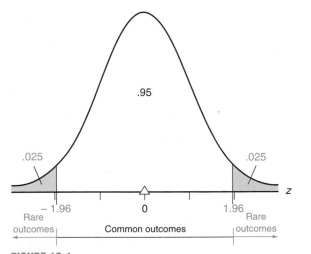

FIGURE 12.1
One possible model for determining common and rare outcomes.

part of Section 7.6.) Accordingly, the probability of a randomly selected z score anywhere between ± 1.96 equals .95. Because it should happen about 95 times out of 100, this is a fairly *common* outcome signifying that nothing special may be happening in the underlying population. On the other hand, because the standard normal curve indicates that .025 is the proportion of total area above a z score of $+1.96$, and also .025 is the proportion of total area below a z score of -1.96, then the probability of a randomly selected z score anywhere beyond either $+1.96$ or -1.96 equals .05 (from .025 + .025, thanks to the addition rule). Because it should happen only about 5 times in 100, this is a *rare* outcome signifying that something special probably is happening in the underlying population.

At this point, you're not expected to understand the rationale behind the preceding perspective, but merely that, once identified with a particular outcome, the proportion of area under a curve will be interpreted as the probability of that outcome. Furthermore, because the probability of an outcome has important implications for the underlying population, probabilities play a key role in inferential statistics.

*Exercise 12.4   Referring to the standard normal table (Table A, Appendix D), find the probability that a randomly selected z score will be

(a) above 1.96.

(b) below -1.96.

(c) either above 1.96 or below -1.96.

(d) between -1.96 and 0 or between 0 and 1.96.

(e) either above 2.58 or below -2.58.

(f) either above 3.30 or below -3.30.

Answers on page 504

Summary
. . . .

The probability of an outcome specifies the proportion of times that this outcome is likely to occur.

Whenever you must find the probability of sets of mutually exclusive outcomes connected with the word *or,* use the addition rule: Add together the separate probabilities of each of the mutually exclusive outcomes to find the probability that any one of these outcomes will occur. Whenever outcomes aren't mutually exclusive, the addition rule must be adjusted for the overlap between outcomes.

Whenever you must find the probability of sets of independent outcomes connected with the word *and,* use the multiplication rule: Multiply together the separate probabilities of each of the independent outcomes to find the probability that these outcomes will occur together. Whenever outcomes are dependent, the multiplication rule must be adjusted by using the conditional probability of the second outcome, given the occurrence of the first outcome.

In inferential statistics, proportions of area under various theoretical curves are interpreted as probabilities.

Important Terms
. . . .

Probability	**Independent outcomes**
Mutually exclusive outcomes	**Multiplication rule**
Addition rule	**Conditional probability**

REVIEW EXERCISES

*12.5 The probability of a boy being born equals .50, or $\frac{1}{2}$, as does the probability of a girl being born. For a randomly selected family with two children, what's the probability of

(a) two boys, that is, a boy and a boy? (Reminder: Before using either the addition or multiplication rule, satisfy yourself that the various outcomes are either mutually exclusive or independent, respectively.)

(b) two girls?

(c) either two boys or two girls?
   Answers on page 504

12.6 Assume the same probabilities as in Exercise 12.5. For a randomly selected family with three children, what's the probability of

(a) three boys?

(b) three girls?

(c) either three boys or three girls?

(d) neither three boys nor three girls? *Hint:* Exercise 12.6(d) can be answered indirectly by first finding the opposite of the specified outcome, then subtracting from one.

12.7 A startling discovery was made by a mathematician working in operations research for the British Bomber Command during World War II (Freeman Dyson, *Disturbing the Universe* [New York: Harper & Row, 1981]). He was set to figuring the odds of survival of the bomber crews sent on night raids over Germany. A crewman had three chances in ten of completing a standard tour of duty (thirty missions).

(a) What was the probability of any crewman surviving a standard tour of duty?

(b) What was the probability of any crewman *not* surviving a standard tour of duty?

(c) Survivors of the first tour of duty were assigned to a second tour of duty (another thirty missions). If, as Dyson's data suggested, the same probabilities existed for each tour of duty—in other words, successive tours of duty can be treated as independent outcomes—what was the probability of a crewman surviving two tours of duty?

(d) What was the probability of a crewman surviving the first tour but not the second?

(e) If the survivors of two tours were assigned to a third tour of duty, what was the probability of a crewman surviving three tours of duty?

*12.8 In *Against All Odds,* the TV series on statistics, Bruce Hoadley, a statistician at Bell Communication Research, discusses the catastrophic failure of the *Challenger* shuttle in 1986. Hoadley estimates that there was a *failure* probability of .02 for each of the six O-rings (designed to prevent the escape of potentially explosive burning gases from the joints of the segmented rocket boosters).

(a) What was the *success* probability of *each* O-ring?

(b) Given that the six O-rings function independently of each other, what was the probability that *all* six O-rings would succeed, that is, perform as designed? In other words, what was the success probability of the first O-ring and the second O-ring, and the third O-ring, and so forth?

(c) Given that you know the probability that all six O-rings would succeed (from the previous question), what was the probability that at least one O-ring would fail? *Hint:* Use your answer to the previous question to solve this problem.

(d) Given the abysmal failure rate revealed by your answer to the previous question, why, you might wonder, was this space mission even attempted? According to Hoadley, missile engineers thought that a secondary set of O-rings would function independently of the primary set of O-rings. If true and if the failure probability of each of the secondary O-rings was the same as that for each primary O-ring (.02), what would be the probability that *both* the primary and secondary O-rings fail at a particular joint? *Hint:* Concentrate on the present question, ignoring your answers to previous questions.

(e) In fact, under conditions of low temperature, as on the morning of the *Challenger* catastrophe, both primary and secondary O-rings lost their flexibility, and whenever the primary O-ring failed, its associated secondary O-ring also failed. Under these conditions, what would be the *conditional* probability of a secondary O-ring failure, *given* the failure of its associated primary O-ring? Note: Any probability, including a conditional probability, can vary between zero and one.

Answers on page 505

12.9 A sensor is used to monitor the performance of a nuclear reactor. The sensor accurately reflects the state of the reactor with a probability of .97. But with a probability of .02, it gives a false alarm (by reporting excessive radiation even though the reactor is performing normally), and with a probability of .01, it misses excessive radiation (by failing to report excessive radiation even though the reactor is performing abnormally).

(a) What is the probability that a sensor will give an incorrect report, that is, either a false alarm or a miss?

(b) To reduce costly shutdowns caused by false alarms, management introduces a second completely independent sensor, and the reactor is only shut down when both sensors report excessive radiation. (According to this perspective, solitary reports of excessive radiation should be viewed as false alarms and ignored, because both sensors provide accurate information much of the time.) What is the new probability that the reactor will be shut down because of simultaneous false alarms by both the first and second sensors?

(c) Being more concerned about failures to detect excessive radiation, someone who lives near the nuclear reactor proposes an entirely different strategy: shut down the reactor whenever either sensor reports excessive radiation. (According to this point of view, even a solitary report of excessive radiation should trigger a shut down, because a failure to detect excessive radiation is potentially catastrophic.) If this policy were adopted, what is the new probability that excessive radiation will be missed simultaneously by both the first and second sensors?

*12.10 Let's assume that the *regular* or *unconditional* probability of being left-handed in the general population equals .10. Furthermore, let's assume that the *conditional* probability of being left-handed, given that a person is a major league baseball player, equals .18.

(a) Being left-handed and a major league baseball player are independent outcomes. (True or false?)

(b) Justify your answer.

Answers on page 505

12.11 Among 100 couples who had undergone marital counseling, 60 couples described their relationships as improved, and among this latter group, 45 couples had children.

(a) What is the probability of randomly selecting a couple who described their relationship as improved?

(b) What is the conditional probability of randomly selecting a couple with children, given that their relationship was described as improved?

(c) What is the conditional probability of randomly selecting a couple without children, given that their relationship was described as improved?

(d) Multiply two probabilities to find the probability of randomly selecting a couple that described their relationship as improved and who had children.

CHAPTER 13

Sampling Distribution of the Mean

13.1   AN EXAMPLE

13.2   CREATING A SAMPLING DISTRIBUTION FROM SCRATCH

13.3   SOME IMPORTANT SYMBOLS

13.4   MEAN OF ALL SAMPLE MEANS ($\mu_{\bar{X}}$)

13.5   STANDARD ERROR OF THE MEAN ($\sigma_{\bar{X}}$)

13.6   SHAPE OF THE SAMPLING DISTRIBUTION

13.7   WHY THE CENTRAL LIMIT THEOREM WORKS

13.8   OTHER SAMPLING DISTRIBUTIONS

Summary

Important Terms

Review Exercises

....
13.1 AN EXAMPLE

As was mentioned in Exercise 8.8 on page 130, there's a good chance that you've taken the Scholastic Assessment Test (SAT), and you might even remember your scores for your verbal and math aptitudes. On a nationwide basis, the verbal scores for all college-bound students during a recent year were distributed around a mean of 500 with a standard deviation of 110. An investigator at a university wishes to test the claim that, on the average, the SAT verbal scores for local freshmen equals the national average of 500. His task would be straightforward if in fact the verbal scores for all local freshmen were readily available. Then, after calculating the mean score for all local freshmen, a direct comparison would indicate whether, on the average, local freshmen score below, at, or above the national average.

Given our current concern, let's assume that for some reason, it's not possible to obtain scores for the entire freshman class. Instead, SAT verbal scores are obtained for a random sample of 100 students from the local population of freshmen, and the mean score for this sample equals 533. If each sample were an exact replica of the population, generalizations from the sample to the population also would be most straightforward. Having observed a mean verbal score of 533 for a sample of 100 freshmen, we could have concluded, without even a pause, that the mean verbal score for the entire freshman class also equaled 533 and, therefore, exceeded the national average.

Concept of a Sampling Distribution

Random samples rarely represent the underlying population exactly. Even a mean verbal score of 533 could originate, just by chance, from a population of freshmen whose mean equals the national average of 500. Accordingly, generalizations from a single sample to a population are much more tentative. Indeed, generalizations are based not merely on the single sample mean of 533 but also on its distribution—a distribution of sample means for all possible random samples. Representing the statistician's model of random outcomes,

....

Sampling distribution of the mean

Probability distribution of means for all possible random samples of a given size from some population.

> the *sampling distribution of the mean* refers to the probability distribution of means for all possible random samples of a given size from some population.

In effect, this distribution describes the entire spectrum of sample means that could occur just by chance and thereby provides a frame of reference for generalizing from a single sample mean to a population mean. In other words, the sampling distribution of the mean allows us to determine whether, among a set of random possibilities, the one observed sample mean can be viewed as a *common* outcome or as a *rare* outcome. If the sample mean of 533 can be viewed as a *common* outcome in the sampling distribution of the mean, then the difference between 533 and 500 isn't large enough to signify that anything special, relative to the national average, is happening in the underlying population—the mean verbal score for the entire freshman class could be the same as the national average. On the other hand, if the sample mean of 533 can be viewed as a *rare* outcome in the sampling distribution of the mean, then the difference between 533 and 500 is large enough to signify that something special, relative to the national average, probably is happening in the underlying population—more specifically, that the mean verbal score for the entire freshman class probably exceeds the national average.

All Possible Random Samples

When attempting to generalize from a single sample mean to a population mean, we must consult the sampling distribution of the mean. In the present case, this distribution is based on *all possible* random samples, each of size 100, that can be taken from the local population of freshmen. "All possible random samples" refers not to the number of samples of size 100 required to *survey completely* the local population of freshmen but to the number of different ways in which a *single* sample of size 100 can be selected from this population.

"All possible random samples" tends to be a huge number. For instance, if the local population contained at least 1000 freshmen, the total number of possible random samples, each of size 100, would be astronomical in size. The 301 digits in this number would, by comparison, dwarf even the national debt. Even with the aid of a computer, it would be a horrendous task to construct this sampling distribution from scratch, itemizing each mean for all possible random samples.

Preview

Fortunately, statistical theory supplies us with considerable information about the sampling distribution of the mean, as will be discussed in the remainder of this chapter. Armed with this information about sampling distributions, we'll return to the current example in the next chapter and test the claim that the mean verbal score for the local population of freshmen equals the national average of 500. Only at that point—and not at the end of this chapter—should you expect to understand the role of sampling distributions in practical applications.

13.2 CREATING A SAMPLING DISTRIBUTION FROM SCRATCH

Let's establish precisely what constitutes a sampling distribution by creating one from scratch under highly simplified conditions. Imagine some ridiculously small population of four observations with values of 2, 3, 4, and 5, as shown in **Figure 13.1.** Next, itemize all possible random samples, each of size two, that could be taken from this population. There are four possibilities on the first draw from the population and also four possibilities on the second draw from the population, as indicated in **Table 13.1.**[1] The two sets of possibilities combine to yield a total of sixteen possible samples. At this point, remember, we're clarifying the notion of a sampling distribution of the mean. In

---

[1]Ordinarily, a single observation is sampled only once, that is, sampling is *without replacement*. If employed with the present, highly simplified example, however, sampling without replacement would magnify an unimportant technical adjustment.

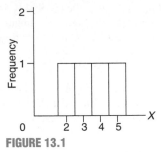

FIGURE 13.1

Group of miniature population.

Table 13.1 ALL POSSIBLE SAMPLES OF SIZE TWO FROM MINIATURE POPULATION		
All Possible Samples	**Mean ($\overline{X}$)**	**Probability**
(1) 2,2	2.0	$\frac{1}{16}$
(2) 2,3	2.5	$\frac{1}{16}$
(3) 2,4	3.0	$\frac{1}{16}$
(4) 2,5	3.5	$\frac{1}{16}$
(5) 3,2	2.5	$\frac{1}{16}$
(6) 3,3	3.0	$\frac{1}{16}$
(7) 3,4	3.5	$\frac{1}{16}$
(8) 3,5	4.0	$\frac{1}{16}$
(9) 4,2	3.0	$\frac{1}{16}$
(10) 4,3	3.5	$\frac{1}{16}$
(11) 4,4	4.0	$\frac{1}{16}$
(12) 4,5	4.5	$\frac{1}{16}$
(13) 5,2	3.5	$\frac{1}{16}$
(14) 5,3	4.0	$\frac{1}{16}$
(15) 5,4	4.5	$\frac{1}{16}$
(16) 5,5	5.0	$\frac{1}{16}$

practice, only one random sample, not sixteen possible samples, would be taken from the population; sample size would be very small relative to population size, and, of course, not all observations in the population would be known.

For each of the sixteen possible samples, Table 13.1 also lists a sample mean (found by adding the two observations and dividing by 2) and its probability of occurrence (expressed as $\frac{1}{16}$, because each of the sixteen possible samples is equally likely). When cast into a relative frequency or probability distribution, as in **Table 13.2,** the sixteen sample means constitute the sampling

**Table 13.2
SAMPLING
DISTRIBUTION OF
THE MEAN
(SAMPLES OF SIZE
TWO FROM MINIA-
TURE POPULATION)**

SAMPLE MEAN ($\overline{X}$)	PROBA- BILITY
5.0	$\frac{1}{16}$
4.5	$\frac{2}{16}$
4.0	$\frac{3}{16}$
3.5	$\frac{4}{16}$
3.0	$\frac{3}{16}$
2.5	$\frac{2}{16}$
2.0	$\frac{1}{16}$

distribution of the mean, previously defined as the probability distribution of means for all possible random samples of a given size from some population. Not all probabilities are equal in Table 13.2, you'll notice, since some values of the sample mean occur more than once among the sixteen possible samples. For instance, a sample mean value of 3.5 appears among four of sixteen possibilities and has a probability of $\frac{4}{16}$.

Probability of a Particular Sample Mean

The distribution in Table 13.2 can be consulted to determine the probability of obtaining a particular sample mean or set of sample means. The probability of a randomly selected sample mean of 5.0 equals $\frac{1}{16}$ or .0625. According to the addition rule for mutually exclusive outcomes, described in Chapter 12, the probability of a randomly selected sample mean of either 5.0 or 2.0 equals $\frac{1}{16} + \frac{1}{16} = \frac{2}{16} = .1250$. Subsequent applications in inferential statistics will always entail, in some form or another, probability statements based on sampling distributions similar to that for the mean.

Review

FIGURE 13.2 summarizes the previous discussion. It depicts the emergence of the sampling distribution of the mean from the set of all possible (sixteen) samples of size two, based on the miniature population of four observations. Familiarize yourself with this figure, as it will be referred to again in subsequent sections.

*Exercise 13.1 Imagine a very simple population consisting of only five observations: 2, 4, 6, 8, 10.

(a) List all possible samples of size two. (*Hint:* There are five possibilities (2, 4, 6, 8, 10) on the first draw from the population and also the same five possibilities (2, 4, 6, 8, 10) on the second draw from the population).

(b) Construct a relative frequency table showing the sampling distribution of the mean.

Answers on page 505

....

13.3 SOME IMPORTANT SYMBOLS

Having established precisely what constitutes a sampling distribution under highly simplified conditions, let's introduce the special symbols that identify the mean and the standard deviation of the sampling distribution of the mean. Table 13.3 also lists the corresponding symbols for the sample and the population. To facilitate your understanding of the next few chapters, it would be wise to memorize these symbols.

You're already acquainted with the English letters $\overline{X}$ and S, representing the mean and the standard deviation of any sample, and also the Greek letters μ (mu) and σ (sigma), representing the mean and standard deviation of any population. New are the Greek letters $\mu_{\overline{X}}$ (mu sub X-bar) and $\sigma_{\overline{X}}$ (sigma sub X-bar), representing the mean and the standard deviation, respectively, of the

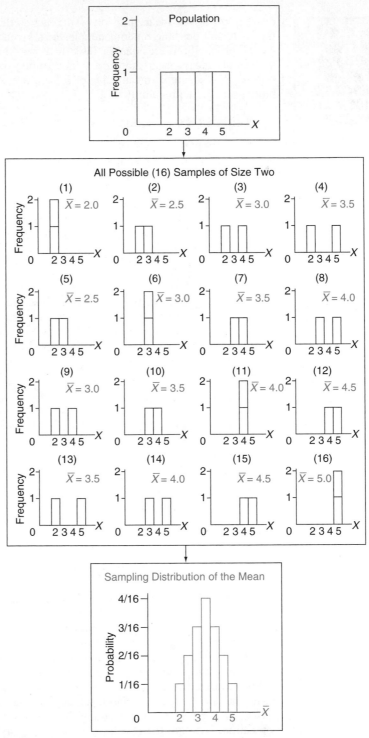

FIGURE 13.2

Emergence of the sampling distribution of the mean from all possible samples.

> **Table 13.3**
> ## SYMBOLS FOR MEAN AND STANDARD DEVIATION OF THREE TYPES OF DISTRIBUTIONS
>
> | TYPE OF DISTRIBUTION | MEAN | STANDARD DEVIATION | |
> |---|---|---|---|
> | Sample | $\bar{X}$ | S | |
> | Population | μ | σ | |
> | Sampling distribution of the mean | $\mu_{\bar{X}}$ | $\sigma_{\bar{X}}$ | $\left(\begin{array}{l}\text{standard error}\\\text{of the mean}\end{array}\right)$ |

sampling distribution of the mean. To minimize confusion, the latter term, $\sigma_{\bar{X}}$, is often referred to as the *standard error of the mean,* or simply as the *standard error.*

Significance of Greek Letters

Note that Greek letters are used to describe characteristics of both populations and sampling distributions, suggesting a common feature. Both types of distribution deal with all possibilities, that is, with *all possible observations,* as in the case of the population, or with the *means of all possible random samples,* as in the case of the sampling distribution of the mean.

With this background, let's focus on three important characteristics—the mean, standard deviation, and shape—of the sampling distribution of the mean. In subsequent chapters, these three characteristics will form the basis for applied work in inferential statistics.

*Exercise 13.2 List the special symbol for the mean of the population __(a)__ , mean of the sampling distribution of the mean __(b)__ , mean of the sample __(c)__ , standard error of the mean __(d)__ , standard deviation of the sample __(e)__ , and standard deviation of the population __(f)__ .

Answers on page 505

....

13.4 MEAN OF ALL SAMPLE MEANS ($\mu_{\bar{X}}$)

The distribution of sample means, itself, has a mean.

The mean of the sampling distribution of the mean *always* equals the mean of the population.

Expressed in symbols, we have

Mean of sampling distribution of the mean ($\mu_{\bar{x}}$)

The mean of all sample means always equals the population mean.

> **MEAN OF SAMPLING DISTRIBUTION**
> $$\mu_{\bar{X}} = \mu \qquad (13.1)$$

where $\mu_{\bar{X}}$ represents the mean of the sampling distribution, and μ represents the mean of the population.

Interchangeable Means

Because the mean of all sample means ($\mu_{\bar{X}}$) always equals the mean of the population (μ), these two terms are used interchangeably in inferential statistics. Any claims about the population mean can be transferred directly to the mean of the sampling distribution, and vice versa. If the mean verbal score for the local population of freshmen equals the national average of 500, as claimed, then the mean of the sampling distribution also automatically will equal 500. By the same token, it's permissible to view the one observed sample mean of 533 as a deviation from either the mean of the sampling distribution or the mean of the population. It should be apparent, therefore, that *whether an expression involves either $\mu_{\bar{X}}$ or μ, it reflects, at most, a difference in emphasis on either the sampling distribution or the population, respectively, rather than any difference in numerical value.*

Explanation

Although important, it's not particularly startling that the mean of all sample means equals the population mean. As can be seen in Figure 13.2, samples are not exact replicas of the population, and most sample means are either larger or smaller than the population mean (equal to 3.5 in Figure 13.2). By taking the mean of all sample means, however, you effectively neutralize chance differences between sample means and retain a value equal to the population mean.

*Exercise 13.3 Indicate whether the following statements, referring to the mean of all sample means, $\mu_{\bar{X}}$, are true or false.

(a) It always equals the value of a particular sample mean.

(b) It equals 100 if, in fact, the population mean equals 100.

(c) It usually equals the value of a particular sample mean.

(d) It is the mean of the sampling distribution of the mean.

(e) It usually doesn't equal the value of a particular sample mean.

(f) It is interchangeable with the population mean.

Answers on page 505

13.5 STANDARD ERROR OF THE MEAN ($\sigma_{\bar{X}}$)

The distribution of sample means also has a standard deviation, referred to as the standard error of the mean.

The standard error of the mean equals the standard deviation of the population divided by the square root of the sample size.

Expressed in symbols,

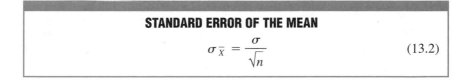

STANDARD ERROR OF THE MEAN

$$\sigma_{\bar{X}} = \frac{\sigma}{\sqrt{n}} \qquad\qquad (13.2)$$

---

where $\sigma_{\bar{X}}$ represents the standard error of the mean; σ represents the standard deviation of the population; and, as usual, n represents the sample size.

Special Type of Standard Deviation

Standard error of the mean ($\sigma_{\bar{x}}$)

Being the standard deviation of the sampling distribution of the mean, it's a rough measure of the average amount by which sample means deviate from the population mean.

The **standard error of the mean** serves as a special type of standard deviation that measures variability in the sampling distribution. It supplies us with a standard, much like a yardstick, that describes the amount by which sample means deviate from the mean of the sampling distribution. (Because these deviations occur just by chance, whenever a random sample is not an exact replica of the population, these deviations are often referred to as *errors*— hence the term *standard error of the mean*.)

You might find it helpful to think of the standard error of the mean as a rough measure of the average amount by which sample means deviate from the mean of the sampling distribution (or the population mean).

Insofar as the shape of the distribution of sample means approximates a normal curve, as described in the next section, about 68 percent of all sample means deviate less than one standard error from the mean of the distribution of sample means, whereas only about 5 percent of all sample means deviate more than two standard errors from the mean of this distribution.

Effect of Sample Size

A most important implication of Formula 13.2 is that whenever the sample size equals two or more, the variability of the sampling distribution of the mean is less than that in the population. A modest demonstration of this effect appears in Figure 13.2, in which the means of all possible samples cluster closer to the population mean (equal to 3.5 in Figure 13.2) than do the four observations in the population. A more dramatic demonstration occurs with larger sample sizes. Earlier in this chapter, for instance, 110 was given as the value of σ, the population standard deviation for verbal SAT scores. Much smaller is the variability in the sampling distribution of mean verbal scores, each based on samples of 100 freshmen. According to Formula 13.2,

$$\sigma_{\bar{X}} = \frac{\sigma}{\sqrt{n}} = \frac{110}{\sqrt{100}} = \frac{110}{10} = 11$$

In the present example, there is a tenfold reduction in variability, from 110 to 11, when our focus shifts from the population to the sampling distribution.

According to Formula 13.2, any increase in sample size translates into a smaller standard error and, therefore, into a *new* sampling distribution with less variability. In other words, with a larger sample size, sample means cluster more closely about the mean of the sampling distribution (and about the population mean), allowing more precise generalizations from samples to populations.

Explanation

It's not surprising that variability should be less in distributions of sample means than in populations. After all, the population standard deviation reflects variability among *individual observations,* and it's directly affected by any relatively large or small observations within the population. On the other hand, the standard error of the mean reflects variability among *sample means,* each of which represents a collection of individual observations. The appearance of relatively large or small observations within a particular sample tends to affect the sample mean only slightly, because of the stabilizing presence in the same sample of other, more moderate observations or extreme observations in the opposite direction. This stabilizing effect becomes even more pronounced with larger sample sizes.

*Exercise 13.4 Indicate whether the following statements, referring to the standard error of the mean, are true or false.

(a) It roughly measures the average amount by which sample means deviate from the population mean.

(b) It measures variability in a particular sample.

(c) It increases in value with larger sample sizes.

(d) It equals the population standard deviation.

(e) It equals 5, given that $\sigma = 40$ and $n = 64$.

(f) It is the standard deviation of the sampling distribution of the mean.
Answers on page 505

13.6 SHAPE OF THE SAMPLING DISTRIBUTION

A product of statistical theory,

Central limit theorem

A statement that the shape of the sampling distribution of the mean will approximate a normal curve if the sample size is sufficiently large.

the *central limit theorem,* in its simplest form, states that the shape of the sampling distribution of the mean will approximate a normal curve if the sample size is sufficiently large.

According to this theorem, it doesn't matter whether the shape of the parent population is normal, positively skewed, negatively skewed, or some nameless, bizarre shape, as long as the sample size is sufficiently large. What constitutes "sufficiently large" depends on the shape of the parent population. If the shape of the parent population is normal, then any sample size (even a sample size of one) will be sufficiently large. Otherwise, depending on the degree of non-normality in the parent population, a sample size between 25 and 100 is sufficiently large.

Examples

For a population with a non-normal shape, as in the top of Figure 13.2, the shape of the sampling distribution reveals a preliminary drift toward normality—that is, a shape having a peak in the middle with tapered flanks on either side—even for very small samples of size 2. Essentially this same drift describes the shape of the sampling distributions, also based on samples of size 2, for two other populations with non-normal shapes, as shown in **Figure 13.3.** When the sample size equals 25, the shapes of these two sampling distributions closely approximate normality, as suggested at the bottom of Figure 13.3.

Earlier in this chapter, 533 was given as the mean verbal score for a random sample of 100 freshmen. Because this sample size satisfies the requirements of the central limit theorem, we can view the sample mean of 533 as originating from a sampling distribution whose shape approximates a normal curve, even though we lack information about the shape of the population of verbal scores for the entire freshman class. It will be possible, therefore, to make precise

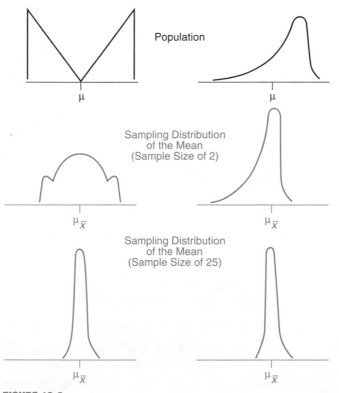

FIGURE 13.3
Effect of central limit theorem.

statements about this sampling distribution, as described in Chapter 14, by referring to the table for the standard normal curve.

13.7 WHY THE CENTRAL LIMIT THEOREM WORKS

In a normal curve, you'll recall, intermediate values are the most prevalent, and more extreme values, either larger or smaller, occupy the tapered flanks. Why, when the sample size is large, does the sampling distribution approximate a normal curve, even though the parent population might be non-normal?

Many Sample Means with Intermediate Values

When the sample size is large, it's *most likely* that any single sample will contain the full spectrum of small, intermediate, and large scores from the parent population, *whatever its shape*. The calculation of a mean for this type of sample tends to neutralize or dilute the effects of any extreme scores, and the sample mean emerges with some intermediate value. Accordingly, intermediate values prevail in the sampling distribution, and they cluster around a peak frequency representing the most common or modal value of the sample mean, as suggested at the bottom of Figure 13.3.

Few Sample Means with Extreme Values

To account for the rarer sample mean values in the tails of the sampling distribution, focus on those relatively infrequent samples that, just by chance, contain less than the full spectrum of scores from the parent population. Sometimes, because of the relatively large number of extreme scores in a particular direction, the calculation of a mean only slightly dilutes their effect, and the sample mean emerges with some more extreme value. The likelihood of obtaining extreme sample mean values declines with the extremity of the value, producing the smoothly tapered, slender tails that characterize a normal curve.

*Exercise 13.5 Indicate whether the following statements, referring to the central limit theorem, are true or false.

(a) It states that, with sufficiently large sample sizes, the shape of the population is normal.

(b) It states that, regardless of sample size, the shape of the sampling distribution of the mean is normal.

(c) It states that, with sufficiently large sample sizes, the shape of the sampling distribution of the mean is normal.

(d) It requires that the shape of the population be known.

(e) It ensures that the shape of the sampling distribution of the mean equals the shape of the population.

(f) It applies to the shape of the sampling distribution—not to the shape of the population and also not to the shape of the sample.

Answers on page 506

13.8 OTHER SAMPLING DISTRIBUTIONS

For the Mean

There are many different sampling distributions of means. Because a new sampling distribution is created by a switch to a new population, the population should, therefore, always be specified for any particular sampling distribution of the mean. For any one population, there are as many different sampling distributions of the mean as there are possible sample sizes. Although each of these sampling distributions has the same mean, the value of the standard error always differs and depends upon the size of the sample.

For Other Measures

Furthermore, there are many different sampling distributions for measures other than a single mean. For instance, there are sampling distributions for medians, proportions, standard deviations, variances, and correlations, as well as for differences between pairs of means, pairs of proportions, and so forth.

INTERNET DEMONSTRATION

Go to the Web site for this book **(http://darwin.cwru.edu/~witte/statistics)** and click on **Sampling Distributions** to create from scratch the sampling distribution of the mean and also to observe how this distribution changes with shifts in population shape and sample size.

Summary

The notion of a sampling distribution is the most important concept in inferential statistics. The sampling distribution of the mean is defined as the probability distribution of means for all possible random samples of a given size from some population.

Statistical theory pinpoints three important characteristics of the sampling distribution of the mean:

1. The mean of the sampling distribution equals the mean of the population.

2. The standard deviation of the sampling distribution, that is, the standard error of the mean, equals the standard deviation of the population, divided by the square root of the sample size. An important implication of this formula is that a larger sample size translates into a sampling distribution of the mean with a smaller variability, allowing more precise generalizations from samples to populations. The standard error of the mean serves as a rough measure of the average amount by which sample means deviate from the mean of the sampling distribution (or the population mean).

3. According to the central limit theorem, the shape of the sampling distribution will approximate a normal curve if the sample size is sufficiently large. Depending on the degree of non-normality in the parent population, a sample size of between 25 and 100 is sufficiently large.

Any single sample mean can be viewed as originating from a sampling distribution whose (1) mean equals the population mean (whatever its value); whose (2) standard error equals the population standard deviation divided by the square root of the sample size; and whose (3) shape approximates a normal curve (assuming that the sample size satisfies the requirements of the central limit theorem).

Important Terms

Sampling distribution of the mean
Standard error of the mean $(\sigma_{\bar{x}})$
Central limit theorem

Mean of sampling distribution of the mean $(\mu_{\bar{x}})$

REVIEW EXERCISES

13.6 A random sample tends not to be an exact replica of its parent population. This fact has several implications; indicate which are true and which are false.

(a) A random sample of one observation usually is as representative as a random sample of ten observations is.

(b) In practice, more than one random sample often is taken.

(c) Insofar as it misrepresents the parent population, a random sample can cause an erroneous generalization.

(d) Several random samples from the same population usually have the same sample mean.

(e) In practice, the mean of a single random sample is evaluated relative to the means for all possible random samples.

(f) A more representative sample can be obtained by handpicking (rather than randomly selecting) observations.

(g) All possible random samples can include a few samples that are exact replicas of the population, but most samples aren't exact replicas.

13.7 Define the sampling distribution of the mean.

13.8 Indicate whether the following statements, referring to the sampling distribution of the mean, are true or false.

(a) It is always constructed from scratch, even when the populations are large.

(b) It serves as a bridge to aid generalizations from a sample to a population.

(c) It is the same as the sample mean.

(d) It always reflects the shape of the underlying population.

(e) It has a mean that itself always coincides with the population mean.

(f) It is a device used to determine what can happen, just by chance, when samples are random.

(g) It remains unchanged even with shifts to new populations or sample sizes.

(h) It supplies a spectrum of possibilities against which to evaluate the one observed sample mean.

(i) It remains unchanged even with shifts to a new measure, such as the sample median.

(j) It tends to cluster more closely about the population mean with increases in sample size.

(k) It is based on the notion that when the samples are random, chance dictates the composition of any particular sample.

13.9 Someone claims that because the mean of the sampling distribution equals the population mean, any single sample mean must also equal the population mean. Any comment?

13.10 (a) A random sample of size 144 is taken from the local population of grade-school children. Each child estimates the number of hours per week spent watching TV. At this point, what can be said about the sampling distribution of the mean?

(b) Assume that a standard deviation, σ, of 8 hours describes the TV estimates for the local population of schoolchildren. At this point, what can be said about the sampling distribution of the mean?

(c) Assume that a mean, μ, of 21 hours does describe the TV estimates for the local population of schoolchildren. Now what can be said about the sampling distribution of the mean?

(d) Roughly speaking, the sample means in the sampling distribution of the mean should deviate, on the average, about _____ hours from the mean of the sampling distribution and from the mean of the population.

(e) About 95 percent of the sample means in this sampling distribution should be between _____ hours and _____ hours.

CHAPTER
14

Introduction to Hypothesis Testing: The *z* Test

14.1   TESTING A HYPOTHESIS ABOUT SAT SCORES
14.2   *z* TEST FOR A POPULATION MEAN
14.3   STEP-BY-STEP PROCEDURE
14.4   STATEMENT OF THE RESEARCH PROBLEM
14.5   NULL HYPOTHESIS (H_0)
14.6   ALTERNATIVE HYPOTHESIS (H_1)
14.7   DECISION RULE
14.8   CALCULATIONS
14.9   DECISION
14.10   INTERPRETATION

Summary

Important Terms

Review Exercises

14.1 TESTING A HYPOTHESIS ABOUT SAT SCORES

In the previous chapter, we postponed a test of the hypothesis that the mean SAT verbal score for all local freshmen equals the national average of 500. Now, given a mean verbal score of 533 for a random sample of 100 freshmen, let's test the hypothesis that, with respect to the national average, nothing special is happening in the local population. Insofar as an investigator usually suspects just the opposite—namely, that something special *is* happening in the local population—he hopes to reject the hypothesis that nothing special is happening, henceforth referred to as the "null" hypothesis (and defined more formally in a later section).

Hypothesized Sampling Distribution

If the null hypothesis is true, then the distribution of sample means—that is, the sampling distribution of the mean for all possible random samples, each of size 100, from the local population of freshmen—will be centered about the national average of 500. (Remember that the mean of the sampling distribution always equals the population mean.) In **Figure 14.1,** this sampling distribution is referred to as the *hypothesized* sampling distribution, because its mean equals 500, the hypothesized mean verbal score for the local population of freshmen. In other words, this distribution is hypothesized because it's centered about the null hypothesized value of 500.

Anticipating the key role of the hypothesized sampling distribution in our hypothesis test, let's spotlight two more properties of this distribution:

1. In Figure 14.1, several vertical lines appear, at intervals of size 11, on either side of the hypothesized population mean of 500. These intervals reflect the size of the standard error of the mean, $\sigma_{\bar{X}}$. To verify this fact,

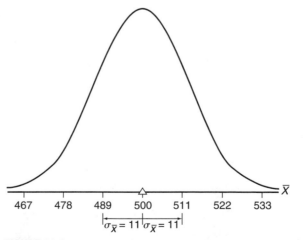

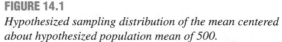

FIGURE 14.1

Hypothesized sampling distribution of the mean centered about hypothesized population mean of 500.

originally demonstrated in Chapter 13, substitute 110 for the population standard deviation, σ, and 100 for the sample size, n, in Formula 13.2 to obtain

$$\sigma_{\bar{X}} = \frac{\sigma}{\sqrt{n}} = \frac{110}{\sqrt{100}} = \frac{110}{10} = 11$$

2. Also notice that the shape of the hypothesized sampling distribution in Figure 14.1 approximates a normal curve, because the sample size of 100 is large enough to satisfy the requirements of the central limit theorem. Eventually, with the aid of normal curve tables, we'll be able to construct boundaries that determine whether sample means qualify as common or rare outcomes under the null hypothesis.

The null hypothesis that the population mean for the freshman class equals 500 is *tentatively* assumed to be true. It is tested by determining whether the one observed sample mean qualifies as a common outcome or a rare outcome in the hypothesized sampling distribution of Figure 14.1.

Common Outcomes

An observed sample mean qualifies as a *common* outcome if the difference between its value and that of the hypothesized population mean is small enough to be viewed as a probable outcome under the null hypothesis.

That is, a sample mean qualifies as a common outcome if it doesn't deviate too far from the hypothesized population mean but appears to emerge from the dense concentration of possible sample means in the middle of the sampling distribution. *A common outcome signifies a lack of evidence that, with respect to the null hypothesis, something special is happening in the underlying population.* Because now there is no compelling reason for rejecting the null hypothesis, it is retained.

Rare Outcomes

An observed sample mean qualifies as a *rare* outcome if the difference between its value and the hypothesized value is too large to be reasonably viewed as a probable outcome under the null hypothesis.

That is, a sample mean qualifies as a rare outcome if it deviates too far from the hypothesized mean and appears to emerge from the sparse concentration of possible sample means in either tail of the sampling distribution. *A rare outcome signifies that, with respect to the null hypothesis, something special probably is happening in the underlying population.* Because now there are grounds for suspecting the null hypothesis, it is rejected.

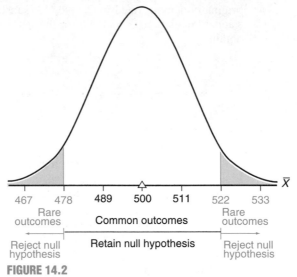

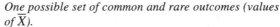

467   478   489   500   511   522   533
Rare   Rare
outcomes   Common outcomes   outcomes

Reject null   Retain null hypothesis   Reject null
hypothesis   hypothesis

FIGURE 14.2
One possible set of common and rare outcomes (values of $\overline{X}$).

Boundaries for Common and Rare Outcomes

Superimposed on the hypothesized sampling distribution in **Figure 14.2** is one possible set of boundaries for common and rare outcomes, expressed in values of $\overline{X}$. (Techniques for constructing these boundaries are described in Section 14.7.) If the one observed sample mean is located between approximately 478 and 522, it will qualify as a common outcome under the null hypothesis and the null hypothesis will be retained. If, however, the one observed sample mean is greater than 522 or less than 478, it will qualify as a rare outcome under the null hypothesis and the null hypothesis will be rejected. Because the observed sample mean of 533 does exceed 522, the null hypothesis is rejected. On the basis of the present test, it's unlikely that the sample of 100 freshmen, with a mean verbal score of 533, originates from a population whose mean equals the national average of 500, and, therefore, the investigator can conclude that the mean verbal score for the local population of freshmen probably differs from (exceeds) the national average.

Sampling distribution of z

The distribution of z values that would be obtained if a value of z were calculated for each sample mean for all possible random samples of a given size from some population.

14.2 z TEST FOR A POPULATION MEAN

For the hypothesis test with SAT verbal scores, it's customary to base the test not on the hypothesized sampling distribution of $\overline{X}$ shown in Figure 14.2, but rather on its counterpart, the hypothesized sampling distribution of z shown in **Figure 14.3.** Now z represents a variation on the familiar standard score, and it displays all of the properties of standard scores described in Chapters 6 and 8. Furthermore, like the sampling distribution of $\overline{X}$, the **sampling distribution of z** *represents the distribution of z values that would be obtained if a value of z*

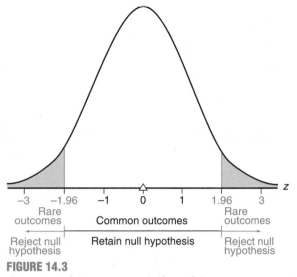

FIGURE 14.3

Common and rare outcomes (values of z*).*

were calculated for each sample mean for all possible random samples of a given size from some population.

The conversion from $\overline{X}$ to z yields a distribution that approximates the standard normal curve in Table A of Appendix D, because, as indicated in Figure 14.3, the original hypothesized population mean (500) emerges as a z score of 0 and the original standard error of the mean (11) emerges as a z score of 1. The shift from $\overline{X}$ to z eliminates the original units of measurement and standardizes the hypothesis test across all situations without, however, affecting the test results.

Reminder: Converting a Raw Score to z

To convert a raw score into a standard score (also described in Chapter 6), express the raw score as a distance from its mean (by subtracting the mean from the raw score), and then split this distance into standard deviation units (by dividing by the standard deviation). Expressing this definition as a word formula, we have

$$Standard\ score = \frac{raw\ score\ -\ mean}{standard\ deviation}$$

in which, of course, the standard score indicates the deviation of the raw score in standard deviation units, above or below the mean.

Converting a Sample Mean to z

The z for the present situation emerges as a slight variation of this word formula: Replace the *raw score* with the one observed sample mean; replace the *mean* with the mean of the sampling distribution, that is, the hypothesized population mean μ_{hyp}; and replace the *standard deviation* with the standard error of the mean $\sigma_{\overline{X}}$. Now

---

z RATIO FOR SINGLE POPULATION MEAN

$$z = \frac{\overline{X} - \mu_{hyp}}{\sigma_{\overline{X}}} \tag{14.1}$$

---

where z indicates the deviation of the observed sample mean in standard error units, above or below the hypothesized population mean.

To test the hypothesis for SAT verbal scores, we must determine the value of z from Formula 14.1. Given a sample mean of 533, a hypothesized population mean of 500, and a standard error of 11, we find

$$z = \frac{533 - 500}{11} = \frac{33}{11} = 3$$

The observed z of 3 exceeds the value of 1.96 specified in the hypothesized sampling distribution in Figure 14.3. Thus the observed z qualifies as a rare outcome under the null hypothesis, and the null hypothesis is rejected. The results of this test with z are the same as those for the original hypothesis test with $\overline{X}$.

Assumptions of z Test

When a hypothesis test evaluates how far the observed sample mean deviates, in standard error units, from the hypothesized population mean, as in the present example, *it's referred to as a z test* or, more accurately, as a **z test for a population mean.** This z test is accurate only when (1) the population is normally distributed or the sample size is large enough to satisfy the requirements of the central limit theorem and (2) the population standard deviation is known. In the present example, the z test is appropriate because the sample size of 100 is large enough to satisfy the central limit theorem and the population standard deviation is known to be 110.

....

z Test for a population mean

A hypothesis test that evaluates how far the observed sample mean deviates, in standard error units, from the hypothesized population mean.

---

*Exercise 14.1   Calculate the value of the z test for each of the following situations:

(a) $\overline{X} = 566$; $\sigma = 30$; $n = 36$; $\mu_{hyp} = 560$

(b) $\overline{X} = 24$; $\sigma = 4$; $n = 64$; $\mu_{hyp} = 25$

(c) $\overline{X} = 82$; $\sigma = 14$; $n = 49$; $\mu_{hyp} = 75$

(d) $\overline{X} = 136$; $\sigma = 15$; $n = 25$; $\mu_{hyp} = 146$

Answers on page 506

....

14.3 STEP-BY-STEP PROCEDURE

Having been exposed to some of the more important features of hypothesis testing, let's take a detailed look at the test for SAT verbal scores. The test procedure lends itself to a step-by-step description, beginning with a brief statement of the problem that inspired the test and ending with an interpretation of the test results. Refer to the boxed step-by-step procedure while reading the next several sections. Whenever appropriate, this format will be used in the remainder of the book.

HYPOTHESIS TEST SUMMARY: z TEST FOR A POPULATION MEAN
(SAT Verbal Scores)

Research Problem:
Does the mean SAT verbal score for all local freshmen differ from the national average of 500?

Statistical Hypotheses:

$$H_0: \mu = 500$$
$$H_1: \mu \neq 500$$

Decision Rule:
Reject H_0 at the .05 level of significance if z equals or is more positive than 1.96 or if z equals or is more negative than -1.96.

Calculations:
Given

$$\overline{X} = 533; \quad \mu_{hyp} = 500; \quad \sigma_{\overline{x}} = \frac{\sigma}{\sqrt{n}} = \frac{110}{\sqrt{100}} = 11$$

$$z = \frac{533 - 500}{11} = 3$$

Decision:
Reject H_0 at the .05 level of significance because $z = 3$, which is more positive than 1.96.

Interpretation:
The mean SAT verbal score for all local freshmen does not equal—it exceeds—the national average of 500.

14.4 STATEMENT OF THE RESEARCH PROBLEM

The formulation of a research problem often represents the most crucial and exciting phase of an investigation. Indeed, the mark of a skillful investigator is to focus on an important research problem that can be answered. Do children from broken families score lower on tests of personal adjustment? Do aggressive TV cartoons incite more aggressive behavior in preschool children? Does profit sharing increase the productivity of employees? Because of our emphasis on hypothesis testing, research problems appear in this book as finished products, usually in the first one or two sentences of a new example.

14.5 NULL HYPOTHESIS (H_0)

Once the problem has been described, it must be translated into a statistical hypothesis regarding some population characteristic. Abbreviated as H_0, the null hypothesis becomes the focal point for the entire test procedure (even though we usually hope to reject it). In the test with SAT verbal scores, the null hypothesis asserts that, with respect to the national average of 500, nothing special is happening to the mean verbal score for the local population of freshmen. An equivalent statement, in symbols, reads

$$H_0: \mu = 500$$

Null hypothesis (H_0)

A statistical hypothesis that usually asserts that nothing special is happening with respect to some characteristic of the underlying population.

where H_0 represents the null hypothesis and μ is the population mean for the local freshman class.

Generally speaking, **the null hypothesis, H_0,** is a statistical hypothesis that usually asserts that nothing special is happening with respect to some characteristic of the underlying population. Because the hypothesis testing procedure requires that the hypothesized sampling distribution of the mean be centered about a single number (500), the null hypothesis always includes an equality about one number ($H_0: \mu = 500$). Furthermore, the null hypothesis always makes a precise statement about a characteristic of the population, never about a characteristic of the sample. (Remember, the purpose of a hypothesis test is to determine whether a particular outcome, such as an observed sample mean, could have reasonably originated from a population with the hypothesized characteristic.)

Finding the Single Number for H_0

The single number actually used in H_0 varies, of course, from problem to problem. Even for a given problem, this number could originate from any of several sources. For instance, it could be based on available information about some relevant population other than the target population, as in the present example in which 500 reflects the mean SAT verbal scores for all college-bound students during a recent year. It also could be based on some existing standard or theory—for example, that the mean verbal score for the current population of local freshmen should equal 540 because that happens to be the mean score achieved by all local freshmen during the last five years.

If, as sometimes happens, it's impossible to identify a meaningful null hypothesis, don't try to salvage the situation with arbitrary numbers. Instead, use another entirely different technique, known as estimation, which is described in Chapter 17.

14.6 ALTERNATIVE HYPOTHESIS (H_1)

In the present example, the alternative hypothesis asserts that, with respect to the national average of 500, something special is happening to the mean verbal score for the local population of freshmen (because the mean for the local population doesn't equal the national average of 500). An equivalent statement, in symbols, reads

$$H_1: \mu \neq 500$$

Alternative hypothesis (H_1)

The opposite of the null hypothesis.

---

where H_1 represents the alternative hypothesis, μ is the population mean for the local freshman class, and $\neq$ indicates "is not equal to."

The **alternative hypothesis, H_1,** *asserts the opposite of the null hypothesis.* A decision to retain the null hypothesis implies a lack of support for the alternative hypothesis, and a decision to reject the null hypothesis implies support for the alternative hypothesis.

As will be described in Chapter 15, the alternative hypothesis may assume any one of three different forms, depending on the perspective of the investigator. In its present form, H_1 specifies a *range* of possible values about the *single* number (500) that appears in H_0.

Research hypothesis

Usually identified with the alternative hypothesis, this is the informal hypothesis or hunch that inspires the entire investigation.

As also will become increasingly apparent, H_1 usually is identified with the **research hypothesis,** *the informal hypothesis or hunch that, by implying the presence of something special in the underlying population, serves as inspiration for the entire investigation.* "Something special" might be, as in the current example, a deviation from a national average, or it could be, as in later chapters, a deviation from some control condition produced by a new teaching method, a weight-reduction diet, or a self-improvement workshop. In any event, it's this research hypothesis—and certainly not the null hypothesis—that supplies the motive behind an investigation.

*Exercise 14.2   Indicate what's wrong with each of the following statistical hypotheses:

(a) $H_0: \mu = 155$
$\quad H_1: \mu \neq 160$

(b) $H_0: \overline{X} = 241$
$\quad H_1: \overline{X} \neq 241$

*Exercise 14.3 First using words, then symbols, identify the null hypothesis for each of the following situations. (Don't concern yourself about the precise form of the alternative hypothesis at this point.)

(a) A school administrator wishes to determine whether sixth grade boys in her school district differ, on the average, from the national norms of 6.2 push-ups for sixth grade boys.

(b) A consumer group investigates whether, on the average, the true weights of packages of ground beef sold by a large supermarket chain differ from the specified 16 ounces.

(c) A marriage counselor wishes to determine whether, during a standard conflict-resolution session, his clients differ, on the average, from the average of 11 verbal interruptions reported for "healthy couples."

Answers on page 506

14.7 DECISION RULE

Decision rule

Specifies precisely when H_0 should be rejected (because the observed z qualifies as a rare outcome).

A **decision rule** *specifies precisely when H_0* should be rejected (because the observed z qualifies as a rare outcome). There are many possible decision rules. A very common one, already introduced in Figure 14.3, specifies that H_0 should be *rejected* if the observed z equals or is more positive than 1.96 or if the observed z equals or is more negative than -1.96. In other words, H_0 should be *retained* if the observed z falls between ± 1.96.

Critical z Scores

Critical z score

A z score that separates common from rare outcomes and hence dictates whether H_0 should be retained or rejected.

Figure 14.4 indicates that z scores of ± 1.96 define the boundaries for the middle .95 of the total area (1.00) under the hypothesized sampling distribution for z. Derived from the normal curve table, as you can verify by checking Table A, Appendix D, these two z scores *separate common from rare outcomes and hence dictate whether H_0 should be retained or rejected.* Because of their vital role in the decision about H_0, these scores are referred to as **critical z scores.**

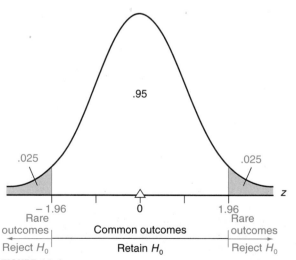

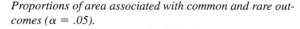

FIGURE 14.4
Proportions of area associated with common and rare outcomes ($\alpha = .05$).

Level of Significance (α)

Figure 14.4 also indicates the proportion (.025 + .025 = .05) of the total area under the sampling distribution of z that is identified with rare outcomes. This proportion is often referred to as the level of significance of the statistical test and is symbolized by the Greek letter α (alpha). In the present example, the level of significance, α, equals .05.

The **level of significance** (α) *indicates the degree of rarity required of an observed outcome in order to reject the null hypothesis* (H_0). For instance, the .05 level of significance indicates that H_0 should be rejected if the observed z could have occurred just by chance with a probability of only .05 (one chance out of twenty) *or less.*

Level of significance (α)

The degree of rarity required of an observed outcome in order to reject the null hypothesis (H_0).

14.8 CALCULATIONS

We can use information from the sample to calculate a value for z. As has been noted previously, use Formula 14.1 to convert the observed sample mean of 533 into a z of 3.

14.9 DECISION

Either retain or reject H_0, depending on the location of the observed z value relative to the critical z values specified in the decision rule. According to the present rule, H_0 should be rejected at the .05 level of significance because the observed z of 3 equals or exceeds the critical z of 1.96. In other words, because the observed z of 3 is a rare outcome (that is, an unlikely outcome from a population with the hypothesized characteristic), we reject H_0.

Retain or Reject H_0?

If you're ever confused about whether to retain or reject H_0, recall the logic behind the hypothesis test. You want to reject H_0 only if the observed value of z qualifies as a rare outcome, because it deviates too far into the tails of the sampling distribution. In other words, you want to reject H_0 only if the observed value of z equals or is more positive than the upper critical z (1.96) or if the observed value of z equals or is more negative than the lower critical z (-1.96). Before deciding, you might find it helpful to sketch the hypothesized sampling distribution, along with the critical z values and shaded rejection regions, and then to use some mark, such as an arrow ($\uparrow$), to designate the location of the observed value of z (3) along the z scale. If this mark is located in the shaded rejection region—or farther out than this region, as in Figure 14.4—then H_0 should be rejected.

*Exercise 14.4 For each of the following situations, indicate whether H_0 should be retained or rejected and justify your answer by specifying the precise relationship between observed and critical z scores. (Remember, you might find it helpful to locate z in a sketch showing both the sampling distribution of z and the decision rule.) Should H_0 be retained or rejected given a hypothesis test with critical z scores of ± 1.96, and

(a) $z = 1.74$ **(b)** $z = 0.13$ **(c)** $z = -2.51$

Answers on page 506

14.10 INTERPRETATION

Finally, interpret the decision in terms of the original research problem. In the present example, it can be concluded that, since the null hypothesis was rejected, the mean SAT verbal score for the local freshman class probably differs from the national average of 500.

Although not a strict consequence of the present test, a more specific conclusion is possible. Since the sample mean of 533 (or its equivalent z of 3) falls in the *upper* rejection region of the hypothesized sampling distribution, it can be concluded that the population mean SAT verbal score for all local freshman probably *exceeds* the national average of 500. By the same token, if the observed sample mean or its equivalent z had fallen in the *lower* rejection region of the hypothesized sampling distribution, it could have been concluded that the population mean for all local freshmen probably is *below* the national average.

If the observed sample mean or its equivalent z had fallen in the retention region of the hypothesized sampling distribution, it would have been concluded (somewhat weakly, as discussed in Section 15.2 of the next chapter) that there is no evidence that the population mean for all local freshman differs from the national average of 500.

*Exercise 14.5 According to the American Psychological Association, members with a doctorate and a full-time teaching appointment earn, on the average, $51,500 per year, with a standard deviation of $3,000. An investigator wishes to determine whether $51,500 also describes the mean salary for all female members with a doctorate and a full-time teaching appointment. Salaries are obtained for a random sample of 100 women from this population, and the mean salary equals $50,300.

(a) Someone claims that the observed difference between $50,300 and $51,500 is large enough by itself to support the conclusion that female members earn less. Explain why it's important to conduct a hypothesis test.

(b) The investigator wishes to conduct a hypothesis test for what population?

(c) What's the null hypothesis, H_0?

(d) What's the alternative hypothesis, H_1?

(e) Specify the decision rule, using the .05 level of significance.

(f) Calculate the value of z. (Remember to convert the standard deviation to a standard error.)

(g) What's your decision about H_0?

(h) Using words, interpret this decision in terms of the original problem.

Answers on pages 506–507

Summary
....

To test a hypothesis about the population mean, a single observed sample mean is viewed within the context of a hypothesized sampling distribution, itself centered about the hypothesized population mean. If the sample mean appears to emerge from the dense concentration of possible sample means in the middle of the hypothesized sampling distribution, it qualifies as a common outcome, and the null hypothesis should be retained. On the other hand, if the sample mean appears to emerge from the sparse concentration of possible sample means in the extremities of the hypothesized sampling distribution, it qualifies as a rare outcome, and the null hypothesis should be rejected.

Hypothesis tests are based not on the sampling distribution of $\overline{X}$, expressed in original units of measurement, but on its counterpart, the sampling distribution of z. Referred to as the z test for a single population mean, this test is appropriate only when (1) the population is normally distributed or the sample size is large enough to satisfy the central limit theorem and (2) the population standard deviation is known.

When testing a hypothesis, adopt a step-by-step procedure, beginning with a statement of the research problem and ending with an interpretation of the test results:

State the research problem. Using words, state the problem to be resolved by the investigation.

Identify the statistical hypotheses. The statistical hypotheses consist of a null hypothesis (H_0) and an alternative (or research) hypothesis (H_1). The null hypothesis supplies the value about which the hypothesized sampling distribution is centered. Depending on the outcome of the hypothesis test, H_0 will either be retained or rejected. Insofar as H_0 implies that nothing special is happening in the underlying population, the investigator usually hopes to reject it in favor of H_1, the research hypothesis. In the present chapter, the statistical hypotheses take the form

$$H_0: \mu = \text{some number}$$
$$H_1: \mu \neq \text{some number}$$

(Two other possible forms for statistical hypotheses will be described in Chapter 15.)

Specify a decision rule. This rule indicates precisely when H_0 should be rejected. The exact form of the decision rule depends on several factors, to be discussed in the next chapter. In any event, H_0 is rejected whenever the observed z deviates from 0 as far as, or farther than, the critical z does.

The level of significance indicates how rare an observed z must be (assuming that H_0 is true) before H_0 can be rejected.

Calculate the value of the observed z. Express the one observed sample mean as an observed z, using Formula 14.1.

Make a decision. Either retain or reject H_0 at the specified level of significance, justifying this decision by noting the relationship between observed and critical z scores.

Interpret the decision. Using words, interpret the decision in terms of the original research problem. Rejection of the null hypothesis supports the research hypothesis, while retention of the null hypothesis fails to support the research hypothesis.

Important Terms

Sampling distribution of *z*	Research hypothesis
z Test for a population mean	Decision rule
Null hypothesis (*H_0*)	Critical *z* score
Alternative hypothesis (*H_1*)	Level of significance (α)

REVIEW EXERCISES

*14.6 For the population at large, the Wechsler Intelligence Scale is designed to yield a normal distribution of test scores with a mean of 100 and a standard deviation of 15. School district officials wonder whether, on the average, an IQ score of 100 describes the intellectual aptitudes of all students in their district. Accordingly, Wechsler IQ scores are obtained for a random sample of 25 of their students, and the mean IQ is found to equal 105. Using the step-by-step procedure described in this chapter, test the null hypothesis at the .05 level of significance.

Answers on page 507

14.7 According to a recent government survey, the daily one-way commute distance of U.S. workers averages 13 miles with a standard deviation of 13 miles. An investigator wishes to determine whether the national average describes the mean commute distance for all workers in the Chicago area. Commute distances are obtained for a random sample of 169 workers from this area, and the mean distance is found to be 15.5 miles. Test the null hypothesis at the .05 level of significance.

14.8 Supply the missing word(s) in the following statements:
If the one observed sample mean can be viewed as a __(a)__ outcome under the hypothesis, H_0 will be __(b)__ Otherwise, if the one observed sample mean can be viewed as a __(c)__ outcome under the hypothesis, H_0 will be __(d)__.

The pair of z scores that separates common and rare outcomes is referred to as __(e)__ z scores. Within the hypothesized sampling distribution, the proportion of area allocated to rare outcomes is referred to as the __(f)__ and is symbolized by the Greek letter __(g)__.

When based on the sampling distribution of z, the hypothesis test is referred to as a __(h)__ test. This test is appropriate if sample size is sufficiently large to satisfy the __(i)__ and if the __(j)__ is known.

CHAPTER
15

More about Hypothesis Testing

15.1   HYPOTHESIS TESTING: AN OVERVIEW

15.2   STRONG OR WEAK DECISIONS

15.3   WHY THE RESEARCH HYPOTHESIS ISN'T TESTED DIRECTLY

15.4   ONE-TAILED AND TWO-TAILED TESTS

15.5   CHOOSING A LEVEL OF SIGNIFICANCE (α)

Summary

Important Terms

Review Exercises

15.1 HYPOTHESIS TESTING: AN OVERVIEW

Why Hypothesis Tests?

There is a crucial link between hypothesis tests and the need of investigators, whether pollsters or researchers, to generalize beyond existing data. If the 100 freshmen in the SAT example of the previous chapter had represented not a sample, but a *census* of the entire freshman class, there wouldn't have been any need to generalize beyond existing data, and it would have been inappropriate to conduct a hypothesis test. The observed difference between the *population* mean of 533 and the national average of 500, by itself, would have been sufficient grounds for concluding that the mean SAT verbal score for all local freshmen exceeds the national average. Indeed, *any* observed difference in favor of the local freshmen, regardless of size, would have supported this conclusion.

If we need to generalize beyond the 100 freshmen to a larger local population, as was actually the case, the observed difference between 533 and 500 can't be interpreted at face value. The basic problem is that if a second random sample of 100 freshmen were taken, its sample mean probably would differ, just by chance, from that for the first sample. Accordingly, chance must be considered when we attempt to decipher the observed difference between 533 and 500 (as either a common or rare outcome).

Importance of the Standard Error

To evaluate the effect of chance, statisticians use the concept of a sampling distribution, that is, the concept of the sample means for all possible random outcomes. A key element in this concept is the standard error of the mean, a measure of the average amount by which sample means differ, just by chance, from the population mean. Dividing the observed difference $(533 - 500)$ by the standard error (11) to obtain a value of z (3) locates the original observed difference along a z scale of either common outcomes (easily attributable to chance) or rare outcomes (not easily attributable to chance). If, when expressed as z, the ratio of the observed difference to the standard error is small enough to be easily attributed to chance, we should retain H_0. Otherwise, if the ratio of the observed difference to the standard error is too large to be easily attributed to chance, as in the SAT example in which this ratio equals 3, we should reject H_0.

Before generalizing beyond the existing data, we must always measure the effect of chance; that is, we must obtain a value for the standard error. To appreciate the vital role of the standard error in the SAT example, increase its value from 11 to 33, and note that even though the observed difference remains the same $(533 - 500)$, we now would retain, not reject H_0, because now z would equal 1 (rather than 3) and be less than the critical z of 1.96.

Possibility of Incorrect Decisions

Having made a decision about the null hypothesis, we never know absolutely whether that decision is correct or incorrect, unless, of course, we survey the entire population. Even if H_0 is true (and, therefore, the normal curve model for random outcomes is also true), there's a *slight* possibility that, just by chance, the one observed z actually originates from one of the shaded rejection regions of the hypothesized distribution of z, thus causing the true H_0 to be rejected. This type of incorrect decision—rejecting a true H_0—is referred to as a type I error. The type I error is described in more detail in Chapter 16.

On first impulse, it might seem desirable to abolish the shaded rejection regions in the hypothesized sampling distribution to ensure that a true H_0 never is rejected. A most unfortunate consequence of this strategy, however, is that no H_0, not even a radically false H_0, ever would be rejected. This second type of incorrect decision—retaining a false H_0—is referred to as a type II error. The type II error also is described in more detail in Chapter 16.

Minimizing Incorrect Decisions

Traditional hypothesis-testing procedures, such as the one illustrated in **Figure 15.1,** tend to minimize both types of incorrect decisions. *If H_0 is true,* there is a high probability that the observed z will qualify as a common outcome under the hypothesized sampling distribution and that the true H_0 will be retained. (This probability equals the proportion of white area (.95) in the hypothesized sampling distribution in Figure 15.1.) On the other hand, *if H_0 is seriously false,* because the hypothesized population mean differs considerably from the true population mean, there is also a high probability that the observed z will qualify as a rare outcome under the hypothesized distribution and that the false H_0 will be rejected. (This probability can't be determined from Figure 15.1, because, in this case, the hypothesized sampling distribution doesn't actually reflect the true sampling distribution. More about this in Chapter 16.)

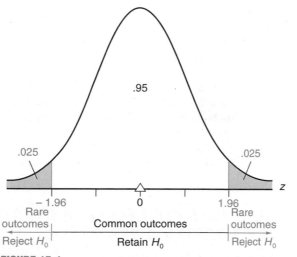

FIGURE 15.1

Proportions of area associated with common and rare outcomes ($\alpha = .05$).

Even though we never really know whether a particular decision is correct or incorrect, it's reassuring that in the long run, *most* decisions will be correct—assuming the null hypotheses are *either true or seriously false*.

15.2 STRONG OR WEAK DECISIONS

Retaining H_0 Is a *Weak* Decision

There are subtle but important differences in the interpretation of decisions to retain H_0 and to reject H_0. H_0 is retained whenever, on the assumption that H_0 is true, the observed z qualifies as a common outcome. Therefore, H_0 *could* be true. But the same observed result also would qualify as a common outcome when the original value in H_0 (500) is replaced with a slightly different value. Thus the retention of H_0 must be viewed as a relatively weak decision. Because of this weakness, many statisticians prefer to describe this decision as simply a *failure to reject H_0* rather than as the retention of H_0. In any event, the retention of H_0 can't be interpreted as proving H_0 to be true. If H_0 had been retained in the present example, it would have been appropriate to conclude not that the mean verbal score for all local freshmen equals the national average, but that the mean verbal score *could* equal the national average, as well as many other possible values in the general vicinity of the national average.

Rejecting H_0 Is a *Strong* Decision

On the other hand, H_0 is rejected whenever the observed z qualifies as a rare outcome—one that could have occurred just by chance with a probability of .05 or less—on the assumption that H_0 is true. This suspiciously rare outcome implies that H_0 is probably false (and that H_1 is probably true). Therefore, the rejection of H_0 can be viewed as a strong decision. When H_0 was rejected in the present example, it was appropriate to report a definitive conclusion that the mean verbal score for all local freshmen probably exceeds the national average.

To summarize,

the decision to retain H_0 implies not that H_0 is probably true, but only that H_0 *could* be true, whereas the decision to reject H_0 implies that H_0 is *probably* false (and that H_1 is *probably* true).

Because most researchers hope to reject H_0 in favor of H_1, the relative weakness of the decision to retain H_0 usually doesn't pose a serious problem.

15.3 WHY THE RESEARCH HYPOTHESIS ISN'T TESTED DIRECTLY

Even though H_0, the null hypothesis, is the focus of a statistical test, it's usually of secondary concern to the investigator. Nevertheless, there are several reasons why, although of primary concern, the research hypothesis is identified with H_1 and tested indirectly.

Lacks Necessary Precision

The research hypothesis, but not the null hypothesis, lacks the necessary precision to be tested directly.

To be tested, a hypothesis must specify a single number about which the hypothesized sampling distribution can be constructed. *Because it specifies a single number, the null hypothesis, rather than the research hypothesis, is tested directly.* In the SAT example, the null hypothesis specifies that a precise value (the national average of 500) describes the mean for the current population of interest (all local freshmen). Typically, the research hypothesis violates this requirement. It merely specifies that some inequality exists between the hypothesized value (500) and the mean for the current population of interest (all local freshmen).

Supported by Strong Decision to Reject

Logical considerations also argue for the indirect testing of the research hypothesis and the direct testing of the null hypothesis.

Because the research hypothesis is identified with the alternative hypothesis, the decision to reject the null hypothesis, should it be made, will provide *strong* support for the research hypothesis, whereas the decision to retain the null hypothesis, should it be made, will provide, at most, *weak* support for the null hypothesis.

As was mentioned previously, the decision to reject the null hypothesis is stronger than the decision to retain it. It makes sense, therefore, to identify the research hypothesis with the alternative hypothesis. If, as hoped, the data favor the research hypothesis, the test will generate strong support for your "hunch": It's *probably* true. If the data don't favor the research hypothesis, the hypothesis test will generate, at most, weak support for the null hypothesis: It *could* be true. *Weak support for the null hypothesis is of little consequence, as this hypothesis — that nothing special is happening in the population — usually serves only as a convenient testing device.*

15.4 ONE-TAILED AND TWO-TAILED TESTS

In this section and the next, let's consider some techniques that can make the hypothesis test more responsive to special conditions. These techniques should be selected before the data are collected.

Two-tailed Test

Generally, the alternative hypothesis, H_1, is the complement of the null hypothesis, H_0. Under typical conditions, the form of H_1 resembles that shown for the SAT example, namely,

$$H_1: \mu \neq 500$$

This alternative hypothesis says that the null hypothesis should be rejected if the mean verbal score for the population of local freshmen differs in either direction from the national average of 500. An observed z will qualify as a rare outcome if it deviates either too far below or too far above the national average. Panel A of **Figure 15.2** shows rejection regions that are associated with both tails of the hypothesized sampling distribution. The corresponding decision rule, with its pair of critical z scores of ± 1.96, is referred to as a **two-tailed** or **nondirectional test.**

One-tailed Test (Lower Tail Critical)

Now let's assume that the research hypothesis for the investigation of SAT verbal scores was based on complaints from instructors about the poor preparation of local freshmen. Let's assume also that if the investigation supports these complaints, a remedial program will be instituted. Under these circumstances, the investigator might prefer a hypothesis test that is specially designed to detect only whether the population mean verbal score for all local freshmen *is less* than the national average.

This alternative hypothesis reads

$$H_1: \mu < 500$$

It reflects a concern that the null hypothesis should be rejected only if the population mean verbal score for all local freshmen is less than the national average of 500. Accordingly, an observed z triggers the decision to reject H_0 only if z deviates too far below the national average. Panel B of Figure 15.2 illustrates a rejection region that is associated with only the lower tail of the hypothesized sampling distribution. The corresponding decision rule, with its critical z of -1.65, is referred to as a **one-tailed** or **directional test** *with the lower tail critical.* Use Table A in Appendix D to verify that if the critical z equals -1.65, then .05 of the total area under the distribution of z has been allocated to the lower rejection region. Notice that the level of significance, α, equals .05 for this one-tailed test and also for the original two-tailed test.

Extra Sensitivity of One-tailed Tests

This new one-tailed test is extra sensitive to any drop in the mean for the population of local freshmen below the national average. If H_0 is false because a drop has occurred, then the observed z will be more likely to deviate below the national average. As can be seen in panels A and B of Figure 15.2, an observed deviation in the direction of concern—below the national average—is more likely to penetrate the broader rejection region for the one-tailed test than that for the two-tailed test. Therefore the decision to reject a *false H_0* (in favor of the research hypothesis) is more likely to occur in the one-tailed test than in the two-tailed test.

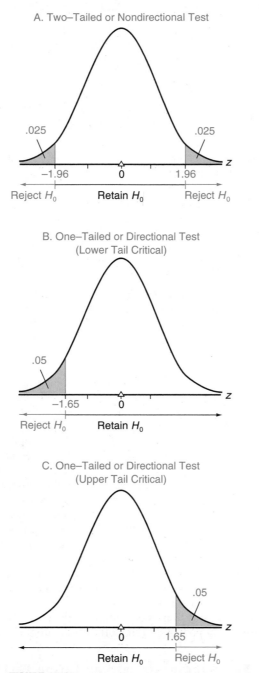

FIGURE 15.2
Three different types of tests ($\alpha = .05$).

One-tailed Test (Upper Tail Critical)

Panel C of Figure 15.2 illustrates a one-tailed or directional test *with the upper tail critical*. This one-tailed test is the mirror image of the previous test. Now the alternative hypothesis reads

$$H_1: \mu > 500$$

and its critical z equals 1.65. This test is specially designed to detect only whether the population mean verbal score for all local freshmen *exceeds* the national average. For example, the research hypothesis for this investigation might have been inspired by an alumnus who will donate a large sum of money if it can be demonstrated that, on the average, the SAT verbal scores of all local freshmen exceed the national average.

One or Two Tails?

Before a hypothesis test, if there's a concern that the true population mean differs from the hypothesized population mean *only* in a particular direction, use the appropriate one-tailed or directional test for extra sensitivity. Otherwise, use the more customary two-tailed or nondirectional test.

Having committed yourself to a one-tailed test with its single rejection region, you must retain H_0, regardless of how far the observed z deviates from the hypothesized population mean in the direction of "no concern." For instance, if a one-tailed test with the lower tail critical had been used with the data for 100 freshmen from the SAT example, H_0 would have been retained because, even though the observed z equals an impressive value of 3, it deviates in the direction of no concern—in this case, above the national average. Clearly, a one-tailed test should be adopted only when there is absolutely no concern about deviations, even very large deviations, in one direction. If there is the slightest concern about these deviations, use a two-tailed test.

Ideally, the selection of a one- or two-tailed test should be made before the data are collected, but in any event, it shouldn't be influenced by the value of the observed z. Never "peek" at the value of the observed z to determine whether to locate the rejection region for a one-tailed test in the upper or the lower tail of the distribution of z. To qualify as a one-tailed test, the location of the rejection region must reflect the investigator's concern only about deviations in a particular direction *before any inspection of the data*. Indeed, the investigator should be able to muster a compelling reason, based on an understanding of the research hypothesis, to support the direction of the one-tailed test.

New Null Hypothesis for One-tailed Tests

When tests are one tailed, a complete statement of the null hypothesis also should include all possible values of the population mean in the direction of no concern. For example, given a one-tailed test with the lower tail critical, such as $H_1: \mu < 500$, the complete null hypothesis should be stated as $H_0: \mu \geq 500$

instead of H_0: $\mu = 500$. By the same token, given a one-tailed test with the upper tail critical, such as H_1: $\mu > 500$, the complete null hypothesis should be stated as H_0: $\mu \leq 500$.

If you think about it, the complete H_0 describes all of the population means that could be true if a one-tailed test results in the retention of the null hypothesis. For instance, if a one-tailed test with the lower tail critical results in the retention of H_0: $\mu \geq 500$, the complete H_0 accurately reflects the fact that not only $\mu = 500$ could be true, but also any other value of the population mean in the direction of no concern, that is, $\mu > 500$, could be true. (Remember, when the test is one-tailed, even a very deviant result in the direction of no concern—possibly reflecting a μ much larger than 500—still will trigger the decision to retain H_0.) Henceforth, whenever a one-tailed test is employed, write H_0 to include values of the population mean in the direction of no concern—*even though the single number in the complete H_0 identified by the equality sign is the one value about which the hypothesized sampling distribution is centered and, therefore, the one value actually used in the hypothesis test.*

*Exercise 15.1 Each of the following statements could represent the point of departure for a hypothesis test. Given only the information in each statement, would you use a two-tailed (or nondirectional) test, a one-tailed (or directional) test with the lower tail critical, or a one-tailed (or directional) test with the upper tail critical? Indicate your decision by specifying the appropriate H_0 and H_1. Furthermore, whenever you conclude that the test is one tailed, indicate the precise word (or words) in the statement that justifies the one-tailed test.

(a) To increase rainfall, extensive cloud-seeding experiments are to be conducted, and the results are to be compared with a baseline figure of 0.54 inch of rainfall (for comparable periods when cloud seeding wasn't done).

(b) Public health statistics indicate, we'll assume, that American males gain an average of 23 pounds during the twenty-year period after age forty. An ambitious weight-reduction program, spanning twenty years, is being tested on a random sample of forty-year-old men.

(c) When untreated during their lifetimes, cancer-susceptible mice have an average life span of 134 days. To determine the effects of a potentially life-prolonging (and cancer-retarding) drug, the average life span is determined for a randomly selected group of mice that receives this drug.

*Exercise 15.2 For each of the following situations, indicate whether H_0 should be retained or rejected. (Remember, you might find it helpful to locate z in a sketch showing both the hypothesized sampling distribution and the decision rule.)

Given a one-tailed test, lower tail critical with $\alpha = .01$, and

(a) $z = -2.34$ **(b)** $z = -5.13$ **(c)** $z = 4.04$

Given a one-tailed test, upper tail critical with $\alpha = .05$, and

(d) $z = 2.00$ **(e)** $z = -1.80$ **(f)** $z = 1.61$

Answers on pages 507–508

15.5 CHOOSING A LEVEL OF SIGNIFICANCE (α)

The level of significance indicates how rare an observed z must be before H_0 can be rejected. To reject H_0 at the .05 level of significance implies that the observed z would have occurred, just by chance, with a probability of only .05 (one chance out of twenty) *or less*.

The level of significance also spotlights an inherent risk in hypothesis testing, the risk of rejecting a true H_0. When the level of significance equals .05, there is a probability of .05 that, even though H_0 is true, the observed z will stray into the rejection region and cause the true H_0 to be rejected.

Which Level of Significance?

When the rejection of a true H_0 is particularly serious, a smaller level of significance can be selected. For example, the .01 level of significance implies that before H_0 can be rejected, the observed z must achieve a degree of rarity equal to .01 (one chance out of one hundred) *or less;* it also limits, to a probability of .01, the risk of rejecting a true H_0. The .01 level might be used in a hypothesis test in which the rejection of a true H_0 would cause the introduction of a costly new remedial education program, even though the population mean verbal score for all local freshmen really equals the national average. An even smaller level of significance, such as the .001 level, might be used when the rejection of a true H_0 would have horrendous consequences—for instance, the treatment of serious illnesses, such as AIDS, exclusively with a new, very expensive drug that not only is worthless but also has severe side effects.

Although many different levels of significance are possible, most tables for hypothesis tests are geared to the .05 and .01 levels. In this book, the level of significance will be specified for you. But in real-life applications, you, as an investigator, might have to select a level of significance. *Unless there are obvious reasons for selecting either a larger or a smaller level of significance, use the customary .05 level*—the largest level of significance reported in most professional journals.

When testing hypotheses with the z test, you may find it helpful to refer to **Table 15.1,** which lists the critical z values for one- and two-tailed tests at the .05 and .01 levels of significance. These z values were obtained from Table A, Appendix D.

*Exercise 15.3 Specify the decision rule for each of the following situations (referring to Table 15.1 to find critical z Values):

   (a) a two-tailed test with $\alpha = .05$

   (b) a one-tailed test, upper tail critical, with $\alpha = .01$

   (c) a one-tailed test, lower tail critical, with $\alpha = .05$

   (d) a two-tailed test with $\alpha = .01$
     Answers on page 508

Table 15.1
CRITICAL *z* VALUES

TYPE OF TEST	LEVEL OF SIGNIFICANCE (α)	
	.05	.01
Two-tailed or nondirectional test	±1.96	±2.58
(H_0: μ = some number)		
(H_1: $\mu \neq$ some number)		
One-tailed or directional test, lower tail critical	−1.65	−2.33
(H_0: $\mu \geq$ some number)		
(H_1: $\mu <$ some number)		
One-tailed or directional test, upper tail critical	+1.65	+2.33
(H_0: $\mu \leq$ some number)		
(H_1: $\mu >$ some number)		

INTERNET DEMONSTRATION

Go to the Web site for this book **(http://darwin.cwru.edu/~ witte/statistics)** and click on **Hypothesis Test** to explore how decisions about the null hypothesis vary with the form of the alternative hypothesis and the level of significance.

Summary

Chance must be considered when we make a decision about the null hypothesis (H_0) by deciphering whether an observed difference reflects a common or rare outcome. Even though we never know whether a particular decision about the null hypothesis is correct or incorrect, it's reassuring that, in the long run, most decisions will be correct, assuming that the null hypotheses are either true or seriously false.

The decision to retain H_0 is weak; it implies only that H_0 *could* be true, whereas the decision to reject H_0 is strong; it implies that H_0 is *probably* false (and that H_1 is *probably* true).

Although the research hypothesis, rather than the null hypothesis, is of primary concern, the research hypothesis is usually identified with the alternative hypothesis and tested indirectly for two reasons: (1) because it lacks the necessary precision, and (2) because of logical considerations, based on the fact that rejecting the null hypothesis is a stronger decision than retaining the null hypothesis.

Use a more sensitive one-tailed test only when, before an investigation, there's an exclusive concern about deviations in a particular direction. Otherwise, use a two-tailed test.

Select the statistical hypotheses from among the following three possibilities:

$$H_0: \mu = \text{some number}$$
$$H_1: \mu \neq \text{some number}$$

The above hypotheses reflect a concern that the true population mean *will not equal* some hypothesized number, and it translates into a decision rule for a two-tailed or nondirectional test.

$$H_0: \mu \geq \text{some number}$$
$$H_1: \mu < \text{some number}$$

The above hypotheses reflect a concern only that the true population mean *will fall below* the hypothesized number, and it translates into a decision rule for a one-tailed or directional test with the lower tail critical.

$$H_0: \mu \leq \text{some number}$$
$$H_1: \mu > \text{some number}$$

The above hypotheses reflect a concern only that the true population mean *will exceed* the hypothesized number, and it translates into a decision rule for a one-tailed or directional test with the upper tail critical.

Unless there are obvious reasons for selecting either a larger or smaller level of significance, use the customary .05 level.

Important Terms

Two-tailed or nondirectional test   **One-tailed or directional test**

REVIEW EXERCISES

15.4 An assembly line at a candy plant is designed to yield two-pound boxes of assorted candies whose weights in fact follow a normal distribution with a mean of 33 ounces and a standard deviation of .30 ounce. A random sample of 36 boxes from the production of the most recent shift reveals a mean weight of 33.09 ounces. (Incidentally, if you think about it, this is an exception to the usual situation where the investigator hopes to reject the null hypothesis.)

(a) Describe the population being tested.

(b) Using the customary procedure, test the null hypothesis at the .05 level of significance.

(c) Someone uses a one-tailed test, upper tail critical, because the sample mean of 33.09 exceeds the hypothesized population mean of 33. Any comment?

15.5 It has been suggested that the null hypothesis always states a claim about a population, *never about a sample.* Someone insists, neverthe-less, on testing a null hypothesis that the population mean equals the observed value of the sample mean. What's wrong with this procedure?

15.6 Why not conclude that the null hypothesis is true when, as a result of a hypothesis test, the null hypothesis is retained?

15.7 Why is the research hypothesis not tested directly?

15.8 Read again the problem described in Exercise 14.5 on page 238.

(a) What form should H_0 and H_1 take if the investigator is concerned only with salary discrimination against female members?

(b) If this hypothesis test supports the conclusion of salary discrimination against female members, a costly class-action suit will be initiated against American colleges and universities. Under these circumstances, do you recommend using the .05 or the .01 level of significance? Why?

15.9 Supply the missing word(s) in the following statements:

If, before a hypothesis test, there's concern about deviations only in a particular direction, use a **(a)** or **(b)** test. Otherwise, use a **(c)** or **(d)** test.

If H_0 is rejected at the .05 level of significance, this implies that if H_0 were **(e)**, the observed z could have occurred just by chance with a probability of only **(f)** or less.

When the rejection of a true H_0 is particularly serious, use the **(g)** level of significance or even the **(h)** level of significance.

CHAPTER 16

Controlling Type I and Type II Errors

16.1   TESTING A HYPOTHESIS ABOUT VITAMIN C

16.2   FOUR POSSIBLE OUTCOMES

16.3   IF H_0 REALLY IS TRUE

16.4   IF H_0 REALLY IS FALSE BECAUSE OF A *LARGE* EFFECT

16.5   IF H_0 REALLY IS FALSE BECAUSE OF A *SMALL* EFFECT

16.6   INFLUENCE OF SAMPLE SIZE

16.7   SELECTION OF SAMPLE SIZE

16.8   POWER CURVES

Summary

Important Terms

Review Exercises

16.1 TESTING A HYPOTHESIS ABOUT VITAMIN C

Personal experience, loosely conducted experiments, appeals to authority—all have been used to justify the daily ingestion of vitamin C. A researcher wishes to test a hunch that vitamin C increases the intellectual aptitude of high school students. After being randomly selected from some large school district, each of 36 students takes a daily dose of 60 milligrams of vitamin C (the commonly recommended dosage) for a period of two months before being tested for IQ.

Ordinarily, IQ scores for all students in this school district approximate a normal distribution with a mean of 100 and a standard deviation of 15. According to the null hypothesis, a mean of 100 still would describe the distribution of IQ scores even if all of the students in the district were to receive the vitamin C treatment. Furthermore, given the researcher's exclusive concern about detecting only any deviation of the population mean *above* 100, the null hypothesis takes the form appropriate for a one-tailed test with the upper tail critical, namely,

$$H_0: \mu \leq 100$$

The rejection of H_0 would support H_1, the research hypothesis that something special is happening in the underlying population (because vitamin C increases intellectual aptitude), namely,

$$H_1: \mu > 100$$

z Test Is Appropriate

To determine whether the sample mean IQ for the 36 students qualifies as a common or a rare outcome under the null hypothesis, a z test will be used. The z test for a population mean is appropriate because, for IQ scores, the population standard deviation is known to be 15 and the shape of the population is known to be normal.

Two Groups Would Have Been Better

Although poorly designed, the present experiment supplies a perspective that will be most useful in later chapters. A better-designed experiment would contrast the IQ scores for the group of subjects who receive vitamin C with the IQ scores for the group of subjects who receive fake vitamin C—thereby controlling for the "placebo effect," a self-induced improvement in performance caused solely by the subject's awareness of being treated in a special way. In other words, a better-designed experiment would evaluate the relationship between an independent variable (real or fake vitamin C) and a dependent variable (IQ scores), as defined in Section 1.9. Hypothesis tests for experiments with two groups are described in Chapters 19 and 20.

Summarized in the following box are those features of the hypothesis test that can be identified before the collection of any data for the vitamin C experiment.

HYPOTHESIS TEST SUMMARY: *z* TEST FOR A POPULATION MEAN (*PRIOR* TO VITAMIN C EXPERIMENT)

Research Problem:

Does the daily ingestion of vitamin C cause an increase, on the average, among IQ scores of all students in the school district?

Statistical Hypotheses:

$$H_0: \mu \leq 100$$
$$H_1: \mu > 100$$

Decision Rule:

Reject H_0 at the .05 level of significance if z equals or is more positive than 1.65.

Calculations:

$$\sigma_{\overline{X}} = \frac{\sigma}{\sqrt{n}} = \frac{15}{\sqrt{36}} = \frac{15}{6} = 2.5$$

16.2 FOUR POSSIBLE OUTCOMES

Table 16.1 summarizes the four possible outcomes of any hypothesis test. Before testing a hypothesis, we must be concerned about all four possible outcomes because we don't know whether H_0 is true or false—that is why we are testing the hypothesis. If, unknown to us, H_0 really is true, a well-designed hypothesis test will tend to confirm this fact; that is, it will cause us to retain H_0 and conclude that H_0 could be true. To conclude otherwise, as is always a slight possibility even with a well-designed test, reflects a type I error. On the other hand, if, unknown to us, H_0 really is *seriously* false, a well-designed hypothesis test also will tend to confirm this fact; that is, it will cause us to reject H_0 and conclude that H_0 is false. To conclude otherwise, as is always a slight possibility, reflects a type II error.

Four Possible Outcomes of the Vitamin C Experiment

It's instructive to describe the four possible outcomes in Table 16.1 in terms of the vitamin C experiment.

Table 16.1 POSSIBLE OUTCOMES OF HYPOTHESIS TEST		
DECISION	**STATUS OF H_0**	
	True H_0	**False H_0**
Retain H_0	(1) Correct decision	(3) Type II error (miss)
Reject H_0	(2) Type I error (false alarm)	(4) Correct decision

1. If H_0 really is true (because vitamin C doesn't cause an increase in the population mean IQ), then *it's a correct decision to retain the true H_0.* In this case, the researcher would conclude correctly that there's no evidence that vitamin C increases IQ.

2. If H_0 is really true, then *it's a type I error to reject the true H_0* and conclude that vitamin C increases IQ when, in fact, it doesn't. Type I errors are sometimes called *false alarms* because, as with their firehouse counterparts, they trigger wild-goose chases after something that doesn't exist. For instance, a type I error might encourage a batch of worthless experimental efforts to discover precisely what dosage level of vitamin C maximizes the nonexistent "increase" in IQ.

3. If H_0 really is false (because vitamin C really causes an increase in the population mean IQ), then *it's a type II error to retain the false H_0* and conclude that there's no evidence that vitamin C increases IQ when, in fact, it does. Type II errors are sometimes called *misses* because they fail to detect a potentially important relationship, such as that between vitamin C and IQ.

4. If H_0 really is false, then *it's a correct decision to reject the false H_0* and conclude that vitamin C increases IQ.

Type I error

Rejecting a true null hypothesis.

Type II error

Retaining a false null hypothesis.

Importance of Null Hypothesis

Refer to Table 16.1 when, as in the following exercise, you must describe the four possible outcomes for a particular hypothesis test. To avoid confusing the type I and II errors, first identify the null hypothesis, H_0. Typically, *the null hypothesis asserts that there is no effect, thereby contradicting the research hypothesis.* In the present case, contrary to the research hypothesis, the null hypothesis (H_0: $\mu \leq 100$) assumes that vitamin C has no positive effect on IQ.

Decisions Usually Are Correct

When generalizing beyond existing observations, there's always the possibility of a type I or a type II error, and we never can be absolutely certain of having made the correct decision. At best, we can use a test procedure that *usually* produces a correct decision when H_0 is either true or seriously false. This claim will be examined in the context of the vitamin C experiment, assuming first that H_0 really is true and then that H_0 really is false. Although you might view this approach as hopelessly theoretical, *since we never actually know* *whether H_0 really is true or false,* read the next few sections carefully, for they have important implications for any hypothesis test.

*Exercise 16.1

(a) List the four possible outcomes for any hypothesis test.

(b) Under the U.S. Criminal Code, a defendant is presumed innocent until proven guilty. Viewing a criminal trial as a hypothesis test (with H_0 specifying that the defendant is innocent), describe each of the four possible outcomes.

Answer on page 508

16.3 IF H_0 REALLY IS TRUE

Assume that H_0 really is true because vitamin C really doesn't increase the population mean IQ. In this case, we need be concerned only about either retaining or rejecting a true H_0 (the two leftmost outcomes in Table 16.1). It's instructive to view these two possible outcomes in terms of the sampling distribution in **Figure 16.1.** Centered about a value of 100, the hypothesized sampling distribution in Figure 16.1 reflects the properties of the projected one-tailed test for vitamin C. If H_0 really is true—and this is a crucial point— the *hypothesized* sampling distribution also can be viewed as the *true* sampling distribution (from which the one observed sample mean actually originates). Therefore, the one observed sample mean (or z) in the experiment can be viewed as being randomly selected from the hypothesized distribution.

Probability of a Type I Error

When, just by chance, a randomly selected sample mean originates from the small, shaded portion of the sampling distribution in Figure 16.1, its z value equals or exceeds 1.65; hence H_0 is rejected. Because H_0 really is true, this is

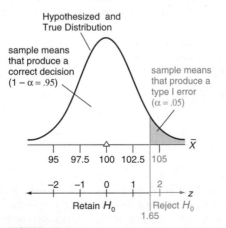

FIGURE 16.1

Hypothesized and true sampling distribution when H_0 is true (because vitamin C causes no increase in IQ).

an incorrect decision or type I error—a false alarm, announced as evidence that vitamin C increases IQ, even though it really doesn't. The probability of a type I error simply equals **alpha (α),** the level of significance. (The level of significance, remember, indicates the proportion of the total area of the sampling distribution in the rejection region for H_0.) In the present case, the probability of a type I error equals .05, as indicated in Figure 16.1.

Probability of a Correct Decision

When, just by chance, a randomly selected sample mean originates from the large white portion of the sampling distribution in Figure 16.1, its z value is less than 1.65 and H_0 is retained. Because H_0 really is true, this is a correct decision—announced as a lack of evidence that vitamin C increases IQ. The probability of a correct decision equals $1 - \alpha$, that is, .95.

Reducing the Probability of a Type I Error

If H_0 really is true, the present test will produce a correct decision with a probability of .95 and a type I error with a probability of .05.[1] If a false alarm has serious consequences, the probability of a type I error can be reduced to .01 or even to .001 simply by using the .01 or .001 levels of significance, respectively. One of these levels of significance might be preferred for the vitamin C test if, for instance, a false alarm could cause the adoption of an expensive program to supply worthless vitamin C to all students in the district and, perhaps, the creation of an accelerated curriculum to accommodate the fictitious increase in intellectual aptitude.

True H_0 Usually Retained

If H_0 really is true, the probability of a type I error, α, equals the level of significance, and the probability of a correct decision equals $1 - \alpha$.

Because values of .05 or less are usually selected for α, we can conclude that if H_0 really is true, correct decisions will occur much more frequently than type I errors will.

*Exercise 16.2 In order to eliminate the type I error, someone decides to use the .00 level of significance. What's wrong with this tactic?

   Answer on page 508

....

[1] Strictly speaking, if H_0: $\mu \leq 100$ really is true, the true sampling distribution also could be centered about some value less than 100, in the direction of no concern. In this case, the consequences of the hypothesis test would be even more favorable than previously suggested. Essentially, because the true sampling distribution would be shifted to the left of the one shown in Figure 16.1, while the retention and rejection regions remain the same, the type I error would have a smaller probability than .05, and a correct decision would have a larger probability than .95.

16.4 IF H_0 REALLY IS FALSE BECAUSE OF A *LARGE* EFFECT

Next assume that H_0 really is false because vitamin C actually increases the population mean by not just a few points, but *by many points*, let's say by ten points. Using the vocabulary of researchers, we also could describe this increase as a ten-point effect, because *any difference between a true and a hypothesized population mean* is referred to as an **effect.** If H_0 really is false, because of the relatively large ten-point effect of vitamin C on IQ, we need be concerned only about either retaining or rejecting a false H_0 (the two right most outcomes in Table 16.1). Let's view each of these two possible outcomes in terms of the sampling distributions in Figure 16.2.

Effect

Any difference between a true and a hypothesized population mean.

Hypothesized sampling distribution

Centered about the hypothesized population mean, this distribution is used to generate the decision rule.

True sampling distribution

Centered about the true population mean, this distribution produces the one observed mean (or z).

Hypothesized Sampling Distribution

It's essential to distinguish between the *hypothesized* sampling distribution and the *true* sampling distribution shown in **Figure 16.2.** Centered about the hypothesized population mean of 100, the **hypothesized sampling distribution** serves as the parent distribution for the familiar decision rule with a critical z of 1.65 for the projected one-tailed test. Once the decision rule has been identified, attention shifts from the hypothesized sampling distribution to the true sampling distribution.

True Sampling Distribution

Centered about the true population mean of 110 (which reflects the ten-point effect, that is, $100 + 10 = 110$), the **true sampling distribution** serves as the parent distribution for the one randomly selected sample mean (or z) that will be observed in the experiment. Viewed relative to the decision rule (based on the hypothesized sampling distribution), the one randomly selected sample mean (originating from the true sampling distribution) dictates whether we retain or reject the false H_0.

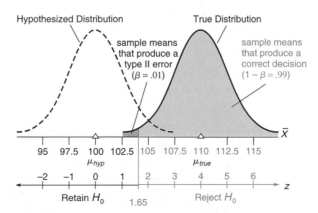

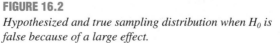

FIGURE 16.2

Hypothesized and true sampling distribution when H_0 is false because of a large effect.

Low Probability of a Type II Error for a *Large* Effect

When, just by chance, a randomly selected sample mean originates from the very small black portion of the true sampling distribution of the mean, its z value is less than 1.65 and, therefore, in compliance with the decision rule, H_0 is retained. Because H_0 really is false, this is an incorrect decision or type II error—a miss, announced as a lack of evidence that vitamin C increases IQ, even though, in fact, it does. With the aid of tables for the normal curve, it can be demonstrated that in the present case, the probability of a type II error, symbolized by the Greek letter **beta (β),** equals .01.

(The present argument doesn't require that you know how to calculate this probability of .01 or those given in the remainder of the chapter. In brief, these probabilities represent areas under the true sampling distribution found by re-expressing the critical z as a deviation from the true population mean [110] rather than from the hypothesized population mean [100] and referring to the normal curve table, Table A, Appendix D.)

- - - - - - - - - - - - - - - - - - - -

Beta (β)

The probability of a type II error, that is, the probability of retaining a false null hypothesis.

High Probability of a Correct Decision for a *Large* Effect

When, just by chance, a sample mean originates from the large shaded portion of the true sampling distribution, its z value equals or exceeds 1.65, and H_0 is rejected. Because H_0 really is false, this is a correct decision—announced as evidence that vitamin C increases IQ. In the present case, the probability of a correct decision, symbolized as $1 - \beta$, equals .99.

Review

If H_0 really is false, because vitamin C has a large ten-point effect on the population mean IQ, the projected one-tailed test will do quite well. There's a high probability of .99 that a correct decision will be made and a probability of only .01 that a type II error will be committed. This conclusion, when combined with that for the previous section, justifies the earlier claim that hypothesis tests tend to produce correct decisions when either H_0 really is true or H_0 really is false because of a large effect.

*Exercise 16.3 Indicate whether the following statements, *all referring to Figure 16.2,* are true or false:

(a) The assumption that H_0 really is false is depicted by the separation of the hypothesized and true distributions.

(b) In practice, when actually testing a hypothesis, we wouldn't know that the true population mean equals 110.

(c) The one observed sample mean is viewed as originating from the hypothesized sampling distribution.

(d) A correct decision would be made if the one observed sample mean has a value of 103.

Answers on page 508

16.5 IF *H₀* REALLY IS FALSE BECAUSE OF A *SMALL* EFFECT

The projected hypothesis test doesn't fare nearly as well if H_0 really is false because vitamin C increases the population mean IQ by *only a few points*, let's say by only three points. Once again, as indicated in **Figure 16.3,** there are two different distributions of sample means: the *hypothesized* sampling distribution centered about the hypothesized population mean of 100 and the *true* sampling distribution centered about the true population mean of 103 (which reflects the three-point effect, that is, $100 + 3 = 103$). After the decision rule has been constructed with the aid of the hypothesized sampling distribution, attention shifts to the true sampling distribution from which the one randomly selected sample mean actually will originate.

Low Probability of a Correct Decision for a *Small* Effect

Viewed relative to the decision rule, the true sampling distribution supplies two types of randomly selected sample means: those that produce a type II error because they originate from the black sector and those that produce a correct decision because they originate from the shaded sector. Because of the small three-point effect, the true and hypothesized population means are much closer in Figure 16.3 than in Figure 16.2. As a result, the entire true sampling distribution in Figure 16.3 is shifted toward the retention region for the false H_0, and proportionately more of this distribution is black. Now the projected one-tailed test performs more poorly; there is a fairly high probability of .67 that a type II error will be committed and a low probability of .33 that the correct decision will be made. (Remember, you needn't determine these normal curve probabilities in order to understand the argument.)

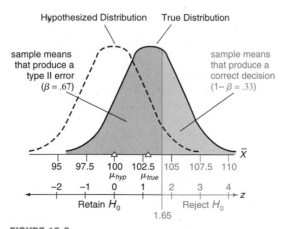

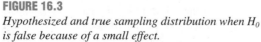

FIGURE 16.3

Hypothesized and true sampling distribution when H_0 is false because of a small effect.

Rejection of False H_0 Depends on Size of Effect

If H_0 really is false, the probability of a type II error, β, and the probability of a correct decision, $1 - \beta$, depend on the size of the effect, that is, the difference between the true and the hypothesized population means. The smaller the effect is, the higher the probability of a type II error and the lower the probability of a correct decision will be.

If you think about it, this conclusion isn't particularly surprising. If H_0 really is false, there must be some effect. The smaller this effect is, the less likely it is that it will be detected (by correctly rejecting the false H_0), and the more likely it is that it will be missed (by erroneously retaining the false H_0). As will be described in the next section, if it's important to detect even a relatively small effect, the probability of a correct decision can be raised to any desired value by increasing the sample size.

*Exercise 16.4 Indicate whether the following statements, *all referring to Figure 16.3,* are true or false:

(a) The value of the true population mean (103) dictates the location of the true sampling distribution.

(b) The critical value of z (1.65) is based on the true sampling distribution.

(c) Because the hypothesized population mean of 100 really is false, it would be impossible to observe a sample mean value less than or equal to 100.

(d) A correct decision would be made if the one observed sample mean has a value of 105.

Answers on page 508

....

16.6 INFLUENCE OF SAMPLE SIZE

Ordinarily, the researcher might not be too concerned about the low detection rate of .33 for the relatively small three-point effect of vitamin C on IQ. Under special circumstances, however, this low detection rate might be unacceptable. For example, previous experimentation might have established that vitamin C has many positive effects, including the reduction of common colds, and no apparent negative side effects. Furthermore, huge quantities of vitamin C might be available at no cost to the school district. The establishment of one more positive effect, even a fairly mild one such as a small increase in the population mean IQ, might clinch the case for supplying vitamin C to all students in the district. Therefore the investigator might wish to use a test procedure for which, when H_0 really is false because of a small effect, the detection rate is appreciably higher than .33.

To increase the probability of detecting a false H_0, increase the sample size.

Assuming that vitamin C still has only a small three-point effect on IQ, let's check the properties of the projected one-tailed test when the sample size increases from 36 to 100 students. Recall the formula for the standard error of the mean, $\sigma_{\bar{X}}$, namely,

$$\sigma_{\overline{X}} = \frac{\sigma}{\sqrt{n}}$$

For the original experiment with its sample size of 36,

$$\sigma_{\overline{X}} = \frac{15}{\sqrt{36}} = \frac{15}{6} = 2.5$$

whereas for the new experiment with its sample size of 100,

$$\sigma_{\overline{X}} = \frac{15}{\sqrt{100}} = \frac{15}{10} = 1.5$$

Clearly, any increase in sample size causes a reduction in the standard error of the mean.

Consequences of Reducing Standard Error

As can be seen by comparing Figures 16.3 and **16.4,** the reduction of the standard error from 2.5 to 1.5 has two important consequences:

1. It shrinks the upper retention region back toward the hypothesized population mean of 100.
2. It shrinks the entire true sampling distribution toward the true population mean of 103.

The net result is that among randomly selected sample means for 100 students, fewer sample means (.36) produce a type II error because they originate from the black sector, and more sample means (.64) produce a correct decision—that is, more lead to the detection of a false H_0—because they originate from the shaded sector.

An obvious implication is that the standard error can be reduced to any desired value merely by increasing the sample size. To cite an extreme case, when the sample size equals 10,000 students (!), the standard error drops to 0.15. In this case, the upper retention region shrinks to the immediate vicinity of the hypothesized population mean of 100, and the entire true sampling distribution of the mean shrinks to the immediate vicinity of the true population

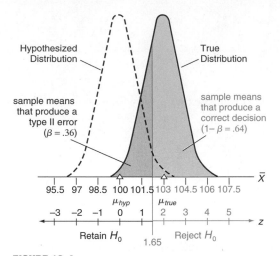

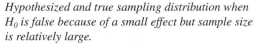

FIGURE 16.4

Hypothesized and true sampling distribution when
H_0 is false because of a small effect but sample size
is relatively large.

mean of 103. The net result is that a type II error hardly ever is committed, and the small three-point effect virtually always is detected.

16.7 SELECTION OF SAMPLE SIZE

Samples Can Be Too Large

At this point, you might think that the sample size always should be as large as possible in order to maximize the detection of a false H_0. Not so. An excessively large sample size produces an extra sensitive hypothesis test that detects even a very small effect that, from almost any perspective, lacks importance. For example, an excessively large sample size could cause H_0 to be rejected, even though vitamin C actually increases the population mean IQ by only one-half point. Because from almost any perspective, this very small effect lacks importance, most researchers would just as soon miss it; that is, most would just as soon retain this false H_0. Thus, before an experiment, a wise researcher attempts to select a sample size that, because it isn't excessively large, minimizes the detection of a small, unimportant effect.

Samples Can Be Too Small

On the other hand, the sample size can be too small. An unduly small sample size will produce an insensitive hypothesis test (with a large standard error) that will miss even a very large, important effect. For example, an unduly small sample size can cause H_0 to be retained, even though vitamin C actually increases the population mean IQ by 15 points. Before an experiment, a wise researcher also attempts to select a sample size that, because it isn't unduly small, maximizes the detection of a large, important effect.

Neither Too Large nor Too Small

For the purposes of most researchers, a sample size in the hundreds is excessively large and one of less than about five is unduly small. There remains, of course, considerable latitude for sample size selection between these rough extremities. Statisticians supply researchers with charts, often referred to as *power curves*, to help select the appropriate sample size for a particular experiment.

*Exercise 16.5 Comment critically on the following experimental reports:

(a) Using a group of four subjects, an investigator announces that H_0 was retained at the .05 level of significance.

(b) Using a group of 600 subjects, a researcher reports that H_0 was rejected at the .05 level of significance.

Answers on page 509

16.8 POWER CURVES

Power (1 − β)

The probability of detecting a particular effect.

Power curves

Show how the likelihood of detecting any possible effect varies with different sample sizes.

The **power** of a hypothesis test *equals the probability of detecting a particular effect*, that is, of rejecting a false H_0. In other words, **power (1 − β)** is the complement of the probability of a type II error (β). In Figures 16.2, 16.3, and 16.4, power equals the proportion of shaded area in each of the true sampling distributions.

Basically, **power curves** *show how the likelihood of detecting any possible effect*—ranging from very small to very large—*varies with different sample sizes*. To locate the appropriate sample size for a projected experiment, the researcher must answer two questions: (1) What is the smallest effect that, *if present*, merits being detected and (2) what is a reasonable detection rate for this effect? (Section 21.10 describes a rough rule of thumb for small, medium, and large effects.)

Need Not Predict True Effect Size

The use of power curves does not require the researcher to predict the *true* effect size—an impossible task—but merely specify the smallest effect that, *if present*, merits detection. If, unknown to the researcher, the true effect size actually is larger than the specified effect, the true detection rate actually will be larger than the specified detection rate—because more of the true sampling distribution overlaps the rejection region for the false H_0 than does the sampling distribution for the specified effect. In other words, a more important effect is even more likely to be detected. On the other hand, if the true effect size actually is smaller than the specified effect, the entire process works in reverse but still to the researcher's advantage, because an unimportant effect, which the researcher would just as soon miss, is even less likely to be detected.

Advantages of Power Curves

Once the researcher has specified both the smallest important effect and a reasonable detection rate, the appropriate sample size can be obtained by consulting charts of power curves. Because the use of power curves depends on several factors, including the researcher's judgment and the availability of local resources, two equally competent researchers could select different sample sizes for the same experiment. Nonetheless, in the hands of a judicious researcher,

> **the use of power curves represents a distinct improvement over the arbitrary selection of sample size, for they help identify a sample size that, being neither unduly small nor excessively large, produces a hypothesis test with the proper sensitivity.**[2]

*Exercise 16.6 A researcher consults a chart of power curves to determine the sample size required to detect an eight-point effect with a probability of .90. What happens to this detection rate of .90—will it actually be *smaller,* the *same,* or *larger*—if, unknown to her, the true effect actually equals

(a) twelve points?

(b) five points?

> *Answers on page 509*

INTERNET DEMONSTRATION

Go to the Web site for this book **(http://darwin.cwru.edu/~witte/statistics)** and click on **Power** to determine how power–the probability of detecting a false null hypothesis–varies with the size of the effect and also with the size of the sample.

Summary

There are four possible outcomes for any hypothesis test:

1. If H_0 really is true, it's a correct decision to retain the true H_0.
2. If H_0 really is true, it's a type I error to reject the true H_0.

---

[2]For an excellent introduction to power curves, see E. Minium, R. Clarke, and T. Coladarci, *Elements of Statistical Reasoning*, 2nd ed. (New York: Wiley, 1999), chapter 17. For more detailed information, see J. Cohen, *Statistical Power Analysis*, 2nd ed. (Hillsdale, NJ: Erlbaum, 1988).

3. If H_0 really is false, it's a type II error to retain the false H_0.

4. If H_0 really is false, it's a correct decision to reject the false H_0.

When generalizing beyond the existing data, there's always the possibility of a type I or a type II error. At best, a hypothesis test tends to produce a correct decision when either H_0 really is true or H_0 really is false because of a large effect.

If H_0 really is true, the probability of a type I error, α, equals the level of significance, and the probability of a correct decision equals $1 - \alpha$.

If H_0 really is false, the probability of a type II error, β, and the probability of a correct decision, $1 - \beta$, depends on the size of the effect—that is, the difference between the true and the hypothesized population means. The larger the effect is, the lower the probability of a type II error and the higher the probability of a correct decision will be.

To increase the probability of detecting a false H_0, even a false H_0 that reflects a very small effect, increase the sample size.

It's desirable to select a sample size that, being neither unduly small nor excessively large, produces a hypothesis test with the proper sensitivity. Power curves are available to help the researcher select the appropriate sample size for a particular experiment.

Important Terms

Type I error	**True sampling distribution**
Type II error	**Beta (β)**
Alpha (α)	**Power ($1 - \beta$)**
Effect	**Power curves**
Hypotheized sampling distribution	

REVIEW EXERCISES

*16.7 Recalling the vitamin C experiment described in this chapter, you could describe the null hypothesis in both symbols and words as follows:

$H_0: \mu \leq 100$, that is, vitamin C doesn't increase IQ

Following the format of Table 16.1 and being as specific as possible, you could describe the four possible outcomes of the vitamin C experiment as follows:

	STATUS OF H_0	
DECISION	TRUE H_0	FALSE H_0
Retain H_0	*Correct Decision:* Conclude that there is no evidence that vitamin C increases IQ when in fact it doesn't.	*Type II Error:* Conclude that there is no evidence that vitamin C increases IQ when in fact it does.
Reject H_0	*Type I Error:* Conclude that vitamin C increases IQ when in fact it doesn't.	*Correct Decision:* Conclude that vitamin C increases IQ when in fact it does.

Using the answer for the vitamin C experiment as a model, specify the null hypothesis and the four possible outcomes for each of the following exercises:

*(a) Exercise 15.1 (a) on page 251

Answer on page 509

(b) Exercise 15.1(b)

(c) Exercise 15.1(c)

16.8 We must be concerned about four possible outcomes *before* conducting a hypothesis test.

(a) Assuming that the test already has been conducted and *the hypothesis already has been retained*, must we still be concerned about four possible outcomes?

(b) Assuming that the test already has been conducted and *the hypothesis already has been rejected*, must we still be concerned about four possible outcomes?

16.9 Using the .05 level of significance, a researcher retains H_0. There is, he concludes, a probability of .95 that H_0 is true. Please comment.

16.10 In another study, a researcher rejects H_0 at the .01 level of significance. There is, she concludes, a probability of .99 that H_0 is false. Comments?

16.11 If you conduct a series of hypothesis tests, each at the .05 level of significance, you'll commit a type I error about five times in a hundred. Comments?

16.12 For a projected one-tailed test, lower tail critical, at the .05 level of significance, construct two rough graphs. Each graph should show the sector in the true sampling distribution that produces a type II error and the sector that produces a correct decision. One graph should reflect the case when H_0 really is false because the true population mean is *slightly* *less* than the hypothesized population mean, and the other graph should reflect the case when H_0 really is false because the true population mean is *appreciably less* than the hypothesized population mean. *Hint:* First identify the decision rule for the hypothesized population mean, and then draw the true sampling distribution for each case.

16.13 **(a)** If you're particularly concerned about the type I error, how should a projected hypothesis test be modified?

(b) If you're particularly concerned about the type II error, how should a projected hypothesis test be modified?

CHAPTER 17

Estimation

17.1   ESTIMATING μ FOR SAT SCORES
17.2   POINT ESTIMATE FOR μ
17.3   CONFIDENCE INTERVAL FOR μ
17.4   WHY CONFIDENCE INTERVALS WORK
17.5   CONFIDENCE INTERVAL FOR μ BASED ON z
17.6   INTERPRETATION OF A CONFIDENCE INTERVAL
17.7   LEVEL OF CONFIDENCE
17.8   EFFECT OF SAMPLE SIZE
17.9   HYPOTHESIS TESTS OR CONFIDENCE INTERVALS?
17.10   CONFIDENCE INTERVAL FOR POPULATION PERCENT
17.11   OTHER TYPES OF CONFIDENCE INTERVALS

Summary

Important Terms

Review Exercises

17.1 ESTIMATING μ FOR SAT SCORES

In Chapter 14, an investigator was concerned about detecting any difference between the mean SAT verbal score for all local freshmen and the national average. As has been seen, this concern translates into a hypothesis test, and with the aid of a z test, it was concluded that the mean for the local population exceeds the national average. Given a concern about the national average, this conclusion is most informative—it might even create some joy among local university officials. However, the same SAT investigation could have been prompted by a wish merely to *estimate* the value of the local population mean rather than to *test a hypothesis* based on the national average. This new concern translates into an estimation problem, and with the aid of point estimates and confidence intervals, known sample characteristics can be used to estimate the unknown population mean SAT verbal score for all local freshmen.

17.2 POINT ESTIMATE FOR μ

Point estimate

A single value that represents some unknown population characteristic, such as the population mean.

A *point estimate* for μ specifies a single value that represents the unknown population mean.

This is the most straightforward type of estimate. If a random sample of 100 local freshmen reveals a sample mean SAT verbal score of 533, then 533 will be the point estimate of the unknown population mean for all local freshmen. The best single point estimate for the unknown population mean is simply the observed value of the sample mean.

A Basic Deficiency

Although straightforward, simple, and precise, point estimates suffer from a basic deficiency: They tend to be inaccurate. Because of sampling variability, it's unlikely that a single sample mean will equal the population mean. By the same token, it's unlikely that a point estimate, such as 533, will coincide with the population mean SAT verbal score for all local freshmen. Furthermore, point estimates convey no indication of this inaccuracy, causing statisticians to supplement point estimates with another, more realistic type of estimate, known as interval estimates or confidence intervals.

*Exercise 17.1 A random sample of 200 graduates of U.S. colleges reveals a mean annual income of $62,600. What is the best estimate of the unknown mean annual income for all graduates of U.S. colleges?

Answer on page 509

Confidence interval

A range of values that, with a known degree of certainty, includes an unknown population characteristic, such as a population mean.

17.3 CONFIDENCE INTERVAL FOR μ

A *confidence interval* for μ specifies a range of values that, with a known degree of certainty, includes the unknown population mean.

For instance, using the following techniques, the SAT investigator might claim, *with 95 percent confidence*, that the interval between 511.44 and 554.56 includes the value of the unknown population mean SAT verbal score for all local freshmen. To be 95 percent confident signifies that if many of these intervals were constructed for a long series of samples, approximately 95 percent would include the population mean verbal score for all local freshmen. In the long run, 95 percent of these confidence intervals are true because they include the unknown population mean. The remaining 5 percent are false because they fail to include the unknown population mean.

17.4 WHY CONFIDENCE INTERVALS WORK

In order to understand confidence intervals, you must view them within the context of three important properties of the sampling distribution of the mean (encountered previously in Chapter 13 under slightly different circumstances). For the sampling distribution from which the sample mean of 533 originates, as shown in **Figure 17.1,** the three important properties are as follows:

1. The mean of the sampling distribution equals the unknown population mean for all local freshmen, whatever its value, because the mean of this sampling distribution always equals the population mean.
2. The standard error of the sampling distribution equals the value (11) obtained from dividing the population standard deviation (110) by the square root of the sample size ($\sqrt{100}$).
3. The shape of the sampling distribution approximates a normal distribution because the sample size of 100 satisfies the requirements of the central limit theorem.

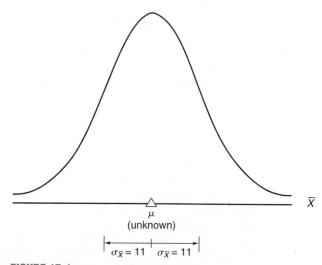

FIGURE 17.1
Sampling distribution of the mean (SAT verbal scores).

A Series of Confidence Intervals

In practice, only one sample mean is actually taken from this sampling distribution and used to construct a single 95 percent confidence interval. However, imagine taking not just one but a series of randomly selected sample means from this sampling distribution. Because of sampling variability, these sample means tend to differ among themselves. For each sample mean, construct a 95 percent confidence interval by adding 1.96 standard errors to the sample mean and subtracting 1.96 standard errors from the sample mean; that is, use the expression

$$\overline{X} \pm 1.96\, \sigma_{\overline{X}}$$

to obtain a 95 percent confidence interval for each sample mean.

True Confidence Intervals

Why, according to statistical theory, do 95 percent of these confidence intervals include the unknown population mean? As indicated in **Figure 17.2,** because the sampling distribution is normal, 95 percent of all sample means are within 1.96 standard errors of the unknown population mean. In other words, 95 percent of all sample means deviate less than *1.96 standard errors from the unknown population mean.* Therefore, when sample means are expanded into confidence intervals—by adding 1.96 standard errors to the sample mean and by subtracting 1.96 standard errors from the sample mean—95 percent of all possible confidence intervals are true because they include the unknown population mean. To illustrate this point, fifteen of the sixteen sample means shown in Figure 17.2 are within 1.96 standard errors of the unknown population mean. The corresponding fifteen confidence intervals, shown in black, have ranges that span the broken line for the population mean, thereby qualifying as true intervals because they include the value of the unknown population mean.

False Confidence Intervals

Five percent of all confidence intervals fail to include the unknown population mean. As indicated in Figure 17.2, 5 percent of all sample means (2.5 percent in each tail) deviate more than 1.96 standard errors from the unknown population mean. Therefore, when sample means are expanded into confidence intervals—by adding 1.96 standard errors to the sample mean and by subtracting 1.96 standard errors from the sample mean—5 percent of all possible confidence intervals are false because they fail to include the unknown population mean. To illustrate this point, only one of the sixteen sample means shown in Figure 17.2 is not within 1.96 standard errors of the unknown population mean. The resulting confidence interval, shown in color, has a range that doesn't span the broken line for the population mean, thereby being designated as a false interval because it fails to include the value of the unknown population mean.

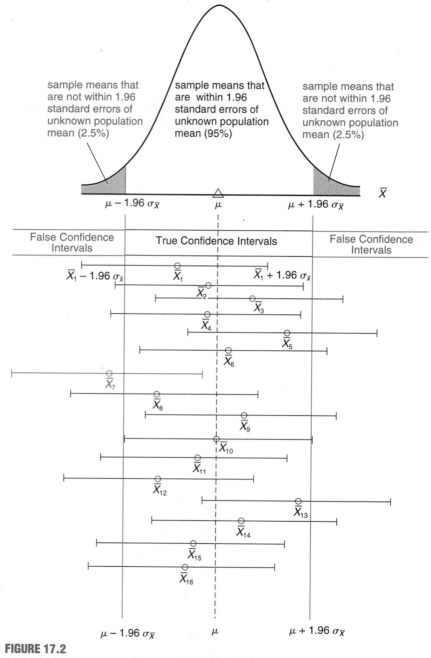

FIGURE 17.2

A series of 95 percent confidence intervals (emerging from sampling distribution).

17.5 CONFIDENCE INTERVAL FOR μ BASED ON z

To determine the previously reported confidence interval of 511.44 to 554.56 for the unknown mean SAT verbal score of all local freshmen, use the following general expression:

CONFIDENCE INTERVAL FOR μ (BASED ON z)

$$\overline{X} \pm (z_{\text{conf}})(\sigma_{\overline{X}})$$

(17.1)

---

where $\overline{X}$ represents the sample mean; z_{conf} represents a number from the standard normal table that satisfies the confidence specifications for the confidence interval; and $\sigma_{\overline{X}}$ represents the standard error of the mean.

Given that the sample mean SAT verbal score, $\overline{X}$, equals 533, that z_{conf} equals 1.96 (from the standard normal tables, where z scores of ± 1.96 define the middle 95 percent of the area under the normal curve), and that the standard error, $\sigma_{\overline{X}}$, equals 11, Formula 17.1 becomes

$$533 \pm (1.96)(11) = 533 \pm 21.56 = \begin{cases} 554.56 \\ 511.44 \end{cases}$$

where 554.56 and 511.44 represent the upper and lower limits of the confidence interval. Now it can be claimed, with 95 percent confidence, that the interval between 511.44 and 554.56 includes the value of the unknown mean SAT verbal score for all local freshmen.

Two Assumptions

The use of Formula 17.1 to construct confidence intervals assumes that the population standard deviation is known and that the population is normal or that the sample size is sufficiently large—at least 25—to satisfy the requirements of the central limit theorem.

*Exercise 17.2 Reading achievement scores are obtained for a group of fourth graders. A score of 4.0 indicates a level of achievement appropriate for fourth grade; a score below 4.0 indicates underachievement; and a score above 4.0 indicates overachievement. Furthermore, let's assume that the population standard deviation equals 0.4. A random sample of 64 fourth graders reveals a mean achievement score of 3.82.

(a) Construct a 95 percent confidence interval for the unknown population mean. (Remember to convert the standard deviation to a standard error.)

(b) Interpret this confidence interval. In other words, do you find any consistent evidence either of overachievement or of underachievement among the possible values of the population mean?

Answers on page 509

17.6 INTERPRETATION OF A CONFIDENCE INTERVAL

A 95 percent confidence claim reflects a long-term performance rating for an extended series of confidence intervals. If a series of confidence intervals is constructed to estimate the same population mean, as in Figure 17.2, approximately 95 percent of these intervals should include the population mean. In practice, only one confidence interval, not a series of intervals, is constructed, and that one interval is either true or false—because it either includes the population mean or fails to include the population mean. Of course, *we never really know whether a particular confidence interval is true or false* unless the entire population is surveyed. However,

> **when the level of confidence equals 95 percent or more, we can be *reasonably confident* that the one observed confidence interval includes the true population mean.**

For instance, we can be *reasonably confident* that the true population mean SAT verbal score for all local freshmen is neither less than 511.44 nor more than 554.56. That's the same as being *reasonably confident* that the true population mean SAT verbal score for all local freshmen is between 511.44 and 554.56.

*Exercise 17.3 Before taking the Graduate Record Exam (GRE), a random sample of college seniors received special training on how to take the GRE. After analyzing their subsequent scores on the GRE, the investigator reported a dramatic gain, relative to the national average of 500, as indicated by a 95 percent confidence interval of 507 to 527. Are the following interpretations true or false?

(a) Every subject who received special training scored between 507 and 527.

(b) The interval from 507 to 527 refers to possible values of the sample mean.

(c) About 95 percent of all subjects scored between 507 and 527.

(d) The interval from 507 to 527 refers to possible values of the population mean for all students who undergo special training.

(e) The true population mean definitely is between 507 and 527.

(f) This *particular* interval describes the population mean about 95 percent of the time.

(g) In the long run, a series of intervals similar to this one would describe the population mean about 95 percent of the time.

(h) In practice, we never really know whether the interval from 507 to 527 is true or false.

(i) We can be reasonably confident that the population mean is between 507 and 527.

Answers on pages 509–510

Level of confidence

The percent of time that a series of confidence intervals includes the unknown population characteristic, such as the population mean.

17.7 LEVEL OF CONFIDENCE

The **level of confidence** *indicates the percent of time that a series of confidence intervals includes the unknown population characteristic, such as the population mean.* Any level of confidence may be assigned to a confidence interval merely by substituting an appropriate value for z_{conf} in Formula 17.1. For instance, to construct a 99 percent confidence interval from the data for SAT verbal scores, first consult Table A in Appendix D to verify that z_{conf} values of ± 2.58 define the middle 99 percent of the total area under the normal curve. Then substitute numbers for symbols in Formula 17.1 to obtain

$$533 \pm (2.58)(11) = 533 \pm 28.38 = \begin{cases} 561.38 \\ 504.62 \end{cases}$$

It can be claimed, *with 99 percent confidence,* that the interval between 504.62 and 561.38 includes the value of the unknown mean SAT verbal score for all local freshmen. This implies that in the long run, 99 percent of these confidence intervals will include the unknown population mean.

Effect on Width of Interval

Notice that the 99 percent confidence interval of 504.62 to 561.38 for the previous data is wider and, therefore, less precise than is the corresponding 95 percent confidence interval of 511.44 to 554.56. The shift from a 95 percent to a 99 percent level of confidence requires an increase in the value of z_{conf} from 1.96 to 2.58. This increase, in turn, causes a wider, less precise confidence interval. Any shift to a higher level of confidence always produces a wider, less precise confidence interval, unless offset by an increase in sample size, as mentioned in the next section.

Choosing a Level of Confidence

Although many different levels of confidence have been used, 95 percent and 99 percent are the most prevalent. Generally, a larger level of confidence, such as 99 percent, should be reserved for those situations in which a false interval might have particularly serious consequences, for example, the widely publicized failure of national opinion pollsters to predict the election of President Harry Truman in 1948.

17.8 EFFECT OF SAMPLE SIZE

The larger the sample size is, the smaller the standard error and, hence, the more precise (narrower) the confidence interval will be. Indeed, as the sample size grows larger, the standard error will approach zero, and the confidence interval will shrink to a point estimate.

Selection of Sample Size

As with hypothesis tests, sample size can be selected according to specifications established before the investigation. To generate a confidence interval that possesses some desired precision (width), yet complies with the desired level of confidence, refer to formulas for sample size in other statistics books. Valid use of these formulas requires that before the investigation, the population standard deviation be either known or estimated.

*Exercise 17.4 On the basis of a random sample of 120 adults, a pollster reports, with 95 percent confidence, that between 58 to 72 percent of all Americans believe in life after death.

(a) If this interval is too wide, what, if anything, can be done with the existing data to obtain a narrower confidence interval?

(b) What can be done to obtain a narrower 95 percent confidence interval if another similar investigation is being planned?

Answers on page 510

INTERNET DEMONSTRATION

Go to the Web site for this book **(http://darwin.cwru.edu/~witte/statistics)** and click on **Confidence Intervals** to determine whether an extended series of 95 percent and 99 percent confidence intervals cover a known population mean the specified percent of times.

17.9 HYPOTHESIS TESTS OR CONFIDENCE INTERVALS?

Ordinarily, data are used either to test a hypothesis or to construct a confidence interval, but not both. Traditionally, hypothesis tests have been preferred to confidence intervals in the behavioral sciences, and that emphasis is reflected in this book. As a matter of fact, however, *confidence intervals tend to be more informative than hypothesis tests are.*

Hypothesis tests merely indicate whether an effect is present, whereas confidence intervals indicate the possible size of the effect.

For the vitamin C experiment proposed in Chapter 16, a hypothesis test merely would indicate whether vitamin C has an effect on IQ scores, whereas a 95 percent confidence interval would indicate the possible size of the effect of vitamin C on IQ scores—for instance, we could claim, with 95 percent confidence, that the interval between 102 and 112 includes the true population mean IQ for students who receive vitamin C. In other words, the true effect of vitamin C is probably somewhere between 2 and 12 IQ points (above the null hypothesized value of 100).

When to Use Confidence Intervals

If the primary concern is whether an effect is present—as is often the case in relatively new research areas—use a hypothesis test. For example, given that a social psychologist is uncertain whether the consumption of alcohol by witnesses compromises the accuracy of their subsequent recall of some event, it would be appropriate to use a hypothesis test. Otherwise, given that previous research clearly demonstrates alcohol-induced inaccuracies in witnesses' testimonies, a new investigator might use a confidence interval to estimate the possible mean rate of these inaccuracies.

Indeed, you should consider using a confidence interval whenever a hypothesis test triggers the rejection of the null hypothesis. For example, referring again to the vitamin C experiment proposed in Chapter 16, after it's been established (by rejecting the null hypothesis) that vitamin C has an effect on IQ scores, it makes sense to estimate, with a 95 percent confidence interval, that the interval between 102 and 112 describes the possible size of that effect, namely, an increase (above 100) of between 2 and 12 IQ points.

17.10 CONFIDENCE INTERVAL FOR POPULATION PERCENT

Let's briefly describe a type of confidence interval—that for population percents or proportions—often encountered in the media. For example, a recent news release reported that among a random or "scientific" sample of 1500 adult Americans, 74 percent favor some form of capital punishment. Furthermore, the **margin of error** equals ±3 percent, given that we wish to be 95 percent confident of our results. Rephrased slightly, this is the same as claiming, with 95 percent confidence, that the interval between 71 and 77 percent (from 74 ± 3) includes the true percent of Americans who favor some form of capital punishment.

Essentially, this 95 percent confidence interval originates from the following expression:

$$\text{sample percent} \pm (1.96)(\text{standard error of the percent})$$

where 1.96 comes from the standard normal curve and the standard error of the percent is analogous to the standard error of the mean.[1] Otherwise, all of the previous comments about confidence intervals for population means apply to confidence intervals for population percents or proportions. Thus, in the present case, we can be reasonably certain that the true population percent is between 71 percent and 77 percent.

Sample Size and Margin of Error

Often encountered in national polls, the huge sample of 1500 Americans guarantees a relatively small, ±3 percent margin of error (by reducing the size of the standard error). If, in the pollster's judgment, a larger margin of error would have

---

> **Margin of error**
>
> *That which is added to and subtracted from some sample value, such as the sample proportion or sample mean, to obtain the limits of a confidence interval.*

---

[1] A proportion (or a percent, which is merely 100 times a proportion) is a special type of mean where, after all observations have been coded as either zero or one, the ones are added and divided by the total number of observations. Therefore, although not emphasized in this book, the standard error of the proportion (or percent) could be obtained from the formula for the standard error of the mean.

been acceptable, smaller samples could have been used. For instance, if a ± 5 percent margin of error would have been acceptable, a random sample of about 500 adults could have been used, whereas if a ± 10 percent margin of error would have been acceptable, about 100 adults could have been used.

Pollsters Use Larger Samples

In any event, pollsters often use samples in the hundreds, or even the thousands, to produce narrower, more precise confidence intervals. When contrasted with the much smaller samples of most researchers, these larger samples reflect several factors, including the relative cheapness of pollsters' observations, often just a randomly dialed phone call away, as well as the notion that samples can't ever be too large in surveys—although they can be too large in experiments, as was discussed in Section 16.7.

A Final Caution

When based on randomly selected respondents, confidence intervals reflect only one kind of error—the *statistical* error due to sampling variability. There are other kinds of *nonstatistical* errors that could compromise the value of a confidence interval. For example, the previous estimate that 71 to 77 percent of all Americans favor capital punishment might have been inflated by adding to a neutral question, such as "Do you favor capital punishment?" a biased phrase, such as "in view of the recent epidemic of murders of innocent children?" Or the previous interval might fail to reflect the targeted population of all adult Americans because the random sample actually originated from a severely limited population, such as that for only registered voters, or because of a failure to minimize, through a concerted follow-up effort, the substantial minority of non-respondents (often as high as 30 percent) among the original random sample, whose attitudes toward capital punishment might differ appreciably from those who responded when first contacted. In the absence of this kind of background information, reports of confidence intervals should be interpreted cautiously.

*Exercise 17.5 In a recent scientific sample of about 900 adult Americans, 80 percent favor some form of stricter gun control, with a margin of error of ± 4 percent for a 95 percent confidence interval. In other words, the 95 percent confidence interval equals 76 to 84 percent. Indicate whether the following interpretations are true or false:

(a) The interval from 76 to 84 percent refers to possible values of the sample percent.

(b) The interval from 76 to 84 percent refers to possible values of the population percent for all adult Americans.

(c) The true population percent is between 76 and 84 percent.

(d) In the long run, a series of intervals similar to this one would fail to include the population percent about 5 percent of the time.

(e) In practice, we never really know whether the interval from 76 to 84 percent is true or false.

(f) We can be reasonably confident that the population percent is between 76 and 84 percent.

Answers on page 510

17.11 OTHER TYPES OF CONFIDENCE INTERVALS

Confidence intervals can be constructed not only for population means and percents but also for differences between two population means, as discussed in subsequent chapters. Although not discussed in this book, confidence intervals also can be constructed for other characteristics of populations, including variances and correlation coefficients.

Summary

Rather than test a hypothesis about a single population mean, you might choose to estimate this population characteristic, using a point estimate and a confidence interval.

In point estimation, a single sample characteristic, such as a sample mean, is used to estimate the corresponding population characteristic. Point estimates ignore sampling variability and, therefore, tend to be inaccurate.

Confidence intervals specify ranges of values that, in the long run, include the unknown population characteristic, such as the mean, a certain percent of the time. For instance, given a 95 percent confidence interval, then, in the long run, approximately 95 percent of all of these confidence intervals are true because they include the unknown population characteristic. Confidence intervals work because they are products of sampling distributions.

Any level of confidence can be assigned to a confidence interval, but the 95 percent and 99 percent levels are the most prevalent. Given one of these levels of confidence, then, even though we can never know whether a particular confidence interval is true or false, we can be "reasonably confident" that a particular interval actually includes the unknown population characteristic.

Narrower, more precise confidence intervals are produced by lower levels of confidence (for example, 95 percent rather than 99 percent) and by larger sample sizes.

Confidence intervals tend to be more informative than hypothesis tests are. Hypothesis tests merely indicate whether an effect is present, whereas confidence intervals indicate the possible size of the effect. Whenever appropriate—including whenever the null hypothesis has been rejected—consider using confidence intervals.

Confidence intervals for population percents or proportions are similar, both in origin and interpretation, to confidence intervals for population means.

Important Terms

Point estimate	**Level of confidence**
Confidence interval	**Margin of error**

REVIEW EXERCISES

*17.6 In Exercise 14.5 on page 238, it was concluded that the mean salary among the population of female members of the American Psychological Association is less than that ($51,500) for all comparable members who have a doctorate and teach full time.

(a) Given a population standard deviation of $3,000, and a sample mean salary of $50,300 for a random sample of 100 female members, construct a 99 percent confidence interval for the mean salary for all female members.

(b) Interpret this confidence interval. In other words, is there any consistent evidence that the mean salary for all female members falls below $51,500, the mean salary for all members?

Answers on page 510

17.7 In Exercise 15.4 on page 000, instead of testing a hypothesis, you might prefer to construct a confidence interval for the mean weight of all two-pound boxes of candy during a recent production shift.

(a) Given a population standard deviation of .30 ounce, and a sample mean weight of 33.09 ounces for a random sample of 36 candy boxes, construct a 95 percent confidence interval.

(b) Interpret this interval, given the manufacturer's desire to produce boxes of candy that on the average exceed 32 ounces.

17.8 It's tempting to claim that once a particular 95 percent confidence interval has been constructed, it includes the unknown population characteristic with a probability of .95. What's wrong with this claim?

*17.9 Imagine that one of the following 95 percent confidence intervals is based on an experiment dealing with the effect of vitamin C on IQ scores:

95% CONFIDENCE INTERVAL	LOWER LIMIT	UPPER LIMIT
1	102	108
2	95	104
3	112	115
4	90	111
5	91	98

(a) Which confidence interval is the most precise?

(b) Which most strongly supports the conclusion that vitamin C *increases* IQ scores?

(c) Which is the least precise?

(d) Which implies the largest sample size?

(e) Which most strongly supports the conclusion that vitamin C *decreases* IQ scores?

(f) Which would most likely stimulate the investigator to conduct an additional experiment, using larger sample sizes?

Answers on page 510

17.10 Unlike confidence intervals, hypothesis tests require that some predetermined population value be used to evaluate sample values. Can you think of any other differences between hypothesis tests and confidence intervals?

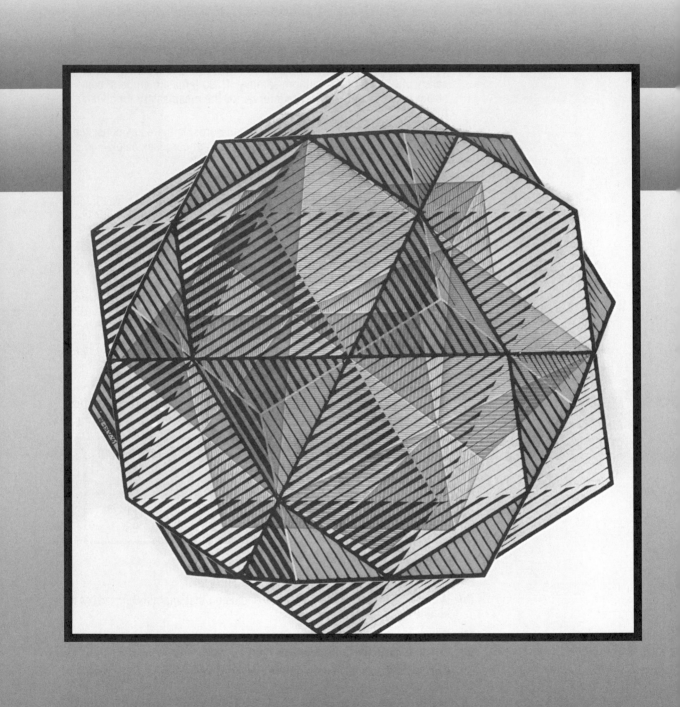

CHAPTER
18

t Test for One Sample

18.1   GAS MILEAGE INVESTIGATION
18.2   *t* SAMPLING DISTRIBUTION
18.3   *t* TABLES
18.4   *t* RATIO
18.5   HYPOTHESIS TESTS: A COMMON THEME
18.6   CONFIDENCE INTERVALS FOR μ BASED ON *t*
18.7   ASSUMPTIONS

DETAILS

18.8   ESTIMATING THE POPULATION STANDARD DEVIATION
18.9   ESTIMATING THE STANDARD ERROR ($s_{\bar{x}}$)
18.10 CALCULATIONS FOR *t* TEST FOR ONE
       SAMPLE (GAS MILEAGE INVESTIGATION)
18.11 DEGREES OF FREEDOM

Summary

Important Terms

Review Exercises

18.1 GAS MILEAGE INVESTIGATION

Federal law may eventually specify that new automobiles must average, for instance, 45 miles per gallon of gasoline. Because, of course, it is impossible to test all new cars, tests will be based on random samples from the entire production of each car model. If a hypothesis test indicates substandard performance, the manufacturer will be penalized 50 dollars per car for the entire production.

In these tests, the null hypothesis states that with respect to the mandated mean of 45 miles per gallon, nothing special is happening in the population for some car model—that is, there is no substandard performance and the population mean equals (or exceeds) the mandated 45 miles per gallon. The alternative hypothesis reflects a concern that the population mean is less than 45 miles per gallon. Symbolically, the two statistical hypotheses read

$$H_0: \mu \geq 45$$
$$H_1: \mu < 45$$

From most perspectives, including the manufacturer's, a type I error (a stiff penalty, even though the car complies with the standard) is very serious. Accordingly, it's appropriate to replace the customary .05 level of significance with a smaller value, for instance, the .01 level. To simplify computations, the projected one-tailed test is based on data from only six randomly selected cars.

For reasons that will become apparent, a new hypothesis test for a population mean, the *t* test, must replace the *z* test described in Chapter 14. Although we'll have more to say about the new *t* test in subsequent sections, spend a few minutes familiarizing yourself with the step-by-step summary for the gas mileage investigation, noting the considerable similarities between it and previous summaries of hypothesis tests with the *z* test.

18.2 *t* SAMPLING DISTRIBUTION

Sampling distribution of t

The distribution of t values that would be obtained if a value of t were calculated for each sample mean for all possible random samples of a given size from some population.

Like the sampling distribution of *z*, the **sampling distribution of** *t represents the distribution of* t *values that would be obtained if a value of* t *were calculated for each sample mean for all possible random samples of a given size from some population.* In the early 1900s, William Gosset discovered the sampling distribution of *t* and subsequently reported his achievement under the pen name of Student. Actually, Gosset discovered not just one but an entire family of *t* sampling distributions (or "Student's" distributions). Each *t* distri-bution is associated with a special number referred to as *degrees of freedom* (discussed in more detail in Section 18.11). When testing a hypothesis regarding the population mean, as in the current example, the number of degrees of freedom, abbreviated *df*, always equals the sample size minus one. Symbolically,

---

DEGREES OF FREEDOM (SINGLE SAMPLE)
$$df = n - 1 \qquad (18.1)$$

For instance, because the gas mileage investigation involves 6 cars, the corresponding t test is based on a sampling distribution with 5 degrees of freedom (from $df = 6 - 1$).

HYPOTHESIS TEST SUMMARY: *t* TEST FOR A POPULATION MEAN
(Gas Mileage Investigation)

Research Problem:

Does the mean gas mileage for some population of cars drop below the legally required minimum of 45 miles per gallon?

Statistical Hypotheses:

$$H_0: \mu \geq 45$$
$$H_1: \mu < 45$$

Decision Rule:

Reject H_0 at the .01 level of significance if t equals or is more negative than -3.365 (from Table B in Appendix D, given $df = n - 1 = 6 - 1 = 5$).

Calculations:

Given $\overline{X} = 43$, $s_{\overline{X}} = 0.89$

then

$$t = \frac{43 - 45}{0.89} = -2.25 \text{ (See Table 18.1 for computations.)}$$

Decision:

Retain H_0 at .01 level of significance because $t = -2.25$ is less negative than -3.365

Interpretation:

The population mean gas mileage *could* equal 45 or more miles per gallon; the manufacturer shouldn't be penalized.

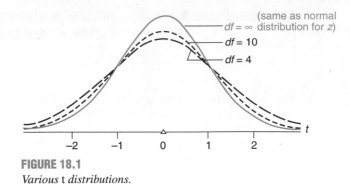

FIGURE 18.1
Various t *distributions.*

Compared to the Standard Normal Distribution

Figure 18.1 shows three *t* distributions. When there are an infinite (∞) number of degrees of freedom, the distribution of *t* is the same as the standard normal distribution of *z*. Notice that even with only 4 to 10 degrees of freedom, a *t* distribution shares several properties with the normal distribution. All *t* distributions are symmetrical, unimodal, and bell shaped, with a dense concentration that peaks in the middle (when *t* equals 0) and tapers off both to the right of the middle (as *t* becomes more positive) and to the left of the middle (as *t* becomes more negative). *The inflated tails of the* t *distribution, particularly apparent with small values of* df, *constitute the most important difference between* t *and* z *distributions.*

18.3 *t* TABLES

General Properties

To save space, tables for *t* distributions concentrate only on the critical values of *t* that correspond to the more common levels of significance. Table B of Appendix D lists the critical *t* values for either one- or two-tailed hypothesis tests at the .05, .01, and .001 levels of significance. All listed critical *t* values are positive and originate from the upper half of each distribution. Because of the symmetry of the *t* distribution, you can obtain the corresponding critical *t* values for the lower half of each distribution merely by placing a negative sign in front of any entry in the table.

Finding Critical *t* Values

To find a critical *t* in Table B of Appendix D, read the entry in the cell intersected by the row for the correct number of degrees of freedom and the column for the test specifications. For example, to find the critical *t* for the gas mileage investigation, first go to the right-hand panel for a one-tailed test, then locate the row corresponding to 5 degrees of freedom (from $df = n - 1 = 6 - 1 = 5$), and then locate the column for a one-tailed test at the .01 level of significance. The intersected cell specifies 3.365. A negative sign must be placed in front of 3.365, because the hypothesis test requires the lower tail to be critical.

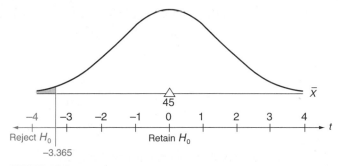

FIGURE 18.2

Hypothesized sampling distribution of t *(gas mileage investigation).*

Thus, -3.365 is the critical t for the gas mileage investigation, and the corresponding decision rule is illustrated in **Figure 18.2**, where the distribution of t is centered about the null hypothesized value.

If the gas mileage investigation had involved a two-tailed test (still at the .01 level with 5 degrees of freedom), then the left-hand panel for a two-tailed test would have been appropriate, and the intersected cell would have specified 4.032. A positive sign and a negative sign would have to be placed in front of 4.032, because both tails are critical. In this case, ± 4.032 would have been the pair of critical t values.

Missing *df* in Table

If the desired number of degrees of freedom doesn't appear in the *df* column of Table B in Appendix D, use the row in the table with the next smallest number of degrees of freedom. For example, when 36 degrees of freedom are specified, use the information from the row for 30 degrees of freedom. (The customary procedure of always rounding off to the next smallest *df* produces a slightly larger critical t that makes the null hypothesis slightly more difficult to reject. Therefore, it defuses potential disputes about borderline decisions by investigators with a vested interest in rejecting the null hypothesis.)

*Exercise 18.1 Find the critical *t* values for the following hypothesis tests:

(a) two-tailed test, $\alpha = .05$, *df* = 12

(b) one-tailed test, lower tail critical, $\alpha = .01$, *df* = 19

(c) one-tailed test, upper tail critical, $\alpha = .05$, *df* = 38

(d) two-tailed test, $\alpha = .01$, *df* = 48

   Answers on page 510

18.4 *t* RATIO

Almost always, as in the gas mileage investigation, *the population standard deviation is unknown and must be estimated from the sample.* The subsequent shift from the standard error of the mean, $\sigma_{\bar{X}}$, to its estimate, $s_{\bar{X}}$ has an important effect on the entire hypothesis test for a population mean. The familiar z test, which presumes that the ratio

$$z = \frac{\overline{X} - \mu_{hyp}}{\sigma_{\overline{X}}}$$

is normally distributed, must be replaced by a new t test, which presumes that the ratio

<table>
<tr><td colspan="2">t RATIO FOR SINGLE POPULATION MEAN</td></tr>
<tr><td>$$t = \frac{\overline{X} - \mu_{hyp}}{s_{\overline{X}}}$$</td><td>(18.2)</td></tr>
</table>

is distributed according to the t sampling distribution with $n - 1$ degrees of freedom, as described in Section 18.2.

For the gas mileage investigation, given that the sample mean gas mileage, $\overline{X}$, equals 43; that the hypothesized population mean, μ_{hyp}, equals 45; and that the estimated standard error, $s_{\overline{X}}$, equals 0.89 (from Table 18.1), Formula 18.2 becomes

$$t = \frac{43 - 45}{0.89} = -2.25$$

with $df = 5$. Because the observed value of t (-2.25) is less negative than the critical value of t (-3.365), the null hypothesis is retained, and we can conclude that the auto manufacturer shouldn't be penalized, because the mean gas mileage for the population cars *could* equal the mandated 45 miles per gallon.

Greater Variability of t Ratio

As has been noted, the tails of the sampling distribution for t are more inflated than are those for z, particularly when the sample size is small.[1] Consequently, to accommodate the greater variability of t, the critical t value must be larger than the corresponding critical z value. For example, given the one-tailed

---

[1]Essentially, the inflated tails are caused by extra variability in the denominator of t. For a complete explanation, see D. C. Howell, *Statistical Methods for Psychology*, 4th ed. (Belmont, CA: Duxbury, 1997), Chapter 7.

---

t Ratio

A replacement for the z ratio whenever the unknown population standard deviation must be estimated.

test at the .01 level of significance for the gas mileage investigation, the critical value for t (-3.365) is much larger than that for z (-2.33).

*Exercise 18.2 A consumers' group suspects that a large food chain makes extra money by supplying less than the specified weight of 16 ounces in its standard one-pound packages of ground beef. Given that a random sample of ten packages yields a mean of 14.7 ounces and an estimated standard error of the mean of 0.26 ounces, use the customary step-by-step procedure to test the null hypothesis at the .05 level of significance with t.

Answer on pages 510–511

18.5 HYPOTHESIS TESTS: A COMMON THEME

The remainder of this book discusses an alphabet-soup variety of tests—z, t, F, U, T, and H—for an assortment of situations. Notwithstanding the new formulas with their special symbols,

> **all of these hypothesis tests represent variations on the same theme: If some observed characteristic, such as the mean for a random sample, qualifies as a rare outcome under the null hypothesis, the hypothesis will be rejected. Otherwise, the hypothesis will be retained.**

To determine whether an outcome is rare, the observed characteristic is converted to some new value, such as t, and compared with critical values from the appropriate sampling distribution. Generally, if the observed value equals or exceeds a positive critical value (or if it equals or is more negative than a negative critical value), the outcome will be viewed as rare and the null hypothesis will be rejected.

18.6 CONFIDENCE INTERVALS FOR μ BASED ON *t*

Under slightly different circumstances, you might wish to estimate the unknown mean gas mileage for the population of cars, rather than test a hypothesis based on 45 miles per gallon. For example, there might be no legally required minimum of 45 miles per gallon, but merely a desire on the part of the manufacturer to estimate the mean gas mileage for a population of cars—possibly as a first step toward the design of a new, improved version of the current model.

When the population standard deviation is unknown and, therefore, must be estimated, as in the present case, t replaces z in the new formula for a confidence interval:

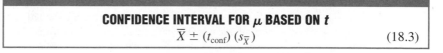

CONFIDENCE INTERVAL FOR μ **BASED ON** *t*
$$\overline{X} \pm (t_{\text{conf}}) (s_{\overline{X}}) \tag{18.3}$$

where $\overline{X}$ represents the sample mean; t_{conf} represents a number (distributed with $n - 1$ degrees of freedom) from the t tables, which satisfies the confidence specifications for the confidence interval; and s represents the estimated standard error of the mean, defined later in Formula 18.6 on page 299.

Finding t_{conf}

To find the appropriate value for t_{conf} in Formula 18.3, refer to Table B, Appendix D. Read the entry from the cell intersected by the row for the correct number of degrees of freedom and the column for the confidence specifications. In the present case, if a 95 percent confidence interval is desired, first locate the row corresponding to 5 degrees of freedom (from $df = n - 1 = 6 - 1 = 5$), and then locate the column for the 95 percent level of confidence, that is, the column heading identified with a single asterisk. (A double asterisk identifies the column for the 99 percent level of confidence.) The intersected cell specifies that a value of 2.571 be entered in Formula 18.6.[2]

Given this value for t_{conf}, as well as (from Table 18.1 on page 300) values of 43 for $\overline{X}$, the sample mean gas mileage, and 0.89 for $s_{\overline{X}}$, the estimated standard error, Formula 18.3 becomes

$$43 \pm (2.571)(0.89) = 43 \pm 2.29 = \begin{cases} 45.29 \\ 40.71 \end{cases}$$

It can be claimed, with 95 percent confidence, that the interval between 40.71 and 45.29 includes the true mean gas mileage for all of the cars in the population.

Interpretation

The interpretation of this confidence interval is the same as that based on z. In the long run, 95 percent of all confidence intervals, similar to the one just discussed, will include the unknown population mean. Although we never really know whether this particular confidence interval is true or false, we can be *reasonably confident* that the true mean for the entire population of cars is neither less than 40.71 miles per gallon nor more than 45.29 miles per gallon.

*Exercise 18.3 A consumers' group concludes (from Exercise 18.2 on page 295) that, in spite of the claims of the large food chain, the mean weight of its "one-pound" packages of ground beef drops below the specified 16 ounces.

....

[2] Specifications for confidence intervals are taken from the left-hand panel of Table B because the symmetrical limits of a confidence interval are analogous to a two-tailed hypothesis test. Both cases require that a specified number of standard errors be added and subtracted relative to either the value of the null hypothesis (to obtain the upper and lower critical values for a two-tailed hypothesis test) or the value of the sample mean (to obtain the upper and lower limits of a confidence interval).

(a) Construct a 95 percent confidence interval for the true weight of all "one-pound" packages of ground beef.

(b) Interpret this confidence interval.
Answers on page 511

18.7 ASSUMPTIONS

Whether testing hypotheses or constructing confidence intervals for population means, use *t* rather than *z* whenever, *as almost always is the case*, the population standard deviation is unknown. Strictly speaking, when using *t*, you must assume that the underlying population is normally distributed. Even if this normality assumption is violated, *t* retains much of its accuracy as long as sample size isn't too small. If a very small sample (less than about 10) is being used *and* you believe that the sample originates from a non-normal population—possibly because of a pronounced positive or negative skew among the observations in the sample—it would be wise to increase the size of the sample before testing a hypothesis or constructing a confidence interval.

DETAILS

18.8 ESTIMATING THE POPULATION STANDARD DEVIATION

If the population standard deviation is unknown, it must be estimated from the sample. This seemingly simple complication has important implications for hypothesis testing—indeed, it's the reason why the *z* test must be replaced by the *t* test, and why the estimated standard error, $s_{\bar{X}}$, replaces $\sigma_{\bar{X}}$ in Formula 18.2 on page 294.

One Possible Estimate (*S*)

One possible estimate of the unknown population standard deviation, σ, is the sample standard deviation, *S*, as defined in Formula 5.1:

$$S = \frac{\sqrt{\Sigma (X - \bar{X})^2}}{n}$$

This formula appears to provide a straightforward solution to our problem. Simply calculate *S* from Formula 5.1 and use that value to estimate σ. Unfortunately, Formula 5.1 is designed for *descriptive* purposes, as in Part 1 of this book, rather than for *estimation* purposes, as in Part 2. As a matter of fact, *S* tends to underestimate σ. This occurs because the divisor, n, in Formula 5.1 is slightly too large, because the sample contains only $n - 1$ (not n) valid bits of information *about variability in the population.* More about this in Section 18.11.

A More Accurate Estimate (s)

A more accurate estimate of σ can be obtained by replacing n with $n - 1$ in the denominator of the formula for S. Denoted by the small letter s, this estimate of σ is defined as

---

SAMPLE STANDARD DEVIATION FOR INFERENTIAL STATISTICS (DEFINITION FORMULA)

$$s = \frac{\sqrt{\Sigma(X - \bar{X})^2}}{n - 1} \tag{18.4}$$

Sample standard deviation for inferential statistics (s)

The version of the sample standard deviation, with $n - 1$ in its denominator, that is used in statistical inference to estimate the unknown population standard deviation.

*Always use this version of the **sample standard deviation, s,** to estimate the unknown population standard deviation, σ.* The value of s can be determined either from the preceding definition formula or from the following computation formula:

SAMPLE STANDARD DEVIATION FOR INFERENTIAL STATISTICS (COMPUTATION FORMULA)

$$s = \sqrt{\frac{n\Sigma X^2 - (\Sigma X)^2}{n(n - 1)}} \tag{18.5}$$

This formula can be derived from Formula 5.3, the original computation formula for S, by replacing one n term in the denominator with $n - 1$. Part I of Table 18.1 (in Section 18.10) shows that s equals 2.19, given values of 40, 44, 46, 41, 43, and 44 miles per gallon for the six cars in the gas mileage investigation described at the start of this chapter.

18.9 ESTIMATING THE STANDARD ERROR ($s_{\bar{X}}$)

Now s replaces σ in the formula for the standard error of the mean. Instead of

$$\sigma_{\bar{X}} = \frac{\sigma}{\sqrt{n}}$$

we have

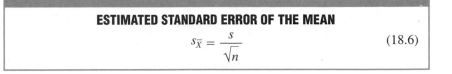

ESTIMATED STANDARD ERROR OF THE MEAN

$$s_{\bar{X}} = \frac{s}{\sqrt{n}} \qquad\qquad (18.6)$$

where $s_{\bar{X}}$ represents the estimated standard error of the mean; s is defined in Formulas 18.4 or 18.5; and n equals the sample size. This new version of the standard error, the **estimated standard error of the mean** ($s_{\bar{X}}$) *is used whenever the unknown population standard deviation must be estimated.* Part II of Table 18.1 shows that s equals 0.89 for the gas mileage investigation.

Estimated standard error of the mean ($S_{\bar{X}}$)

The version of the standard error of the mean that is used whenever the unknown population standard deviation must be estimated.

---

*Exercise 18.4 Calculate **(a)** the mean and **(b)** the estimated standard error of the mean for the random sample of ten "one-pound" packages of ground beef, referred to in Exercise 18.2, given the following weights in ounces: 16, 15, 14, 15, 14, 15, 16, 14, 14, 14.

Answers on page 511

18.10 CALCULATIONS FOR *t* TEST FOR ONE SAMPLE (GAS MILEAGE INVESTIGATION)

Table 18.1 on page 300 shows all of the calculations for a *t* test for the gas mileage investigation.

18.11 DEGREES OF FREEDOM

The notion of degrees of freedom is used throughout the remainder of this book.

Degrees of freedom (df)

The number of values free to vary, given one or more mathematical restrictions on a sample of observed values used to estimate *some unknown population characteristic.*

Degrees of freedom (df) **refer to the number of values that are free to vary, given one or more mathematical restrictions on a sample of observed values used to** *estimate* **some unknown population characteristic.**

Typically, when used to estimate some unknown population characteristic, not all observed values within the sample are free to vary. For example, the gas mileage data consist of six values: 40, 44, 46, 41, 43, and 44. Nevertheless, the *t* test for these data has only five degrees of freedom because only five of these six observed values are free to vary and, therefore, provide valid information for purposes of estimation. Let's look at this more closely, remembering that the *concept of degrees of freedom is introduced only because we are using observations in a sample to estimate some unknown characteristic of the population.*

Why One *df* Is Lost

The loss of one degree of freedom occurs, without fanfare, when the sample standard deviation, *s*, is used to *estimate* the unknown population standard deviation, σ. The missing degree of freedom can be detected most easily by referring to Formula 18.4 for *s*, that is,

> ## Table 18.1
> ## CALCULATIONS FOR *t* TEST (GAS MILEAGE INVESTIGATION)
>
> ### I. FINDING $\overline{X}$ AND s
> **(a)** Computational sequence:
> Assign a value to n **1**
> Sum all X scores **2**
> Substitute numbers in formula **3** and solve for $\overline{X}$
> Square each X score **4**, one at a time, and then add all squared X scores **5**
> Substitute numbers into the formula **6** and solve for s.
>
> **(b)** Data and computations:
>
> | | **4** |
> |---|---|
> | **X** | **X²** |
> | 40 | 1600 |
> | 44 | 1936 |
> | 46 | 2116 |
> | 41 | 1681 |
> | 43 | 1849 |
> | 44 | 1936 |
> | **1** $n = 6$ **2** $\Sigma X = 258$ | **5** $\Sigma X^2 = 11118$ |
>
> **3** $\overline{X} = \dfrac{\Sigma X}{n} = \dfrac{258}{6} = 43$
>
> **6** $s = \sqrt{\dfrac{n(\Sigma X^2) - (\Sigma X)^2}{n(n-1)}} = \sqrt{\dfrac{6(11118) - (258)^2}{6(6-1)}}$
>
> $= \sqrt{\dfrac{66708 - 66564}{30}} = \sqrt{\dfrac{144}{30}} = \sqrt{4.8} = 2.19$
>
> ### II. FINDING $s_{\overline{X}}$
> **(a)** Computational sequence:
> Substitute numbers obtained above in formula **7** and solve for $s_{\overline{X}}$
> **(b)** Computations:
>
> **7** $s_{\overline{X}} = \dfrac{s}{\sqrt{n}} = \dfrac{2.19}{\sqrt{6}} = \dfrac{2.19}{2.45} = 0.89$
>
> ### III. FINDING THE OBSERVED t
> **(a)** Computational sequence:
> Assign value to μ_{hyp} **8**, the hypothesized population mean.
> Substitute numbers obtained above in formula **9** and solve for t.
> **(b)** Computations:
>
> **8** $\mu_{hyp} = 45$
>
> **9** $t = \dfrac{\overline{X} - \mu_{hyp}}{s_{\overline{X}}} = \dfrac{43 - 45}{0.89} = \dfrac{-2}{0.89} = -2.25$

$$s = \sqrt{\frac{\Sigma(X - \overline{X})^2}{n - 1}}$$

To estimate σ from the gas mileage data, this formula specifies that the six original gas mileage values, X, be expressed, one at a time, as positive or negative distances from their sample mean, $\overline{X}$. At this point, a subtle mathematical restriction causes the loss of one degree of freedom. It's always true, as demonstrated in **Table 18.2** for the gas mileage data, that *the sum of all values, expressed as deviations from their mean, equals zero.* (If you're skeptical, verify this property for any set of values. Also, you might refer to Section 4.3, in which this property of the mean was first discussed.) Given *any* five of the six deviations in Table 18.2, the remaining deviation is not free to vary; its value must comply with the restriction that the sum of all deviations must equal zero. For instance, given the first five deviations in Table 18.2, the remaining deviation must equal 1, and therefore the remaining (sixth) observation must equal 44, that is, the mean of 43 combined with the deviation of 1. By the same token, given the last five deviations in Table 18.2, the remaining deviation must equal -3, and therefore the remaining (first) observation must equal 40, that is, the mean of 43 combined with the deviation of -3.

A slightly different perspective may further clarify the notion of degrees of freedom. The population standard deviation reflects the deviations of observations from the population mean, $X - \mu$. To estimate the unknown population standard deviation, it would be most efficient to take a random sample of these deviations, but usually this is impossible because the population mean is unknown. Typically, therefore, the unknown population mean, μ, must be replaced with the sample mean, $\overline{X}$, and a random sample of these deviations, $X - \overline{X}$, must be used to estimate the population deviations, $X - \mu$. Even though there are n deviations in the sample, only $n - 1$ of the sample deviations are free to vary, if the sum of the n deviations from *their own sample mean* always equals zero. In other words, only $n - 1$ of the

---

Table 18.2
DEMONSTRATION: $\Sigma(X - \overline{X}) = 0$ (GAS MILEAGE DATA)

X	$X - \overline{X}$	*Positive Deviations*	*Negative Deviations*
40	$40 - 43$		-3
44	$44 - 43$	1	
46	$46 - 43$	3	
41	$41 - 43$		-2
43	$43 - 43$	0	
44	$44 - 43$	1	
		$\overline{5}$	$\overline{-5}$

Therefore, $\Sigma(X - \overline{X}) = 5 + (-5) = 5 - 5 = 0$

sample deviations are free to supply valid information about population deviations; one bit of valid information—one degree of freedom—has been lost because the known sample mean is used to replace the unknown population mean.

S Less Accurate Than s

As has been noted, the original version of the sample standard deviation, S, tends to underestimate the population standard deviation, σ. This occurs because, even though there are only $n - 1$ independent deviations (estimates of variability) in the numerator, n is retained in the denominator. A more accurate estimate of σ is obtained when the denominator term reflects the number of independent deviations—that is, the number of degrees of freedom—in the numerator, as in formula for s, in which the denominator term equals $n - 1$.

Other Mathematical Restrictions

In subsequent sections, we shall encounter other mathematical restrictions, and sometimes more than one degree of freedom will be lost. In any event, however, the degrees of freedom always indicate the number of values free to vary, given one or more mathematical restrictions on a set of values used to estimate some unknown population characteristic.

*Exercise 18.5 As a first step toward modifying his study habits, Phil keeps daily records of his study time.

(a) During the first two weeks, Phil's mean study time equals 20 hours per week. If he studied 22 hours during the first week, how many hours did he study during the second week?

(b) During the first four weeks, Phil's mean study time equals 21 hours. If he studied 22 hours during the first week, 18 hours during the second week, and 21 hours during the third week, how many hours did he study during the fourth week?

(c) Given the information in (a) and (b), indicate how many degrees of freedom are associated with each situation.

(d) Describe the mathematical restriction that causes a loss of degrees of freedom in the previous situations.

Answers on page 511

Summary
. . . .

When the population standard deviation, σ, is unknown, it must be estimated with the sample standard deviation, s. By the same token, the standard error of the mean, $\sigma_{\bar{X}}$, then must be estimated with $s_{\bar{X}}$. Under these circumstances, t rather than z should be used to test a hypothesis or to construct a confidence interval for the population mean.

The t ratio, as defined in Formula 18.2, is distributed with $n - 1$ degrees of freedom, and the critical t values are obtained from Table B, Appendix D. Because of the inflated tails of t sampling distributions—particularly when the sample size is small—the critical t values are larger (either positive or negative) than the corresponding critical z values.

When using t, you must assume that the underlying population is normally distributed. Violations of this assumption are important only when the observations in small samples appear to originate from non-normal populations.

Degrees of freedom (*df*) refer to the number of values free to vary, given one or more mathematical restrictions on a set of values used to estimate some population characteristic. When s is used to estimate σ, only $n - 1$ of the deviations in the numerator of s are free to vary, because of the mathematical restriction that the sum of deviations from the sample mean always must equal zero. The loss of one degree of freedom requires that $n - 1$ be used in the denominator term for s.

Important Terms

Sampling distribution of *t*

***t* ratio**

Sample standard deviation for inferential statistics (*s*)

Estimated standard error of the mean ($s_{\bar{X}}$)

Degrees of freedom (*df*)

REVIEW EXERCISES

Reminder about Computational Accuracy

Whenever necessary, round numbers two places to the right of the decimal, using the rounding procedure described in Section 7 of Appendix A. Otherwise, if your preliminary computations are based on numbers with more than two digits to the right of the decimal point—possibly because you're using the full capacity of a calculator or computer—*slight* differences occasionally will appear between your answers and those in Appendix C.

*18.6 A library system lends books for periods of 21 days. This policy is being reevaluated in view of a possible new loan period that could be either longer or shorter than 21 days. To aid in this decision, book-lending records were consulted to determine the loan periods actually used by the patrons. A random sample of eight records revealed the following loan periods in days: 21, 15, 12, 24, 20, 21, 13, 16. Test the null hypothesis with t, using the .05 level of significance.

Answer on page 511

18.7 It's well established that lab rats require, on the average, 32 trials before reaching a criterion of three consecutive errorless trials in a complex water maze. To determine whether a mildly adverse stimulus has any effect on performance, a sample of seven lab rats were given a mild electrical shock just before each trial. They required the following number of trials before reaching the criterion: 35, 38, 39, 33, 31, 32, 36.

(a) Test the null hypothesis with t, using the .05 level of significance.

(b) Construct a 95 percent confidence interval for the true number of trials required to learn the water maze.

(c) Interpret this confidence interval.

*18.8 Is the temperature of earth getting warmer because, instead of escaping into outer space, heat is trapped by so-called greenhouse gas emissions, such as carbon dioxide, in the Earth's atmosphere? An issue of *Science News* (January 17, 1998) describes globally averaged temperatures based on thousands of land and sea stations. Although the long-term average temperature, in degrees Fahrenheit, equals 61.7, temperatures for each of the seven most recent years were above this average. Listed in chronological order, these yearly averages were 62.2, 61.9, 61.9, 62.1, 62.3, 62.1, 62.5.[3]

(a) Using t and the .01 level of significance, test the null hypothesis that the temperature of the Earth is not getting warmer. *Hint*: Simplify computations by setting the null hypothesized value of 61.7 equal to 0, and reexpressing the seven remaining temperatures as deviations above 61.7. For example, reexpress 62.2 as a deviation of .5 (above 61.7).

(b) If appropriate (because the null hypothesis has been rejected), construct a 99 percent confidence interval and interpret this interval.

Answers on page 512

18.9 A tire manufacturer wishes to determine whether, on the average, a brand of steel-belted radial tires provides more than 50,000 miles of wear. A random sample of 36 tires yielded a sample mean, $\overline{X}$, of 52,100 miles and a sample standard deviation, s, of 2,500 miles.

(a) Use t to test the null hypothesis at the .01 level of significance.

(b) If appropriate (because the null hypothesis has been rejected), construct a 99 percent confidence interval, and interpret this interval.

18.10 Assume that on the average, healthy young adults dream 90 minutes each night, as inferred from several measures, including rapid eye movement. An investigator wishes to determine whether the consumption of alcohol just before sleep affects the amount of dream time. After consuming a standard amount of alcohol, dream time is monitored for each of 28 healthy young adults in a random sample. Results show a sample mean, $\overline{X}$, of 88 minutes and a sample standard deviation, s, of 9 minutes.

(a) Use t to test the null hypothesis at the .05 level of significance.

(b) If appropriate (because the null hypothesis has been rejected), construct a 95 percent confidence interval, and interpret this interval.

18.11 In the gas mileage test described in this chapter, would you prefer a smaller or a larger sample size if you were

(a) the car manufacturer? Why?

(b) a vigorous prosecutor for a federal regulatory agency? Why?

....

[3]Thanks to Rob Quayle of the National Climatic Data Center for providing the actual numerical values used in this exercise.

18.12 Even though the population standard deviation is unknown, an investigator uses z rather than the more appropriate t to test a hypothesis at the .05 level of significance.

(a) Does .05 describe the true level of significance? If not, is the true level of significance larger or smaller than .05?

(b) Does the value of the critical z coincide with that of the true critical t? If not, is the true critical t value larger or smaller than that of the critical z?

18.13 When discussing degrees of freedom, it was suggested that the most efficient way to estimate the population standard deviation, σ, is to take a random sample of the deviations $X - \mu$. Ordinarily this is impossible because the value of the population mean, μ, is unknown. If the value of μ were known, however, how many degrees of freedom would be associated with an estimate of σ based on n of these deviations $(X - \mu)$?

CHAPTER 19

t Test for Two Independent Samples

19.1   BLOOD-DOPING EXPERIMENT
19.2   TWO INDEPENDENT SAMPLES
19.3   TWO HYPOTHETICAL POPULATIONS
19.4   STATISTICAL HYPOTHESES
19.5   SAMPLING DISTRIBUTION OF $\overline{X}_1 - \overline{X}_2$
19.6   MEAN OF THE SAMPLING DISTRIBUTION OF $\overline{X}_1 - \overline{X}_2$
19.7   STANDARD ERROR OF THE SAMPLING DISTRIBUTION OF $\overline{X}_1 - \overline{X}_2$
19.8   *z* TEST
19.9   *t* RATIO
19.10   CONFIDENCE INTERVAL FOR $\mu_1 - \mu_2$
19.11   ASSUMPTIONS

DETAILS

19.12   ESTIMATING THE STANDARD ERROR ($s_{\overline{X}_1 - \overline{X}_2}$)
19.13   CALCULATIONS FOR *t* TEST FOR TWO INDEPENDENT SAMPLES (BLOOD-DOPING EXPERIMENT)

Summary

Important Terms

Review Exercises

19.1 BLOOD-DOPING EXPERIMENT

During the 1998 Tour de France, the world's best-known bicycle race, some cyclists were expelled for attempting to enhance their performance by "blood doping" with the synthetic hormone erythropoietin, or EPO, that stimulates the production of oxygen-bearing (and fatigue-inhibiting) red blood cells. An investigator wants to determine whether this type of blood doping increases the endurance of athletes under controlled laboratory conditions. (We'll ignore the very real medical, ethical, and legal issues raised by this type of experimentation.) Volunteer athletes from the local track team are randomly assigned to one of two groups: a blood-doped group, (X_1), that receives a prescribed amount of EPO and a non-blood-doped group, (X_2), that receives a comparable amount of a harmless neutral substance. Subsequently, after an appropriate interval of time has elapsed, each athlete runs on a rapid treadmill until exhausted. Total time on the treadmill is used as the measure of endurance.

19.2 TWO INDEPENDENT SAMPLES

Two independent samples

Observations in one sample aren't paired, on a one-to-one basis, with observations in the other sample.

In the present chapter, the two samples are assumed to be independent. ***Two independent samples*** *occur if the observations in one sample aren't paired, on a one-to-one basis, with observations in the other sample.* The discussion in the present chapter for two independent samples should be contrasted with that in the next chapter for two matched samples (when the investigator creates pairs of athletes by matching them for body weight.)[1]

19.3 TWO HYPOTHETICAL POPULATIONS

The subjects in the blood-doping experiment originate from a very limited real population: all volunteer athletes from the local track team. This is hardly an inspiring target for statistical inference. A standard remedy is to characterize the sample of athletes *as if* it were a random sample from a much larger hypothetical population, loosely defined as "all similar volunteer athletes who could conceivably participate in the experiment." Strictly speaking, there are two hypothetical populations, one defined for the endurance scores of athletes who are blood doped with EPO and the other for the endurance scores of athletes who are not blood doped. These two populations are cited in the null hypothesis, and as noted in Chapter 11, any generalizations to hypothetical populations must be viewed as provisional conclusions based on the wisdom of the researcher rather than on any logical or statistical necessity. Only additional experimentation can resolve whether a given experimental finding merits the generality assigned to it by the researcher.

---

[1]Don't confuse "independent samples" with "independent variable." *Independent samples* refers to the absence of any pairing or matching across the two samples. *Independent variable* refers to the variable manipulated by the investigator, that is, in the above experiment, either real or fake blood doping.

Difference between Population Means: Blood Doping

The difference between population means reflects the **effect** of blood doping with EPO on endurance. If blood doping has little or no effect on endurance, then the endurance scores would tend to be about the same for both populations of athletes, and the difference between populations means would be close to zero. But if blood doping does facilitate endurance, the scores for the blood-doped population would tend to exceed those for the non-blood-doped population, and the difference between population means would be positive: The stronger the facilitative effect of blood doping on endurance, the larger the positive difference between population means. If blood doping hinders endurance, the endurance scores for the blood-doped population would tend to be exceeded by those for the non-blood-doped population, and the difference between population means would be negative.

19.4 STATISTICAL HYPOTHESES

Null Hypothesis

According to the null hypothesis for the blood-doping experiment, nothing special is happening because blood doping with EPO doesn't facilitate endurance, that is, either there is no difference between the means for the two populations or the difference between population means is negative in opposition to blood doping. An equivalent statement in symbols reads

$$H_0: \mu_1 - \mu_2 \leq 0$$

where H_0 represents the null hypothesis, and μ_1 and μ_2 represent the mean endurance scores for the blood-doped and non-blood-doped populations, respectively.

Alternative Hypothesis

The investigator wants to reject the null hypothesis only if blood doping increases endurance scores. Given this perspective, the alternative (or research) hypothesis should specify that the difference between population means is positive in favor of blood doping. An equivalent statement in symbols reads

$$H_1: \mu_1 - \mu_2 > 0$$

where H_1 represents the alternative hypothesis, and as above, μ_1 and μ_2 represent the mean endurance scores for the blood-doped and non-blood-doped populations, respectively. This directional alternative hypothesis translates into a one-tailed test with the upper tail critical.

Two Other Possible Alternative Hypotheses

Although not appropriate for the current experiment, two other possible alternative hypotheses exist. Another directional hypothesis, expressed as

$$H_1: \mu_1 - \mu_2 < 0$$

translates into a one-tailed test with the lower tail critical, whereas a nondirectional hypothesis, expressed as

$$H_1: \mu_1 - \mu_2 \neq 0$$

translates into a two-tailed test. As emphasized in Section 15.4, a directional alternative hypothesis should be used when there's a concern *only* about differences in a particular direction.

*Exercise 19.1 Specify both the null and alternative hypotheses for each of the following studies. Remember that a directional alternative hypothesis is appropriate only when a word or phrase justifies an exclusive concern about population mean differences in a particular direction. The group that receives the special treatment customarily is identified with μ_1.

(a) After randomly assigning migrant children to two groups, a school psychologist wishes to determine whether there is a difference in the mean reading comprehension scores between the group that receives a special bilingual reading program and the group that receives a traditional reading program.

(b) On further reflection, the school psychologist decides that, because of the extra expense of the special bilingual reading program, she wishes to reject the null hypothesis only if there is evidence that reading comprehension scores are improved, on the average, by the special reading program.

(c) A health researcher wishes to determine whether, on the average, cigarette consumption is reduced for smokers who chew caffeine gum. Accordingly, smokers in attendance at an anti-smoking workshop are randomly assigned to two groups—one that chews caffeine gum and one that doesn't—and their daily cigarette consumption is monitored for six months after the workshop.

(d) A political scientist wishes to determine whether males and females tend to differ about the amount of money that, in their opinion, should be spent on national defense. After being informed about the size of the current U.S. defense budget, in billions of dollars, randomly selected males and females are asked to indicate the percent by which they would alter this amount—for example, −8 percent for an 8 percent reduction, 0 percent for no change, 4 percent for a 4 percent increase.

Answers on page 512

19.5 SAMPLING DISTRIBUTION OF $\overline{X}_1 - \overline{X}_2$

In previous chapters, when the null hypothesis for a single population mean was tested, considerable attention was devoted to the single sample mean, $\overline{X}$, and its sampling distribution. In the present chapter, when the null hypothesis for the difference between two population means is being tested, we focus on the **sampling distribution of sample means differences,** $\overline{X}_1 - \overline{X}_2$.

Definition

Sampling distribution of sample means differences, $\overline{X}_1 - \overline{X}_2$
Differences between sample means based on all possible pairs of random samples—of given sizes—from two underlying populations.

In practice, of course, there's only one observed difference between means, $\overline{X}_1 - \overline{X}_2$. To determine whether this one difference qualifies as a common or a rare outcome under the null hypothesis, it is viewed as originating from the sampling distribution of the differences between means, that is, the sampling distribution of sample mean differences, $\overline{X}_1 - \overline{X}_2$, for all possible pairs of random samples—of given sizes—from the two underlying populations.

Not Constructed from Scratch

Because all the possible pairs of random samples usually translate into a huge number of possibilities—often of astronomical proportions—the sampling distribution of $\overline{X}_1 - \overline{X}_2$ is not constructed from scratch. As with the sampling distribution of $\overline{X}$ described in Chapter 13, statistical theory must be relied on for information about the mean and standard error for this new sampling distribution.

19.6 MEAN OF THE SAMPLING DISTRIBUTION OF $\overline{X}_1 - \overline{X}_2$

In the one-sample case, the mean of the sampling distribution of $\overline{X}$ equals the population mean. Likewise, in the two-sample case, *the mean of the sampling distribution of* $\overline{X}_1 - \overline{X}_2$ *equals the difference between population means.*

If you think about it, this conclusion isn't particularly startling. Because of sampling variability, it's unlikely that the one observed difference between sample means will equal the difference between population means. Instead, it's likely that just by chance, this one observed difference will be either larger or smaller than the difference between population means. However, because not just one, but all possible differences between sample means contribute to the mean of the sampling distribution of $\overline{X}_1 - \overline{X}_2$, the effects of sampling variability are neutralized, and the mean of the sampling distribution of $\overline{X}_1 - \overline{X}_2$ equals the difference between population means. Accordingly, these two terms are used interchangeably. Any claims about the difference between population means can be transferred directly to the mean of the sampling distribution of $\overline{X}_1 - \overline{X}_2$.

Standard error of $\overline{X}_1 - \overline{X}_2$
A rough measure of the average amount by which any difference between sample means deviates from the difference between population means.

19.7 STANDARD ERROR OF THE SAMPLING DISTRIBUTION OF $\overline{X}_1 - \overline{X}_2$

The sampling distribution of $\overline{X}_1 - \overline{X}_2$ also has a standard deviation, referred to as the standard error of the difference between sample means. The **standard error of** $\overline{X}_1 - \overline{X}_2$ *is a rough measure of the average amount by which any* *sample mean difference deviates, just by chance, from the mean of the sampling distribution of $\overline{X}_1 - \overline{X}_2$ (or the difference between population means).*

Effect of Increased Sample Sizes

The size of standard error of $\overline{X}_1 - \overline{X}_2$—much like that of the original standard error described in Section 13.5—becomes smaller with increases in sample sizes. As a result, the values of $\overline{X}_1 - \overline{X}_2$ tend to cluster closer to the mean of the sampling distribution of $\overline{X}_1 - \overline{X}_2$ (and the difference between population means), allowing more precise generalizations from samples to populations.

19.8 z TEST

The hypothesis test for blood doping will be based not on the hypothesized sampling distribution of $\overline{X}_1 - \overline{X}_2$, but on its counterpart, the hypothesized sampling distribution of t. There also is a z test for two population means, but its use requires that both population standard deviations be known. In practice, this information is rarely available, and therefore, the z test for two population means is hardly ever appropriate. No further description of the z test will be given in this or the next chapter.

19.9 t RATIO

In the blood-doping experiment, the null hypothesis can be tested with the following ratio:

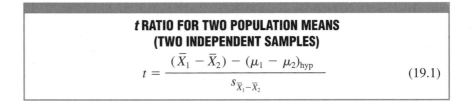

**t RATIO FOR TWO POPULATION MEANS
(TWO INDEPENDENT SAMPLES)**

$$t = \frac{(\overline{X}_1 - \overline{X}_2) - (\mu_1 - \mu_2)_{\text{hyp}}}{s_{\overline{X}_1 - \overline{X}_2}} \qquad (19.1)$$

which complies with a t sampling distribution with degrees of freedom equal to the sum of the two sample sizes minus two, that is, $n_1 + n_2 - 2$, for reasons discussed in Section 19.12. (This sampling distribution represents the distribution of t values that would be obtained if a value of t were calculated for each pair among all possible pairs of random samples of given sizes, n_1 and n_2, from the two underlying populations.) In Formula 19.1, $\overline{X}_1 - \overline{X}_2$ represents the one observed difference between sample means; $(\mu_1 - \mu_2)_{\text{hyp}}$ represents the hypothesized difference (of zero) between population means; and $s_{\overline{X}_1 - \overline{X}_2}$ represents the estimated standard error, as defined later in Formulas 19.3 and 19.4 on page 317.

HYPOTHESIS TEST SUMMARY: *t* TEST FOR TWO POPULATION MEANS: INDEPENDENT SAMPLES
(Blood-doping Experiment)

Research Problem:

Does the population mean endurance score for athletes who are blood doped with EPO exceed that for athletes who are not blood doped?

Statistical Hypotheses:

$$H_0: \mu_1 - \mu_2 \leq 0$$
$$H_1: \mu_1 - \mu_2 > 0$$

Decision Rule:

Reject H_0 at the .05 level of significance if t is equal to or more positive than 1.860 (from Table B, Appendix D, given $df = n_1 + n_2 - 2 = 5 + 5 - 2 = 8$).

Calculations:

$$t = \frac{(7 - 6) - (0)}{1.10} = 0.91 \quad \text{(See Table 19.1 on pages 319–320}$$
for all computations.)

Decision:

Retain H_0 at .05 level of significance because $t = 0.91$ is less positive than 1.860.

Interpretation:

The difference between population means *could* equal zero (or less); there is no evidence that blood doping with EPO increases the endurance scores of athletes.

Spend a few minutes familiarizing yourself with the step-by-step hypothesis test summary shown above for the current blood-doping experiment, which involves two groups of five athletes each. Note the considerable similarities between it and previous summaries of hypothesis tests. Also note that **Figure 19.1** illustrates the decision rule for the current test. As usual, the sampling distribution of *t* is centered about the null hypothesized value of zero.

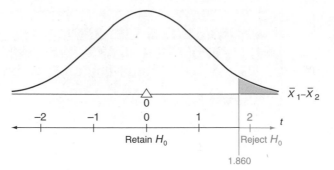

FIGURE 19.1
Hypothesized sampling distribution of t *(blood-doping experiment).*

19.10 CONFIDENCE INTERVAL FOR $\mu_1 - \mu_2$

Let's assume that the experimenter is not interested in determining merely whether blood doping increases endurance. Instead, he wishes to estimate, with the aid of a confidence interval, the possible sizes of the effect of blood doping on endurance, *whatever that effect might be*. Under these circumstances, a confidence interval must be constructed for the difference between population means, $\mu_1 - \mu_2$.

Confidence Interval for

$\mu_1 - \mu_2$

A range of values that, in the long run, includes the unknown difference between population means a certain percent of the time.

> **Confidence intervals for $\mu_1 - \mu_2$ specify ranges of values that, in the long run, include the unknown difference between population means a certain percent of the time.**

For instance, using the following techniques, the investigator of blood doping might claim, with 95 percent confidence, that the interval between -1.54 and 3.54 minutes includes the true difference between population means (due to blood doping with EPO).

Given that two samples are independent, as in the present blood-doping experiment, a confidence interval for $\mu_1 - \mu_2$ can be constructed from the following expression:

CONFIDENCE INTERVAL FOR $\mu_1 - \mu_2$
(TWO INDEPENDENT SAMPLES)

$$\overline{X}_1 - \overline{X}_2 \pm (t_{\text{conf}})(s_{\overline{X}_1 - \overline{X}_2}) \qquad (19.2)$$

---

where $\overline{X}_1 - \overline{X}_2$ represents the difference between sample means; t_{conf} represents a number (distributed with $n_1 + n_2 - 2$ degrees of freedom) from the *t* tables, which satisfies the confidence specifications; and $s_{\overline{X}_1 - \overline{X}_2}$ represents the estimated standard error defined in Formula 19.3.

Finding t_{conf}

To find the appropriate value of t_{conf} in Formula 19.2, refer to Table B, Appendix D, and follow the same procedure as described in Section 18.6 for one sample. For example, if a 95 percent confidence interval is desired for the blood-doping experiment, first locate the row corresponding to 8 degrees of freedom (from $df = n_1 + n_2 - 2 = 5 + 5 - 2 = 8$, given five subjects per group) and then locate the column for the 95 percent level of confidence, that is, the column heading identified with a single asterisk. The intersected cell specifies that 2.306 be entered for t_{conf} in Formula 19.2.

Given this value for t_{conf}, and (from Table 19.1 on pages 319–320) values of 1 for $\overline{X}_1 - \overline{X}_2$ (the difference between sample means) and 1.10 for $s_{\overline{x}_1 - \overline{x}_2}$ (the estimated standard error), Formula 19.2 becomes

$$1 \pm (2.306)(1.10) = 1 \pm 2.54 = \begin{cases} 3.54 \\ -1.54 \end{cases}$$

Now it can be claimed, with 95 percent confidence, that the interval between -1.54 and 3.54 includes the true difference between population means.

Interpreting Confidence Intervals for $\mu_1 - \mu_2$

Notice that the numbers in this confidence interval refer to *differences* between population means, and the signs are particularly important because they indicate the *direction* of these differences. Otherwise, the interpretation of a confidence interval for $\mu_1 - \mu_2$ is the same as that for μ. In the long run, 95 percent of all confidence intervals, similar to the one just stated, will include the unknown difference between population means. Although we never really know whether this particular confidence interval is true or false, we can be *reasonably confident* that the true difference between population means (due to blood doping with EPO) is neither less than -1.54 minutes nor more than 3.54 minutes.

Because both positive and negative differences appear in this confidence interval, no single interpretation is possible. The appearance of negative differences, such as -1.54, indicates that blood doping might actually hinder endurance, whereas the appearance of positive differences, such as 3.54, indicates that blood doping might facilitate endurance. The inclusion of a zero difference indicates that blood doping may have no effect whatsoever on endurance.[2]

....

[2]Because of the common statistical origins of confidence intervals and hypothesis tests, the appearance of a zero difference in a 95 percent confidence interval always signifies that the null hypothesis would have been retained if the *same data* were used to conduct a comparable hypothesis test—in this case, a *two-tailed* test at the .05 level of significance. Although the previous hypothesis test for blood doping wasn't strictly comparable, because it was a one-tailed rather than a two-tailed test, the appearance of a zero difference in the above confidence interval is, nevertheless, consistent with the previous decision to retain the null hypothesis for blood doping.

In spite of these contradictory possibilities, the relatively large upper limit of 3.54 minutes might stimulate additional research on blood doping with EPO. Any new experiment on blood doping should use a larger sample size in order to produce a narrower, more precise confidence interval. If the signs of the two new limits are either both positive or both negative, the new confidence interval would permit a single interpretation for all possibilities.

*Exercise 19.2 Imagine that one of the following 95 percent confidence intervals is based on the blood-doping experiment with two independent groups of subjects. (Remember, each of these numbers reflects a possible *difference,* in minutes, between the two population means, and the sign of the difference is crucial.)

95% CONFIDENCE INTERVAL	LOWER LIMIT	UPPER LIMIT
1	−3.45	4.25
2	1.89	2.21
3	−1.54	−0.32
4	0.21	1.53
5	−2.53	1.78

(a) Which confidence interval is most precise?

(b) Which most strongly supports the conclusion that blood doping *facilitates* endurance?

(c) Which is the least precise?

(d) Which implies the largest sample sizes?

(e) Which most strongly supports the conclusion that blood doping *hinders* endurance?

(f) Which would most likely stimulate the investigator to conduct an additional experiment using larger sample sizes?

Answers on page 512

19.11 ASSUMPTIONS

Whether testing a hypothesis or constructing a confidence interval, *t* assumes that both underlying populations are normally distributed with equal variances. You needn't be too concerned about violations of these assumptions, particularly if both sample sizes are equal and each is fairly large (greater than about 10). Otherwise, in the *unlikely* event that you observe conspicuous departures from normality or equality of variances in the data for the two groups, consider the following possibilities: (1) Increase sample sizes (to minimize the effect of any non-normality); (2) equate sample sizes (to minimize the effect of unequal population variances); (3) use a slightly less sensitive, more complex version of *t* (designed for unequal variances, alluded to in Section 22.5, and described more fully in some advanced books, such as D. C. Howell's *Statistical Meth-*

ods for Psychology, 4th ed. [Belmont, CA: Duxbury, 1997], chapter 7); or (4) use a less sensitive, but more assumption-free test, such as the Mann-Whitney *U* test described in Chapter 25.

DETAILS

19.12 ESTIMATING THE STANDARD ERROR ($s_{\bar{X}_1 - \bar{X}_2}$)

Ordinarily, before obtaining a value for *t*, you'll have to compute the value of the estimated standard error, $s_{\bar{X}_1 - \bar{X}_2}$, using the following formula:

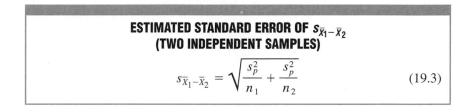

$$
\textbf{ESTIMATED STANDARD ERROR OF } s_{\bar{X}_1 - \bar{X}_2}
$$
$$
\textbf{(TWO INDEPENDENT SAMPLES)}
$$
$$
s_{\bar{X}_1 - \bar{X}_2} = \sqrt{\frac{s_p^2}{n_1} + \frac{s_p^2}{n_2}} \tag{19.3}
$$

where s_p^2 represents the pooled variance estimate, defined in Formula 19.4, as follows, and n_1 and n_2 are the two sample sizes.

The Pooled Variance (s_p^2)

Because the present *t* test assumes that the two population variances are equal, *the variance common to both populations can be estimated most accurately by combining the two sample variances to obtain the* **pooled variance estimate,** s_p^2.

To find the pooled variance estimate, s_p^2, combine the two sample variances, s_1^2 and s_2^2, on the basis of their respective degrees of freedom, $n_1 - 1$ and $n_2 - 1$, as follows:

> **Pooled variance estimate (S_p^2)**
>
> *The most accurate estimate of the population variance (assumed to be the same for both populations) based on a combination of two sample variances.*

$$
\textbf{POOLED VARIANCE ESTIMATE}
$$
$$
s_p^2 = \frac{(n_1 - 1)s_1^2 + (n_2 - 1)s_2^2}{(n_1 - 1) + (n_2 - 1)} \tag{19.4}
$$

where s_1^2 and s_2^2 are the variances of the two samples, and n_1 and n_2 are the two sample sizes. The values of s_1^2 and s_2^2 can be found by squaring each of the sample standard deviations, s_1 and s_2, obtained from either Formula 18.4 or Formula 18.5.

Value of s_p^2 between s_1^2 and s_2^2

Multiplying each sample variance by its degrees of freedom, as in Formula 19.4, ensures that the contribution of each sample variance is proportionate to its degrees of freedom. For instance, if s_1^2 contains twice as many degrees of freedom as s_2^2 does, then the value of s_1^2 also will count twice as much as s_2^2 will in determining the final value of s_p^2. Viewed in this fashion, s_p^2 can be characterized as the mean of s_1^2 and s_2^2, once these estimates have been adjusted for their degrees of freedom. Accordingly, if the values of s_1^2 and s_2^2 are different, s_p^2 will always assume some intermediate value. If sample sizes (and, therefore, degrees of freedom) are equal, the value of s_p^2 will be exactly midway between s_1^2 and s_2^2, as it is in the blood-doping experiment, where the value of s_p^2 (3) is exactly midway between s_1^2 (3.5) and S_2^2 (2.5). (See Table 19.1 in Section 19.13.)

Two Degrees of Freedom Lost

The degrees of freedom for the pooled variance estimate equal the sum of the two sample sizes minus two, that is, $n_1 + n_2 - 2$. Two degrees of freedom are lost because the observations in each of the two samples are expressed as deviations from their respective sample means. For this reason, the corresponding t ratio has a sampling distribution with $n_1 + n_2 - 2$ degrees of freedom.

*Exercise 19.3 Using Table B in Appendix D, find the critical t values for each of the following hypothesis tests:

(a) two-tailed test; $\alpha = .05$; $n_1 = 12$; $n_2 = 11$

(b) one-tailed test, upper tail critical; $\alpha = .05$; $n_1 = 15$; $n_2 = 13$

(c) one-tailed test, lower tail critical; $\alpha = .01$; $n_1 = n_2 = 25$

(d) two-tailed test; $\alpha = .01$; $n_1 = 8$; $n_2 = 10$

  Answers on page 512

19.13 CALCULATIONS FOR *t* TEST FOR TWO INDEPENDENT SAMPLES (BLOOD-DOPING EXPERIMENT)

Table 19.1 lists the various computations that produce a t of 0.91 for the blood-doping experiment. For computational convenience, the present test is based on very small samples of only five endurance scores per group. Ordinarily, the test should be based on larger samples whose sizes are determined by consulting the power curves alluded to in Chapter 16. Also, for computational convenience, the endurance scores have been rounded to the nearest minute even though, in practice, they surely would reflect more precise measurement.

*Exercise 19.4 A psychologist wants to determine the effect of instructions on the time required to solve a mechanical puzzle. Each of 20 volunteers is given the same mechanical puzzle to be solved as rapidly as possible. Before the task, the subjects are randomly assigned in equal numbers to receive two different sets of instructions. One group is told that the task is difficult (X_1), and the other group is told that the task is easy (X_2). The score for each subject reflects the time in minutes required to solve the puzzle. Use a t to test the null hypothesis at the .05 level of significance for the solution times listed at the top of page 321.

Table 19.1
CALCULATIONS FOR *t* TEST: TWO INDEPENDENT SAMPLES
(BLOOD-DOPING EXPERIMENT)

I. FINDING SAMPLE MEANS AND VARIANCES, $\overline{X}_1$, $\overline{X}_2$, s_1^2, AND s_2^2

(a) Computational sequence:

Assign a value to n_1 **1**.

Sum all X_1 scores **2**.

Substitute numbers in formula **3** and solve for $\overline{X}_1$.

Square each X_1 score **4**, one at a time, and then add all squared X_1 scores **5**.

Substitute numbers into formula **6** and solve for s_1^2.

Repeat this entire computational sequence for n_2 and X_2 and solve for $\overline{X}_2$ and s_2^2.

(b) Data and computations:

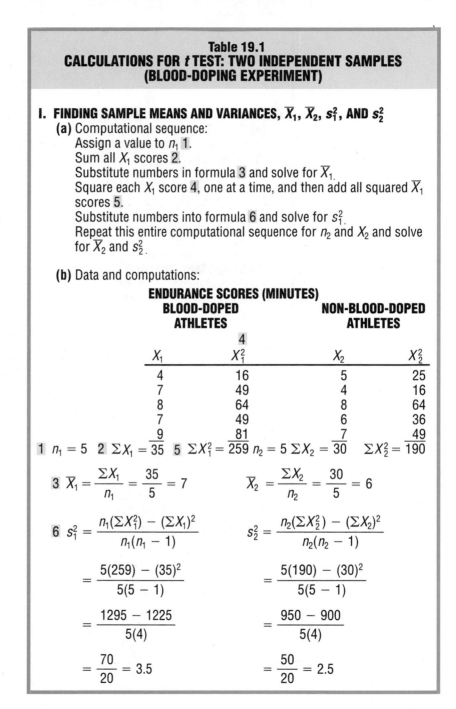

ENDURANCE SCORES (MINUTES)

BLOOD-DOPED ATHLETES		NON-BLOOD-DOPED ATHLETES	
	4		
X_1	X_1^2	X_2	X_2^2
4	16	5	25
7	49	4	16
8	64	8	64
7	49	6	36
9	81	7	49

1 $n_1 = 5$ **2** $\Sigma X_1 = 35$ **5** $\Sigma X_1^2 = 259$ $n_2 = 5$ $\Sigma X_2 = 30$ $\Sigma X_2^2 = 190$

3 $\overline{X}_1 = \dfrac{\Sigma X_1}{n_1} = \dfrac{35}{5} = 7$ $\qquad$ $\overline{X}_2 = \dfrac{\Sigma X_2}{n_2} = \dfrac{30}{5} = 6$

6 $s_1^2 = \dfrac{n_1(\Sigma X_1^2) - (\Sigma X_1)^2}{n_1(n_1 - 1)}$ $\qquad$ $s_2^2 = \dfrac{n_2(\Sigma X_2^2) - (\Sigma X_2)^2}{n_2(n_2 - 1)}$

$= \dfrac{5(259) - (35)^2}{5(5 - 1)}$ $\qquad$ $= \dfrac{5(190) - (30)^2}{5(5 - 1)}$

$= \dfrac{1295 - 1225}{5(4)}$ $\qquad$ $= \dfrac{950 - 900}{5(4)}$

$= \dfrac{70}{20} = 3.5$ $\qquad$ $= \dfrac{50}{20} = 2.5$

Table 19.1
CALCULATIONS FOR *t* TEST: TWO INDEPENDENT SAMPLES
(BLOOD-DOPING EXPERIMENT) (cont.)

II. FINDING POOLED VARIANCE, s_p^2

(a) Computational sequence:
Substitute numbers obtained above in formula [7] and solve for s_p^2.

(b) Computations:

$$[7]\ s_p^2 = \frac{(n_1 - 1)\,s_1^2 + (n_2 - 1)\,s_2^2}{(n_1 - 1) + (n_2 - 1)} = \frac{(5 - 1)\,3.5 + (5 - 1)\,2.5}{(5 - 1) + (5 - 1)}$$

$$= \frac{(4)\,3.5 + (4)\,2.5}{(4) + (4)} = \frac{14 + 10}{8} = \frac{24}{8} = 3$$

III. FINDING STANDARD ERROR, $s_{\bar{X}_1 - \bar{X}_2}$

(a) Computational sequence:
Substitute numbers obtained above in formula [8] and solve for $s_{\bar{X}_1 - \bar{X}_2}$.

(b) Computations:

$$[8]\ s_{\bar{X}_1 - \bar{X}_2} = \sqrt{\frac{s_p^2}{n_1} + \frac{s_p^2}{n_2}} = \sqrt{\frac{3}{5} + \frac{3}{5}}$$

$$= \sqrt{\frac{6}{5}} = \sqrt{1.2} = 1.10$$

IV. FINDING THE OBSERVED *t* RATIO

(a) Computational sequence:
Substitute numbers obtained above in formula [9], as well as a value of 0 for the expression $(\mu_1 - \mu_2)_{\text{hyp}}$, and solve for *t*.

(b) Computations:

$$[9]\ t = \frac{(\bar{X}_1 - \bar{X}_2) - (\mu_1 - \mu_2)_{\text{hyp}}}{s_{\bar{X}_1 - \bar{X}_2}} = \frac{(7 - 6) - 0}{1.10} = \frac{1}{1.10} = 0.91$$

*Exercise 19.4 continued

SOLUTION TIMES	
"DIFFICULT" TASK	**"EASY" TASK**
5	13
20	6
7	6
23	5
30	3
24	6
9	10
8	20
20	9
12	12

Answer on pages 512–513

Summary

Statistical hypotheses for the difference between two population means must be selected from among the following three possibilities:
Nondirectional:

$$H_0: \mu_1 - \mu_2 = 0$$
$$H_1: \mu_1 - \mu_2 \neq 0$$

Directional, lower tail critical:

$$H_0: \mu_1 - \mu_2 \geq 0$$
$$H_1: \mu_1 - \mu_2 < 0$$

Directional, upper tail critical:

$$H_0: \mu_1 - \mu_2 \leq 0$$
$$H_1: \mu_1 - \mu_2 > 0$$

Tests of this null hypothesis require that you deal with the sampling distribution of the difference between sample means, $\overline{X}_1 - \overline{X}_2$. Two important properties of this sampling distribution are as follows:

1. Its mean equals the difference between population means.
2. Its standard error roughly measures the average amount by which any difference between sample means deviates from the difference between population means.

Actual hypothesis tests are based on the t ratio, as defined in Formula 19.1 for two independent samples. This t ratio has a sampling distribution with $n_1 + n_2 - 2$ degrees of freedom.

A confidence interval also can be constructed for the differences between population means. A single interpretation is possible only if the two limits of the confidence interval share similar signs, either both positive or both negative.

The use of *t* assumes that both underlying populations are normally distributed with equal variances. Except under rare circumstances, you needn't be concerned about violations of these assumptions.

Important Terms
. . . .

Two independent samples

Effect

Sampling distribution of sample mean differences, $\overline{X}_1 - \overline{X}_2$

Standard error of $\overline{X}_1 - \overline{X}_2$ ($s_{\overline{x}_1 - \overline{x}_2}$)

Pooled variance estimate (s_p^2)

Confidence interval for $\mu_1 - \mu_2$

REVIEW EXERCISES

19.5   To test compliance with authority, a classical experiment in social psychology requires that subjects administer increasingly painful electric shocks to seemingly helpless victims who agonize in an adjacent room. Each subject earns a score between 0 and 25, depending on the point at which the subject refuses to comply with authority—an experimenter who orders the administration of increasingly intense shocks. A score of 0 signifies the subject's unwillingness to comply at the very outset, and a score of 25 signifies the subject's willingness to comply completely with orders.

Ignore the ethical issues raised by this type of experiment, and assume that you are curious about the effect of a "committee atmosphere" on compliance with authority. In one condition, shocks are administered only after an affirmative decision by the committee, consisting of one real subject and two experimental confederates who act as subjects but, in fact, merely go along with the decision of the real subject. In the other condition, shocks are administered only after an affirmative decision by a solitary real subject.

Six subjects are randomly assigned to the committee condition (X_1) and six subjects are randomly assigned to the solitary condition (X_2). A compliance score is obtained for each subject. Use *t* to test the null hypothesis at the .05 level of significance.

COMPLIANCE SCORES	
COMMITTEE	SOLITARY
2	3
5	8
20	7
15	10
4	14
10	0

19.6 Let's return to the investigator, first described in Chapter 16, who wants to determine whether daily doses of vitamin C increase intellectual aptitude. Now a total of 70 high school students are randomly divided into two groups of 35 each, designated to receive daily doses of 60 milligrams of either vitamin C or fake vitamin C. After two months of daily doses, IQ scores are obtained. For the subjects who received vitamin C, the sample mean IQ ($\overline{X}_1$) equals 110, and for the subjects who received fake vitamin C, the sample mean IQ ($\overline{X}_2$) equals 108. The estimated standard error equals 1.80.

(a) Using t, test the null hypothesis at the .01 level of significance.

(b) If appropriate (because the null hypothesis has been rejected), construct a 99 percent confidence interval for the true population mean difference, and interpret this interval.

*19.7 Is the performance of college students affected by grading policy? In the same introductory biology class, a total of 40 student volunteers were randomly assigned, in equal numbers, to take the course for either letter grades or a simple pass/fail. At the end of the academic term, the mean achievement score for the letter grade students ($\overline{X}_1$) equaled 86.2, and the mean achievement score for pass/fail students ($\overline{X}_2$) equaled 81.6. The estimated standard error was 1.50.

(a) Use t to test the null hypothesis at the .05 level of significance.

(b) How would the above hypothesis test change if the roles of X_1 and X_2 were reversed—that is, if X_1 were identified with pass/fail students and X_2 were identified with letter grade students?

(c) If appropriate (because the null hypothesis was rejected), construct a 95 percent confidence interval for the true population mean difference, and interpret this interval.

(d) Most students would doubtless prefer to select their favorite grading policy rather than be randomly assigned to a particular grading policy. Why not, therefore, replace random assignment with self-selection?

*19.8 An investigator wishes to determine whether alcohol consumption causes a deterioration in the performance of automobile drivers. Before the driving test, subjects drank a glass of orange juice that, in the case of one group of subjects, was laced with two ounces of vodka. Performance was measured by the number of errors made on a driving simulator. One hundred twenty volunteer subjects were randomly assigned, in equal numbers, to the two groups. For the subjects who drank the vodka-laced orange juice, the mean number of errors ($\overline{X}_1$) equaled 26.4, and for the subjects who drank the regular orange juice, the mean number of errors ($\overline{X}_2$) equaled 18.6. The estimated standard error equaled 2.4. Use t to test the null hypothesis at the .05 level of significance.

Answers on page 514

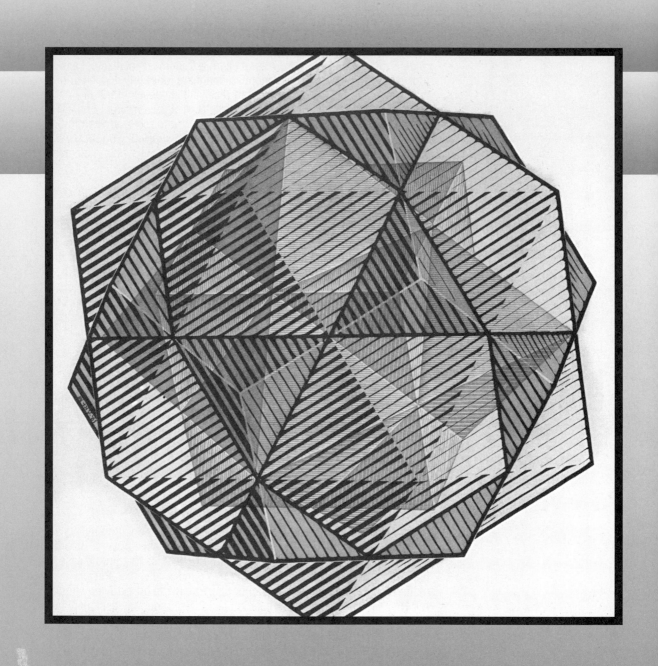

CHAPTER 20

t Test for Two Matched Samples

TWO POPULATION MEANS

20.1   MATCHING PAIRS OF ATHLETES IN THE BLOOD-DOPING EXPERIMENT
20.2   TWO MATCHED SAMPLES
20.3   DIFFERENCE SCORES (*D*)
20.4   STATISTICAL HYPOTHESES
20.5   SAMPLING DISTRIBUTION OF $\overline{D}$
20.6   *t* RATIO
20.7   CONFIDENCE INTERVAL FOR μ_D
20.8   TO MATCH OR NOT TO MATCH?
20.9   USING THE SAME SUBJECTS IN BOTH GROUPS (REPEATED MEASURES)
20.10   ASSUMPTIONS
20.11   THREE *t* TESTS FOR POPULATION MEANS: AN OVERVIEW

DETAILS

20.12   ESTIMATING THE STANDARD ERROR ($s_{\overline{D}}$)
20.13   CALCULATIONS FOR *t* TEST FOR TWO MATCHED SAMPLES (BLOOD-DOPING EXPERIMENT)

POPULATION CORRELATION COEFFICIENT

20.14   *t* TEST FOR THE GREETING CARD EXCHANGE
20.15   ASSUMPTIONS
20.16   A LIMITATION

Summary

Important Terms

Review Exercises

TWO POPULATION MEANS

20.1 MATCHING PAIRS OF ATHLETES IN THE BLOOD-DOPING EXPERIMENT

Preliminary studies of blood doping with EPO, as described in Chapter 19, might reveal an unexpected phenomenon: Lightweight athletes in *both* the blood-doped and non-blood-doped groups have better endurance scores than heavier athletes. This factor complicates the search for the effect of blood doping on endurance scores.

In subsequent experiments, therefore, it might be advantageous to pair athletes with similar body weights. Before collecting the data, the athletes could be matched for body weight, beginning with the two lightest and ending with the two heaviest. Once a member of a given pair has been randomly assigned to either the blood-doped or the non-blood-doped group, the other member of that pair is automatically assigned to the remaining group. Thereafter, the athletes are treated in the same way as in the original blood-doping experiment; that is, endurance scores are obtained both for athletes who are blood doped with EPO and for athletes who receive a harmless neutral substance.

20.2 TWO MATCHED SAMPLES

Because the athletes of each pair are matched for body weight, the pairs of endurance scores will tend to be similar, and so the statistical test should be altered to reflect this new dependency between the scores for *pairs* of blood-doped and non-blood-doped athletes. **Two matched samples** occur whenever, as in the current case, *each observation in one sample is paired, on a one-to-one basis, with a single observation in the other sample.* This new test for two matched samples is relatively straightforward, and, under appropriate circumstances, it's preferred to the test for two independent samples described in the previous chapter.

Two matched samples

Each observation in one sample is paired, on a one-to-one basis, with a single observation in the other sample.

*Exercise 20.1 Indicate whether each of the following studies involves two independent samples or two matched samples.

(a) Estimates of weekly TV-viewing time of third grade girls are compared with those of third grade boys.

(b) Annual incomes of husbands are compared with those of their wives.

(c) SAT scores of freshmen who plan to major in engineering are compared with the scores of freshmen who plan to major in social sciences.

(d) Problem-solving skills of recognized scientists are compared with those of recognized artists, given that individuals have been matched for IQ.

Answers on page 514

20.3 DIFFERENCE SCORES (D)

When using t to test the null hypothesis for the current blood-doping experiment, computations can be simplified by working directly with the difference between pairs of endurance scores, that is, by working directly with the

DIFFERENCE SCORE	
$D = X_1 - X_2$	(20.1)

where D is the **difference score**, and X_1 and X_2 are the endurance scores for pairs of blood-doped and non-blood-doped athletes, respectively. Essentially, the use of difference scores converts a two-sample problem with X_1 and X_2 scores into a one-sample problem with D scores.

20.4 STATISTICAL HYPOTHESES

Null Hypothesis

When converted to difference scores, the original pair of populations — one for scores of blood-doped athletes and the other for scores of non-blood-doped athletes — becomes a single population of difference scores, and the null hypothesis can be expressed in terms of this new population. If blood doping with EPO has either no consistent effect or a negative effect on endurance scores when athletes are paired for body weight, the population mean of all difference scores, μ_D, should equal zero or less. In symbols, an equivalent statement reads

$$H_0: \mu_D \leq 0$$

Alternative Hypothesis

As before, the investigator wants to reject the null hypothesis only if blood doping actually increases endurance scores. An equivalent statement in symbols reads

$$H_1: \mu_D > 0$$

This directional alternative hypothesis translates into a one-tailed test with the upper tail critical.

Two Other Possible Alternative Hypotheses

Although not appropriate for the current experiment, there are two other possible alternative hypotheses. Another directional hypothesis, expressed as

$$H_1: \mu_D < 0$$

translates into a one-tailed test with the lower tail critical, and a nondirectional hypothesis, expressed as

$$H_1: \mu_D \neq 0$$

translates into a two-tailed test.

20.5 SAMPLING DISTRIBUTION OF $\overline{D}$

The sample mean of the difference scores, $\overline{D}$, varies from sample to sample and has its own sampling distribution. The mean of the sampling distribution of $\overline{D}$ equals the difference between population means.

Insofar as there is a dependency between paired observations, the standard error for two matched samples is smaller than that for two independent samples. This is a desirable consequence since a smaller standard error translates into a hypothesis test that is more likely to detect a false null hypothesis.

20.6 t RATIO

The null hypothesis can be tested with the following ratio:

---

t RATIO FOR TWO POPULATION MEANS (TWO MATCHED SAMPLES)

$$t = \frac{\overline{D} - \mu_{Dhyp}}{s_{\overline{D}}}$$

(20.2)

---

that has a t sampling distribution with $n - 1$ degrees of freedom, where n equals the number of difference scores. In Formula 20.2, $\overline{D}$ represents the sample mean of the difference scores; $\mu_{D_{hyp}}$ represents the hypothesized mean (of zero) for all difference scores in the population; and $s_{\overline{D}}$ represents the estimated standard error of the mean of the difference scores, as defined later in Formula 20.4.

Spend a few minutes familiarizing yourself with the step-by-step hypothesis test summary for the current blood-doping experiment, which involves six pairs of athletes matched for body weight.

According to the present test, the null hypothesis can be rejected, and there is evidence that when athletes are matched for body weight, blood doping increases endurance. It's important to mention the matching procedure in any

HYPOTHESIS TEST SUMMARY: *t* TEST FOR TWO POPULATION MEANS: MATCHED SAMPLES (BLOOD-DOPING EXPERIMENT)

Research Problem:

Does the population mean endurance score for athletes who are blood doped with EPO exceed that for athletes who aren't blood doped, given that athletes are matched for body weight?

Statistical Hypotheses:

$$H_0 : \mu_D \leq 0$$
$$H_1 : \mu_D > 0$$

Decision Rule:

Reject H_0 at the .05 level of significance if *t* equals or is more positive than 2.015 (from Table B, Appendix D, given that $df = n - 1 = 6 - 1 = 5$).

Calculations:

$$t = \frac{2 - 0}{0.68} = 2.94 \text{ (See Table 20.2 on page 337 for all}$$
computations).

Decision:

Reject H_0 at .05 level of significance because $t = 2.94$ is more positive than 2.015.

Interpretation:

There is evidence that when athletes are matched for body weight, blood doping with EPO increases endurance.

conclusion. When matching was absent, as in the previous test of blood doping with two independent samples, the null hypothesis was retained, not rejected. In effect, the matching procedure eliminates one source of variability among endurance scores—the variability due to differences in body weight—that otherwise inflates the standard error term and causes an increase in β, the probability of a type II error.

20.7 CONFIDENCE INTERVAL FOR μ_D

Given that two samples are matched, as when athletes were matched for body weight in the blood-doping experiment, a confidence interval for μ_D can be constructed from the following expression:

CONFIDENCE INTERVAL FOR μ_D (TWO MATCHED SAMPLES)

$$\overline{D} \pm (t_{\text{conf}})(s_{\overline{D}}) \qquad\qquad (20.3)$$

---

where $\overline{D}$ represents the sample mean of the difference scores; t_{conf} represents a number (distributed with $n - 1$ degrees of freedom) from the t tables, which satisfies the confidence specifications; and $s_{\overline{D}}$ represents the estimated standard error defined in Formula 20.4.

Finding t_{conf}

To find the appropriate value of t_{conf} in Formula 20.3, refer to Table B, Appendix D, and follow the same procedure as described previously for two independent samples. If a 95 percent confidence interval is desired for the blood-doping experiment with matched athletes, first locate the row corresponding to 5 degrees of freedom (since there are 6 pairs of athletes matched for body weight, $df = n - 1 = 6 - 1 = 5$), and then locate the column for the 95 percent level of confidence, that is, the column heading identified with a single asterisk. The intersected cell specifies a value of 2.571 to be entered in Formula 20.3.

Given this value for t_{conf}, and (from Table 20.2 on page 337) values of 2 for $\overline{D}$ (the sample mean of the difference scores) and 0.68 for $s_{\overline{D}}$ (the estimated standard error), Formula 20.3 becomes

$$2 \pm (2.571)(0.68) = 2 \pm 1.75 = \begin{cases} 3.75 \\ 0.25 \end{cases}$$

It can be claimed, with 95 percent confidence, that the interval between 0.25 and 3.75 includes the true difference between population means.

Interpreting Confidence Intervals for μ_D

In this case, because both limits have similar (positive) signs, a single interpretation describes all of the possibilities included in the confidence interval. The appearance of only positive differences indicates that when athletes are matched for body weight, blood doping with EPO facilitates endurance. Furthermore, we can be *reasonably confident* that on the average, the true facilitative effect is neither less than 0.25 minute nor more than 3.75 minutes.

Comparing Confidence Intervals for Two Experiments

It's instructive to compare the present confidence interval with the previous one of -1.54 to 3.54 minutes for two independent samples. The narrower, more precise interval for two matched samples reflects a reduction in the estimated standard error caused by matching for body weight (which, you'll recall, eliminates the variability in endurance scores due to differences in body weight). In general, there's no guarantee that matching always will produce narrower, more precise confidence intervals. As suggested in the next section, therefore, match only when a "worthy" variable has been identified.

*Exercise 20.2 Imagine that one of the following 95 percent confidence intervals is based on the blood-doping experiment with two matched groups of subjects.

95% CONFIDENCE INTERVAL	LOWER LIMIT	UPPER LIMIT
1	−3.45	4.25
2	1.89	2.21
3	−1.54	−0.32
4	0.21	1.53
5	−2.53	1.78

(a) Which confidence interval is least precise?

(b) Which most strongly supports the conclusion that blood doping *hinders* endurance?

(c) Which is the most precise?

(d) Which implies the largest number of matched subjects?

(e) Which most strongly supports the conclusion that blood doping *facilitates* endurance?

(f) Which would most likely stimulate the investigator to conduct an additional experiment using a larger number of matched subjects?

Answers on page 514

20.8 TO MATCH OR NOT TO MATCH?

Matching Reduces Degrees of Freedom

A shift in the unit of analysis from original observations, as in the case of two independent samples, to the differences between *pairs* of original observations, as in the case of two matched samples, causes the degrees of freedom to be reduced by a factor of one-half. For example, when 30 subjects are used in an experiment with two independent samples, the t test has 28 degrees of freedom (from $15 + 15 - 2$), but when the same number of subjects is sorted into 15 pairs in an experiment with two matched samples, the t test has only half as many degrees of freedom, that is, 14 (from $15 - 1$). Given a one-tailed test, upper tail critical, at the .05 level of significance, the first t test with 28 degrees of freedom has a critical t of 1.701, and the second t test with only 14 degrees of freedom has a slightly larger critical t of 1.761. Thus, in the absence of effective matching, the net effect might be a less sensitive hypothesis test—that is, a hypothesis test with a larger critical t—because of the wasted degrees of freedom.

Matching Reduces Generality of Conclusion

When samples are matched, any conclusion applies only to a population with matching restrictions. In the most recent hypothesis test, there is evidence that blood doping increases endurance scores *only* in a population of athletes *who are matched for body weight.* There is no basis for assuming that blood doping also will produce a demonstrable increase in endurance scores under less-controlled circumstances—when athletes aren't matched for body weight.

Matching Can Increase Sensitivity of Test

Matching is desirable only when some uncontrolled variable appears to have considerable impact on the variable being measured, the dependent variable. Appropriate matching reduces variability—and the estimated standard error—and, therefore, produces a more sensitive hypothesis test.

Finding an Uncontrolled Variable Worthy of Matching

To identify a variable worthy of matching, you should familiarize yourself with all previous research in the area and conduct pilot studies. Use matching only if, before the full-fledged investigation, you are able to detect an uncontrolled variable that, when identified, aids your interpretation of preliminary findings.

20.9 USING THE SAME SUBJECTS IN BOTH GROUPS (REPEATED MEASURES)

As a special case, two matched samples might involve using the same subjects in both samples. If, as in Exercise 20.9 at the end of this chapter, the same athletes were used in both the blood-doped and non-blood-doped groups of a blood-doping experiment, any variability due to "individual differences" would be eliminated from the difference scores used in the statistical analysis. Often referred to as **repeated measures,** *because each subject is measured more than once,* this technique controls the groups not only for body weight but also for any other differences among individual subjects, such as physical strength, age, sex, experience, attitude, and so forth. Ideally, because of the removal of a major source of uncontrolled variability due to individual differences, any remaining differences between pairs of endurance scores would be due primarily to blood doping—a most desirable consequence.

Repeated measures

Whenever the same subject is measured more than once.

Don't Repeat Measurements If Effects Linger

Unfortunately, the attractiveness of this design sometimes fades upon closer inspection. For instance, because each athlete performs twice, once in the blood-doped condition and once in the non-blood-doped condition, sufficient time must elapse between these two conditions to eliminate any lingering effects due to blood doping. If there is any concern that these effects can't be eliminated, use each subject in only one condition.

Counterbalancing

Otherwise, when subjects do perform double duty in both conditions, *it's customary to randomly assign half of the subjects to experience the two conditions in a particular order*—say, first the blood-doped and then the non-blood-doped condition—*while the other half of the subjects experience the two conditions in the reverse order.* Known as **counterbalancing,** this adjustment eliminates a potential bias in favor of one condition merely because most subjects happen to experience it first (or second).

....

Counterbalancing

Reversing the order of conditions for equal numbers of all subjects.

20.10 ASSUMPTIONS

In the present context, *t* assumes that the population of difference scores is normally distributed. You needn't be too concerned about violations of this assumption as long as sample size is fairly large (greater than about ten pairs). Otherwise, in the unlikely event that you encounter conspicuous departures from normality, consider either increasing the sample size or using the less sensitive but more assumption-free Wilcoxon *t* test described in Chapter 25.

20.11 THREE *t* TESTS FOR POPULATION MEANS: AN OVERVIEW

Previous chapters have described three *t* tests for population means, and their more distinctive features are summarized in **Table 20.1.** Given a hypothesis test for one or two population means, a *t* test is appropriate if, as usually is the case, the population standard deviation must be estimated. In practice, you must be able to decide whether to use a *t* test for one sample, two independent samples, or two matched samples. This decision is fairly straightforward if you proceed, step by step, as follows:

One or Two Samples?

First, establish whether there are one or two samples. If there is only one sample, because the study deals with a single set of observations, then, of course, you needn't search any further; the appropriate *t* is that for one sample.

Are the Two Samples Paired?

Second, if there are two samples, establish whether there is any pairing. If each observation is paired, on a one-to-one basis, with a single observation in the other sample (because of matching or repeated measures), then the appropriate *t* is that for two matched samples.

Finally, if there are two samples but no evidence of pairing among individual observations, then the appropriate *t* is that for two independent samples.

Examples

Let's illustrate this strategy by identifying the appropriate *t* test for several different studies where, with the aid of radar guns, investigators clock the speeds of randomly selected motorists on a dangerous section of a state

Table 20.1
SUMMARY OF *t* TESTS FOR POPULATION MEANS

TYPE OF SAMPLE	SAMPLE MEAN	NULL HYPOTHESIS*	STANDARD ERROR	*t* RATIO	DEGREES OF FREEDOM
One sample	$\bar{X}$	$H_0: \mu =$ some number	$s_{\bar{X}}$ (Formula 18.6)	$\dfrac{\bar{X} - \mu_{hyp}}{s_{\bar{X}}}$	$n - 1$
Two Independent Samples (No pairing)	$\bar{X}_1 - \bar{X}_2$	$H_0: \mu_1 - \mu_2 = 0$	$s_{\bar{X}_1 - \bar{X}_2}$ (Formula 19.3)	$\dfrac{(\bar{X}_1 - \bar{X}_2) - (\mu_1 - \mu_2)_{hyp}}{s_{\bar{X}_1 - \bar{X}_2}}$	$n_1 + n_2 - 2$
Two Matched Samples (Pairing)	$\bar{D}$	$H_0: \mu_D = 0$	$s_{\bar{D}}$ (Formula 20.4)	$\dfrac{\bar{D} - \mu_{D_{hyp}}}{s_{\bar{D}}}$	$n - 1$ (where n refers to pairs of observations)

*For two-tailed test.

highway. Use the recommended strategy to arrive at your own answer before reading the one in the book.

Study A

Research Problem: Clocked speeds of randomly selected motorists are compared with the posted speed limit of 55 miles per hour.

Answer: Because there is a single set of observations, the appropriate *t* test is that for one sample (where, incidentally, the null hypothesis equals 55 miles per hour).

Study B

Research Problem: Clocked speeds of randomly selected trucks are compared with clocked speeds of randomly selected cars.

Answer: Because there are two sets of observations (speeds for trucks and speeds for cars), there are two samples. Because there is no indication of pairing among individual observations, the appropriate *t* test is that for two independent samples.

Study C

Research Problem: Clocked speeds of randomly selected motorists are compared at two different locations: one mile before and one mile after a large sign listing the number of motor fatalities on that stretch of highway during the previous year.

Answer: Because there are two sets of observations (speeds before and speeds after the sign), there are two samples. Furthermore, because each observation in one sample (the speed of a particular motorist one mile before the sign) is paired with a single observation in the other sample (the speed of the same motorist one mile after the sign), the appropriate *t* test is that for two matched samples.

Beginning with the next set of exercises, you will be exposed to a variety of studies for which you must identify the appropriate statistical test. By following a step-by-step procedure, such as the one begun here, you'll be able to make this identification not only for textbook studies, but also for those encountered in everyday practice.

*Exercise 20.3 Each of the following studies requires a *t* test for one or more population means. Specify whether the appropriate *t* test is for one sample, two independent samples, or two matched samples.

(a) College students are randomly assigned to undergo either behavioral therapy or Gestalt therapy. After twenty therapeutic sessions, each student earns a score on a mental health questionnaire.

(b) A researcher wishes to determine whether attendance at a day-care center increases the scores of three-year-old children on a motor skill test. Random assignment dictates which member from each of twenty pairs of twins attends the day-care center and which member stays at home.

(c) One hundred college freshmen are randomly assigned to sophomore roommates having either similar or dissimilar vocational goals. At the end of their freshman year, these one hundred freshmen's GPAs are to be analyzed on the basis of the previous distinction.

(d) According to the U.S. Department of Health, the average 16-year-old male can do 23 push-ups. A physical education instructor finds that in his school district, 30 randomly selected 16-year-old males can do an average of 28 push-ups.

(e) A child psychologist assigns aggression scores to each of ten children during two 60-minute observation periods separated by an intervening exposure to a series of violent TV cartoons.

Answers on page 514

INTERNET DEMONSTRATION

Go to the Web site for this book **(http://darwin.cwru.edu/~witte/statistics)** and click on **Experiments** to simulate outcomes for either two independent or two matched samples, given specified differences between population means and varying sample sizes.

DETAILS

20.12 ESTIMATING THE STANDARD ERROR ($s_{\overline{D}}$)

The estimated standard error, $s_{\overline{D}}$, can be obtained from the following expression:

ESTIMATED STANDARD ERROR (TWO MATCHED SAMPLES)

$$s_{\overline{D}} = \frac{s_D}{\sqrt{n}} \qquad (20.4)$$

---

where s_D represents the sample standard deviation for the observed difference scores, as defined next in Formula 20.5, and n equals the number of difference scores.

Finding the Sample Standard Deviation (s_D)

The required sample standard deviation, s_D, can be obtained from the following expression:

SAMPLE STANDARD DEVIATION (DIFFERENCE SCORES)

$$s_D = \sqrt{\frac{n\Sigma D^2 - (\Sigma D)^2}{n(n-1)}} \qquad (20.5)$$

---

where D represents the difference scores defined in Formula 20.1, and n equals the number of difference scores.

Except for a change in notation from X to D, the previous formulas for two matched samples are exactly the same as their counterparts for one sample in Chapter 18. By the same token, the computational procedures for difference scores should seem familiar, as they already were encountered for the original scores in Chapter 18.

20.13 CALCULATIONS FOR _t_ TEST FOR TWO MATCHED SAMPLES (BLOOD-DOPING EXPERIMENT)

The present test is based on the endurance scores for only six pairs of athletes (matched for body weight), as shown in Table 20.2. This table also shows the various computations that produce a _t_ of 2.94, which leads to the rejection of the null hypothesis.

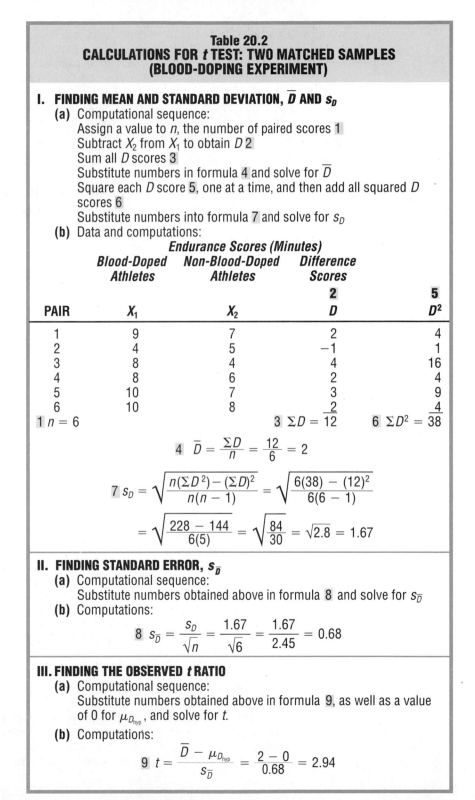

Table 20.2
CALCULATIONS FOR t TEST: TWO MATCHED SAMPLES (BLOOD-DOPING EXPERIMENT)

I. FINDING MEAN AND STANDARD DEVIATION, $\overline{D}$ AND s_D

(a) Computational sequence:
Assign a value to n, the number of paired scores **1**
Subtract X_2 from X_1 to obtain D **2**
Sum all D scores **3**
Substitute numbers in formula **4** and solve for $\overline{D}$
Square each D score **5**, one at a time, and then add all squared D scores **6**
Substitute numbers into formula **7** and solve for s_D

(b) Data and computations:

Endurance Scores (Minutes)

PAIR	Blood-Doped Athletes X_1	Non-Blood-Doped Athletes X_2	Difference Scores **2** D	**5** D^2
1	9	7	2	4
2	4	5	−1	1
3	8	4	4	16
4	8	6	2	4
5	10	7	3	9
6	10	8	2	4
1 $n = 6$			**3** $\Sigma D = 12$	**6** $\Sigma D^2 = 38$

$$\textbf{4}\quad \overline{D} = \frac{\Sigma D}{n} = \frac{12}{6} = 2$$

$$\textbf{7}\quad s_D = \sqrt{\frac{n(\Sigma D^2) - (\Sigma D)^2}{n(n-1)}} = \sqrt{\frac{6(38) - (12)^2}{6(6-1)}}$$

$$= \sqrt{\frac{228 - 144}{6(5)}} = \sqrt{\frac{84}{30}} = \sqrt{2.8} = 1.67$$

II. FINDING STANDARD ERROR, $s_{\overline{D}}$

(a) Computational sequence:
Substitute numbers obtained above in formula **8** and solve for $s_{\overline{D}}$

(b) Computations:

$$\textbf{8}\quad s_{\overline{D}} = \frac{s_D}{\sqrt{n}} = \frac{1.67}{\sqrt{6}} = \frac{1.67}{2.45} = 0.68$$

III. FINDING THE OBSERVED t RATIO

(a) Computational sequence:
Substitute numbers obtained above in formula **9**, as well as a value of 0 for $\mu_{D_{hyp}}$, and solve for t.

(b) Computations:

$$\textbf{9}\quad t = \frac{\overline{D} - \mu_{D_{hyp}}}{s_{\overline{D}}} = \frac{2 - 0}{0.68} = 2.94$$

*Exercise 20.4 An investigator wants to test still another claim for vitamin C, namely, that it reduces the frequency of common colds. To eliminate the variability due to different family environments, pairs of children from the same family are randomly assigned either to a group that receives vitamin C or to a group that receives fake vitamin C. At the end of the study, after an entire school year, each child has a score that reflects the total number of days ill because of colds, as determined from daily inspections by the school nurse. The following scores were obtained for ten pairs of children:

	DAYS ILL DUE TO COLDS	
PAIR NO.	VITAMIN C (X_1)	FAKE VITAMIN C (X_2)
1	2	3
2	5	4
3	7	9
4	0	3
5	3	5
6	7	7
7	4	6
8	5	8
9	1	2
10	3	5

Using *t*, test the null hypothesis at the .05 level of significance.

Answers on page 515

POPULATION CORRELATION COEFFICIENT

20.14 *t* TEST FOR THE GREETING CARD EXCHANGE

In Chapter 9, .80 describes the sample correlation coefficient, *r*, between the number of cards given and the number of cards received by five friends. Any conclusions about the correlation coefficient in the underlying population—for instance, the population of all friends—must consider chance sampling variability, as described by the sampling distribution of *r*.

Null Hypothesis

Let's view the greeting card data for the five friends as a random sample of pairs of observations from the population of all friends. Then it's possible to test the null hypothesis that the *population correlation coefficient*, symbolized by the Greek letter ρ (rho), equals zero. In other words, it's possible to test the hypothesis that in the population of all friends, there is no correlation between the number of cards given and the number of cards received.

Focus on Relationship Instead of Group Difference

These five pairs of observations also can be viewed as two matched samples, because each observation in one sample is paired with a single observation in the other sample. Now, however, we wish to determine whether there is a *relationship* between the number of cards given and received, not whether there is a *mean difference* between the number of cards given and received. Accordingly, the appropriate measure is the correlation coefficient, not the difference between two sample means, and the appropriate *t* test is for the population correlation coefficient, not the difference between population means.

t Ratio

A new *t* test can be used to determine whether an *r* of .80 qualifies as a common or a rare outcome in the *t* sampling distribution. To obtain a value for the *t* ratio, use the following formula:

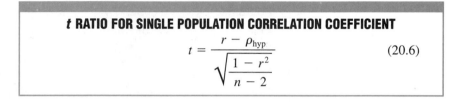

t RATIO FOR SINGLE POPULATION CORRELATION COEFFICIENT

$$t = \frac{r - \rho_{\text{hyp}}}{\sqrt{\dfrac{1 - r^2}{n - 2}}} \qquad (20.6)$$

where *r* refers to the sample correlation coefficient (Formula 9.2); ρ_{hyp} refers to the hypothesized population correlation coefficient (which always must be equal to 0); and *n* refers to the number of pairs of observations. The expression in the denominator represents the standard error of the sample correlation coefficient. As implied by the term at the bottom of this expression, the sampling distribution of *t* has $n - 2$ degrees of freedom. When pairs of observations are represented as points in a scatterplot, *r* presumes that the cluster of points approximates a straight line. Two degrees of freedom are lost because the points are free to vary only about some straight line that, itself, always depends on two points.

HYPOTHESIS TEST SUMMARY:
t TEST FOR A POPULATION CORRELATION COEFFICIENT
(GREETING CARD EXCHANGE)

Problem:
Could there be a correlation between the number of cards given and the number of cards received for the population of all friends?

Statistical Hypotheses:

$$H_0: \rho = 0$$
$$H_1: \rho \neq 0$$

Decision Rule:
Reject H_0 at the .05 level of significance if *t* equals or is more positive than 3.182 or if *t* equals or is more negative than -3.182 (from Table B in Appendix D, given that $df = n - 2 = 5 - 2 = 3$).

Calculations:
Given that $r = 0.80$ and $n = 5$:

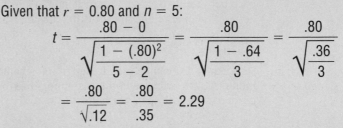

$$t = \frac{.80 - 0}{\sqrt{\dfrac{1 - (.80)^2}{5 - 2}}} = \frac{.80}{\sqrt{\dfrac{1 - .64}{3}}} = \frac{.80}{\sqrt{\dfrac{.36}{3}}}$$

$$= \frac{.80}{\sqrt{.12}} = \frac{.80}{.35} = 2.29$$

Decision:
Retain H_0 at .05 level of significance because $t = 2.29$ is less positive than 3.182.

Interpretation:
The population correlation coefficient *could* equal zero, and there might not be any relationship between the number of cards given and the number of cards received in the population of friends.

Importance of Sample Size

According to the present hypothesis test, the population coefficient *could* equal zero. This conclusion might seem surprising, given that an *r* of .80 was observed for the greeting card exchange. When the value of *r* is based on only 5 pairs of observations, as in the present example, its sampling variability is huge, and, in fact, an *r* of .88 would be required to reject the null hypothesis. Ordinarily, a serious investigation would use a larger sample size—preferably one that, with the aid of power curves, reflects the investigator's judgment about what constitutes the smallest important effect (or correlation).

*Exercise 20.5 A random sample of 27 California taxpayers reveals an *r* of .43 between years of education and annual income. Use *t* to test the null hypothesis at the .05 level of significance that there is no relationship between educational level and annual income for the population of California taxpayers.

Answer on page 515

20.15 ASSUMPTIONS

When using the *t* test for the population correlation coefficient, you must assume that the relationship between the two variables, *X* and *Y*, can be described with a straight line and that the sample originates from a *normal bivariate population.* The latter term means that the separate population distributions for each variable (X and Y) should be normal. When these assumptions are suspect—for instance, if the observed distribution for one variable appears to be extremely non-normal—the test results are only approximate and should be interpreted accordingly.

20.16 A LIMITATION

The present t test can't be used to test the hypothesis that the population correlation coefficient ρ equals some number other than zero. When ρ is not equal to zero, the sampling distribution of r is skewed and can't be adequately approximated by the symmetrical sampling distribution of t. A better approximation is supplied by Fisher's r-to-z transformation, as described in more advanced statistics books. Incidentally, the Fisher transformation also should be used to construct confidence intervals for population correlation coefficients (whose values usually don't equal zero).

Summary

Two samples are matched whenever each observation in one sample is paired, on a one-to-one basis, with a single observation in the other sample. The use of difference scores transforms the original pair of populations into a single population of difference scores.

The statistical hypotheses must be selected from among the following three possibilities, where μ_D represents the population mean for all difference scores:

Nondirectional:

$$H_0: \mu_D = 0$$
$$H_1: \mu_D \neq 0$$

Directional, lower tail critical:

$$H_0: \mu_D \geq 0$$
$$H_1: \mu_D < 0$$

Directional, upper tail critical:

$$H_0: \mu_D \leq 0$$
$$H_1: \mu_D > 0$$

The distribution of $\overline{D}$ is very similar to its counterpart, the distribution of $\overline{X}_1 - \overline{X}_2$, for two independent samples. A major difference is that the standard error is smaller when the two samples are matched.

The t ratio for two matched samples has a sampling distribution with $n - 1$ degrees of freedom, given that n equals the number of paired observations.

A confidence interval also can be constructed for μ_D. A single interpretation is possible only if the limits of the interval have similar signs, either both positive or both negative.

Inappropriate matching is costly and should be avoided. When matching is appropriate, it reduces the size of the estimated standard error—a most desirable consequence. Only match when, before the full-fledged investigation, you are able to detect a heretofore uncontrolled variable that aids the interpretation of preliminary findings.

As a special case, two matched samples may involve using the same subjects in both samples. Often referred to as repeated measures, this technique eliminates any variability due to individual differences. There is the possibility, however, that the performance in one condition might be contaminated by the subject's prior experience with the other condition. If this possibility can't be eliminated, use each subject in only one condition.

When using t for two matched samples, you must assume that the population of difference scores is normally distributed. You needn't be too concerned about violations of this assumption as long as sample sizes are relatively large.

To test the hypothesis that the population correlation coefficient equals zero, use a new t ratio (defined in Formula 20.6) with $n - 2$ degrees of freedom.

When using t to test whether a population correlation coefficient differs from zero, you must assume that the relationship between X and Y is linear and that the population distributions for X and Y are normally distributed.

Important Terms
. . . .

Two matched samples

Difference score (D)

Repeated measures

Counterbalancing

REVIEW EXERCISES

*20.6 An educational psychologist wants to check claims that daily meditation will improve the academic achievement of its practitioners. To control the experiment for academic aptitude, pairs of college students with similar grade point averages (GPA's) are randomly assigned to either a group that receives daily training in meditation, or a group that doesn't receive training in meditation. At the end of the experiment, which lasts for one semester, the following GPAs were reported for the seven pairs of participants:

	GPAs	
PAIR NO.	**MEDITATION (X_1)**	**NO MEDITATION (X_2)**
1	4.00	3.75
2	2.67	2.74
3	3.65	3.42
4	2.11	1.67
5	3.21	3.00
6	3.60	3.25
7	2.80	2.65

Using t, test the null hypothesis at the .01 level of significance.
Answer on pages 515–516

20.7 A public health investigator wishes to determine whether a new anti-smoking film actually reduces the daily consumption of cigarettes by heavy smokers after they see it. The mean daily cigarette consumption is calculated for each of eight heavy smokers during the month *before* the film presentation and also during the month *after* the film presentation, with the following results:

	MEAN DAILY CIGARETTE CONSUMPTION	
SMOKER NO.	**BEFORE FILM (X_1)**	**AFTER FILM (X_2)**
1	28	26
2	29	27
3	31	32
4	44	44
5	35	35
6	20	16
7	50	47
8	25	23

This illustrates the important case in which repeated measurements are obtained for the same subject. *Note:* Because X_1 and X_2 are identified with the earlier and later scores, respectively, a positive difference score $(D = X_1 - X_2)$ reflects a *decline* in cigarette consumption. Remember this when deciding on the form of the alternative hypothesis, H_1.

(a) Using t, test the null hypothesis at the .05 level of significance.

(b) If appropriate (because the null hypothesis was rejected), construct a 95 percent confidence interval for the true population mean for all difference scores, and interpret this interval.

(c) What might be done to improve the design of this experiment?

20.8 A manufacturer of a gas additive claims that it improves gas mileage under virtually any kind of driving conditions. A random sample of 30 drivers test this claim by determining their gas mileage for a full tank of gas that contains the additive (X_1) and for a full tank of gas that doesn't contain the additive (X_2). The sample mean difference, $\bar{D}$, equals 2.12 miles (in favor of the additive), and the estimated standard error equals 1.50 miles.

(a) Using t, test the null hypothesis at the .05 level of significance.

(b) Are there any special precautions that should be taken with the present experimental design?

*20.9 In a classic study of blood doping, which predates the existence of the blood-doping drug EPO, Melvin Williams of Old Dominion University actually injected extra oxygen-bearing red cells into the subjects' bloodsteam just prior to the treadmill test. Twelve long-distance runners were tested in five-mile runs on treadmills. Essentially, two running times were obtained for each athlete, once in the blood-doped condition after the injection of two pints of blood and once in the non-blood-doped condition after the injection of a comparable amount of fake blood (a harmless red saline solution). The presentation of real and fake blood was counterbalanced, with half of the athletes unknowingly receiving the real blood first and then the fake blood, and the other half receiving the reverse. Their running times, measured to the nearest second, were as follows:

RUNNER	NON-BLOOD-DOPED (MIN:SEC)	BLOOD-DOPED (MIN:SEC)	DIFFERENCE (SECONDS)
1	27:05	26:48	17
2	31:01	28:55	126
3	27:32	27:15	17
4	30:10	29:34	36
5	33:13	33:39	− 26
6	29:13	28:28	45
7	33:01	30:28	153
8	29:41	29:53	− 12
9	33:34	31:26	128
10	33:57	31:42	135
11	26:47	26:22	25
12	28:13	28:41	− 28

Source: *New York Times* (May 4, 1980).

Note: Positive difference scores signify that blood doping has a facilitative effect, that is, the athlete is running faster when blood doped.

(a) Using t, test the null hypothesis at the .05 level of significance.

(b) Would you have arrived at the same decision about the null hypothesis if the difference scores had been reversed by subtracting the non-blood-doped scores from the blood-doped scores?

(c) If appropriate, construct a 95 percent confidence interval for the true effect of blood doping, and interpret this interval.

(d) Why is it important to counterbalance the presentation of blood-doped and non-blood-doped conditions?

(e) Comment on the wisdom of testing each subject twice—once with real blood and once with fake blood—during a single twenty-four-hour period. (Williams actually used much longer intervals in his study.)

Answers on pages 516–517

20.10 A researcher randomly assigns college freshmen to either of two experimental conditions. Because both groups consist of college freshmen, someone claims that it's appropriate to use a *t* test for two matched samples. Please comment.

20.11 Although samples are actually matched, an investigator ignores this fact in the statistical analysis and uses a *t* test for two independent samples. On the assumption that matching involves a relevant variable, how will this mistake affect the probability of a type II error?

20.12
(a) When done properly, what's accomplished by matching?
(b) What's the cost of matching?
(c) What makes a variable worthy of matching?
(d) What serious complication can negate any advantages gained by using the same subjects in both conditions?

20.13 A random sample of 38 statistics students from a large statistics class reveals an *r* of $-.24$ between their test scores on a statistics exam and the amounts of time they spent taking the exam. Test the null hypothesis with *t*, using the .01 level of significance.

CHAPTER 21

Beyond Hypothesis Tests: *p*-Values and Effect Size

p-VALUES

21.1   DEFINITION
21.2   FINDING APPROXIMATE *p*-VALUES
21.3   READING *p*-VALUES REPORTED BY OTHERS
21.4   MERITS OF LESS STRUCTURED (*p*-VALUE) APPROACH
21.5   LEVEL OF SIGNIFICANCE OR *p*-VALUE?
21.6   A NOTE ON USAGE
21.7   COMPUTER OUTPUT

EFFECT SIZE

21.8   STATISTICALLY SIGNIFICANT RESULTS
21.9   SQUARED POINT BISERIAL CORRELATION, r_{pb}^2
21.10 SMALL, MEDIUM, OR LARGE EFFECT?
21.11 A RECOMMENDATION

Summary

Important Terms

Review Exercises

p-VALUES

21.1 DEFINITION

Most investigators adopt a less structured approach to hypothesis testing than that described in this book. The null hypothesis is neither retained nor rejected, but viewed with *degrees of suspicion*, depending on the degree of rarity of the observed value of *t* or, more generally, the test result. Instead of subscribing to a single *predetermined* level of significance, the investigator waits until *after* the test result actually has been observed, then assigns a probability, known as a *p*-value, representing the degree of rarity attained by the test result.

> **The *p-value* for a test result represents the degree of rarity of that result, given that the null hypothesis is true. Smaller *p*-values tend to discredit the null hypothesis and to support the research hypothesis.**

Strictly speaking, the *p*-value indicates the degree of rarity of the observed test result when combined with all potentially *more deviant* test results. In other words, the *p*-value represents the proportion of area, beyond the observed result, in the tail of the sampling distribution, as shown in **Figure 21.1** by the shaded sectors for two

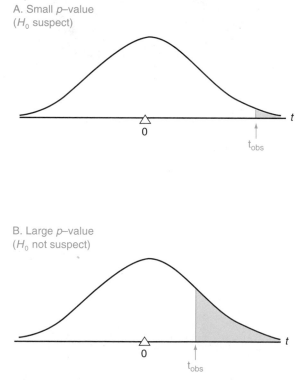

A. Small *p*–value
(H_0 suspect)

B. Large *p*–value
(H_0 not suspect)

FIGURE 21.1
Shaded sectors showing small and large p-*values.*

different test results. In the top panel of Figure 21.1, a relatively deviant (from zero) observed *t* is associated with a small *p*-value that makes the null hypothesis suspect, whereas in the bottom panel, a relatively nondeviant observed *t* is associated with a large *p*-value that doesn't make the null hypothesis suspect.[1]

21.2 FINDING APPROXIMATE *p*-VALUES

p-*Value*

The degree of rarity of a test result, given that the null hypothesis is true.

Table B, Appendix D, can be used to find *approximate p*-values, that is, *p*-values involving an inequality, such as $p < .05$ or $p > .05$. To aid in the identification of these approximate *p*-values, a color-coded outline has been superimposed over the entries for *t* in Table B, Appendix D (as well as over the entries of most subsequent tables in Appendix D). Once you've located the observed *t* relative to the tabular entries, simply follow the vertical line upward to identify the correct approximate *p*-value.

Let's find the approximate *p*-value for the *t* of 0.91 for the original blood-doping experiment involving two independent samples of 5 athletes per group, described on page 313. First, identify the row in Table B for a one-tailed test with 8 degrees of freedom (from $5 + 5 - 2$). The three entries in this row, 1.860, 2.896, and 4.501, serve as benchmarks for degrees of rarity corresponding to *p*-values of .05, .01, and .001, respectively. Because the observed *t* of 0.91 doesn't deviate from the null hypothesized value of 0 as far as the first entry of 1.860, follow the vertical line, to the left of 1.860, upward to $p > .05$. From most perspectives, this is a fairly large *p*-value: The test result is not particularly rare; it could have occurred just by chance with a probability of more than .05, given that H_0 is true. In other words, little support has been mustered for the research hypothesis. This conclusion is consistent with the original decision to retain H_0, when a more structured hypothesis test at the .05 level of significance was conducted for the same data in Section 19.9.

Other Examples of Finding Approximate *p*-Values

It's instructive to find the approximate *p*-value for the previous hypothesis test, assuming a more deviant test result, such as an observed *t* of 3.20. The same three entries from Table B (1.860, 2.896, and 4.501) serve as benchmarks. Now because the observed *t* lies between 2.896 and 4.501, follow the intervening vertical line upward to $p < .01$. From most perspectives, this is a small *p*-value: The test result is a rare event; it could have occurred just by chance with a probability less than .01. In this case, considerable support would have been mustered for the research hypothesis.

We could follow essentially the same procedure if the same observed *t* of 3.20 had been obtained for a two-tailed test. First, identify the row in Table B,

[1]Figure 21.1 illustrates *one-tailed p*-values that are appropriate whenever the investigator has an exclusive interest only in deviations in a particular direction, as with a one-tailed hypothesis test. Otherwise, as with a two-tailed hypothesis test, *two-tailed p*-values are appropriate. Although not shown in Figure 21.1, two-tailed *p*-values would require equivalent shaded areas to be located in *both* tails of the sampling distribution, and the resulting two-tailed *p*-value would actually be twice as large as the corresponding one-tailed *p*-value.

Appendix D, for a two-tailed-test with 8 degrees of freedom. The three entries in this row, 2.306, 3.355, and 5.041, serve as benchmarks. Because the observed t of 3.20 lies between 2.306 and 3.355, follow the intervening vertical line upward to $p < .05$. Although this p-value also would have supported the research hypothesis, it isn't as small as the p-value ($p < .01$) for the same observed t when a more sensitive one-tailed test was used.

*Exercise 21.1 Find the approximate p-value for each of the following test results:

(a) one-tailed test, upper tail critical; $df = 12$; $t = 4.61$

(b) one-tailed test, lower tail critical; $df = 19$; $t = -2.41$

(c) two-tailed test; $df = 15$; $t = 3.76$

(d) two-tailed test; $df = 42$; $t = 1.305$

(e) two-tailed test; $df = 58$; $t = 1.76$

(f) one-tailed test, upper tail critical; $df = 11$; $t = -4.23$ (Be careful!)

(g) one-tailed test, lower tail critical; $df = 28$; $t = -2.06$
    Answers on page 517

21.3 READING *p*-VALUES REPORTED BY OTHERS

A single research report might describe a batch of tests with a variety of approximate p-values, such as $p < .05$, $p < .01$, and $p < .001$, and, if the test failed to support the research hypothesis, $p > .05$. You must attend carefully to the direction of the inequality symbol. For example, the test result supports the research hypothesis when $p < .05$, but not when $p > .05$.

Exact *p*-Values Generated by Computers

As illustrated in many of the computer outputs in this book, including those later in this chapter, when statistical tests are performed by computers, with their capacity to obtain *exact* p-values (or values of Sig. in the case of SPSS), reports contain many different p-values, such as $p = .03$, $p = .27$, and $p = .009$. Even though more precise equalities replace inequalities, exact p-values listed on computer outputs are interpreted the same as those read from tables. For example, it's still true that $p = .03$ describes a rare test result, whereas $p = .27$ describes a result that isn't particularly rare. Incidentally, sometimes you'll see even a very rare $p = .000$, which, however, doesn't signify that p actually equals zero—an impossibility, because the t sampling distribution extends outward to infinity—but merely that rounding off causes the disappearance of nonzero digits from the reported p-value.

21.4 MERITS OF LESS STRUCTURED (*p*-VALUE) APPROACH

This less structured approach does have merit. Having eliminated the requirement that the null hypothesis be either retained or rejected, you can postpone a decision until sufficient evidence has been mustered, possibly from a series of investigations. This perspective is very attractive when test results are border-line. For instance, imagine a hypothesis test in which the null hypothesis is retained, even though the observed t of 1.80 is only slightly less deviant than the critical t of 1.860 for the .05 level of significance. Given the less structured approach, an investigator might, with the aid of a computer, establish that $p = .06$ for the observed t. Reporting the borderline result, with $p = .06$, implies at least some support for the research hypothesis.

Weakness of Less Structured (p-Value) Approach

One weakness of this less structured approach is that in the absence of a firm commitment to either retain or reject the null hypothesis according to some pre-determined level of significance, it's difficult to deal with the important notions of type I and type II errors. For this reason, a more structured approach to hypoth-esis testing will continue to be featured in the remaining chapters of this book, although not to the exclusion of the important approach involving p-values.

21.5 LEVEL OF SIGNIFICANCE OR p-VALUE?

A final word of caution. Don't confuse the level of significance with a p-value, even though both originate from the same column headings of Table B, Appendix D. Specified *before* the test result has been observed, the level of significance describes a degree of rarity that, if attained subsequently by the test result, triggers the decision to reject H_0. Specified *after* the test result has been observed, a p-value describes the most impressive degree of rarity actually attained by the test result.

Testing H_0 with a p-Value

You needn't drop a personal preference for a more structured hypothesis test, with a predetermined level of significance, just because a research report contains only p-values. For instance, any p-value less than .05, such as $p < .05$, $p = .03$, $p < .01$, or $p < .001$, implies that, with the same data, H_0 would have been rejected at the .05 level of significance. By the same token, any p-value greater than .05, such as $p > .05$, $p < .10$, $p < .20$, or $p = .18$ implies that, with the same data, H_0 would have been retained at the .05 level of significance.

*Exercise 21.2 Indicate which member of each of the following pairs of p-values describes the *more rare* test result:

(a₁) $p > .05$ **(a₂)** $p < .05$

(b₁) $p < .001$ **(b₂)** $p < .01$

(c₁) $p < .05$ **(c₂)** $p < .01$

(d₁) $p < .10$ **(d₂)** $p < .20$

(e₁) $p = .04$ **(e₂)** $p = .02$

*Exercise 21.3 Treating each of the p-values in the previous exercise sepa-rately, indicate those that would cause you to reject the null hypothesis at the .05 level of significance.

Answers on page 517

21.6 A NOTE ON USAGE

Published reports of statistical tests usually are brief, often consisting of only an interpretive comment plus a parenthetical statement that summarizes the statistical analysis and, almost invariably, includes a *p*-value. A published report of the original hypothesis test for blood doping in Chapter 19 might read as follows:

There is a lack of evidence that, on the average, blood doping increases endurance scores [$t(8) = 0.91$, $p > .05$].

Or

The difference between mean endurance scores for the two groups of athletes isn't statistically significant [$t(8) = 0.91$, $p > .05$].

The parenthetical statement indicates that a *t* based on 8 degrees of freedom was found to equal 0.91. Because the *p*-value of more than .05 reflects a fairly common test result, given that the null hypothesis is true, this result fails to support the research hypothesis, as implied in the interpretative statements.

If a more deviant *t* of 3.20 had been observed in the original hypothesis test for blood doping, a published report might have read as follows:

There is evidence that, on the average, blood doping increases endurance scores [$t(8) = 3.20$, $p < .01$].

Or

There is a statistically significant difference between the mean endurance scores in favor of blood-doped athletes [$t(8) = 3.20$, $p < .01$].

Now, because the *p*-value of less than .01 reflects a rare test result, given that the null hypothesis is true, this result supports the research hypothesis, as implied in the interpretative statements.

*Exercise 21.4 Recall that in Exercise 19.4 on page 318, a psychologist wants to determine the effect of instructions on the time required by subjects to solve the same mechanical puzzle. For two independent samples of ten subjects per group, solution times were 6.8 minutes longer, on the average, for subjects given "difficult" instructions than for subjects given "easy" instructions. A *t* ratio of 2.15 culminated in the rejection of the null hypothesis. In one sentence, indicate how these results might be described in a published report.

Answers on page 517

21.7 COMPUTER OUTPUT

Table 21.1 shows an SAS output for the original hypothesis test for blood doping with two independent samples, as summarized on page 353.

Table 21.1
SAS OUTPUT: *t* TEST FOR ENDURANCE SCORES

The SAS System
21:38 Wednesday, December 8, 1999
t Test Procedure

Variable : ENDURE

Group	N	Mean	Std Dev	Std Error
blood	5	7.00000000	1.87082869	0.83666003
nonblood	5	6.00000000	1.58113883	0.70710678

Variances	T	DF	Prob > |T|
Unequal	0.9129	7.8	0.3888
Equal	1 0.9129	8.0	0.3880

2 For HO : Variances are equal, $F' = 1.40$ DF = (4,4)
Prob > $F' = 0.7523$

Comments:

1 *Compare value of* t *with that given in Table 19.1 on page 320. Report the results for the customary* t *test (discussed in this book) that assumes* equal *variances rather than the more generalized* t *test (not discussed in this book) that accommodates* unequal *variances, unless, as explained in comment 2 below the assumption of equal population variances has been rejected. In SAS outputs, PROB > |T| identifies the two-tailed p-value for the current* t *test, that is,* $p = 0.3880$ *(Divide this value by 2 to obtain* p = 0.1940, *the more appropriate one-tailed p-value for the current* t *test.)*
2 *The F' (or F) test for equal population variances or, as it's often called, "homogeneity of variance." The F' value of 1.40 is found by dividing the square of the larger standard deviation (1.87) (1.87) by the square of the smaller standard deviation (1.58) (1.58). When PROB > F', the p-value for F', is too small—say, less than about .10—there is a possibility that the population variances aren't equal and, therefore, that results for the* t *test with unequal variances should be reported because, if this were the case, these results would have been more accurate. (Because the F' test responds to any non-normality, as well as to unequal population variances, some investigators prefer other tests, such as Levene's test in the SPSS output for Exercise 21.5, as a screening device before reporting* t *results based on unequal variances. For more information about both the* t *test that accommodates unequal population variances and Levene's test for equal population variances, see D. C. Howell,* Statistical Methods for Psychology, *4th ed. (Belmont, CA: Duxbury, 1997), chapter 7.*

*Exercise 21.5 The following SPSS output is based on the compliance scores in Review Exercise 19.5 on page 322.

SPSS OUTPUT: *t* TEST FOR COMPLIANCE SCORES

***t*-Test**

Group Statistics

	Variable	N	Mean	Std Deviation	Std. Error Mean
SCORES	Committee	6	9.3333	7.0333	2.8713
	Solitary	6	7.0000	4.9800	2.0331

		Levene's Test for Equality of Variances		t-test for Equality of Means					95% Confidence Interval of the Difference	
		F	Sig.	t	df	Sig. (2-tailed)	Mean Difference	Std. Error Difference	Lower	Upper
SCORES	Equal variances assumed	1.224	.294	.663	10	.522	2.3333	3.5182	−5.5057	10.1724
	Equal variances not assumed			.663	9.006	.524	2.3333	3.5182	−5.6245	10.2912

SPSS OUTPUT: *t* TEST FOR COMPLIANCE SCORES (Continued)

Independent Samples Test

Note: In SPSS outputs, *p*-values are listed under Sig.

(a) Which *t* test results should be reported—those based on the customary *t* test that assumes equal population variances, or those based on a more generalized *t* test that accommodates unequal population variances? *Hint:* See comment 2 in Table 21.1 on page 353.

(b) Specify the value of *t*, along with its degrees of freedom and *p*-value.

(c) Does the test result reach the .05 level of significance?

(d) In words, indicate the precise meaning of *p* = .522 for the *t* value reported in the output.

Answers on page 517

EFFECT SIZE

21.8 STATISTICALLY SIGNIFICANT RESULTS

It's important that you accurately interpret the findings of others—often reported as "statistically significant" results. Tests of hypotheses often are referred to as tests of significance, and test results are described as having *statistical significance* (if the null hypothesis has been rejected) or as not having statistical significance (if the null hypothesis has been retained).

Don't assume that statistical significance is the same as importance.

Statistical significance *indicates merely that the null hypothesis is probably false.* It doesn't indicate whether the null hypothesis is false because of a huge

Statistical significance

Not an indication of importance, but merely that the null hypothesis is probably false.

difference between population means or because of only a slight difference between population means.

Beware of Excessively Large Sample Sizes

Statistical significance that lacks importance is often caused by using excessively large sample sizes. With a large sample—for instance, imagine a blood-doping experiment involving two independent samples with 200 athletes per group—even a very small, unimportant **effect** (*difference between population means*) will be detected (because of the small standard error), and the test will be reported as having statistical significance.

Check for Importance

Viewed in this way, statistical significance merely indicates that an observed effect, such as an observed difference between the sample means, is sufficiently large, relative to the standard error, to be viewed as a rare outcome. (Statistical significance also implies that, if the experiment were repeated with new subjects, a *similar* observed effect probably would be obtained.) One way to gauge the importance of a statistically significant result *based on large sample sizes*, such as the blood-doping experiment with 200 subjects per group alluded to previously, is to use a measure, analogous to the squared correlation coefficient discussed in Section 10.8, that estimates—*without being swayed by large sample sizes*—the proportion of variance in endurance scores explained by whether the subject is blood doped.

21.9 SQUARED POINT BISERIAL CORRELATION, r_{pb}^2

This new measure focuses not on differences between group means, but on the correlation between pairs of observations, once subjects' endurance scores have been paired with arbitrary numerical codes, such as 0 or 1, depending on whether they were members of the blood-doped or control group. Once arbitrary numerical codes have been introduced, this "new" correlation coefficient, briefly introduced as a *point biserial* correlation coefficient in Section 9.10, is computationally indistinguishable from the familiar Pearson correlation coefficient. For all practical purposes, the **squared point biserial correlation coefficient** (r_{pb}^2) also can be interpreted similarly, that is, *as the proportion (from 0 to 1) of variance in the dependent variable that is predictable from, or explained by the independent variable.* Furthermore, if the study is a well-designed experiment, the squared point biserial correlation can be interpreted as the proportion of variance in the dependent variable that is *caused* by the independent variable.

You might find it helpful to interpret the value of the squared point biserial correlation, r_{pb}^2, in terms of the discussion of predictive accuracy in Section 10.8. Recall that the squared correlation coefficient, r^2, indicates the relative improvement in predictive accuracy, expressed as a proportion of the original error variance, when the repetitive prediction of $\overline{Y}$ is replaced by a series of Y' predictions based on the least squares equation. In the present context, r_{pb}^2 indicates the relative improvement in accuracy (for predicting, one at a time, each of the observed endurance scores, X, for all subjects), when the repetitive prediction of the mean for *all* subjects, $\overline{X}$, is replaced by more specialized

---

Effect

Difference between population means.

Squared point biserial correlation coefficient (r_{pb}^2)

The proportion of variance in the dependent variable that can be explained by the independent variable.

predictions of either of the two sample means, $\overline{X}_1$ or $\overline{X}_2$, depending on whether a particular subject was blood doped or not. In effect, r_{pb}^2 indicates the proportion of the total variance of endurance scores that is predictable from the relationship between endurance scores and the presence or absence of blood doping.

Happily, instead of actually calculating from scratch the squared point biserial correlation coefficient, r_{pb}^2, we merely need to solve the following simple formula:

PROPORTION OF EXPLAINED VARIANCE (TWO SAMPLES)

$$r_{pb}^2 = \frac{t^2}{t^2 + df} \tag{21.1}$$

---

where t^2 is the square of the obtained value of t, and df refers to the number of degrees of freedom associated with the significant t test. If the significant t involves two independent samples, as in the present example, $df = n_1 + n_2 - 2$. Otherwise, if the significant t involves two matched samples, $df = n - 1$. In either case, notice that the df term in the denominator of Formula 21.1 adjusts the value of r_{pb}^2 downward to compensate for the otherwise inflationary effect of large sample sizes on the value of t (and claims of statistical significance).

Let's look more closely at how we might use this formula to evaluate the importance of a statistically significant t based on large sample sizes. Pretend that a statistically significant t of 2.0 was reported for the blood-doping experiment with a large sample size of 200 subjects per group, that is $df = 200 + 200 - 2 = 398$. Substituting numbers in Formula 21.1, we obtain

$$r_{pb}^2 = \frac{2^2}{2^2 + 398} = \frac{4}{402} = .01$$

This small value of .01 for r_{pb}^2 suggests that only 1 percent of the variance in endurance scores is explained by (or caused by) whether subjects are blood doped. The remaining 99 percent of variance of endurance scores is not explained by the presence or absence of blood doping.

21.10 SMALL, MEDIUM, OR LARGE EFFECT?

One rough rule of thumb, suggested by Cohen, is that the estimated effect (estimated difference between population means) is small (and could lack importance) if r_{pb}^2 is in the vicinity of .01; the estimated effect is medium (and could have some importance) if r_{pb}^2 is in the general vicinity of .06; and

Table 21.2 COHEN'S GUIDELINES FOR EFFECT SIZE	
r^2_{pb}	EFFECT
.01	Small
.06	Medium
.14	Large

the estimated effect is large (and probably has importance) if r^2_{pb} is in the vicinity of, or exceeds, .14 (See Table 21.2).[2] Using this rule of thumb, the estimated effect of .01 would be judged to be small and could lack importance.

A Complication

Beware of special circumstances that, if present, render Cohen's guidelines irrelevant. Even a very small effect might be judged to be important because of the particular phenomena being studied. As a far-fetched example, if blood doping provided some protection against AIDS, then even an r^2_{pb} of .01 might be viewed as very important. The widely cited Physicians' Health Study involving more than 20,000 physicians, as reported in the *New England Journal of Medicine* (January 28, 1988) supplies a more realistic example. Even though, with a measure analogous to r^2_{pb}, only a minuscule one-tenth of 1 percent of heart attacks among participants could be attributed to whether an aspirin was taken each day, this very small effect was judged to be of sufficient importance—because almost twice as many, that is 85 more, control physicians had heart attacks—that the experiment was discontinued, with the recommendation that aspirin therapy be considered by all high-risk individuals in the population.

21.11 A RECOMMENDATION

Calculate (and report) the squared point biserial correlation coefficient—or its more accurate, but less-intuitive competitor, ω^2 (omega-squared), cited in advanced statistics books—whenever you encounter a statistically significant t based on large sample sizes. Not being inflated by large sample sizes, the value of r^2_{pb} provides us with an accurate estimate of effect size.[3] Do not, however, blindly apply any rule of thumb about effect sizes, such as the one stated earlier, without regard to special circumstances that could give considerable importance even to a very small effect.

A Reminder

By using the discussion of power curves in Chapter 16 as a point of departure, you should be able to avoid excessively large sample sizes in your own investigations. More positively, you should be able to specify a sample size that will detect, *with a high probability*, only those differences between population means that you, the investigator, judge to be important.

------

[2]J. Cohen, *Statistical Power Analysis for the Behavioral Sciences*, 2nd ed. (Hillsdale, NJ: Erlbaum, 1988), pp. 24–26. Parenthetical comments about importance were added by the authors.

[3]Being itself a product of chance sampling variability, r^2_{pb} provides us with a *stable* estimate of effect size only when sample sizes are relatively large. Otherwise, when sample sizes are moderate, any estimate based on r^2_{pb} must be viewed as quite speculative. Regardless of sample size, it doesn't make sense to estimate the effect size in the absence of a statistically significant t—that is, in the absence of evidence that there is an effect.

*Exercise 21.6 Statistically significant results were reported for experiment A and experiment B. Although both experiments used the .05 level of significance and obtained identical t values of 2.10, experiment A had only 20 subjects in each of two independent groups, and experiment B had 100 subjects in each of two independent groups.

(a) Which experiment more likely reflects an important effect?

(b) Using Formula 21.1 and Cohen's rule of thumb, estimate the size of the effect—small, medium or large—for each of these experiments.

Answers on page 517

Summary

Rather than subscribing to a predetermined level of significance, many investigators report the p-value for a test result. The p-value indicates the degree of rarity of a test result, given that the null hypothesis is true. Smaller p-values tend to discredit the null hypothesis. Although this approach has merit, the more structured hypothesis testing procedure will be featured in the remainder of this book.

Don't confuse statistical significance with importance. Statistical significance that lacks importance is often caused by the use of excessively large sample sizes. To gauge the importance of a statistically significant result, particularly those based on large sample sizes, first calculate the squared point biserial correlation coefficient, using Formula 21.1. Then apply Cohen's rule of thumb, which identifies r^2_{pb} values in the vicinity of .01, .06, and .14 with small, medium, and large effects, respectively, as well as with increasing degrees of importance.

Important Terms

p-Value

Statistical significance

Effect

Squared point biserial correlation coefficient, r^2_{pb}

REVIEW EXERCISES

21.7 **(a)** A test result is assigned a p-value of $.01 < p < .05$. Precisely what does this signify?

(b) Given that $.01 < p < .05$, would the null hypothesis be retained or rejected at the .05 level of significance?

(c) Given that $.01 < p < .05$, would the null hypothesis be retained or rejected at the .01 level of significance?

*21.8 Refer to Exercise 19.7, described on page 323 and solved on pages 513–514, and complete each of the following assignments:

(a) Specify the p-value for this test result.

(b) State how the test results might appear in a published report.

(c) If the test result is statistically significant, use Formula 21.1 on page 356 and Cohen's rule of thumb to estimate whether the result implies a small, medium, or large effect, even though any estimate based on r_{pb}^2 must be viewed as quite speculative when sample sizes are moderate, as in the present case.

Answers on page 517

21.9 Refer to Exercise 19.8, described on page 323 and solved on page 514, and respond to parts (a), (b), and (c) in Exercise 21.8.

21.10 Refer to Exercise 20.9, described on page 344 and solved on pages 516–517, and respond to parts (a), (b), and (c) in Exercise 21.8. *Note:* The estimate of effect size in (c) must be viewed as *highly* speculative when sample sizes are small, as in the present case.

21.11 After testing several thousand high school seniors, the state department of education reported a statistically significant difference between the mean GPAs for female and male students. Please comment.

V-'53

CHAPTER 22

Analysis of Variance (One Way)

22.1 TESTING A HYPOTHESIS ABOUT RESPONSIBILITY IN CROWDS
22.2 TWO SOURCES OF VARIABILITY
22.3 *F* RATIO
22.4 *F* TEST
22.5 ASSUMPTIONS
22.6 TWO CAUTIONS

DETAILS

22.7 VARIANCE ESTIMATES
22.8 SUM OF SQUARES (*SS*)
22.9 DEGREES OF FREEDOM (*df*)
22.10 MEAN SQUARES (*MS*) AND THE *F* RATIO
22.11 *F* TABLES
22.12 NOTES ON USAGE
22.13 *F* TEST IS NONDIRECTIONAL

BEYOND THE *F* TEST

22.14 SMALL, MEDIUM, OR LARGE EFFECT?
22.15 MULTIPLE COMPARISONS
22.16 SCHEFFÉ'S TEST
22.17 OTHER MULTIPLE COMPARISON TESTS
22.18 COMPUTER OUTPUT

Summary

Important Terms

Review Exercises

22.1 TESTING A HYPOTHESIS ABOUT RESPONSIBILITY IN CROWDS

Do crowds affect our willingness, either positively or negatively, to assume responsibility for the welfare of ourselves and others? For instance, does the presence of other people either facilitate or inhibit our reaction to potentially dangerous smoke seeping from a wall vent? Hoping to answer this question, a social psychologist measures any delay in a subject's alarm reaction (the dependent variable) as smoke gradually fills a waiting room occupied only by the subject, plus "crowds" of either zero, two, or four experimental confederates (the independent variable) who act as regular subjects but, in fact, ignore the smoke.

Null Hypothesis

As usual, the null hypothesis is tested with experimental findings. This null hypothesis states that on the average, the three populations of subjects who are exposed to smoke in rooms with either zero, two, or four confederates will delay an equal amount of time before reporting the smoke to the psychologist. Expressed symbolically, the null hypothesis reads

$$H_0: \; \mu_0 = \mu_2 = \mu_4$$

where μ_0 represents the mean delay before reporting the smoke by the population of subjects who are exposed to smoke in rooms with zero confederates, and where μ_2 and μ_4 represent the mean delays by populations of subjects in rooms with two and four confederates, respectively. Rejection of the null hypothesis implies, most generally, that crowd size affects our willingness to assume responsibility for dealing with a potentially threatening event.

New Test for More Than Two Population Means

Analysis of variance (ANOVA)
An overall test of the null hypothesis for more than two population means.

Resist any urge to test this null hypothesis with *t*, because, as discussed in Section 22.15, the regular *t* test usually can't handle null hypotheses for more than two population means. *When data are quantitative, an overall test of the null hypothesis for more than two population means requires a new statistical procedure known as* **analysis of variance,** *which is often abbreviated as* **ANOVA** (from **AN**alysis **O**f **VA**riance, and pronounced an-OH′-vuh).

One-way ANOVA

One-way ANOVA
The simplest type of analysis of variance that tests whether differences exist among population means categorized by only one factor or independent variable.

Chapter 22 describes situations that can be analyzed with the simplest type of analysis of variance. Often referred to as a **one-way ANOVA,** *this type of analysis tests whether differences exist among population means categorized by only one factor or independent variable,* such as crowd size. The discussion in Chapter 22 for a one-way ANOVA should be contrasted with that in Chapter 23 for a two-way ANOVA.

Later sections treat the computational procedures for ANOVA; the next few sections emphasize the intuitive basis for ANOVA within the context of the social psychologist's experiment in the smoke-filled room.

Two Possible Outcomes

For computational simplicity, let's assume that the social psychologist randomly assigns only three subjects to each of the three groups, to be tested (one subject at a time) in rooms with either zero, two, or four confederates, and she measures the reaction time of each subject to the nearest minute.

Table 22.1 shows two possible experimental outcomes that, when analyzed with ANOVA, produce different decisions about the null hypothesis: It is retained for one outcome but rejected for the other. Using just your intuition, guess which outcome will cause the null hypothesis to be retained and which will cause it to be rejected. Do this before reading further.

Your intuition was correct if you decided that outcome A will cause the null hypothesis to be retained, whereas outcome B will cause the null hypothesis to be rejected.

Mean Differences Still Important

Your decisions for outcomes A and B most likely were based on the relatively small differences between the group means for outcome A and the relatively large differences between group means for outcome B. Observed mean differences have been a major ingredient in previous t tests of the null hypothesis, and these differences are just as important in tests of the null hypothesis involving ANOVA. It's easy to lose sight of this fact because

Table 22.1
TWO POSSIBLE EXPERIMENTAL OUTCOMES:
REACTION TIMES IN MINUTES

OUTCOME A

	GROUPS (NUMBER OF CONFEDERATES)		
	(ZERO)	(TWO)	(FOUR)
	16	9	10
	12	10	12
	11	14	14
Group mean:	13	11	12 Overall mean = 12

OUTCOME B

	GROUPS (NUMBER OF CONFEDERATES)		
	(ZERO)	(TWO)	(FOUR)
	8	11	16
	4	12	18
	3	16	20
Group mean:	5	13	18 Overall mean = 12

observed mean differences appear, somewhat disguised, as one type of variability in ANOVA. It takes extra effort to view ANOVA—with its emphasis on the analysis of several sources of variability—as related to the previous *t* tests. Reminders of this fact appear throughout this chapter.

22.2 TWO SOURCES OF VARIABILITY

Differences between Group Means

First, without worrying about computational details, let's look more closely at one source of variability in outcomes A and B: the differences between group means. Relatively small differences appear between group means of 13, 11, and 12 in outcome A, that is, they show relatively little variability from one mean to the next. As noted in previous chapters, relatively small differences between group means often can be attributed to chance. Even though the null hypothesis is true (because crowd size doesn't affect the subjects' reaction times), group means tend to differ merely as a result of chance sampling variability. In the case of outcome A, the null hypothesis should not be rejected: There appears to be a lack of evidence that crowd size affects the subjects' reaction times in outcome A.

On the other hand, relatively large differences appear between the group means of 5, 13, and 18 for outcome B, that is, they show relatively great variability from one mean to the next. These relatively large differences probably cannot be attributed to chance. Instead they probably indicate that the null hypothesis is false (because crowd size affects the subjects' reaction times). In the case of outcome B, the null hypothesis should be rejected. Now there appears to be evidence of a **treatment effect,** that is, *the existence of at least one difference between the population means categorized by the independent variable* (crowd size).

Treatment effect

The existence of at least one difference between the population means categorized by the independent variable.

Variability between groups

Variability among scores of subjects who, being in different groups, receive different experimental treatments.

Variability within groups

Variability among scores of subjects who, being in the same group, receive the same experimental treatment.

Variability within Groups

A *definitive* decision about the null hypothesis requires that differences between group means be viewed as one source of variability that, when adjusted appropriately, can be compared with a second source of variability. In particular, an estimate of **variability between groups,** that is, *the variation among scores of subjects who, being in different groups, receive different experimental treatments*, must be compared with another, completely independent estimate of **variability within groups,** that is, *the variation among scores of subjects who, being in the same group, receive the same experimental treatment.* As will be seen,

the more that the variability between groups exceeds the variability within groups, the more suspect will be the null hypothesis.

Let's focus on the second source of variability—the variability within groups for subjects treated similarly. Referring to Table 22.1, focus on the differences among the scores of 16, 12, and 11 for the three subjects who are treated similarly in the first group. Continue this procedure, one group at a time, to obtain an overall impression of variability within groups for all three groups in outcome A and for all three groups in outcome B. Notice the relative stability of the differences among the three scores within each of the various groups, regardless of whether the group happens to be in outcome A or outcome B. For instance, one crude measure of variability, the range, equals either 4 or 5 for each group shown in Table 22.1.

A key point is that the variability within each group depends entirely on the scores of subjects treated similarly (exposed to the same crowd size), and it never involves the scores of subjects treated differently (exposed to different crowd sizes). In contrast with the variability between groups, the variability within groups never reflects the presence of a treatment effect. Regardless of whether the null hypothesis is true or false, the variability within groups reflects only **random error,** that is, *the combined effects* (on the scores of individual subjects) *of all uncontrolled factors,* such as individual differences among subjects, slight variations in experimental conditions, and errors in measurement. In ANOVA, the within-group estimate often is referred to simply as the *error term,* and it is analogous to the pooled variance estimate (s_p^2) in the t test for two independent samples.

Random error

The combined effects (on the scores of individual subjects) of all uncontrolled factors.

---

*Exercise 22.1 Imagine a simple experiment with three groups, each containing four observations. For each of the following outcomes, indicate whether variability between groups is present and also whether variability within groups is present. *Note:* You needn't do any calculations, with the possible exception of an occasional group mean, in order to answer this question.

(a)	GROUP 1	GROUP 2	GROUP 3
	8	8	8
	8	8	8
	8	8	8
	8	8	8

(b)	GROUP 1	GROUP 2	GROUP 3
	8	4	12
	8	4	12
	8	4	12
	8	4	12

(c)	GROUP 1	GROUP 2	GROUP 3
	4	6	5
	6	6	7
	8	10	9
	14	10	11

(d)	GROUP 1	GROUP 2	GROUP 3
	6	11	20
	8	12	18
	8	14	23
	10	15	25

Answers on page 518

22.3 *F* RATIO

In previous chapters, the null hypothesis has been tested with a *t* ratio. In the two-sample case, *t* reflects the ratio between the observed difference between the two sample means in the numerator and the estimated standard error in the denominator. For three or more samples, the null hypothesis is tested with a new ratio, the *F* ratio. Essentially, *F* reflects the ratio of the observed differences between all sample means (measured as variability between groups) in the numerator and the estimated error term or pooled variance estimate (measured as variability within groups) in the denominator term, that is,

---

F RATIO

$$F = \frac{Variability\ between\ groups}{Variability\ within\ groups} \tag{22.1}$$

Like *t*, *F* has its own family of sampling distributions that can be consulted, as described in Section 22.11, to test the null hypothesis. The resulting test is known as an *F* test.

22.4 *F* TEST

An *F* test of the null hypothesis is based on the notion that if the null hypothesis really is true, both the numerator and the denominator of the *F* ratio will tend to be about the same, but if the null hypothesis really is false, the numerator will tend to be larger than the denominator.

If the Null Hypothesis Really Is True

If the null hypothesis really is true (because there is no treatment effect due to different crowd sizes), the two estimates of variability (between and within groups) will reflect only random error. In this case,

$$F = \frac{random\ error}{random\ error}$$

Except for chance, estimates in both the numerator and the denominator are similar, and generally, *F* varies about a value of one.

If the Null Hypothesis Really Is False

If the null hypothesis really is false (because there is a treatment effect due to different crowd sizes), both of the estimates still will reflect random error, but that for between groups also will reflect the treatment effect. In this case,

$$F = \frac{random\ error\ +\ treatment\ effect}{random\ error}$$

When the null hypothesis really is false, the presence of a treatment effect tends to cause a chain reaction: The observed differences between group means tend to be large, as also does the variability between groups. Accordingly, the numerator term tends to exceed the denominator term, producing an F whose value is larger than one. When the null hypothesis really is *seriously* false, the sizable treatment effect tends to cause an even more pronounced chain reaction, beginning with very large observed differences between group means and ending with an F whose value tends to be *considerably* larger than one.

*Exercise 22.2 If the null hypothesis really is true, both the numerator and denominator of the F ratio will reflect only __**(a).**__ If the null hypothesis really is false, the numerator of the F ratio will also reflect the __**(b).**__ If the null hypothesis really is seriously false, the value of F tends to be considerably larger than __**(c).**__

Answers on page 518

When Status of Null Hypothesis is Unknown

In practice, of course, we never really know whether the null hypothesis is true or false. Following the usual procedure, we assume the null hypothesis to be true and view the observed F within the context of its hypothesized sampling distribution, as shown in **Figure 21.1.** If, because the differences between group

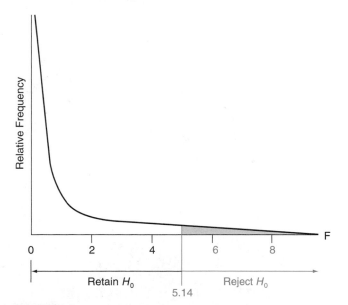

FIGURE 22.1
Hypothesized sampling distribution of F (for 2 and 6 degrees of freedom).

means are relatively small, the observed F appears to emerge from the dense concentration of possible F ratios smaller than the critical F, the experimental outcome will be viewed as a common occurrence, on the assumption that the null hypothesis is true. Therefore, the null hypothesis will be retained. On the other hand, if, because the differences between group means are relatively large, the observed F appears to emerge from the sparse concentration of possible F ratios equal to or greater than the critical F, the experimental outcome will be viewed as a rare occurrence, and the null hypothesis will be rejected. In the latter case, the value of the observed F is presumed to be inflated by a treatment effect.

Test Results for Outcomes A and B

Full-fledged hypothesis tests for outcomes A and B agree with the earlier intuitive decisions. Given the .05 level of significance, the null hypothesis should be retained for outcome A, because the observed F of 0.50 is smaller than the critical F of 5.14. But the null hypothesis should be rejected for outcome B, because the observed F of 21.50 exceeds the critical F. The hypothesis test for outcome B, as summarized in the accompanying box, will be discussed in more detail in later sections of this chapter.

HYPOTHESIS TEST SUMMARY: ONE-WAY F TEST (SMOKE ALARM EXPERIMENT, OUTCOME B)

Research Problem:
On the average, are subjects' reaction times to potentially dangerous smoke affected by crowds of zero, two, or four confederates?

Statistical Hypotheses:

$$H_0: \mu_0 = \mu_2 = \mu_4$$
$$H_1: H_0 \text{ is not true.}$$

Decision Rule:
Reject H_0 at .05 level of significance if F equals or is more positive than 5.14 (from Table C, Appendix D, given $df_{between} = 2$ and $df_{within} = 6$)

Calculations:
$F = 21.50$ (See Table 22.2 on page 371 and Table 22.3 on page 373 for additional details.)

Decision:
Reject H_0 at the .05 level of significance because $F = 21.50$ is more positive than 5.14.

Interpretation:
Crowd size does affect the subjects' mean reaction times to potentially dangerous smoke.

22.5 ASSUMPTIONS

The assumptions for *F* tests in ANOVA are the same as those for *t* tests of mean differences for two independent samples. All underlying populations are assumed to be normally distributed with equal variances. You needn't be too concerned about violations of these assumptions, particularly if all sample sizes are equal and each is fairly large (greater than about 10). Otherwise, in the *unlikely* event that you encounter conspicuous departures from normality or equality of variances, consider various alternatives similar to those discussed in Chapter 19 for the *t* test. More specifically, you might (1) increase sample sizes (to minimize the effect of non-normality); (2) equalize sample sizes (to minimize the effect of unequal population variances); (3) use a more complex version of *F* (designed for unequal population variances); or (4) use a less sensitive but more assumption-free test, such as the Kruskal-Wallis *H* test described in Chapter 25.[1]

22.6 TWO CAUTIONS

The ANOVA techniques described in this book presume that all scores are independent. In other words, the subjects are not matched across groups, and each subject contributes just one score to the overall analysis. Special ANOVA techniques must be used when scores lack independence because, for instance, each subject contributes more than one score.

To simplify computations, unrealistically small sample sizes are used in this and the next chapter. In practice, sample sizes that are either unduly small or excessively large should be avoided, as suggested in Section 16.7.

DETAILS

22.7 VARIANCE ESTIMATES

As its name implies, the analysis of variance uses variance estimates to measure variability between groups and within groups. Introduced in Chapter 5, *variance* is a measure of variability that, because it equals the square of the standard deviation, can be obtained from any formula for the standard deviation by eliminating the square root sign. A variance *estimate* signifies that information from a sample is used to determine the unknown variance of the population.

Variance Estimate for a Single Sample

The most common example of a variance estimate is the square of the sample standard deviation defined in Formula 18.4:

---

[1]For more information about the version of *F* designed for unequal variances, see D. C. Howell, *Statistical Methods for Psychology*, 4th ed. (Belmont, CA: Duxbury, 1997), chapter 11.

$$s^2 = \frac{\Sigma(X-\overline{X})^2}{n-1}$$

This estimate is designed for use with a single sample and therefore can't be used in analysis of variance. It can be used, however, to identify two general features of variance estimates in ANOVA.

1. Sum of Squares in Numerator: The numerator term represents the sum of the squared deviations about the sample mean, $\overline{X}$. In ANOVA, the numerator term of a variance estimate always is the *sum of squares*, that is, *the sum of squared deviations for some set of scores about their mean.*

2. Degrees of Freedom in Denominator: The denominator represents the number of degrees of freedom for these deviations. (Remember, as discussed in Section 18.11, only $n - 1$ of these deviations are free to vary. One degree of freedom is lost because the sum of n deviations about their own mean always must equal zero.) In ANOVA, the denominator term of a variance estimate always is the number of *degrees of freedom*, that is, *the number of deviations in the numerator that are free to vary and, therefore, supply valid information for the purpose of estimation.*

Mean Square

A variance estimate in ANOVA consists of some sum of squares divided by its degrees of freedom.

. . . .

Mean square (MS)

A variance estimate obtained by dividing a sum of squares by its degrees of freedom.

This operation always produces a number equal to the mean of the squared deviations, hence the designation **mean square,** abbreviated as *MS*. In ANOVA, the latter term is the most common, and it will be used in subsequent discussions. A general expression for any variance estimate reads

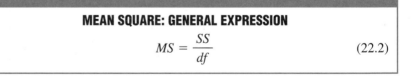

MEAN SQUARE: GENERAL EXPRESSION

$$MS = \frac{SS}{df} \tag{22.2}$$

where *MS* represents the variance estimate; *SS* denotes the sum of squared deviations about their mean; and *df* equals the corresponding number of degrees of freedom. Formula 22.2 should be read as "the mean square equals the sum of squares divided by its degrees of freedom."

The *F* test of the null hypothesis for outcome B will be based on a ratio involving two variance estimates: the mean square for variability between groups and the mean square for variability within groups. Before using these mean squares, we must calculate their sums of squares, as described in Section 22.8, and their degrees of freedom, as described in Section 22.9.

22.8 SUM OF SQUARES *(SS)*

Most of the computational effort in ANOVA is directed toward the various sum of squares terms: the sum of squares for variability between groups, $SS_{between}$; the sum of squares for variability within groups, SS_{within}; and the sum of squares for the total of these two, SS_{total}. Remember, *any **sum of squares** always equals the sum of the squared deviations of some set of scores about their mean.* More specifically,

---
Sum of squares (SS)

The sum of squared deviations of some set of scores about their mean.

- $SS_{between}$ equals the sum of the squared deviations of group means about the overall mean.
- SS_{within} equals the sum of the squared deviations of all scores about their respective group means.
- SS_{total} equals the sum of the squared deviations of all scores about the overall mean.

Calculating the Sums of Squares

When calculating the various SS terms, we do not deal directly with squared deviations. Instead, it's more convenient to use the equivalent expressions listed in the top half of **Table 22.2.** Study these expressions; they comply with a highly predictable computational pattern that once learned is easily remembered:

1. Each SS term consists of two main components, first a positive component and then a second component that is subtracted.

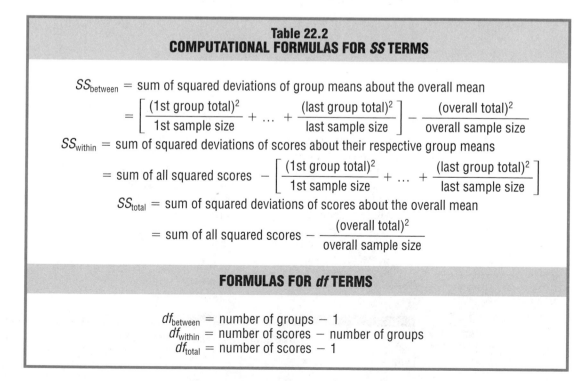

Table 22.2
COMPUTATIONAL FORMULAS FOR *SS* TERMS

$SS_{between}$ = sum of squared deviations of group means about the overall mean

$$= \left[\frac{(\text{1st group total})^2}{\text{1st sample size}} + \cdots + \frac{(\text{last group total})^2}{\text{last sample size}} \right] - \frac{(\text{overall total})^2}{\text{overall sample size}}$$

SS_{within} = sum of squared deviations of scores about their respective group means

$$= \text{sum of all squared scores} - \left[\frac{(\text{1st group total})^2}{\text{1st sample size}} + \cdots + \frac{(\text{last group total})^2}{\text{last sample size}} \right]$$

SS_{total} = sum of squared deviations of scores about the overall mean

$$= \text{sum of all squared scores} - \frac{(\text{overall total})^2}{\text{overall sample size}}$$

FORMULAS FOR *df* TERMS

$df_{between}$ = number of groups − 1
df_{within} = number of scores − number of groups
df_{total} = number of scores − 1

2. Totals replace group means and the overall mean.
3. Each score, whether a total score or an original score, is squared and, in the case of a total score, is divided by its sample size.

Checking for Computational Accuracy

Table 22.3 indicates how to use these computational formulas for the data in outcome B. To minimize computational errors, calculate from scratch each of the three *SS* terms, even though this entails some duplication of effort (not shown in Table 22.3 in order to save space). Then, as an almost foolproof check of your computations, as shown at the bottom of Table 22.3, verify that SS_{total} equals the sum of the various *SS* terms, that is,

---

SUMS OF SQUARES (ONE WAY)

$$SS_{\text{total}} = SS_{\text{between}} + SS_{\text{within}} \qquad (22.3)$$

22.9 DEGREES OF FREEDOM (*df*)

Formulas for the number of degrees of freedom differ for each *SS* term, and, for convenience, the various *df* formulas are listed in the bottom half of Table 22.2. To determine the *df* for any *SS*, simply substitute the appropriate numbers and subtract. For outcome B, which consists of three groups and a total of nine scores,

$$df_{\text{between}} = 3-1 = 2$$
$$df_{\text{within}} = 9-3 = 6$$
$$df_{\text{total}} = 9-1 = 8$$

Remember, the degrees of freedom reflect the number of deviations that are free to vary in the corresponding *SS* term. The value of 2 for df_{between} reflects the loss of one degree of freedom because the three group means are expressed as deviations about the one overall mean. The value of 6 for df_{within} reflects the loss of three degrees of freedom because all nine scores are expressed as deviations about their three respective group means. Finally, the value of 8 for df_{total} reflects the loss of one degree of freedom because all nine scores are expressed as deviations about the one overall mean.

Checking Accuracy of Degrees of Freedom

In ANOVA, the degrees of freedom for SS_{total} always equals the combined degrees of freedom for the remaining *SS* terms, that is,

---

Degrees of freedom (df)

The number of deviations free to vary in any sum of squares term.

DEGREES OF FREEDOM (ONE WAY)

$$df_{\text{total}} = df_{\text{between}} + df_{\text{within}} \qquad (22.4)$$

Table 22.3
CALCULATION OF *SS* TERMS

A. COMPUTATIONAL SEQUENCE

Find each group total and also the overall total for all groups 1.
Substitute numbers into computational formula 2 and solve for $SS_{between}$.
Substitute numbers into computational formula 3 and solve for SS_{within}.
Substitute numbers into computational formula 4 and solve for SS_{total}.
Do computational check 5.

B. DATA AND COMPUTATIONS

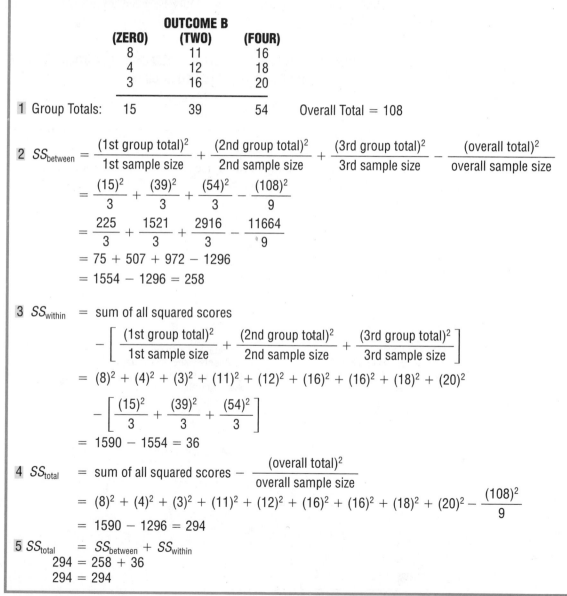

	OUTCOME B	
(ZERO)	(TWO)	(FOUR)
8	11	16
4	12	18
3	16	20

1 Group Totals: 15   39   54   Overall Total = 108

$$2 \quad SS_{between} = \frac{(\text{1st group total})^2}{\text{1st sample size}} + \frac{(\text{2nd group total})^2}{\text{2nd sample size}} + \frac{(\text{3rd group total})^2}{\text{3rd sample size}} - \frac{(\text{overall total})^2}{\text{overall sample size}}$$

$$= \frac{(15)^2}{3} + \frac{(39)^2}{3} + \frac{(54)^2}{3} - \frac{(108)^2}{9}$$

$$= \frac{225}{3} + \frac{1521}{3} + \frac{2916}{3} - \frac{11664}{9}$$

$$= 75 + 507 + 972 - 1296$$

$$= 1554 - 1296 = 258$$

$$3 \quad SS_{within} = \text{sum of all squared scores}$$

$$- \left[\frac{(\text{1st group total})^2}{\text{1st sample size}} + \frac{(\text{2nd group total})^2}{\text{2nd sample size}} + \frac{(\text{3rd group total})^2}{\text{3rd sample size}} \right]$$

$$= (8)^2 + (4)^2 + (3)^2 + (11)^2 + (12)^2 + (16)^2 + (16)^2 + (18)^2 + (20)^2$$

$$- \left[\frac{(15)^2}{3} + \frac{(39)^2}{3} + \frac{(54)^2}{3} \right]$$

$$= 1590 - 1554 = 36$$

$$4 \quad SS_{total} = \text{sum of all squared scores} - \frac{(\text{overall total})^2}{\text{overall sample size}}$$

$$= (8)^2 + (4)^2 + (3)^2 + (11)^2 + (12)^2 + (16)^2 + (16)^2 + (18)^2 + (20)^2 - \frac{(108)^2}{9}$$

$$= 1590 - 1296 = 294$$

$$5 \quad SS_{total} = SS_{between} + SS_{within}$$

$$294 = 258 + 36$$

$$294 = 294$$

This formula can be used to verify that the correct number of degrees of freedom has been assigned to each of the *SS* terms in outcome B.

22.10 MEAN SQUARES (*MS*) AND THE *F* RATIO

Having found the values of the various *SS* terms and their degrees of freedom, we can determine the values of the mean squares for variability between groups and variability within groups and then calculate the value of *F*, as suggested in **Figure 22.2.**

The value of the mean square for variability between groups, $MS_{between}$, is given by the following expression:

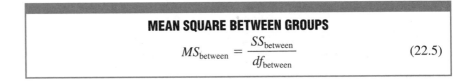

MEAN SQUARE BETWEEN GROUPS

$$MS_{between} = \frac{SS_{between}}{df_{between}} \tag{22.5}$$

$MS_{between}$ reflects the variability between means for groups of subjects who are treated differently. Relatively large values of $MS_{between}$ suggest the presence of a treatment effect.

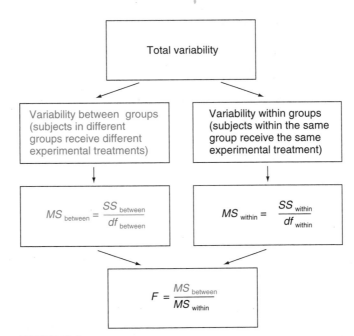

FIGURE 22.2
Sources of variability for ANOVA and the F ratio.

For outcome B,

$$MS_{between} = \frac{258}{2} = 129$$

The value of the mean square for variability within groups, MS_{within} is given by the following expression:

MEAN SQUARE WITHIN GROUPS

$$MS_{within} = \frac{SS_{within}}{df_{within}} \qquad (22.6)$$

MS_{within} reflects the variability among scores for subjects who are treated similarly within each group, pooled across all groups. Regardless of whether a treatment effect is present, MS_{within} measures only random error.

For outcome B,

$$MS_{within} = \frac{36}{6} = 6$$

Finally, Formula 22.1 for F can be rewritten as

F Ratio

Ratio of the between-group mean square (for subjects treated differently) to the within-group mean square (for subjects treated similarly).

F RATIO (ONE WAY)

$$F = \frac{MS_{between}}{MS_{within}} \qquad (22.7)$$

As mentioned previously, if the null hypothesis really is true (because reaction times are not affected by crowd size), the value of F will vary about a value of approximately one, but if the null hypothesis really is false, the value of F will tend to be larger than one.

For outcome B, the null hypothesis is suspect, because

$$F = \frac{129}{6} = 21.50$$

22.11 *F* TABLES

A decision about the null hypothesis requires that the observed *F* be compared with a critical *F*. As has been noted, there's not one but a family of *F* sampling distributions. Any particular *F* sampling distribution is uniquely specified by the pair of degrees of freedom associated with the mean squares in its numerator and denominator.

Critical *F* values for hypothesis tests at the .05 level (light numbers) and the .01 level (dark numbers) are listed in **Table 22.4** (for a few *F* sampling distributions) and in Table C, Appendix D (for the full range of *F* sampling distributions). To read either table, simply find the cell intersected by the column with the degrees of freedom equal to those in the numerator of *F*, $df_{between}$, and by the row with the degrees of freedom equal to those in the denominator of *F*, df_{within}. Table 22.4 illustrates this procedure when, as in outcome B, 2 and 6 degrees of freedom are associated with the numerator and denominator of *F*, respectively. In this case, the column with 2 degrees of freedom and the row with 6 degrees of freedom intersect a cell (shaded in color) that lists a critical *F* value of 5.14 for a hypothesis test at the .05 level of significance. As was anticipated in previous sections, because the observed *F* of 21.50 for outcome B exceeds this critical *F*, the overall null hypothesis can be rejected. There is evidence that crowd size does affect the subjects' reaction times.

Table 22.4
SPECIMEN TABLE FROM TABLE C OF APPENDIX D
CRITICAL VALUES OF *F*
.05 LEVELS OF SIGNIFICANCE (LIGHT NUMBERS)
.01 LEVELS OF SIGNIFICANCE (DARK NUMBERS)

DEGREES OF FREEDOM IN DENOMINATOR	DEGREES OF FREEDOM IN NUMERATOR			
	1	**2**	**3**	**4****
1	161	200	216	
	4052	**4999**	**5403**	
2	18.51	19.00	19.16	
	98.49	**99.01**	**99.17**	
3	10.13	9.55	9.28	
	34.12	**30.81**	**29.46**	
4	7.17	6.94	6.59	
	21.20	**18.00**	**16.69**	
5	6.61	5.79	5.41	
	16.26	**13.27**	**12.06**	
6	5.99	5.14	4.76	
	13.74	**10.92**	**9.78**	
7	5.59	4.47	4.35	
	12.25	**9.55**	**8.45**	
8				
*				
*				
*				

*Exercise 22.3 Find the critical *F* values for the following hypothesis tests:

(a) $\alpha = .05$, $df_{between} = 1$, $df_{within} = 18$

(b) $\alpha = .01$, $df_{between} = 3$, $df_{within} = 56$

(c) $\alpha = .05$, $df_{between} = 2$, $df_{within} = 36$

(d) $\alpha = .05$, $df_{between} = 4$, $df_{within} = 95$

Answers on page 518

22.12 NOTES ON USAGE

ANOVA results are usually reported as shown in **Table 22.5**. "Source" refers to the source of variability: between groups, within groups, and total. Notice the arrangement of column headings from *SS* and *df* to *MS* and *F*. Also notice that the bottom row for total variability contains entries only for *SS* and *df*. Ordinarily, the shaded numbers in parentheses don't appear in ANOVA tables, but in Table 22.5 they show the origin of each *MS* and of *F*. The asterisk in Table 22.5 emphasizes that the observed *F* of 21.50 exceeds the critical *F* of 5.14 and, therefore, causes the null hypothesis to be rejected at the .05 level of significance.

Other Labels

Sometimes ANOVA tables appear with labels other than those shown in Table 22.5. For instance, "Between" might be replaced with "Treatment," as the variability between groups reflects any treatment effect. Or "Between" might be replaced by a description of the actual experimental treatment, such as "Number of Confederates" or, more generally, "Crowd Size." Likewise, "Within" might be replaced with "Error," as variability within groups reflects only the presence of random error.

Published Reports

In addition to an ANOVA table, published reports of a hypothesis test might be limited to an interpretative comment, plus a parenthetical statement that summarizes the statistical test and includes a *p*-value. For example, using the

Table 22.5
ANOVA TABLE (OUTCOME B)

SOURCE	SS	df	MS	F
Between	258	2	$\left(\frac{258}{2} = \right) 129$	$\left(\frac{129}{6} = \right) 21.50*$
Within	36	6	$\left(\frac{36}{6} = \right) 6$	
Total	294	8		

Significant at .05 level.

less structured p-value approach to hypothesis testing described in Section 21.1, an investigator might report the following:

There is evidence that on the average, number of confederates affects the reaction times of subjects to potentially dangerous smoke [F (2,6) = 21.50, $p <$ .01].

The parenthetical statement indicates that an F based on 2 and 6 degrees of freedom was found to equal 21.50. The test result has an approximate p-value of less than .01 because, as can be seen in Table 22.4, the observed F of 21.50 is more positive than the critical F of 10.92 for the .01 level of significance. Furthermore, because the p-value of less than .01 reflects a rare test result, given that the null hypothesis is true, it supports the research hypothesis, as implied in the interpretative statement.

*Exercise 22.4 A common assumption is that sleep deprivation influences aggression. To test this assumption, volunteer subjects are randomly assigned to sleep-deprivation periods of either 0, 24, 48, or 72 hours and are subsequently tested for aggressive behavior in a controlled social situation. Aggressive scores signify the total number of different aggressive behaviors, such as "put downs," arguments, or verbal interruptions, demonstrated by subjects during a test period.

AGGRESSIVE SCORES			
0	24	48	72
0	1	5	7
1	3	4	1
0	2	7	6
3	2	8	9
1	4	6	10
2	6	3	12
4	3	2	8
2	4	5	7

(a) Test the null hypothesis at the .05 level of significance.

(b) Summarize the results with an ANOVA table.

(c) Specify the approximate p-value for these test results.

(d) How might these results appear in a published report?

Answers on page 518

22.13 *F* TEST IS NONDIRECTIONAL

It might seem strange that even though the entire rejection region for the null hypothesis appears only in the upper tail of the F sampling distribution, as in Figure 22.1, *the F test in ANOVA is the equivalent of a nondirectional test.*

Recall that all variations in ANOVA are squared. When squared, all values become positive, regardless of whether the original differences between groups (or group means) are positive or negative. All squared differences between groups have a cumulative positive effect on the observed F and thereby ensure that F is a nondirectional test, even though only the upper tail of its sampling distribution contains the rejection region.

F and t^2

A similar effect could be produced by squaring the t test. When squared, all values of t^2 become positive, regardless of whether the original value for the observed t was positive or negative. Hence, the t^2 test also qualifies as a nondirectional test, even though the entire rejection region appears only in the upper tail of the t^2 sampling distribution. In fact, the values of t^2 and F are identical when both tests are applied to the same data for two independent groups. When only two groups are involved, the t^2 test can be viewed as a special case of the more general F test in ANOVA for two or more groups.

BEYOND THE F TEST

22.14 SMALL, MEDIUM, OR LARGE EFFECT?

Rejection of the overall null hypothesis usually raises some additional questions, such as what is the estimated size—small, medium, or large—of the overall effect? Recall the discussion of Section 21.10, where the squared point biserial correlation coefficient was used to estimate, *independently of sample sizes*, whether a statistically significant t test reflects a small, medium, or large effect. The same type of analysis can be conducted for statistically significant F tests using the squared curvilinear correlation coefficient, symbolized as η^2 and pronounced "eta-squared." Viewed as an extension of the point biserial correlation coefficient to more than two groups, η^2 describes the proportion (from 0 to 1) of the total variance in scores that is predictable from—or if it is a well-designed experiment, caused by—the subjects' membership in one of the various groups defined by the investigator.

The Squared Curvilinear Correlation Coefficient, η^2

More specifically, in terms of the smoke alarm experiment, η^2 indicates the relative improvement in accuracy (for predicting, one at a time, each of the observed reaction times for all subjects), when the repetitive prediction of the overall mean for *all* subjects, $\overline{X}$, is replaced by predictions of one of the three group means, $\overline{X}_0$, $\overline{X}_2$, or $\overline{X}_4$, depending on whether a particular subject had been in crowd sizes of 0, 2, or 4 confederates. In effect, the **squared curvilinear correlation coefficient,** η^2, *indicates the proportion of variance in the dependent variable* (*reaction times*) *explained by the independent variable* (*crowd size*).

Instead of actually calculating η^2 from scratch, we need merely solve the following formula:

Squared curvilinear correlation coefficient, (η^2)

The proportion of variance in the dependent variable that can be explained by the independent variable.

> **PROPORTION OF EXPLAINED VARIANCE (ONE-WAY ANOVA)**
> $$\eta^2 = \frac{SS_{\text{between}}}{SS_{\text{total}}} \tag{22.8}$$

where the two *SS* terms can be obtained from the ANOVA summary table.

Let's look more closely at how we might use this formula to estimate effect size, given the significant *F* for the smoke alarm experiment. Substituting numbers from Table 22.5 for the two *SS* terms, we obtain

$$\eta^2 = \frac{258}{294} = .88$$

This large value of .88 suggests that 88 percent of the variance in reaction times is explained by whether subjects are in crowds of zero, two, or four confederates, whereas only the remaining 12 percent of variance of reaction times is not explained by crowd size.

One rough rule of thumb, suggested by Cohen, is that the estimated effect (estimated differences among population means) is small if η^2 approximates .01; medium if η^2 approximates .06; and large if η^2 approximates .14 or more.[2] Using Cohen's rule of thumb, the estimated effect size for the smoke alarm experiment would be considered spectacularly large, and it reflects the fact that these fictitious data were selected to dramatize the differences due to crowd size. (Also as implied in Section 21.11, this estimated effect size—even if based on real data—would have been highly speculative because of the instability of η^2 when sample sizes are small.)

A Recommendation

Consider calculating η^2 (or its less straightforward, but more accurate competitor, ω^2 [omega-squared], cited in advanced statistics books) whenever you encounter a statistically significant *F*, especially one based on large samples. As mentioned in Section 21.10, however, do not blindly apply Cohen's rule of thumb about effect sizes without regard to special circumstances that could give considerable importance to even a very small effect.

*Exercise 22.5 Given the rejection of the null hypothesis in Exercise 22.4 on page 378, use Formula 22.8 and Cohen's rule of thumb to estimate whether the effect is small, medium, or large (even though this estimate is speculative because of the moderate sample sizes.)

Answer on page 518

---

[2]J. Cohen, *Statistical Power Analysis in the Behavioral Sciences*, 2nd ed. (Hillsdale, NJ: Erlbaum, 1988), pp. 285–87.

22.15 MULTIPLE COMPARISONS

Rejection of the overall null hypothesis indicates only that not all population means are equal. In the case of outcome B, the rejection of H_0 signals the presence of one or more inequalities between the mean reaction times for populations of subjects exposed to crowds of zero, two, and four confederates, that is, between μ_0, μ_2, and μ_4. To pinpoint the one or more differences between pairs of population means that contribute to the rejection of the overall H_0, you must use a test of multiple comparisons. A test of **multiple comparisons** is designed to evaluate not just one but a series of differences between population means. For outcome B, each of the three possible differences between pairs of population means, $\mu_0 - \mu_2$, $\mu_0 - \mu_4$, and $\mu_2 - \mu_4$, should be evaluated.

Multiple comparisons

The series of possible comparisons whenever more than two population means are involved.

---

t Test Not Appropriate

These differences can't be evaluated with a series of regular t tests (except under special circumstances, as alluded to in Section 22.17). Essentially, the regular t test is designed to evaluate a *single* comparison using a pair of observed means, not multiple comparisons using all possible pairs of observed means. Among other complications, the use of multiple t tests increases the probability of a type I error (rejecting a true null hypothesis) beyond that value specified by the level of significance.

A coin-tossing example might clarify this problem. When a fair coin is tossed only once, the probability of heads equals .50—just as when a single t test is to be conducted at the .05 level of significance, the probability of a type I error equals .05. When a fair coin is tossed three times, however, heads can appear not only on the first toss but also on the second or third toss, and hence the probability of heads on *at least one* of the three tosses exceeds .50. By the same token, for a series of three t tests, each conducted at the .05 level of significance, a type I error can be committed not only on the first test but also on the second or third test, and hence the probability of committing a type I error on *at least one* of the three tests exceeds .05. In fact, the cumulative probability of at least one type I error can be as large as .15 for a series of three t tests and even larger for a more extended series of t tests.

22.16 SCHEFFÉ'S TEST

The shortcoming just described does not apply to several specially designed multiple comparison tests, including the test by Scheffé (pronounced Shef-FAY') described in this book. Once the overall null hypothesis has been rejected in ANOVA, **Scheffé's test** *can be used to test each of all possible comparisons, and yet the cumulative probability of at least one type I error never exceeds the specified level of significance.*

Scheffé's test

A multiple comparison test that, regardless of the number of comparisons, never permits the cumulative probability of at least one type I error to exceed the specified level of significance.

Scheffé's Critical Value

One version of Scheffé's test supplies a critical value for evaluating the observed difference between means for any pair of groups, for instance, the groups with zero and two confederates. If the observed difference between means, $\overline{X}_0 - \overline{X}_2$, is either more positive or more negative than Scheffé's critical value, $(\overline{X}_0 - \overline{X}_2)_{\text{crit}}$, the null hypothesis for the comparison of that particular pair of population means can be rejected. To determine the critical value for groups with zero and two confederates, use the following expression, along with numerical data from the ANOVA for outcome B:

$$(\overline{X}_0 - \overline{X}_2)_{\text{crit}} = \pm \sqrt{(df_{\text{between}}) \, (F_{\text{crit}}) \, (MS_{\text{within}}) \left(\frac{1}{n_0} + \frac{1}{n_2} \right)}$$

$$= \pm \sqrt{(2) \, (5.14) \, (6) \left(\frac{1}{3} + \frac{1}{3} \right)} \quad = \pm \sqrt{(2) \, (5.14) \, (6) \left(\frac{2}{3} \right)}$$

$$= \pm \sqrt{41.12} \; = \pm 6.41$$

Because the observed difference between means, $\overline{X}_0 - \overline{X}_2$, equals -8 (from $5 - 13$, as shown in Table 22.1) and is more negative than the critical difference between means, -6.41, the null hypothesis for $\mu_0 - \mu_2$ can be rejected at the .05 level. In other words, there is evidence that the mean reaction times differ between populations of subjects exposed to zero and two confederates.

When sample sizes are unequal, Scheffé's critical value must be calculated for each comparison. When all sample sizes are equal, as in outcome B, the critical mean difference needs to be calculated for only one comparison and then used to evaluate all remaining comparisons. **Table 22.6** summarizes the results of Scheffé's test for each of the three differences between pairs of means in outcome B, using the critical value of ± 6.41.

Interpretation for Outcome B

Table 22.6 reveals that two of the three comparisons reach the .05 level of significance. It can be concluded that the differences between population means for subjects with zero and two confederates and for subjects with zero and four confederates contribute to the original rejection of the overall null hy-

	Table 22.6	
	SUMMARY TABLE FOR SCHEFFÉ'S TEST	
POPULATION COMPARISON	**OBSERVED MEAN DIFFERENCE**	**SCHEFFÉ'S CRITICAL VALUE**
$\mu_0 - \mu_2$	-8^*	± 6.41
$\mu_0 - \mu_4$	-13^*	± 6.41
$\mu_2 - \mu_4$	-5	± 6.41

Significance of 0.5 level.

pothesis. Thus there is evidence that the reaction times of subjects with zero confederates differ—that is, they are shorter than—the reaction times of subjects with either two or four confederates. There is no evidence, however, that the reaction times of subjects with two confederates differ from those of subjects with four confederates.

General Expression

For any comparison of differences between pairs of group means, $\overline{X}_i - \overline{X}_j$, Scheffé's critical value is given by the following expression:

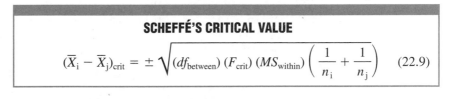

SCHEFFÉ'S CRITICAL VALUE

$$(\overline{X}_i - \overline{X}_j)_{\text{crit}} = \pm \sqrt{(df_{\text{between}})(F_{\text{crit}})(MS_{\text{within}})\left(\frac{1}{n_i} + \frac{1}{n_j}\right)} \quad (22.9)$$

where the subscripts i and j refer to any two groups, and all terms under the square root refer back to the original ANOVA.

Other Types of Comparisons

Scheffé's test should be used only if the overall null hypothesis has been rejected. It is designed to evaluate all possible comparisons of interest, not just differences for pairs of means. For example, when modified appropriately, Scheffé's test can evaluate other types of comparisons, such as the difference between the mean for those subjects with zero confederates and the combined mean for those subjects with two and four confederates.

*Exercise 22.6 After the rejection of the null hypothesis in Exercise 22.4 on page 378, Scheffé's test may be used to infer which pairs of population means differ (and cause the overall null hypothesis to be rejected). Using the .05 level of significance, calculate Scheffé's critical value for the data in Exercise 22.4 and use it to infer which pairs of population means differ.

Answers on page 519

22.17 OTHER MULTIPLE COMPARISON TESTS

That only Scheffé's test has been discussed shouldn't be interpreted as an unqualified endorsement. At least a half dozen other multiple comparison tests are available, and depending on several considerations, any of these could be the most appropriate test for a particular set of comparisons. For instance, when compared to Scheffé's test, none of the other tests protects better against false alarms or type I errors, but, on the other hand, none of them protects worse against misses or type II errors. Depending on the relative seriousness of the type I and II errors, therefore, you might choose to use some other multiple comparison test—possibly even an extension of the *regular t* test for two independent samples, variously referred to as the "*protected t* test" or as the *LSD* (*least significant difference*) test, that reverses the strength and weakness of Scheffé's test.

Selecting a Multiple Comparison Test

We'll not deal with the relatively complex, controversial issue of when, depending on circumstances, a specific multiple comparison test is most appropriate. Indeed, sometimes it isn't even necessary to resolve this issue. With a computer program, such as Minitab, SPSS or SAS, a few keystrokes can initiate not just one, but an entire series of multiple comparison tests. Insofar as the pattern of significant and nonsignificant comparisons remains about the same for all tests—as has often happened in our experience—you simply can report this finding without concerning yourself about *the* most appropriate multiple comparison test for that particular set of comparisons. In those cases where the status of a particular comparison is ambiguous, being designated as significant by some of the multiple comparison tests and as nonsignificant by the remaining tests, this comparison could be reported as having "borderline" significance.

22.18 COMPUTER OUTPUT

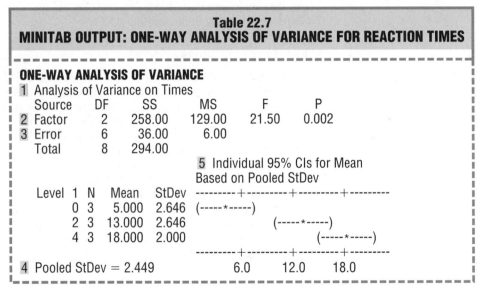

Table 22.7
MINITAB OUTPUT: ONE-WAY ANALYSIS OF VARIANCE FOR REACTION TIMES

ONE-WAY ANALYSIS OF VARIANCE

[1] Analysis of Variance on Times

Source	DF	SS	MS	F	P
[2] Factor	2	258.00	129.00	21.50	0.002
[3] Error	6	36.00	6.00		
Total	8	294.00			

[5] Individual 95% CIs for Mean
Based on Pooled StDev

Level	N	Mean	StDev	
0	3	5.000	2.646	(-----*-----)
2	3	13.000	2.646	(-----*-----)
4	3	18.000	2.000	(-----*-----)

```
                                    ---------+---------+---------+---------
[4] Pooled StDev = 2.449               6.0       12.0      18.0
```

Comments:

[1] *One-way ANOVA for reaction times. Compare with results in Table 22.5*

[2] *Between groups.*

[3] *Within groups.*

[4] *Equal to the square root of the mean square for error (6.00), the pooled standard deviation represents the best estimate of the population standard deviation, given the assumption of equal population variances (and standard deviations).*

[5] *This stack of three confidence intervals can be interpreted as confirming the results for Scheffé's test in Table 22.6. The lack of overlap between the confidence interval for subjects in crowds of zero confederates and either of the other two confidence intervals implies that the corresponding null hypotheses probably are false.*

Exercise 22.7 The following SPSS output is based on the aggressive scores in Exercise 22.4 for subjects who experience various levels of sleep deprivation. It contains both an ANOVA summary table and also a table showing the results for Scheffé's test.

(a) Specify the value of the F ratio.

(b) The SPSS output specifies Sig. as the exact p-value for the F ratio. Precisely what does this tell us?

(c) Does the F ratio attain the .05 level of significance?

SPSS OUTPUT: ONE-WAY ANOVA AND SCHEFFÉ'S TEST FOR AGGRESSIVE SCORES

Oneway

ANOVA

AGGRESS

	Sum of Squares	df	Mean Square	F	Sig.
Between Groups	154.125	3	51.375	10.836	.000
Within Groups	132.750	28	4.741		
Total	286.875	31			

Post Hoc Tests

Multiple Comparisons

Dependent Variable: AGGRESS
Scheffé

(I) DEPRIV	(J) DEPRIV	Mean Difference (I-J)	Std. Error	Sig.	95% Confidence Interval Lower Bound	95% Confidence Interval Upper Bound
0	24	−1.50	1.09	.600	−4.74	1.74
	48	−3.38*	1.09	.038	−6.61	−.14
	72	−5.88*	1.09	.000	−9.11	−2.64
24	0	1.50	1.09	.600	−1.74	4.74
	48	−1.88	1.09	.412	−5.11	1.36
	72	−4.38*	1.09	.005	−7.61	−1.14
48	0	3.38*	1.09	.038	.14	6.61
	24	1.88	1.09	.412	−1.36	5.11
	72	−2.50	1.09	.178	−5.74	.74
72	0	5.88*	1.09	.000	2.64	9.11
	24	4.38*	1.09	.005	1.14	7.61
	48	2.50	1.09	.178	−.74	5.74

The mean difference is significant at the .05 level.

(d) Although the SPSS-produced table of results for Scheffé's test on page 385 has not been discussed heretofore, you'll see that it's not difficult to interpret, once you ignore the slightly confusing redundancy caused by, for example, comparing group 0 with group 24 and also, in the next cell, group 24 with group 0. Whenever an asterisk appears next to the mean difference for any two groups, those two group means differ significantly at the .05 level. Specify precisely which pairs of groups have means that differ significantly at the .05 level. (Notice that, as usual in SPSS, exact p-values are listed under Sig.)

Answers on page 519

Summary
· · · · · · · · · · · ·

When data are quantitative, analysis of variance, abbreviated as ANOVA, tests the null hypothesis for two, three, or more population means by classifying total variability into two independent components: variability between groups and variability within groups. Both components will reflect only random error if the null hypothesis is true, and the resulting F ratio (variability between groups divided by variability within groups) will tend toward a value of approximately one. If the null hypothesis is false, variability between groups will reflect both random error and a treatment effect, whereas variability within groups will still reflect only random error, and the resulting F ratio will tend toward a value greater than one.

Each variance estimate or mean square (MS) is found by dividing the appropriate sum of squares (SS) term by its degrees of freedom (df). In practice, once a value of F has been obtained, it's compared with a critical F in the F sampling distribution. If the observed F equals or is more positive than the critical F, the observed sample mean differences are too large, given that the null hypothesis is true, and therefore the null hypothesis will be rejected. Otherwise, for all smaller observed values of F, the null hypothesis will be retained.

ANOVA results are often summarized in tabular form.

F tests in ANOVA assume that all underlying populations are normally distributed with equal variances. Ordinarily, you needn't be too concerned about violations of these assumptions.

Because all variations in ANOVA are squared, all variance estimates are positive, and the F test is a nondirectional test.

Whenever obtaining a statistically significant F, especially one based on large sample sizes, consider using Formula 22.8 and Cohen's rule of thumb to estimate, independently of sample size, whether the effect is small, medium, or large.

To pinpoint differences between specific pairs of population means that contribute to the rejection of the overall F, use one or more multiple comparison tests, such as the test by Scheffé described in this book. Use these tests only if the overall H_0 has been rejected.

Important Terms
· · · · · · · · · · · · · · · · · ·

Analysis of variance (ANOVA)　　　　　　**Treatment effect**

One-way ANOVA　　　　　　**Variability between groups**

Variability within groups	**F Ratio**
Random error	**Squared curvilinear**
Mean square *(MS)*	**correlation coefficient (η^2)**
Sum of squares *(SS)*	**Multiple comparisons**
Degrees of freedom *(df)*	**Scheffé's test**

REVIEW EXERCISES

22.8 **(a)** Use the data for outcome A in Table 22.1 to test the null hypothesis at the .01 level of significance.

(b) Summarize the results with an ANOVA table, as in Table 22.5.

*22.9 Another social psychologist conducts the smoke alarm experiment described in this chapter. For reasons beyond his control, unequal numbers of "real" subjects occupy the different groups.

(a) Given the following results, test the null hypothesis at the .05 level of significance.

REACTION TIMES IN MINUTES		
ZERO	**TWO**	**TWELVE**
1	4	7
3	7	12
6	5	10
2		9
1		

(b) Summarize the results with an ANOVA table.

(c) If the overall null hypothesis has been rejected, use Scheffé's test and interpret the results.

Answers on pages 519–520

22.10 A third social psychologist modifies the above design to include five different crowd sizes.

(a) Given the following results, test the null hypothesis at the .05 level of significance.

REACTION TIMES IN MINUTES				
ZERO	**TWO**	**FOUR**	**EIGHT**	**TWELVE**
1	4	6	15	20
1	3	1	6	25
3	1	2	9	10
6	7	10	17	10

(b) Summarize the results with an ANOVA table.

(c) If the F test is statistically significant—that is, if the overall null hypothesis has been rejected—estimate whether the overall effect is small, medium, or large (even though this estimate must be viewed as highly speculative because of the small sample sizes).

(d) If appropriate, use Scheffé's test and interpret the results.

22.11 Comment on each of the following statements:
  (a) If the null hypothesis really is true, the F ratio should equal one.
  (b) It's not possible to use a measure of variability, such as the sample variance, to test hypothesis regarding differences between population means.
  (c) If the null hypothesis is rejected, then *all* population means must be different.

*22.12 For some experiment, imagine four possible outcomes, as described in the following ANOVA tables.

A.	SOURCE	SS	df	MS	F
	Between	900	3	300	3
	Within	8000	80	100	
	Total	8900	83		
B.	SOURCE	SS	df	MS	F
	Between	1500	3	500	5
	Within	8000	80	100	
	Total	9500	83		
C.	SOURCE	SS	df	MS	F
	Between	300	3	100	1
	Within	8000	80	100	
	Total	8300	83		
D.	SOURCE	SS	df	MS	F
	Between	300	3	100	1
	Within	400	4	100	
	Total	700	7		

(a) How many groups are in outcome D?
(b) Assuming groups of equal size, what's the size of each group in outcome C?
(c) Which outcome(s) would cause the null hypothesis to be rejected at the .05 level of significance?
(d) Which outcome is based on the largest total number of subjects?
(e) Which outcome provides the least information about a possible treatment effect?

(f) Which outcome would be the least likely to stimulate additional research?

(g) Specify the approximate p-values for each of these outcomes.

(h) For any outcome(s) that cause the null hypothesis to be rejected at the .05 level of significance, use Formula 22.8 and Cohen's rule of thumb to estimate whether effect size(s) are small, medium, or large.

Answers on page 520

22.13 Twenty-three overweight male volunteers are randomly assigned to three different treatment programs designed to produce a weight loss by focusing on either diet, exercise, or the modification of eating behavior. Weight changes were recorded, to the nearest pound, for all participants who completed the two-month experiment. Positive scores signify a weight drop; negative scores, a weight gain.

(a) Using the following results, test the null hypothesis at the .05 level of significance. *Note:* Negative signs should be honored when calculating group and overall totals.

DIET	EXERCISE	BEHAVIOR MODIFICATION
	WEIGHT CHANGES	
3	−1	7
4	8	1
0	4	10
−3	2	0
5	2	18
10	−3	12
3		4
0		6
		5

(b) Summarize the results with an ANOVA table.

(c) Specify the approximate p-value for these test results.

(d) How might these results appear in a published report?

(e) If appropriate, use Scheffé's test and interpret the results.

22.14 The F test describes the ratio of two sources of variability: that for subjects treated differently and that for subjects treated similarly. Is there any sense in which the t test for two independent groups can be viewed likewise?

CHAPTER 23

Analysis of Variance (Two Way)

23.1 TESTING HYPOTHESES ABOUT REACTIONS OF MALES AND FEMALES IN CROWDS

23.2 PRELIMINARY INTERPRETATIONS

23.3 THREE F RATIOS

23.4 INTERACTION

23.5 DESCRIBING INTERACTIONS

23.6 ASSUMPTIONS

23.7 IMPORTANCE OF EQUAL SAMPLE SIZES

23.8 OTHER TYPES OF ANOVA

DETAILS

23.9 VARIANCE ESTIMATES

23.10 SUM OF SQUARES (SS)

23.11 DEGREES OF FREEDOM (df)

23.12 MEAN SQUARES (MS) AND F RATIOS

23.13 F TABLES

23.14 SMALL, MEDIUM, OR LARGE EFFECT?

23.15 MULTIPLE COMPARISONS

23.16 COMPUTER OUTPUT

Summary

Important Terms

Review Exercises

23.1 TESTING HYPOTHESES ABOUT REACTIONS OF MALES AND FEMALES IN CROWDS

Chapter 22 described an attempt to determine the effect of crowd size on the reaction times of subjects—presumably both males and females—to potentially dangerous smoke. If gender were viewed as important, possibly because of the noticeably different reactions of males and females during a pilot study, the social psychologist might design a study with two factors or independent variables: crowd size (either zero, two, or four confederates) and gender (either female or male). Using this two-factor or **two-way ANOVA** design, she still can test the original null hypothesis regarding the effect of crowd size on subjects' reaction times. As a bonus, the psychologist also can test two new null hypotheses: one regarding the effect of gender and the other regarding the combined effect or interaction of crowd size and gender on subjects' reaction times.

For computational simplicity, let's assume that the social psychologist randomly assigns two female subjects to be tested (one at a time) with either zero, two, or four confederates and then repeats the random assignment for an equal number of male subjects. The resulting six groups, each consisting of two subjects, represent all possible combinations of the two factors.

Two-way ANOVA

A more complex type of analysis than one-way ANOVA that tests whether differences exist among population means categorized by two factors or independent variables.

23.2 PRELIMINARY INTERPRETATIONS

Tables

Table 23.1 shows one set of possible outcomes for the two-way study. Although, as indicated in Chapter 22, the actual computations in ANOVA are based on totals, preliminary interpretations can be based on either totals or means. In Table 23.1, the numbers in color represent four different types of means:

Main effect

The effect of a single factor when any other factor is ignored.

1. The mean of the two group means in each column yields the three column means (9, 12, 15). Column means represent the mean reaction times for each crowd size *when gender is ignored.* Any differences among these column means not attributable to chance are referred to as the main effect of crowd size on reaction time. In ANOVA, **main effect**

Table 23.1
OUTCOME OF TWO-WAY EXPERIMENT (REACTION TIMES IN MINUTES)

			CROWD SIZE					
GENDER	ZERO		TWO		FOUR		ROW MEAN	
Female	8 8	8	8 6	7	10 8	9	8	
Male	9 11	10	15 19	17	24 18	21	16	
Column Mean		9		12		15	Overall Mean = 12	

Note: *Numbers in color are means.*

always refers to the effect of a single factor, such as crowd size, when any other factor, such as gender, is ignored.

2. The mean of the three group means in each row yields the two row means (8, 16). Row means represent the mean reaction times for gender when crowd size is ignored. Any difference between these row means not attributable to chance is referred to as the main effect of gender on reaction time.

3. The mean of the reaction times for each group of two subjects yields the six group means (8, 7, 9, 10, 17, 21) for each combination of the two factors. Sometimes referred to as cell means or treatment-combination means, these group means reflect not only the main effects for crowd size and gender described earlier, but more importantly, any effect due to the interaction between crowd size and gender, as described in the following section.

4. Finally, the mean of the three column means—or of the two row means—yields the overall or grand mean (12) for all subjects in the study.

Graphs

Before plunging into a statistical analysis, let's attempt to decipher, with the aid of graphs, the messages embodied in Table 23.1. This type of preliminary interpretation is not designed to replace the statistical analysis, but to supply at least a rough perspective, whenever possible, about those findings that most likely will be supported by the statistical analysis.

Main Effects

The slanted line in panel A of **Figure 23.1** depicts the large differences between column means, that is, between mean reaction times for subjects, regardless of gender, with zero, two, and four confederates. The relatively steep slant of this line suggests that even when chance sampling variability is considered, the null hypothesis for crowd size probably should be rejected. The steeper the slant is, the larger the observed differences between column means and the greater the suspected main effect of crowd size will be. A fairly level line in panel A of Figure 23.1 would have reflected the relative absence of any main effect due to crowd size.

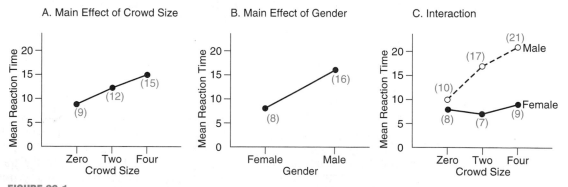

FIGURE 23.1
Graphs of outcomes of two-way experiment.

The slanted line in panel B of Figure 23.1 depicts the large difference between row means, that is, between mean reaction times for females and males, regardless of crowd size. The relatively steep slope of this line suggests that even when chance sampling variability is considered, the null hypothesis for gender probably should be rejected; that is, there probably exists a main effect of gender on reaction time.

Interaction

These preliminary conclusions about main effects must be qualified because of a complication due to the combined effect or interaction of crowd size and gender.

Interaction occurs whenever the effects of one factor are not consistent for all values (or levels) of the second factor.

Panel C of Figure 23.1 depicts the interaction between crowd size and gender. The two nonparallel lines in panel C depict differences between the three group means in the first row and the three group means in the second row, that is, between the mean reaction times of females for different crowd sizes and the mean reaction times of males for different crowd sizes. Although the line for female subjects remains fairly level, that for male subjects is slanted, suggesting that the reaction times of male subjects, but not those of female subjects, are influenced by crowd size. Because the effect of crowd size is not consistent for female and male subjects—portrayed by the apparent nonparallelism between the two lines in panel C of Figure 23.1—the null hypothesis (that there is no interaction between the two factors) probably should be rejected. Section 23.4 contains additional comments about interaction, as well as a more preferred definition of interaction.

Summary of Preliminary Interpretations

To summarize, a nonstatistical evaluation of the graphs of data for the two-way study suggests several preliminary interpretations. Each of the three null hypotheses regarding the effects of crowd size, gender, and the interaction of these factors probably should be rejected. Because of the suspected interaction, however, generalizations about the effects of one factor must be qualified in terms of specific levels of the second factor. Pending the outcome of the statistical analysis, you can speculate that the crowd size probably influences the reaction times of male subjects but not those of female subjects.

*Exercise 23.1 A home economist wishes to determine whether college students prefer a commercially produced pizza with a particular topping (either plain, sausage, salami, or everything) and one type of crust (either thick or thin). One hundred sixty volunteers were randomly assigned to one of the eight groups defined by this two-way experiment. After eating their assigned pizza, the twenty subjects in each group rated their preference on a scale ranging from 0 (inedible) to 10 (the best). The results, in the form of means for groups, rows, and columns, are as follows:

MEAN PREFERENCE SCORES FOR PIZZA AS A FUNCTION OF TOPPING AND CRUST					
			TOPPING		
CRUST	PLAIN	SAUSAGE	SALAMI	EVERYTHING	ROW
Thick	7.2	5.7	4.8	6.1	6.0
Thin	8.9	4.8	8.4	1.3	5.9
Column	8.1	5.3	6.6	3.7	

Construct graphs for each of the three possible effects, and use this information to interpret the outcome of the experiment. (Ordinarily, of course, you would verify these speculations by performing an ANOVA—a task that cannot be performed for these data, because only means are supplied.)

Answers on page 521

23.3 THREE *F* RATIOS

As suggested in **Figure 23.2**, *F* ratios in both a one- and a two-way ANOVA always consist of a numerator, shown in color, that measures some aspect of variability between groups and a denominator, shown in black, that measures variability within groups. In a one-way ANOVA, as you'll recall, a single null hypothesis is tested with one *F* ratio.

In a two-way ANOVA, three different null hypotheses are tested, one at a time, with three *F* ratios: F_{column}, F_{row}, and $F_{interaction}$.

The numerator of each of these three *F* ratios reflects a different aspect of variability between groups: variability between columns (crowd size), variability between rows (gender), and interaction, which is any remaining variability between groups not attributable to either variability between columns (crowd size) or rows (gender).

The numerator terms for the three *F* ratios in the bottom panel of Figure 23.2 estimate random error and, if present, a treatment effect (for subjects treated differently by the investigator, in the case of crowd size, and by "nature," in the case of gender).[1] But the black denominator term always estimates only random error (for subjects treated similarly in the same group).

In practice, a large *F* value is viewed as rare, given that the null hypothesis is true, and, therefore, it leads to the rejection of that particular null hypothesis. Otherwise, the null hypothesis is retained.

---

[1]Strictly speaking, because it can't be manipulated by the investigator, gender fails to qualify as a true independent variable, and, therefore, any conclusion about the causal effect of gender on reaction times is open to discussion.

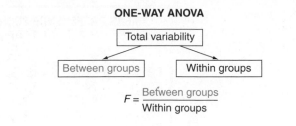

ONE-WAY ANOVA

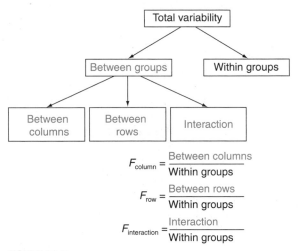

TWO-WAY ANOVA

$$F_{column} = \frac{\text{Between columns}}{\text{Within groups}}$$

$$F_{row} = \frac{\text{Between rows}}{\text{Within groups}}$$

$$F_{interaction} = \frac{\text{Interaction}}{\text{Within groups}}$$

FIGURE 23.2
Sources of variability and F ratios in one- and two-way ANOVAs.

Test Results for Two-Way Study

Hypothesis tests for the results of the smoke alarm study agree with the preliminary interpretations based on graphs. As the adjacent summary shows, each of the three null hypotheses in the study can be rejected at the .05 level of significance. Rejection of the null hypotheses for both main effects indicates that crowd size and gender, taken in turn, influence the reaction times of subjects to smoke. Rejection of the null hypothesis for interaction, however, spotlights the fact that the effect of crowd size on reaction times is different for male and female subjects.

23.4 INTERACTION

Interaction emerges as the most striking feature of a two-way ANOVA. As has been noted previously, two factors will interact if the effects of one factor are not consistent for all of the levels of a second factor. In other words, when two factors are combined, something happens that represents more than a mere

HYPOTHESIS TEST SUMMARY: TWO-WAY *F* TEST
(Smoke Alarm Study)

Research Problem:

Do crowd size and gender, as well as the interaction of these two factors, influence the subjects' mean reaction times to potentially dangerous smoke?

Statistical Hypotheses:

H_0: no main effect due to columns or crowd size
  (or $\mu_0 = \mu_2 = \mu_4$).
H_0: no main effect due to rows or gender
  (or $\mu_{female} = \mu_{male}$).
H_0: no interaction.
H_1: H_0 is not true.
  (Same H_1 accommodates each H_0.)

Decision Rule:

Reject H_0 at the .05 level of significance if F_{column} or $F_{interaction}$ equals or is more positive than 5.14 (from Table C, Appendix D, given 2 and 6 degrees of freedom), and if F_{row} equals or is more positive than 5.99 (given 1 and 6 degrees of freedom).

Calculations:

$$F_{column} = 6.75$$
$$F_{row} = 36.02$$
$$F_{interaction} = 5.25$$

(See Table 23.3 on pages 403–404 for calculations.)

Decision:

Reject all three null hypotheses at the .05 level of significance because $F_{column} = 6.75$ is more positive than 5.14; $F_{row} = 36.02$ is more positive than 5.99; and $F_{interaction} = 5.25$ is more positive than 5.14.

Interpretation:

Both crowd size and gender influence the subjects' mean reaction times to smoke. The interaction indicates that the influence of crowd size depends on whether subjects are males or females. More specifically, it appears that the mean reaction times of males, but not females, increase with the crowd size.

composite of the separate effects of each of the two factors. Standard examples of interaction include the sometimes fatal combination of tranquilizers and alcohol, as well as, in a lighter vein, the taste clash between certain wines and cheeses.

Supplies Valuable Information

Rather than being a complication to be avoided if possible, an interaction often highlights pertinent issues for future research. For example, the interaction between crowd size and gender suggests that subsequent studies might attempt to identify those factors that cause the reaction times of male subjects, but not female subjects, to be relatively sensitive to crowd size. In the process, much might be learned about why some people do or don't assume social responsibility in groups.

Other Examples

The combined effect of crowd size and gender could have differed from that described in panel C of Figure 23.1. Examples of some other possible effects are shown in **Figure 23.3.** The two top panels in Figure 23.3 describe outcomes that, because of their consistency, would cause the retention of the null hypothesis for interaction. The two bottom panels in Figure 23.3 describe outcomes that, because of their inconsistency, probably would cause the rejection of the null hypothesis for interaction.

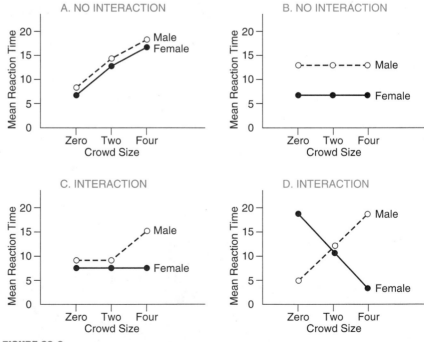

FIGURE 23.3
Some possible outcomes (two-way experiment).

Simple Effects

. . . .

Simple effect

*The effect of one factor at a
single level of another factor.*

The notion of interaction can be clarified further by viewing each line in Figure 23.3 as a simple effect. A **simple effect** *represents the effect of one factor at a single level of the second factor.* Thus, in panel A, there are two simple effects of crowd size, one for males and one for females, and both simple effects are consistent, showing an increase in mean reaction times with larger crowd sizes. Accordingly, the main effect of crowd size can be interpreted without referring to its two simple effects.

Inconsistent Simple Effects

In panel D, on the other hand, the two simple effects of crowd size, one for males and one for females, clearly are inconsistent; the simple effect of crowd size for females shows a decrease in mean reaction times with larger crowd sizes, whereas the simple effect of crowd size for males shows just the opposite — an increase in mean reaction times with larger crowd sizes. Accordingly, the main effect of crowd size — assuming one exists — can't be interpreted without referring to its radically different simple effects.

Simple Effects and Interaction

. . . .

Interaction

*The product of inconsistent
simple effects.*

In Figure 23.3, no interaction is present in panels A and B because their respective simple effects are consistent, as suggested by the parallel lines. Interactions are present in panels C and D because their respective simple effects are inconsistent, as suggested by the diverging or crossed lines. Given the present perspective, **interaction** *can be viewed as the product of inconsistent simple effects.*

*Exercise 23.2 A recent example of interaction from the psychological literature is the tendency of college students, when assigning prison sentences to pictures of "convicted defendants," to judge attractive swindlers more harshly than unattractive swindlers, but to judge attractive robbers less harshly than unattractive robbers.

(a) Construct a graph showing this interaction. As is customary, identify the vertical axis with the dependent variable, the mean prison sentence. For the sake of uniformity, identify the horizontal axis with the two types of criminals (swindlers and robbers), and label the two lines inside the graph as attractive and unattractive.

(b) Assume that, in fact, there is no interaction. Instead, independently of their degree of attractiveness, swindlers are judged more harshly than robbers, and, independently of their crime, unattractive defendants are judged more harshly than attractive defendants. Using the same identifications as in the previous question, construct a graph that depicts this result.

Hint: Because there is no interaction, the two lines in the graph must be parallel, that is, the two simple effects must be consistent.

Answers on pages 521–522

23.5 DESCRIBING INTERACTIONS

The original interaction between crowd size and gender could have been described in two different ways. First, we could have portrayed the inconsistent simple effects of crowd size for females and males by showing panel A of **Figure 23.4** (originally shown in panel C of Figure 23.1). Alternately, we could have portrayed the inconsistent simple effects of gender for zero, two, and four confederates by showing panel B of Figure 23.4. Although different, the configurations in both panels A and B suggest essentially the same interpretation: Crowd size influences the reaction times of male subjects, but not those of female subjects. In cases where one perspective seems to make as much sense as another, it's customary to plot along the horizontal axis the factor with the larger number of levels, as in panel A.

23.6 ASSUMPTIONS

The assumptions for F tests in a two-way ANOVA are similar to those for a one-way ANOVA: All underlying populations (for each treatment combination or group) are assumed to be normally distributed with equal variances. As with the one-way ANOVA, you needn't be too concerned about violations of these assumptions, particularly if all group sizes are equal and each is fairly large (greater than about 10). Otherwise, in the unlikely event that you encounter conspicuous departures from normality or equality of variances, consult a more advanced statistics book.[2]

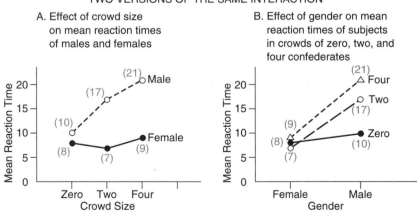

FIGURE 23.4

Two versions of the same interaction. Note: Colored numbers represent group means from Table 23.1.

[2]See G. Keppel, *Design and Analysis: A Researcher's Handbook*, 3rd ed. (Englewood Cliffs, NJ: Prentice-Hall, 1991); or D. C. Howell, *Statistical Methods for Psychology*, 4th ed. (Belmont, CA: Duxbury, 1997).

23.7 IMPORTANCE OF EQUAL SAMPLE SIZES

Insofar as possible, *all groups in two-way studies should have equal sample sizes*. Otherwise, to the degree that sample sizes are unequal and the resulting design lacks balance, problems of interpretation can occur. If you must analyze data based on unequal sample sizes—possibly because of missing subjects, equipment breakdowns, or recording errors—consult a more advanced statistics book. (See footnote, page 400.)

23.8 OTHER TYPES OF ANOVA

One- and two-way studies don't exhaust the possibilities for ANOVA. For instance, you could use ANOVA to analyze the results of a three-way study with three factors or independent variables. Furthermore, regardless of the number of factors, each subject might be measured repeatedly along all levels of one or more factors. Although the basic concepts described in this book transfer almost intact to a wide assortment of more intricate research designs, computational procedures grow more complex, and the interpretation of results often is more difficult. Intricate research designs, requiring the use of complex types of ANOVA, provide the skilled researcher with powerful tools for evaluating complicated situations. Under no circumstances, however, should a study be valued simply because of the complexity of its design and statistical analysis. Try to use the least complex design and analysis that will answer your research question.

DETAILS

23.9 VARIANCE ESTIMATES

There is considerable similarity between measures of variability in one- and two-way ANOVAs. In both cases, every measure of variability consists of a variance estimate or mean square obtained by dividing a particular sum of squares (SS) by its degrees of freedom (df). Let's concentrate first on the various SS terms and then on their associated dfs.

23.10 SUM OF SQUARES (SS)

Once again, most computational effort is directed toward the various SS terms: the column sum of squares, SS_{column}; the row sum of squares, SS_{row}; the interaction sum of squares, $SS_{\text{interaction}}$; the within-group sum of squares, SS_{within}; and the total sum of squares, SS_{total}. Although all of these SS terms can be calculated directly, it's more convenient to calculate $SS_{\text{interaction}}$ indirectly, as shown below, because SS_{total} always equals the sum of the various SS terms, that is,

---

SUMS OF SQUARES (TWO WAY)

$$SS_{\text{total}} = SS_{\text{column}} + SS_{\text{row}} + SS_{\text{interaction}} + SS_{\text{within}} \qquad (23.1)$$

Computational Formulas

The top part of **Table 23.2** lists the various computational formulas for each *SS* term, including the formula for calculating $SS_{interaction}$ indirectly. **Table 23.3** illustrates the application of these formulas to the data for the two-way study. Notice the highly predictable computational pattern first described in Section 22.8. Each score is always squared, and each total, whether for a group, column, or row total, or for the overall total, is always squared and then divided by its sample size.

Table 23.2
COMPUTATIONAL FORMULAS FOR *SS* TERMS

SS_{column} = sum of squared deviations of column means about the overall mean

$$= \frac{(\text{1st column total})^2}{\text{column sample size}} + \ldots + \frac{(\text{last column total})^2}{\text{column sample size}} - \frac{(\text{overall total})^2}{\text{overall sample size}}$$

SS_{row} = sum of squared deviations of row means about the overall mean

$$= \frac{(\text{1st row total})^2}{\text{row sample size}} + \ldots + \frac{(\text{last row total})^2}{\text{row sample size}} - \frac{(\text{overall total})^2}{\text{overall sample size}}$$

SS_{within} = sum of squared deviations of scores about their respective group means

$$= \text{sum of all squared scores} - \left[\frac{(\text{1st group total})^2}{\text{group sample size}} + \ldots + \frac{(\text{last group total})^2}{\text{group sample size}} \right]$$

SS_{total} = sum of squared deviations of scores about the overall mean

$$= \text{sum of all squared scores} - \frac{(\text{overall total})^2}{\text{overall sample size}}$$

$SS_{interaction}$ = sum of squared deviations of group means about the overall mean (not attributable to variations in column and row means)

$$= SS_{total} - [SS_{column} + SS_{row} + SS_{within}]$$

FORMULAS FOR *df* TERMS

$$df_{column} = \text{number of columns} - 1$$

$$df_{row} = \text{number of rows} - 1$$

$$df_{interaction} = (\text{number of columns} - 1)(\text{number of rows} - 1)$$

$$df_{within} = \text{number of scores} - \text{number of groups}$$

$$df_{total} = \text{number of scores} - 1$$

Table 23.3
CALCULATION OF *SS* TERMS

A. COMPUTATIONAL SEQUENCE

Find (and circle) each group total **1**.
Find each column and row total and also the overall total **2**.
Substitute numbers into computational formula **3** and solve for SS_{column}.
Substitute numbers into computational formula **4** and solve for SS_{row}.
Substitute numbers into computational formula **5** and solve for SS_{within}.
Substitute numbers into computational formula **6** and solve for SS_{total}.
Substitute numbers into formula **7** and solve for $SS_{interaction}$.

B. DATA AND COMPUTATIONS

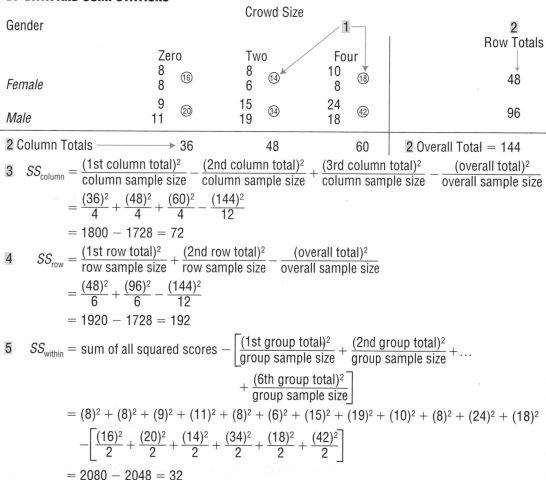

Crowd Size

Gender	Zero	Two	Four	**2** Row Totals
Female	8 8 ⑯	8 6 ⑭	10 8 ⑱	48
Male	9 11 ⑳	15 19 ㉞	24 18 ㊷	96

2 Column Totals ⟶ 36   48   60 | **2** Overall Total = 144

3 $SS_{column} = \dfrac{(\text{1st column total})^2}{\text{column sample size}} - \dfrac{(\text{2nd column total})^2}{\text{column sample size}} + \dfrac{(\text{3rd column total})^2}{\text{column sample size}} - \dfrac{(\text{overall total})^2}{\text{overall sample size}}$

$= \dfrac{(36)^2}{4} + \dfrac{(48)^2}{4} + \dfrac{(60)^2}{4} - \dfrac{(144)^2}{12}$

$= 1800 - 1728 = 72$

4 $SS_{row} = \dfrac{(\text{1st row total})^2}{\text{row sample size}} + \dfrac{(\text{2nd row total})^2}{\text{row sample size}} - \dfrac{(\text{overall total})^2}{\text{overall sample size}}$

$= \dfrac{(48)^2}{6} + \dfrac{(96)^2}{6} - \dfrac{(144)^2}{12}$

$= 1920 - 1728 = 192$

5 $SS_{within} = \text{sum of all squared scores} - \left[\dfrac{(\text{1st group total})^2}{\text{group sample size}} + \dfrac{(\text{2nd group total})^2}{\text{group sample size}} + \cdots \right.$

$\left. + \dfrac{(\text{6th group total})^2}{\text{group sample size}} \right]$

$= (8)^2 + (8)^2 + (9)^2 + (11)^2 + (8)^2 + (6)^2 + (15)^2 + (19)^2 + (10)^2 + (8)^2 + (24)^2 + (18)^2$

$- \left[\dfrac{(16)^2}{2} + \dfrac{(20)^2}{2} + \dfrac{(14)^2}{2} + \dfrac{(34)^2}{2} + \dfrac{(18)^2}{2} + \dfrac{(42)^2}{2} \right]$

$= 2080 - 2048 = 32$

Table 23.3
CALCULATION OF *SS* TERMS *(Continued)*

6 SS_{total} = sum of all squared scores = $\dfrac{(\text{overall total})^2}{\text{overall sample size}}$

= $(8)^2 + (8)^2 + (9)^2 + (11)^2 + (8)^2 + (6)^2 + (15)^2 + (19)^2 +$

$(10)^2 + (8)^2 + (24)^2 + (18)^2 - \dfrac{(144)^2}{12}$

= 2080 − 1728

= 352

7 $SS_{interaction}$ = $SS_{total} - [SS_{column} + SS_{row} + SS_{within}]$

= 352 − [72 + 192 + 32]

= 352 − 296 = 56

Check for Computational Accuracy

To minimize computational errors, repeat all the computations from scratch a second time and proceed only if all the results agree. This is particularly important, because the indirect determination of $SS_{interaction}$ eliminates the value of Formula 23.1 as a computational check.

23.11 DEGREES OF FREEDOM (*df*)

The number of degrees of freedom must be determined for each *SS* term in a two-way ANOVA, and for convenience, the various *df* formulas are listed in the bottom half of Table 23.2. Notice that $df_{interaction}$ represents the product of df_{column} and df_{row}.

The *df* for the present study are as follows:

$$df_{column} = 3 - 1 = 2$$
$$df_{row} = 2 - 1 = 1$$
$$df_{interaction} = (3 - 1)(2 - 1) = 2$$
$$df_{within} = 12 - 6 = 6$$
$$df_{total} = 12 - 1 = 11$$

Check for Accuracy

Recall the general rule that the degrees of freedom for SS_{total} equal the combined degrees of freedom for all remaining *SS* terms, that is,

DEGREES OF FREEDOM (TWO WAY)

$$df_{total} = df_{column} + df_{row} + df_{interaction} + df_{within} \qquad (23.2)$$

This formula can be used to verify that the correct number of degrees of freedom has been assigned to each *SS* term.

23.12 MEAN SQUARES (*MS*) AND *F* RATIOS

Having found values for the various *SS* terms and their *df*, we can determine values for the corresponding *MS* terms and then calculate the three *F* ratios using formulas in **Table 23.4.** Notice that MS_{within}—the estimate of random error—appears in the denominator of each of these three *F* ratios.

The ANOVA results for the two-way study are summarized in **Table 23.5.** The numbers in color (which ordinarily don't appear in ANOVA tables) indicate the origin of each *MS* term and each *F*.

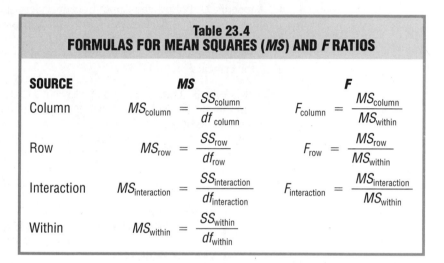

Table 23.4
FORMULAS FOR MEAN SQUARES (*MS*) AND *F* RATIOS

SOURCE	MS	F
Column	$MS_{column} = \dfrac{SS_{column}}{df_{column}}$	$F_{column} = \dfrac{MS_{column}}{MS_{within}}$
Row	$MS_{row} = \dfrac{SS_{row}}{df_{row}}$	$F_{row} = \dfrac{MS_{row}}{MS_{within}}$
Interaction	$MS_{interaction} = \dfrac{SS_{interaction}}{df_{interaction}}$	$F_{interaction} = \dfrac{MS_{interaction}}{MS_{within}}$
Within	$MS_{within} = \dfrac{SS_{within}}{df_{within}}$	

Table 23.5
ANOVA TABLE (TWO-WAY EXPERIMENT)

SOURCE	SS	df	MS	F
Column	72	2	$\left(\dfrac{72}{2} =\right)$ 36	$\left(\dfrac{36}{5.33} =\right)$ 6.75*
Row	192	1	$\left(\dfrac{192}{1} =\right)$ 192	$\left(\dfrac{192}{5.33} =\right)$ 36.02*
Interaction	56	2	$\left(\dfrac{56}{2} =\right)$ 28	$\left(\dfrac{28}{5.33} =\right)$ 5.25*
Within	32	6	$\left(\dfrac{32}{6} =\right)$ 5.33	
Total	352	11		

* Significant .05 level.

Other Labels

Other labels also might have appeared in Table 23.5. For instance, "Column" and "Row" might have been replaced by descriptions of the treatment variables, in this case, "Crowd Size" and "Gender." By the same token, "Interaction" might have been replaced by "Crowd Size × Gender," by "Crowd Size* Gender," or by some abbreviation, such as "CS × G" (for Crowd Size × Gender), and "Within" might have been replaced by "Error."

23.13 *F* TABLES

Each of the three *F* ratios in Table 23.5 exceeds its respective critical *F* ratio. To obtain critical *F* ratios from the *F* sampling distribution, refer to Table C, Appendix D. Follow the same procedure described in Section 22.11 to verify that when 2 and 6 degrees of freedom are associated with the numerator and denominator of *F*, as for F_{column} and $F_{interaction}$, the critical *F* equals 5.14. Also verify that when 1 and 6 degrees of freedom are associated with *F*, as for F_{row}, the critical *F* equals 5.99.

*Exercise 23.3 A social psychologist wishes to determine the effect of TV violence on the subsequent aggressive behavior of first graders. Two first graders are randomly assigned to each of the various combinations of the two factors: the type of TV program (either cartoon or real life) and the amount of viewing time (either 0, 1, 2, or 3 hours). The subjects are then tested in a controlled social situation and assigned an aggression score, reflecting the total number of aggressive behaviors displayed during the test period.

AGGRESSION SCORES OF FIRST GRADERS				
	VIEWING TIME (HOURS)			
TYPE OF PROGRAM	0	1	2	3
Cartoon	0,1	1,0	3,5	6,9
Real Life	0,0	1,1	6,2	6,10

(a) Test the various null hypotheses at the .05 level of significance.

(b) Summarize the results with an ANOVA table.

(c) Specify the approximate *p*-values for the various *F* tests.

(d) Using the format described in Section 22.12, indicate how these results might appear in a published report.

Answers on pages 522–523

23.14 SMALL, MEDIUM, OR LARGE EFFECT?

Recall the discussion of Section 22.14, where the squared curvilinear correlation coefficient, η^2, was used to estimate, independently of sample sizes, whether a statistically significant *F* test reflects a small, medium, or large effect. Precisely the same type of analysis can be conducted for *each* significant *F* in a two-way

ANOVA. For all practical purposes, each η^2 estimates the proportion of the total variance in scores that is predictable from (or caused by, if a well-designed experiment) the factor for columns, by the factor for rows, and by the interaction.

In every case, η^2 is calculated by dividing the appropriate sum of squares (either SS_{column}, SS_{row}, or $SS_{interaction}$) by the total sum of squares (SS_{total}). Using the same rule of thumb as in Section 22.14, the estimated effect for any factor or interaction of factors is small if η^2 approximates .01; medium if η^2 approximates .06; and large if η^2 approximates .14 or more. Once again, consider calculating η^2 whenever you encounter a statistically significant F, especially one based on large sample sizes, but remember that even a very small effect might be important because of special circumstances.

*Exercise 23.4 Referring to Table 23.5 on page 405, estimate whether any significant F reflects a small, medium, or large effect (even though these estimates must be viewed as highly speculative because of the very small sample sizes).

> *Answer on page 523*

23.15 MULTIPLE COMPARISONS

A modification of Scheffe's test for multiple comparisons, described in the previous chapter, may be used to pinpoint important differences between pairs of column or row means whenever the corresponding main effects are statistically significant (and interpretations of these main effects aren't compromised by any inconsistencies associated with a statistically significant interaction). Whenever the interaction is statistically significant, as in the two-way study for the present chapter, special statistical tests, including tests of simple effects, usually are conducted to pinpoint the precise locus of the inconsistencies that cause the interaction. Keppel's *Design and Analysis: A Researcher's Handbook* is an excellent source of information about more detailed analyses of interactions. (See footnote, page 400.)

23.16 COMPUTER OUTPUT

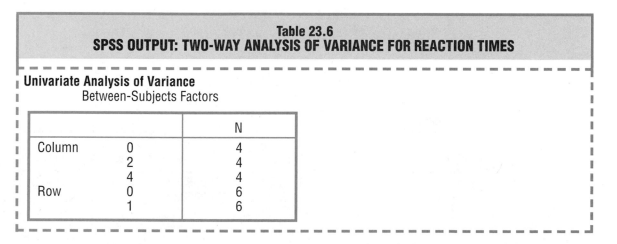

Table 23.6
SPSS OUTPUT: TWO-WAY ANALYSIS OF VARIANCE FOR REACTION TIMES

Univariate Analysis of Variance
Between-Subjects Factors

		N
Column	0	4
	2	4
	4	4
Row	0	6
	1	6

Table 23.6
SPSS OUTPUT: TWO-WAY ANALYSIS OF VARIANCE FOR REACTION TIMES (continued)

1 Tests of Between-Subjects Effects

Dependent Variable: REACTION

	Source	Type II Sum of Squares	df	Mean Square	F	2 Sig.
3	Corrected Model	320.000[a]	5	64.000	12.000	.004
	COLUMN	72.000	2	36.000	6.750	.029
	ROW	192.000	1	192.000	36.000	.001
	COLUMN*ROW	56.000	2	28.000	5.250	.048
4	Error	32.000	6	5.333		
5	Corrected Total	352.000	11			

a. R Squared = .909 (Adjusted R Squared = .833)

Comments:
1 Two-way ANOVA for reaction times. Compare with Table 23.5.
2 Exact p-values.
3 Variability that is attributable to differences between groups, that is, the combined variability attributable to the two main effects and the interaction. (See bottom panel of Figure 23.2.)
4 Within group or error variability.
5 Total variability.

Exercise 23.5 The following SAS output is based on the data in Exercise 23.8 on page 411 for subjects who experience different dosages of vitamin C and different exposures to a sauna. Although its form differs from that for SPSS in Section 23.16, an inspection of the SAS output should reveal the answers to the questions that follow.

SAS OUTPUT: TWO-WAY ANOVA FOR "DAYS OF DISCOMFORT"

The SAS System
12:46 Tuesday, November 23, 1999
Analysis of Variance Procedure

Dependent Variable: DISCOMF

Source	df	Sum of Squares	Mean Square	F Value	Pr > F
Model	11	39.6388889	3.6035354	5.64	0.0002
Error	24	15.3333333	0.6388889		
Corrected Total	35	54.9722222			

SAS OUTPUT: TWO-WAY ANOVA FOR "DAYS OF DISCOMFORT" (continued)					
Source	df	Anova SS	Mean Square	F Value	Pr > F
SAUNA	2	5.0555556	2.5277778	3.96	0.0327
VITAMINC	3	33.6388889	11.2129630	17.55	0.0001
SAUNA*VITAMINC	6	0.9444444	0.1574074	0.25	0.9561

Note: "MODEL" variability in SAS outputs corresponds to "Explained" variability, that is, the combined variability attributable to both main effects and the interaction in SPSS outputs.

(a) How many degrees of freedom are associated with the within group or error term?

(b) What is the F value for vitamin C dosage?

(c) What is the exact p-value for the interaction?

(d) Using the .05 level of significance, indicate your decision about each of the three null hypotheses in this study.

Answer on page 523

Summary

Before any statistical analysis, and particularly before complex analyses such as a two-way ANOVA, it's often helpful to form preliminary impressions by constructing graphs of the various possible effects: slanted lines forecast a possible main effect, and nonparallel lines serve notice of a possible interaction.

In a two-way analysis of variance, three null hypotheses are tested with three different F ratios. The numerator of each F ratio measures a different aspect of variability between groups: variability between columns, variability between rows, and any remaining variability between groups due to interaction. The denominator of each F ratio measures the variability within groups and reflects only random error.

Two factors will interact if their simple effects are inconsistent. Interaction emerges as the most striking feature of a two-way ANOVA.

The assumptions for F tests in a two-way ANOVA are the same as those for one-way ANOVA. Insofar as possible, all groups in two-way studies should have equal sample sizes.

Each measure of variability or mean square (MS) is calculated by dividing the appropriate sum of squares (SS) by its degrees of freedom (df). Once a calculated value of F has been obtained, it's compared with a critical F from the sampling distribution of F. As usual, if the calculated F exceeds the critical F, the corresponding null hypothesis will be rejected.

ANOVA results are often summarized in tabular form.

Whenever obtaining a statistically significant F, especially one based on large sample sizes, consider calculating the squared curvilinear correlation coefficient, η^2, to estimate whether the corresponding effect is small, medium, or large.

Important Terms

Two-way ANOVA   Simple effect
Main effect   Interaction

REVIEW EXERCISES

*23.6 A psychologist randomly assigns eight rats to each group in a two-way experiment designed to determine the effect of food deprivation (either 0, 24, 48, or 72 hours) and reward magnitude (either 1, 2, or 3 food pellets) on their rate of bar pressing.

(a) Specify the various sources of variability and their degrees of freedom.

(b) One possible outcome is that there is a main effect for food deprivation, no main effect for reward magnitude, and no interaction. Viewed in this fashion, there are seven other possible types of outcomes for this experiment. Counting the stated outcome as one possibility, list all eight possibilities.

(c) Among the eight possible outcomes specified in (b), which outcomes would the researcher most prefer?

(d) Among the eight possible outcomes specified in (c), which outcome would the researcher least prefer?

Answers on pages 523–524

23.7 Each of the following (incomplete) ANOVA tables represents some experiment. Determine how many levels of each factor were used; the total number of groups; and, on the assumption that all groups have equal numbers of subjects, the number of subjects in each group. Then, using the .05 level of significance for all hypothesis tests, complete the ANOVA table:

(a) SOURCE	SS	df	MS	F
Column	790	1		
Row	326	2		
Interaction	1887			
Within	14702	60		
Total				

(b) SOURCE	SS	df	MS	F
Treatment A	142	2		
Treatment B	480	2		
A $\times$ B	209			
Error	5030	81		
Total				

(c) For each significant *F* in (a) or (b), estimate whether it reflects a small, medium, or large effect. (See Section 23.14.)

*23.8 A health educator suspects that the "days of discomfort" caused by common colds can be reduced by ingesting large doses of vitamin C and visiting a sauna every day. Using a two-way design, subjects with new colds are randomly assigned to one of four different daily dosages of vitamin C (either 0, 500, 1000, or 1500 milligrams) and to one of three different daily exposures to a sauna (either 0, $\frac{1}{2}$, or 1 hour).

NUMBER OF DAYS OF DISCOMFORT DUE TO COLDS

SAUNA EXPOSURE (HOURS)	VITAMIN C DOSAGE (MILLIGRAMS)			
	0	500	1000	1500
0	6	5	4	2
	4	3	2	3
	5	3	3	2
$\frac{1}{2}$	5	4	3	2
	4	3	2	1
	5	2	3	2
1	4	4	3	1
	3	2	2	2
	4	3	2	1

(a) Using the appropriate sets of means, graph the various possible effects and tentatively interpret the experimental outcomes.

(b) Test the various null hypotheses at the .05 level of significance, using the customary step-by-step procedure.

(c) Summarize the results with an ANOVA table.

(d) Specify the approximate *p*-values for the various *F* tests.

(e) How might these results appear in a published report?

Answers on pages 524–525

23.9 In what sense does a two-way ANOVA use observations more efficiently than a one-way ANOVA does?

23.10 A psychologist employs a two-way experiment to study the combined effect of sleep deprivation and alcohol consumption on the performance of automobile drivers. Before the driving test, the subjects go without sleep for various time periods and then drink a glass of orange juice laced with controlled amounts of vodka. Their performance is measured by the number of errors made on a driving simulator. Two subjects are randomly assigned to each possible combination of sleep deprivation (either

0, 24, 48, or 72 hours) and alcohol consumption (either 0, 1, 2, or 3 ounces), yielding the following results:

NUMBER OF DRIVING ERRORS

ALCOHOL CONSUMPTION (OUNCES)	SLEEP DEPRIVATION (HOURS)			
	0	24	48	72
0	0	2	5	5
	3	4	4	6
1	1	3	6	5
	3	3	7	8
2	3	2	8	7
	5	5	11	12
3	4	4	10	9
	6	7	13	15

(a) Construct graphs for the various effects, and attempt to anticipate the results of the statistical analysis.

(b) Test the various null hypotheses at the .05 level of significance.

(c) Summarize the results with an ANOVA table.

(d) Specify the approximate p-values for the various F tests.

23.11 Does the type of instruction in a college sociology class (either lecture or self-paced) and its grading policy (either letter or pass/fail) influence the performance of students, as measured by the number of quizzes successfully completed during the semester? Six students are randomly assigned to each of the four possible combinations, yielding the following results:

NUMBER OF QUIZZES SUCCESSFULLY COMPLETED

GRADING POLICY	TYPE OF INSTRUCTION	
	LECTURE	SELF-PACED
Letter grades	4	7
	3	8
	5	6
	2	4
	6	10
	5	12
Pass/Fail	8	4
	9	1
	4	3
	5	2
	8	6
	10	8

(a) Graph and interpret the three effects.

(b) Test the various null hypotheses at the .01 level of significance.

(c) Summarize the results with an ANOVA table.

(d) Specify the approximate p-values for the various F tests.

(e) How might these test results appear in a published report?

23.12 In this chapter, all examples of two-way studies involve at least two observations per group (or cell). Would it be possible to perform an ANOVA for a two-way study having only one observation per group?

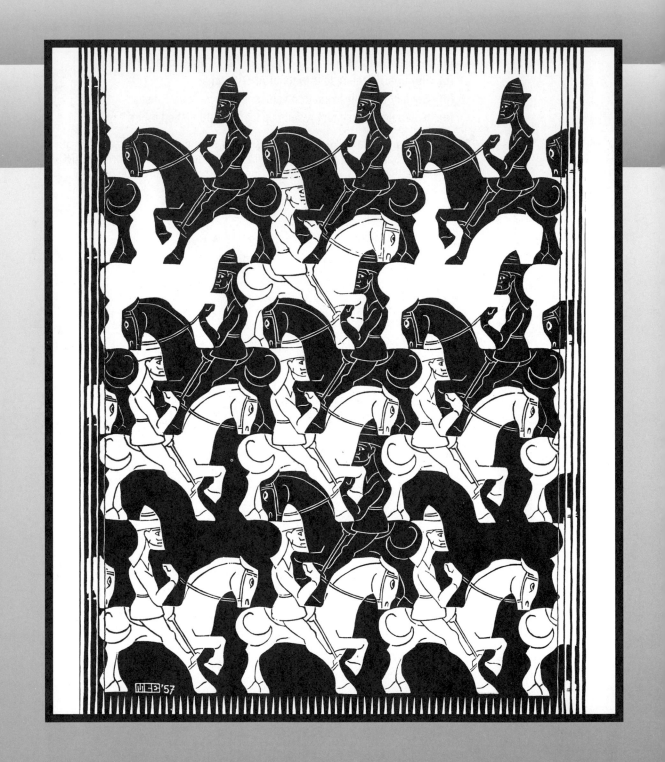

CHAPTER 24

Chi-Square (χ^2) Test for Qualitative Data

ONE-WAY χ^2 TEST

24.1 SURVEY OF BLOOD TYPES
24.2 STATISTICAL HYPOTHESES
24.3 OBSERVED AND EXPECTED FREQUENCIES
24.4 CALCULATION OF χ^2
24.5 χ^2 TABLES AND DEGREES OF FREEDOM
24.6 χ^2 TEST
24.7 χ^2 TEST IS NONDIRECTIONAL

TWO-WAY χ^2 TEST

24.8   LOST-LETTER STUDY
24.9   STATISTICAL HYPOTHESES
24.10 OBSERVED AND EXPECTED FREQUENCIES
24.11 χ^2 TABLES AND DEGREES OF FREEDOM
24.12 χ^2 TEST
24.13 SOME PRECAUTIONS
24.14 CHECKING IMPORTANCE
24.15 COMPUTER OUTPUT

Summary

Important Terms

Review Exercises

ONE-WAY χ^2 TEST

24.1 SURVEY OF BLOOD TYPES

Your blood belongs to one of four genetically determined types: O, A, B, or AB. A recent bulletin issued by a large blood bank claims that these four blood types are distributed in the U.S. population according to the following proportions: .44 are type 0, .41 are type A, .10 are type B, and .05 are type AB. Let's treat this claim as a null hypothesis to be tested with a random sample of 100 students from a large university in California.

A Test for Qualitative Data

When observations are merely classified into various categories—for example, blood types: 0, A, B, and AB; political affiliations: Republican, Democrat, and Independent; ethnic backgrounds: African American, Asian American, European American, and so forth—the data are qualitative, as discussed in Section 1.6. Hypothesis tests for qualitative data require the use of a new test known as the *chi-square* test (symbolized as χ^2 and pronounced "ki square").

One-Way Versus Two-Way

When observations are classified in only one way, that is, classified along a single qualitative variable, as with the four blood types, the test is a **one-way χ^2 test.** Designed to evaluate the adequacy with which observed frequencies are described by hypothesized or expected frequencies, a one-way χ^2 test is also referred to as a *goodness-of-fit* test. Later, when observations are classified in two ways, that is, cross-classified according to two qualitative variables, the test is a two-way χ^2.

One-way χ^2 test

Evaluates whether observed frequencies for a single qualitative variable are adequately described by hypothesized or expected frequencies.

24.2 STATISTICAL HYPOTHESES

Null Hypothesis

For the one-way χ^2 test, the null hypothesis makes a statement about two or more population proportions whose values, in turn, generate the hypothesized or expected frequencies for the test, as described here. Sometimes these population proportions are specified directly, as in the survey of blood types:

$$H_0: P_O = .44; P_A = .41; P_B = .10; P_{AB} = .05$$

where P_O refers to the hypothesized proportion of students with type O blood in the population from which the sample was taken, and so forth. Notice that the values of population proportions always must sum to 1.00.

Other Examples

At other times, you'll have to infer the values of population proportions from verbal statements. Ordinarily, you will not find this too difficult. For example, the null hypothesis that artists are equally likely to be left-handed or right-handed translates into

$$H_0: P_{left} = P_{right} = .50 \text{ (or } \tfrac{1}{2}\text{)}$$

where P_{left} represents the hypothesized proportion of left-handers in the population of artists.

The hypothesis that soda drinkers are equally likely to prefer any one of four different brands (coded 1, 2, 3, and 4) translates into

$$H_0: P_1 = P_2 = P_3 = P_4 = .25 \text{ (or } \tfrac{1}{4}\text{)}$$

where P_1 represents the hypothesized proportion of drinkers who prefer brand 1 in the population of soda drinkers.

Alternative Hypothesis

Because the null hypothesis will be false if population proportions deviate in *any* direction from that hypothesized, the alternative or research hypothesis can be described simply as

$$H_1: H_0 \text{ is false}$$

As usual, the alternative hypothesis indicates that, relative to the null hypothesis, something special is happening in the underlying population, such as, for instance, a tendency for artists to be left-handed or for soda drinkers to prefer one or two brands.

*Exercise 24.1 Specify the null hypothesis for each of the following situations. (Remember, the null hypothesis usually represents a negation of the researcher's hunch or hypothesis.)

(a) A political scientist wants to determine whether voters prefer candidate A more than candidate B for president.

(b) A biologist suspects that, upon being released 10 miles south of their home roost, migratory birds are more likely to fly toward home (north) rather than toward any of the three remaining directions (east, south, or west).

(c) A sociologist believes that crimes are not committed with equal likelihood on each of the seven days of the week.

(d) Being more specific, another sociologist suspects that *proportionately* more crimes are committed during the two days of the weekend (Saturday and Sunday) than during the five days of the week. *Hint:* There are just two (unequal) proportions: one representing the two weekend days and the other representing the five weekdays.

Answers on page 525

24.3 OBSERVED AND EXPECTED FREQUENCIES

If the null hypothesis is true, then, except for chance, the hypothesized proportions should be reflected in the sample. For example, when testing the blood bank's claim with a sample of 100 students, 44 students should have type O (from the product of .44 and 100); 41 should have type A; 10 should have type B; and only 5 should have type AB. In **Table 24.1,** each of these numbers is referred to as an **expected frequency f_e,** that is, *the hypothesized frequency for each category of the qualitative variable if, in fact, the null hypothesis is true.* To obtain the expected frequency for any category, find the product of the hypothesized or expected proportion for that category and the total sample size, namely,

---

Expected frequency (f_e)

The hypothesized frequency for each category, given that the null hypothesis is true.

> **EXPECTED FREQUENCY (ONE-WAY χ^2 TEST)**
> $$f_e = \text{(expected proportion) (total sample size)} \qquad (24.1)$$

Observed frequency (f_o)

The obtained frequency for each category.

---

where f_e represents the expected frequency. In Table 24.1, each expected frequency is paired with an **observed frequency f_o,** *the obtained frequency for each category.*

Evaluating Discrepancies

It's most unlikely that a random sample will exactly reflect the characteristics of its population. Even though the null hypothesis is true, discrepancies will appear between observed and expected frequencies, as in Table 24.1.

Table 24.1
OBSERVED AND EXPECTED FREQUENCIES:
BLOOD TYPES OF 100 STUDENTS

Frequency	BLOOD TYPE				
	O	A	B	AB	Total
Observed (f_o)	38	38	20	4	100
Expected (f_e)	44	41	10	5	100

The crucial question is whether the discrepancies between observed and expected frequencies are small enough to qualify as a common outcome, given that the null hypothesis is true. If so, the null hypothesis should be retained. Otherwise, if the discrepancies are large enough to qualify as a rare outcome, the null hypothesis should be rejected.

24.4 CALCULATION OF χ^2

To determine whether discrepancies between observed and expected frequencies qualify as a common or rare outcome, a value is calculated for χ^2 and compared with its hypothesized sampling distribution. To calculate the value of χ^2, use the following expression:

χ^2 RATIO

$$\chi^2 = \sum \frac{(f_o - f_e)^2}{f_e} \qquad (24.2)$$

---

where f_o denotes the observed frequency, and f_e denotes the expected frequency for each category of the qualitative variable. **Table 24.2** on page 420 illustrates how to use Formula 24.2 to calculate χ^2 for the present example.

Some Properties of χ^2

Notice several features of Formula 24.2. The larger the discrepancies are between the observed and expected frequencies, $f_o - f_e$, the larger the value of χ^2 and, therefore, as will be seen, the more suspect the null hypothesis will be. Because of the squaring of each discrepancy, negative discrepancies become positive, and the value of χ^2 never can be negative. Division by f_e indicates that discrepancies must be evaluated not in isolation, but relative to the size of expected frequencies. For example, a discrepancy of 5 looms more important (and translates into a larger value of χ^2) relative to an expected frequency of 10 than relative to an expected frequency of 100.

24.5 χ^2 TABLES AND DEGREES OF FREEDOM

As with t and F, χ^2 has not one but a family of distributions. Table D of Appendix D supplies critical values from various χ^2 distributions for hypothesis tests at the .10, .05, .01, and .001 levels of significance.

To locate the appropriate row in Table D, first identify the correct number of degrees of freedom. For the one-way test, the degrees of freedom for χ^2 can be obtained from the following expression:

Table 24.2
CALCULATION OF χ^2 (ONE-WAY TEST)

A. COMPUTATIONAL SEQUENCE
Find an expected frequency for each expected proportion **1**.
List observed and expected frequencies **2**.
Substitute numbers in formula **3** and solve for χ^2.

B. DATA AND COMPUTATIONS

1 f_e = (expected proportion)(sample size)
$f_e(O)$ = (.44)(100) = 44
$f_e(A)$ = (.41)(100) = 41
$f_e(B)$ = (.10)(100) = 10
$f_e(AB)$ = (.05)(100) = 5

2
Frequency	O	A	B	AB	Total
f_o	38	38	20	4	100
f_e	44	41	10	5	100

3 $\chi^2 = \Sigma \dfrac{(f_o - f_e)^2}{f_e}$

$$= \frac{(38-44)^2}{44} + \frac{(38-41)^2}{41} + \frac{(20-10)^2}{10} + \frac{(4-5)^2}{5}$$

$$= \frac{(-6)^2}{44} + \frac{(-3)^2}{41} + \frac{(10)^2}{10} + \frac{(-1)^2}{5}$$

$$= \frac{36}{44} + \frac{9}{41} + \frac{100}{10} + \frac{1}{5}$$

$$= .82 + .22 + 10.00 + .20 = 11.24$$

DEGREES OF FREEDOM (ONE-WAY χ^2 TEST)
$$df = C - 1 \qquad\qquad (24.3)$$

where C refers to the total number of categories of the qualitative variable.

Lose One Degree of Freedom

To understand Formula 24.3, focus on the set of observed frequencies for 100 students in Table 24.1. In practice, of course, the observed frequencies of the four (C) categories have equal status, and any combination of four frequencies that sums to 100 is possible. From the more abstract perspective of degrees of freedom, however, only three ($C - 1$) of these frequencies are free to vary because of the mathematical restriction that, when calculating χ^2 for the present data, all observed (or expected) frequencies must sum to 100. Although the observed frequencies of any three of the four categories are free to vary, the frequency of the fourth category must be some number that, when combined with the other three frequencies, will yield a sum of 100. By the same token, if there had been five categories, the frequencies of any four categories would have been free to vary, but not that of the fifth category. For the one-way test, the number of degrees of freedom always equals one less than the total number of categories (C), as indicated in Formula 24.3.

In the present example, in which the categories consist of the four blood types,

$$df = 4 - 1 = 3$$

To find the critical χ^2 for a hypothesis test at the .05 level of significance, locate the cell in Table D of Appendix D intersected by the row for 3 degrees of freedom and the column for the .05 level of significance. This cell lists a value of 7.81 for the critical χ^2.

....

24.6 χ^2 TEST

Following the usual procedure, assume the null hypothesis to be true and view the observed χ^2 within the context of its hypothesized distribution shown in **Figure 24.1.** If, because the discrepancies between observed and expected frequencies are relatively small, the observed χ^2 appears to emerge from the dense concentration of possible χ^2 values smaller than the critical χ^2, the observed outcome would be viewed as a common occurrence, on the assumption that the null hypothesis is true.

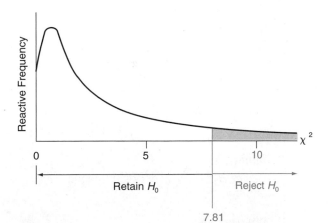

FIGURE 24.1
Hypothesized sampling distribution of χ^2 (3 degrees of freedom).

Therefore, the null hypothesis would be retained. On the other hand, if, because the discrepancies between observed and expected frequencies are relatively large, the observed χ^2 appears to emerge from the sparse concentration of possible χ^2 values equal to or greater than the critical χ^2, the observed outcome would be viewed as a rare occurrence, and the null hypothesis would be rejected.

In fact, because the observed χ^2 of 11.24 is more positive than the critical χ^2 of 7.81, the null hypothesis should be rejected: There is evidence that the distribution of blood types in the student population differs from that claimed for the U.S. population.

Speculations about Particular Discrepancies

Because all discrepancies contribute to the rejection of the null hypothesis, there are no statistical grounds for identifying the largest discrepancy as *the* discrepancy that causes the null hypothesis to be rejected. However, it is

---

HYPOTHESIS TEST SUMMARY: ONE-WAY χ^2 TEST (SURVEY OF BLOOD TYPES)

Research Problem:

Does the distribution of blood types in a population of college students comply with that described (for the U.S. population) in a blood bank bulletin?

Statistical Hypotheses:

$$H_0: P_O = .44; \ P_A = .41; \ P_B = .10; \ P_{AB} = .05$$

(where P_O represents the proportion of type O blood in the population of students, and so forth)

$$H_1: H_0 \text{ is false.}$$

Decision Rule:

Reject H_0 at the .05 level of significance if χ^2 equals or is more positive than 7.81 (from Table D, Appendix D, given $df = C - 1 = 4 - 1 = 3$).

Calculations:

$$\chi^2 = 11.24 \text{ (See Table 24.2 on page 420.)}$$

Decision:

Reject H_0 at .05 level of significance because $\chi^2 = 11.24$ is more positive than 7.81.

Interpretation:

The distribution of blood types in a student population differs from that claimed for the U.S. population.

sometimes helpful to speculate about one or more of the most obvious discrepancies between observed and expected frequencies. As can be seen in Table 24.1, the present survey contains an unexpectedly large number of students with type B blood. A subsequent investigation revealed that the sample included a large number of Asian-American students, a group that has an established high incidence of type B blood. This might explain why the hypothesized distribution of blood types fails to describe that for the population of students from which the random sample was taken. Certainly a random sample should be taken from a much broader spectrum of the general population before questioning the blood bank's claim about the distribution of blood types in the U.S. population.

*Exercise 24.2 A random sample of 90 college students indicates whether they *most* desire love, wealth, power, health, fame, or family happiness.

(a) Using the .05 level of significance and the following results, test the null hypothesis that in the underlying population, the various desires are equally popular.

(b) Specify the approximate *p*-value for this test result.

DESIRES OF COLLEGE STUDENTS

FREQUENCY	LOVE	WEALTH	POWER	HEALTH	FAME	FAMILY HAP.	TOTAL
Observed (f_0)	25	10	5	25	10	15	90

Answers on page 526

24.7 χ² TEST IS NONDIRECTIONAL

The χ^2 test is nondirectional as all discrepancies between observed and expected frequencies are squared. All squared discrepancies have a cumulative positive effect on the value of the observed χ^2 and thereby ensure that χ^2 is a nondirectional test, even though, as illustrated in Figure 24.1, only the upper tail of its distribution contains the rejection region.

INTERNET DEMONSTRATION

Go to the Web site for this book **(http://darwin.cwru.edu/~witte/statistics)** and click on **One-way Chi-square** to simulate outcomes for each of two sets of expected frequencies that either reflect or fail to reflect the known shape of the underlying population.

TWO-WAY χ^2 TEST

So far we have considered the case where observations are classified in terms of only one qualitative variable. Now let's discuss the case where observations are cross-classified in terms of two qualitative variables.

24.8 LOST-LETTER STUDY

The return of addressed, stamped envelopes intentionally "lost" in the vicinity of mailboxes can be used to determine if there is a relationship between two qualitative variables, such as the *type of neighborhood* and the *type of address on returned envelopes*. More specifically, an investigator suspects that the rates of returned letters might reveal some relationship between the type of neighborhood (downtown, suburbia, or campus) and the type of envelope address (personal or business) on returned letters.

A total of 100 letters, half with personal addresses and half with business addresses, are "lost" in *each* of the three types of neighborhoods according to procedures that control for possible contaminating factors, such as mailbox accessibility. Each returned letter is cross-classified on the basis of both its origin and its envelope address, as shown in **Table 24.3.** For instance, 41 letters returned from downtown had personal addresses, whereas 23 of the letters returned from campus had business addresses. When observations are cross-classified according to two qualitative variables, as with the lost-letter study, the test is a two-way χ^2 test.

24.9 STATISTICAL HYPOTHESES

Null Hypothesis

Two-way χ^2 test

Evaluates whether observed frequencies reflect the independence of two qualitative variables.

For the **two-way χ^2 test,** the null hypothesis always makes a statement about the lack of relationship between two qualitative variables in the underlying population. In the present case, it makes a statement about the lack of a relationship between type of neighborhood and the type of address on returned envelopes. This is the same as claiming that, for example, the proportion of returned letters with business addresses doesn't depend on the neighborhood where the letter is lost. Accordingly, the two-way χ^2 test often is referred to as a *test of independence* for the two qualitative variables.

Table 24.3
OBSERVED FREQUENCIES OF RETURNED LETTERS

| | NEIGHBORHOOD | | | |
ENVELOPE ADDRESS	DOWNTOWN	SUBURBIA	CAMPUS	TOTAL
Personal	41	32	47	120
Business	19	38	23	80
Total	60	70	70	200

Although symbolic statements of the null hypothesis are possible, it's much easier to use word descriptions such as

H_0: Type of neighborhood and type of address on returned envelopes are independent;

or as another example,

H_0: Gender and political preference are independent.

If these null hypotheses are true, then among the population of *returned* letters, the type of neighborhood shouldn't change the likelihood or probability that a randomly selected letter has a personal address, and among the population of voters, gender shouldn't change the probability that a randomly selected voter prefers the Democrats. Otherwise, if these null hypotheses are false, type of neighborhood should change the probability that a randomly selected letter has a personal address, and gender should change the probability that a randomly selected voter prefers the Democrats.

Alternative Hypothesis

The alternative or research hypothesis always takes the form

H_1: H_0 is false.

*Exercise 24.3 Specify the null hypothesis for each of the following situations:

(a) A political scientist suspects that there is a relationship between the educational level of adults (grade school, high school, college) and whether they favor right-to-abortion legislation.

(b) A marital therapist believes that groups of clients and nonclients are distinguishable on the basis of whether their parents are divorced.

(c) An organizational psychologist wonders whether employees' annual evaluations, as either satisfactory or unsatisfactory, tend to reflect whether they work day or night shifts.

Answers on page 526

24.10 OBSERVED AND EXPECTED FREQUENCIES

As in the one-way χ^2 test, expected frequencies are calculated on the assumption that the null hypothesis is true, and, depending on the size of the discrepancies between observed and expected frequencies, the null hypothesis is either retained or rejected.

Table 24.4
OBSERVED AND EXPECTED FREQUENCIES OF RETURNED LETTERS

	NEIGHBORHOOD			
ENVELOPE ADDRESS	DOWNTOWN	SUBURBIA	CAMPUS	TOTAL
Personal f_o	41	32	47	120
f_e	36	42	42	
Business f_o	19	38	23	80
f_e	24	28	28	
Total	60	70	70	200

Finding Expected Frequencies from Proportions

Let's look at how expected frequencies are calculated for the two-way χ^2 test. According to the present null hypothesis, type of neighborhood and type of address on returned envelopes are independent. Except for chance, the same proportion of returned letters with personal addresses should be observed for each of the three neighborhoods. Referring to the totals in the margins of Table 24.3, you'll notice that when all three types of neighborhoods are considered together, 120 of the total of 200 returned letters had personal addresses. Therefore, if the null hypothesis is true, 120/200, or .60, should describe the proportion of returned letters with personal addresses in *each* of the three environments. More specifically, among the total of 60 letters returned from downtown, .60 of this total, that is (.60)(60), or 36 letters, should have personal addresses, and 36 is the expected frequency of returned letters with personal addresses from downtown, as indicated in **Table 24.4.** By the same token, among the total of 70 returned letters from suburbia (or from campus), .60 of this total, that is (.60)(70), or 42, is the expected frequency of returned letters with personal addresses from suburbia (or from campus).

As can be verified in Table 24.4, the expected frequencies for returned letters with business addresses can be calculated in the same way, after establishing that when all three neighborhoods are considered together, only 80 of the total of 200 returned letters have business addresses. Now, if the null hypothesis is true, 80/200, or .40, should describe the proportion of returned letters with business addresses from each of the three neighborhoods. For example, among the total of 60 letters returned from downtown, .40 of this total, or 24, will be the expected frequency of returned letters with business addresses from downtown.

Finding Expected Frequencies from Totals

Expected frequencies have been derived from expected proportions in order to spotlight the reasoning behind the test. In the long run, it's more efficient to calculate the expected frequencies directly from the various marginal totals, according to the following formula:

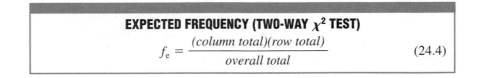

EXPECTED FREQUENCY (TWO-WAY χ^2 TEST)
$$f_e = \frac{(column\ total)(row\ total)}{overall\ total} \qquad (24.4)$$

where f_e refers to the expected frequency for any cell in the table; column total refers to the total frequency for the column occupied by that cell; row total refers to the total frequency for the row occupied by that cell; and overall total refers to the grand total for all columns (or all rows).

Using the marginal totals in Table 24.4, we may verify that Formula 24.4 yields the expected frequencies shown in that table. For example, the expected frequency of returned letters with personal addresses from downtown is

$$f_e = \frac{(60)\,(120)}{200} = \frac{7200}{200} = 36$$

and the expected frequency of returned letters with personal addresses from suburbia is

$$f_e = \frac{(70)\,(120)}{200} = \frac{8400}{200} = 42$$

Before reading on, use Formula 24.4 to obtain the expected frequencies of returned letters for the four remaining cells in Table 24.4.

Having determined the set of expected frequencies, you may use Formula 24.2 to calculate the value of χ^2, as described in **Table 24.5.** Incidentally, for computational convenience, all of the fictitious totals in the margins of Table 24.4 were selected to be multiples of ten. In actual practice the marginal totals are unlikely to be multiples of ten, and consequently, expected frequencies won't always be whole numbers.

24.11 χ^2 TABLES AND DEGREES OF FREEDOM

Locating a critical χ^2 value in Table D of Appendix D requires that you know the correct number of degrees of freedom. For the two-way test, degrees of freedom for χ^2 can be obtained from the following expression:

DEGREES OF FREEDOM (TWO-WAY χ^2 TEST)
$$df = (C - 1)(R - 1) \qquad (24.5)$$

Table 24.5
CALCULATION OF χ^2 (TWO-WAY TEST)

A. COMPUTATIONAL SEQUENCE

Use formula 1 to obtain all expected frequencies from table of observed frequencies.
Construct a table of observed and expected frequencies 2.
Substitute numbers in formula 3 and solve for χ^2.

B. DATA AND COMPUTATIONS

1   $f_e = \dfrac{\text{(column total)(row total)}}{\text{overall total}}$

$f_e \text{ (personal, downtown)} = \dfrac{(60)(120)}{200} = 36$

$f_e \text{ (personal, suburbia)} = \dfrac{(70)(120)}{200} = 42$

$f_e \text{ (personal, campus)} = \dfrac{(70)(120)}{200} = 42$

$f_e \text{ (business, downtown)} = \dfrac{(60)(80)}{200} = 24$

$f_e \text{ (business, suburbia)} = \dfrac{(70)(80)}{200} = 28$

$f_e \text{ (business, campus)} = \dfrac{(70)(80)}{200} = 28$

2

	Downtown	Suburbia	Campus	Total
Personal f_0	41	32	47	120
f_e	36	42	42	
Business f_0	19	38	23	80
f_e	24	28	28	
Total	60	70	70	200

3   $\chi^2 = \sum \dfrac{(f_0 - f_e)^2}{f_e}$

$= \dfrac{(41 - 36)^2}{36} + \dfrac{(32 - 42)^2}{42} + \dfrac{(47 - 42)^2}{42} + \dfrac{(19 - 24)^2}{24} + \dfrac{(38 - 28)^2}{28} + \dfrac{(23 - 28)^2}{28}$

$= 0.69 + 2.38 + 0.60 + 1.04 + 3.57 + 0.89$

$= 9.17$

where C equals the number of categories for the column variable and R equals the number of categories for the row variable. In the present example, which has three columns (downtown, suburbia, and campus) and two rows (personal and business),

$$df = (3 - 1)(2 - 1) = (2)(1) = 2$$

To find the critical χ^2 for a test at the .05 level of significance, locate the cell in Table D of Appendix D intersected by the row for 2 degrees of freedom and the column for the .05 level. In this case, the value of the critical χ^2 equals 5.99.

Explanation for Degrees of Freedom

To understand Formula 24.5, focus on the set of observed frequencies in Table 24.3. In practice, of course, the observed frequencies of the six cells (in the interior of the table) have equal status, and any combination of six frequencies that sum to the various marginal totals is possible. From the more abstract perspective of degrees of freedom, however, only two of these frequencies are free to vary. Why? Because of the mathematical restriction that within each column all observed (or expected) cell frequencies must sum to the column total and that within each row all observed (or expected) cell frequencies must sum to the row total. Although the observed frequencies of any two of the six cells are free to vary, the frequencies of the four remaining cells are determined by the various marginal totals. As indicated in panel A of **Table 24.6,** any frequencies could be assigned to the first two cells in the top row, but the frequency of the third cell in the top row would be determined by the row total, and the frequencies of the three cells in the second row would be determined by their respective column totals.

Another Example

Imagine a study involving three columns and three rows, yielding a total of nine cells with observed frequencies. According to Formula 24.5,

$$df = (3 - 1)(3 - 1) = (2)(2) = 4$$

This implies that after frequencies have been assigned to four of the cells, the frequencies of the five remaining cells are determined by the various marginal totals. As indicated in panel B of Table 24.6, any frequencies could be assigned to the first two cells in the top row and to the first two cells in the middle row. The frequency of the remaining cell within each of these rows would be determined by its row total, and the frequencies of the three cells in the bottom row would be determined by their respective column totals.

Table 24.6
DEGREES OF FREEDOM (TWO-WAY χ^2 TEST)

A. THREE COLUMNS AND TWO ROWS:
$df = (3 - 1)(2 - 1)$
$= 2$

✓	✓	X	120
X	X	X	80
60	70	70	200

B. THREE COLUMNS AND THREE ROWS:
$df = (3 - 1)(3 - 1)$
$= 4$

✓	✓	X	20
✓	✓	X	100
X	X	X	80
60	70	70	200

✓ Cell frequency is free to vary.
X Cell frequency is not free to vary but fixed by marginal total.

24.12 χ^2 TEST

Because the calculated χ^2 of 9.17 is more positive than the critical χ^2 of 5.99, the null hypothesis should be rejected: There is evidence that the type of neighborhood and the type of address on returned envelopes aren't independent. In other words, knowledge about the type of neighborhood supplies extra information about the likelihood that returned letters will have a personal or business address. A comparison of observed and expected frequencies in Table 24.5 suggests that letters with personal addresses are more likely to be returned from either downtown or campus, whereas letters with business addresses are more likely to be returned from suburbia.

HYPOTHESIS TEST SUMMARY TWO-WAY χ^2 TEST (LOST-LETTER STUDY)

Research Problem:
Is there a relationship between the type of neighborhood and the type of address on returned envelopes?

Statistical Hypotheses:
H_0: Type of neighborhood and type of address on returned envelopes are independent.
$$H_1: H_0 \text{ is false.}$$

Decision Rule:
Reject H_0 at .05 level of significance if χ^2 equals or is more positive than 5.99 [from Table D, Appendix D, given that $df = (C - 1)(R - 1) = (3 - 1)(2 - 1) = 2$].

Calculations:
$$\chi^2 = 9.17 \text{ (See Table 24.5 on page 428.)}$$

Decision:
Reject H_0 at .05 level of significance because $\chi^2 = 9.17$ is more positive than 5.99.

Interpretation:
Type of neighborhood and type of address on returned envelopes aren't independent.

Published Reports

A published report of this hypothesis test might be limited to an interpretative comment, plus a parenthetical statement that summarizes the statistical analysis and includes a *p*-value. For example, using the less structured *p*-value approach to hypothesis testing described in Section 21.1, an investigator might report the following:

There is evidence that, among returned letters, the type of address is related to the type of neighborhood [χ^2 (2, $n = 200$) = 9.17, $p < .05$].

The parenthetical statement indicates that a χ^2 based on 2 degrees of freedom and a sample size, n, of 200, was found to equal 9.17. The test result has an approximate p-value less than .05 because, as can be seen in Table D of Appendix D, the observed χ^2 of 9.17 is more positive than the critical χ^2 of 5.99 for the .05 level of significance. Furthermore, because a p-value of less than .05 is a rare event, given that the null hypothesis is true, it supports the research hypothesis, as implied in the interpretative statement.

*Exercise 24.4 A researcher suspects that there might be a relationship, possibly based on genetic factors, between hair color and susceptibility to poison oak. Three hundred volunteer subjects are exposed to a small amount of poison oak and then classified according to their susceptibility (rash or no rash) and their hair color (red, blond, brown, or black), yielding the following frequencies:

HAIR COLOR AND SUSCEPTIBILITY TO POISON OAK					
		HAIR COLOR			
SUSCEPTIBILITY	RED	BLOND	BROWN	BLACK	TOTAL
Rash	10	30	60	80	180
No Rash	20	30	30	40	120
Total	30	60	90	120	300

(a) Test the null hypothesis at the .01 level of significance.

(b) Specify the approximate p-value for this test result.

(c) How might these results appear in a published report?
Answers on pages 526–527

24.13 SOME PRECAUTIONS

Avoid Dependent Observations

The valid use of χ^2 requires that *observations be independent of one another*. One observation should have no influence on another; for instance, when tossing a pair of dice, the appearance of a six spot on one die has no influence on the number of spots displayed on the other die. A violation of independence occurs whenever a single subject contributes more than one observation (or in the two-way case, more than one pair of observations). For example, it would have occurred in the blood type survey if any student's blood type had been counted more than once in the observed frequencies of Table 24.1. It also would have occurred in a preference test for four brands of soda if each subject's preference had been counted more than once, possibly because of a series of taste trials. When considering the use of χ^2, *the total for all observed frequencies never should exceed the total number of subjects.*

Avoid Small Expected Frequencies

The valid use of χ^2 also requires that expected frequencies not be too small. A conservative rule specifies that *all expected frequencies be 5 or more.* Small expected frequencies need not necessarily lead to a statistical dead end; sometimes it's possible to create a larger expected frequency from the combination of smaller expected frequencies (see Exercise 24.13). Otherwise, avoid small expected frequencies by using a larger sample size.

Avoid Extreme Sample Sizes

Although small *expected frequencies* are to be avoided because of technical reasons peculiar to the χ^2 test, a small *sample size* is to be avoided for more general reasons discussed originally in Section 16.7. An unduly small sample size produces a test that tends to miss even a seriously false null hypothesis. By avoiding small expected frequencies, you'll automatically protect the χ^2 test from the more severe cases of small sample size.

An excessively large sample size also is to be avoided. As also discussed in Section 16.7, an excessively large sample size produces a test that tends to detect small unimportant departures from null hypothesized values.

24.14 CHECKING IMPORTANCE

One way to check the importance of a statistically significant two-way χ^2 test based on a large sample size is to use a measure analogous to the squared correlation coefficient, known as the **squared Cramér's phi coefficient,** symbolized as ϕ_c^2. Without being swayed by the large sample size, ϕ_c^2 *very roughly estimates, on a scale from 0 to 1, the proportion of predictability between two qualitative variables.*

Squared Cramér's Phi Coefficient (ϕ_c^2)

Squared Cramér's phi coefficient (ϕ_c^2)

Very rough estimate of the proportion of predictability between two qualitative variables.

Solve for the squared Cramér's phi coefficient using the following formula:

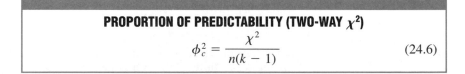

PROPORTION OF PREDICTABILITY (TWO-WAY χ^2)

$$\phi_c^2 = \frac{\chi^2}{n(k-1)}$$

(24.6)

---

where χ^2 is the obtained value of the significant two-way χ^2, n is the sample size (total observed frequency), and k is the smaller of either the C columns or the R rows (or the value of either if the columns and rows are the same).

For the lost-letter study, given a significant χ^2 of 9.17, $n = 200$, and $k = 2$ (from $R = 2$), we can calculate

$$\phi_c^2 = \frac{9.17}{200(2 - 1)} = .05$$

One very rough rule of thumb, suggested by Cohen for correlations, is that the strength (and importance) of the relationship between the two variables is small if ϕ_c^2 approximates .01, medium if ϕ_c^2 approximates .09, and large if ϕ_c^2 approximates or exceeds .25.[1] Using this rule of thumb, the estimated strength of the relationship between neighborhood and letter address for returned letters is somewhere between small and medium, because $\phi_c^2 = .05$.

A Recommendation

Consider calculating ϕ_c^2 whenever a statistically significant two-way χ^2 has been obtained, especially if the observed frequencies number in the hundreds or more. As mentioned in Section 21.11, however, do not blindly apply any rule of thumb about the strength of a relationship without regard to special circumstances that could give considerable importance to even a very weak relationship. For example, the value of ϕ_c^2 actually was only .001 (!) for the important relationship, described in the *New England Journal of Medicine* (January 28, 1988), between the incidence of heart attacks for physicians who did or didn't take a daily aspirin.

*Exercise 24.5 Given the significant χ^2 in Exercise 24.4, use Formula 24.6 to estimate whether the strength of the relationship between hair color and suscepbibility to poison oak is small, medium, or large.

Answer on page 527

INTERNET DEMONSTRATION

Go to the Web site for this book **(http://darwin.cwru.edu/~witte/statistics)** and click on **Two-way Chi-square** to simulate outcomes by manipulating the degree of dependency between two qualitative variables and also by manipulating sample sizes.

---

[1]This is the recommended rule of thumb if χ^2 is based on tables with either two columns or two rows (or both), as often is the case. Otherwise, these values (.01, .09, and .25) probably should be adjusted downward. See Cohen, *Statistical Power Analysis for the Behavioral Sciences*, 2nd ed. (Hillsdale, NJ: Erlbaum, 1988), pp. 224–27.

24.15 COMPUTER OUTPUT

Table 24.7
SAS OUTPUT: TWO-WAY χ^2 TEST FOR LOST-LETTER DATA

The SAS System
12:46 Tuesday, November 23, 1999
TABLE OF ADDRESS BY NGHBRHD

ADDRESS Frequency Percent Row Pct Col Pct	NGHBRHD			
	Downtown	Suburbia	Campus	Total
Personal	**1** 41 20.50 **2** 34.17 68.33	32 16.00 26.67 45.71	47 23.50 39.17 67.14	120 60.00
Business	19 9.50 23.75 31.67	38 19.00 47.50 54.29	23 11.50 28.75 32.86	80 40.00
Total	60 30.00	70 35.00	70 35.00	200 100.00

STATISTICS FOR TABLE OF ADDRESS BY NGHBRHD

Statistic	DF	Value	**4** Prob
3 Chi-Square	2	9.177	0.010
Likelihood Ratio Chi-Square	2	9.116	0.010
Mantel-Haenszel Chi-Square	1	6.663	0.010
Phi Coefficient		0.214	
Contingency Coefficient		0.209	
5 Cramér's V		0.214	

Sample Size = 200

Comments:
1 *Observed frequency.*
2 *The observed frequency expressed as a percent of the grand total; the row total; and the column total.*
3 *Two-way χ^2 test for lost-letter data. Compare with results in Table 24.5.*
4 *Exact p-value.*
5 *An extension of the Pearson r referred to as Cramér's phi coefficient in Section 9.10. A correlation of .214 describes the relationship between the two qualitative variables in this study (given that arbitrary numerical codes have been assigned to the original categories for both variables). The squared Cramér's phi coefficient also is discussed in Section 24.14.*

*Exercise 24.6 The following Minitab output is based on the data for marijuana smoking reported in Exercise 24.12 on page 438.

MINITAB OUTPUT: χ^2 TEST FOR MARIJUANA SMOKING RESPONSES OF TWO GENERATIONS

Chi-Square Test
Expected counts are printed below observed counts

	C1	C2	TOTAL
1	62	43	105
	65.63	39.37	
2	38	17	55
	34.37	20.63	
Total	100	60	160

Chi-Sq = 0.200 + 0.334 + 0.382 + 0.637 = 1.553
DF = 1, p-value = 0.213

(a) What is the observed frequency of Yes responses (coded 1) for the 1990s generation (coded C2)?

(b) What is the expected frequency of No responses (coded 2) for the 1980s generation (coded C1)?

(c) What is the value of χ^2?

(d) What is the exact p-value for χ^2?

Answers on page 527

Summary
. . . .

The chi-square test is designed to test the null hypothesis for qualitative data, expressed as frequencies. For the one-way χ^2 test, the null hypothesis claims that the population distribution complies with a set of hypothesized proportions. For the two-way χ^2 test, the null hypothesis claims that the two qualitative variables are independent (or not related). In either case, the null hypothesis is used to generate a set of expected frequencies, which, in turn, is compared to a corresponding set of observed frequencies.

Essentially, χ^2 reflects the size of the discrepancies between observed and expected frequencies, and the larger the value of χ^2 is, the more suspect the null hypothesis will be.

To obtain critical values for χ^2, Table D of Appendix D must be consulted, with the appropriate number of degrees of freedom for the one- and two-way tests. Because all discrepancies are squared, the χ^2 test is nondirectional, even though only the upper tail of the χ^2 distribution contains the rejection region.

Use of the χ^2 test requires that observations be independent and that expected frequencies be sufficiently large. Unduly small or excessively large sample sizes should be avoided.

Whenever obtaining a statistically significant χ^2, especially one based on a large sample size, consider using Formula 24.6 and Cohen's rule of thumb on page 433 to estimate, independently of sample size, whether the strength of the relationship is small, medium, or large.

Important Terms

One-way χ^2 test

Expected frequency (f_e)

Observed frequency (f_0)

Two-way χ^2 test

Squared Cramér's phi coefficient (ϕ_c^2)

REVIEW EXERCISES

24.7 Randomly selected records of 140 criminals reveal that crimes (for which they were convicted) were committed on the following days of the week:

	DAYS WHEN CRIMES WERE COMMITTED							
Frequency	Mon.	Tue.	Wed.	Thu.	Fri.	Sat.	Sun.	Total
Observed (f_0)	17	21	22	18	23	24	15	140

(a) Using the .01 level of significance, test the null hypothesis that in the underlying population, crimes are equally likely to be committed on any day of the week.

(b) Specify the approximate *p*-value for this test result.

(c) How might these results appear in a published report?

*24.8 Several investigators have reported a tendency for more people to die (from natural causes, such as cancer and strokes) after, rather than before, a major holiday. This post-holiday death peak has been attributed to several factors, including the willful postponement of death until after the holiday, as well as holiday stress and post-holiday depression. Writing in the *Journal of the American Medical Association* (April 11, 1990), Phillips and Smith report that, among a total of 103 elderly California women of Chinese descent who died of natural causes within one week of the Harvest Moon Festival, only 33 died in the week before, whereas 70 died in the week after.

(a) Using the .05 level of significance, test the null hypothesis that in the underlying population, people are equally likely to die either in the week before or in the week after this holiday.

(b) Specify the approximate p-value for this test result.

(c) How might these results appear in a published report?

Answers on page 527

24.9 While playing a coin-tossing game in which you're to guess whether heads or tails will appear, you observe 30 heads in a string of 50 coin tosses.

(a) Test the null hypothesis that this coin is unbiased, that is, that heads and tails are equally likely to appear in the long run.

(b) Specify the approximate p-value for this test result.

24.10 In Chapter 1, Table 1.1 lists the weights of 53 male statistics students. Although students were asked to report their weights to the nearest pound, inspection of Table 1.1 reveals that a disproportionately large number (27) reported weights ending in either a zero or a five. This suggests that many students probably reported their weights rounded to the nearest five or ten pounds rather than to the nearest pound. Using the .05 level of significance, test the null hypothesis that in the underlying population, weights are rounded to the nearest pound. *Hint:* If the null hypothesis is true, two-tenths of all weights should end in either a zero or five, and eight-tenths of all weights should end in a one, two, three, four, six, seven, eight, or nine. Therefore, the situation requires a one-way test with only two categories, and $df = 1$.

*24.11 Students are classified according to religious preference (Buddhist, Jewish, Protestant, Roman Catholic, or Other) and political affiliation (Democrat, Republican, Independent, or Other).

POLITICAL AFFILIATION	RELIGIOUS PREFERENCE AND POLITICAL AFFILIATION RELIGIOUS PREFERENCE					
	BUDDHIST	JEWISH	PROTESTANT	ROM. CATH.	OTHER	TOTAL
Democrat	30	30	40	60	40	200
Republican	10	10	40	20	20	100
Independent	10	10	20	20	40	100
Other	0	0	0	0	100	100
Total	50	50	100	100	200	500

(a) This table consists of five columns and four rows. Identify those cell frequencies that are free to vary and those that are fixed, as in Table 24.6, and determine the number of degrees of freedom for χ^2 in this situation.

(b) Anything suspicious about these observed frequencies?

(c) Using the .05 level of significance, test the null hypothesis that these two variables are independent in the underlying population.

Answers on pages 527–528

24.12 Do two different generations of college students respond similarly to the question "Have you ever smoked marijuana?" The following table shows answers to this question for samples of students who attended statistics classes during the 1980s and 1990s. Using the .05 level of significance, test the null hypothesis that there is no relationship between replies and the two generations of college students.

MARIJUANA SMOKING RESPONSES OF TWO GENERATIONS GENERATION			
RESPONSE	1980s	1990s	TOTAL
Yes	62	43	105
No	38	17	55
Total	100	60	160

24.13 Test the null hypothesis at the .01 level of significance that the distribution of blood types for college students complies with the proportions described in the blood bank bulletin, namely, .44 for O, .41 for A, .10 for B, and .05 for AB. Now, however, assume that the results are available for a random sample of only 60 students. The results are as follows: 27 for O, 26 for A, 4 for B, and 3 for AB. *Note:* The expected frequency for AB, (.05)(60) = 3, is less than 5, the smallest permissible expected frequency. Create a sufficiently large expected frequency by combining B and AB blood types.

24.14 A social scientist cross-classifies the responses of 100 randomly selected people on the basis of gender and whether or not they favor strong gun control laws, to obtain the following:

GENDER AND ATTITUDE TOWARD STRONG GUN CONTROL			
	ATTITUDE		
GENDER	FAVOR	OPPOSE	TOTAL
Male	40	20	60
Female	30	10	40
Total	70	30	100

(a) Using the .05 level of significance, test the null hypothesis for gender and attitude toward gun control.

(b) Specify the approximate *p*-value for the test result.

(c) How might these results appear in a published report?

24.15 To appreciate the impact of large sample size on the value of χ^2, multiply each of the observed frequencies in Exercise 24.14 by ten to obtain the following:

GENDER AND ATTITUDE TOWARD STRONG GUN CONTROL			
	ATTITUDE		
GENDER	FAVOR	OPPOSE	TOTAL
Male	400	200	600
Female	300	100	400
Total	700	300	1000

Note: Even though the sample size has increased by a factor of ten in Exercise 24.15, the proportion of males (and females) who favor gun control remains the same as in Exercise 24.14. In both exercises, gun control is favored by .67 of all males (from $40/60 = 400/600 = .67$) and by .75 of all females (from $30/40 = 300/400 = .75$).

(a) Using the .05 level of significance, again test the null hypothesis for gender and attitude toward gun control.

(b) Specify the approximate *p*-value for this test result.

(c) Given a significant χ^2 for the current analysis, use Formula 24.6 and Cohen's rule of thumb to estimate whether the strength of the relationship between gender and attitude toward gun control is small, medium, or large.

CHAPTER 25

Tests for Ranked Data

25.1   USE ONLY WHEN APPROPRIATE
25.2   A NOTE ON TERMINOLOGY

MANN-WHITNEY U TEST (TWO INDEPENDENT SAMPLES)

25.3   WHY NOT A t TEST?
25.4   STATISTICAL HYPOTHESES FOR U
25.5   CALCULATION OF U
25.6 U TABLES
25.7   DECISION RULE
25.8   DIRECTIONAL TESTS

WILCOXON T TEST (TWO MATCHED SAMPLES)

25.9   WHY NOT A t TEST?
25.10 STATISTICAL HYPOTHESES FOR T
25.11 CALCULATION OF T
25.12 T TABLES
25.13 DECISION RULE

KRUSKAL-WALLIS H TEST (THREE OR MORE INDEPENDENT SAMPLES)

25.14 WHY NOT AN F TEST?
25.15 STATISTICAL HYPOTHESES FOR H
25.16 CALCULATION OF H
25.17 χ^2 TABLES
25.18 DECISION RULE
25.19 H TEST IS NONDIRECTIONAL
25.20 GENERAL COMMENT: TIES

       Summary

       Important Terms

       Review Exercises

25.1 USE ONLY WHEN APPROPRIATE

Use the Mann-Whitney *U*, Wilcoxon *T*, and Kruskal-Wallis *H* tests of this chapter only under appropriate circumstances, that is, (1) when the original data are ranked or (2) when the original data are quantitative but don't appear to originate from normally distributed populations with equal variances.

In the latter case, beware of *non-normality* when the sample sizes are small (less than about ten), and beware of *unequal variances* when the sample sizes are small and unequal.

When the original data are quantitative and the populations appear to be normally distributed with equal variances, use the *t* and *F* tests.

Under these circumstances, the t and F tests are more powerful, that is, they are more likely to detect a false null hypothesis. In other words, they minimize the probability of a type II error.

25.2 A NOTE ON TERMINOLOGY

Nonparametric Tests

Nonparametric tests

Tests, such as U, T, *and* H *of this chapter, that evaluate* entire *population distributions rather than specific population characteristics.*

The U, T, and H tests for ranked data in this chapter, as well as the χ^2 test for qualitative data in the previous chapter, are often referred to as **nonparametric tests.** Parameter refers to any descriptive measure of a population, such as the population mean. Nonparametric tests, such as U, T, and H, evaluate hypotheses for *entire* population distributions, whereas parametric tests, such as t and F in Chapters 18 through 23, evaluate hypotheses for a specific parameter, usually the population mean.

Distribution-Free Tests

Distribution-free tests

Tests—such as U, T, *and* H *of this chapter—that make no assumptions about the form of the population distribution.*

Nonparametric tests also are referred to as **distribution-free tests.** This name highlights the fact that these tests require no assumptions about the precise form of the population distribution. As will be noted, the tests of this chapter, such as U, T, and H, can be conducted without assumptions about the underlying population distributions. In contrast, other types of tests, such as t and F, require populations to be normally distributed with equal variances.

Labels Can Be Misleading

Although widely used, these labels can be misleading. If the two population distributions are assumed to have roughly similar variabilities and shapes, as in the U test for TV-viewing estimates described in the next section (or in the T test described in Section 25.11), we sacrifice the distribution-free status of these tests to gain a more precise parametric test of any differences in central tendency. Consequently, depending on the perspective of the practitioner, the U, T, or H tests might qualify as neither nonparametric nor distribution-free.

Table 25.1 ESTIMATES OF WEEKLY TV-VIEWING TIME (HOURS)	
TV FAVOR- ABLE	TV UNFAVOR- ABLE
12	43
4	14
5	42
20	1
5	2
5	0
10	0
49	

MANN-WHITNEY *U* TEST (TWO INDEPENDENT SAMPLES)

If high school students are asked to estimate the number of hours they spend watching TV each week, are their anonymous replies influenced by whether TV viewing is depicted favorably or unfavorably? More specifically, one-half of the members of a social studies class are selected at random to receive questionnaires that depict TV viewing favorably (as the preferred activity of better students), and the other half of the class receive questionnaires that depict TV viewing unfavorably (as the preferred activity of poorer students). After the replies of several students who responded not with numbers but with words such as "a lot" and "hardly at all" were discarded, the results were listed in **Table 25.1.**

25.3 WHY NOT A *t* TEST?

When taken at face value, it might appear that the estimates in Table 25.1 could be tested with the customary *t* test for two independent samples. But closer inspection reveals a complication. Each group of estimates includes one or two very large values, suggesting that the underlying populations are positively skewed rather than normal. When the sample sizes are small, as in the present experiment, violations of the normality assumption could seriously impair the accuracy of the *t* test by causing the probability of a type I error to differ considerably from that specified in the level of significance.

One remedy is to convert all of the estimates in Table 25.1 into ranks and to analyze the newly ranked data with the Mann-Whitney *U* test for two independent samples. As is true of all tests for ranked data, the *U* test is immune to violations of assumptions about normality and equal variances.

25.4 STATISTICAL HYPOTHESES FOR *U*

For the TV-viewing study, the statistical hypotheses take the form

H_0: population distribution 1 = population distribution 2
H_1: population distribution 1 $\neq$ population distribution 2

in which TV viewing is depicted favorably in population 1 and unfavorably in population 2.

Unspecified Differences

Notice that the null hypothesis equates two *entire* population distributions. Any type of inequality between population distributions, whether caused by differences in central tendency, variability, or shape, could contribute to the rejection of H_0. Strictly speaking, the *rejection of H_0 signifies only that the two populations differ because of some unspecified inequality, or combination of inequalities,* between the original population distributions.

Specified Differences

More precise conclusions are possible if it can be assumed that both population distributions have about equal variabilities and roughly similar shapes, that is, for instance, if both population distributions are symmetrical or if both are similarly skewed. Under these circumstances, the rejection of H_0 signifies that the two population distributions occupy different locations and, therefore, possess different central tendencies (which can be interpreted as a difference between population means or population medians).

··

25.5 CALCULATION OF *U*

Table 25.2 indicates how to convert the estimates in Table 25.1 into ranks. Before assigning numerical ranks to the two groups, coded as groups 1 and 2, list all observations from smallest to largest for the combined groups. Beginning with the smallest estimate, assign the consecutive numerical ranks 1, 2, 3, and so forth, until all the estimates have been converted to ranks. When two or more estimates are the same, assign the mean of the numerical ranks that would have been assigned if the estimates had been different. For example, each of the two estimates of 0 hours receives a rank of 1.5, the mean of the ranks 1 and 2, and each of the three estimates of 5 hours receives a rank of 7, the mean of the ranks 6, 7, and 8.

*Exercise 25.1 Beginning with a rank of 1 for the smallest observation, rank each of the following sets of observations:

(a) 4, 6, 9, 10, 10, 12, 15, 23, 23, 23, 31

(b) 103, 104, 104, 105, 105, 109, 112, 118, 119, 124

(c) 51, 54, 54, 54, 54, 59, 60, 71, 71, 79

Answers on page 528

Preliminary Interpretation

Differences in ranks between groups 1 and 2 are not mentioned in Table 25.2, but it's wise to pause at this point and to form a preliminary impression of any of these differences. The more one group tends to outrank the other, the larger the difference between the mean ranks for the two groups and the more suspect the null hypothesis will be. Because, in Table 25.2, the mean rank for group 1 equals 9 (from 72/8) and the mean rank for group 2 equals 6.86 (from 48/7), there is a tendency for group 1 to outrank group 2. In other words, estimates of weekly TV-viewing time in the TV-favorable group tend to be larger than those in the TV-unfavorable group. It remains to be seen whether this result will cause the null hypothesis to be rejected.

Once all of the observations have been ranked, find the sum of ranks for group 1, R_1, and the sum of the ranks for group 2, R_2. To verify that ranks have been assigned and added correctly, perform the computational check shown in

Table 25.2
CALCULATION OF *U*

A. COMPUTATIONAL SEQUENCE

Identify the sample sizes of group 1, n_1, group 2, n_2, and the combined groups, n **1**.

List observations from smallest to largest for the combined groups **2**.

Assign numerical ranks to the ordered observations for the combined groups **3**.

Sum the ranks for group 1 **4** and for group 2 **5**.

Substitute numbers in formula **6** and verify that ranks have been assigned and added correctly.

Substitute numbers in formula **7** and solve for U_1.

Substitute numbers in formula **8** and solve for U_2.

Set *U* equal to whichever is smaller — U_1 or U_2 **9**.

B. DATA AND COMPUTATIONS

1 $n_1 = 8$; $n_2 = 7$; $n = 8 + 7 = 15$

2 Observations		**3** Ranks	
(1) TV Favorable	**(2)** TV Unfavorable	**(1)** TV Favorable	**(2)** TV Unfavorable
	0		1.5
	0		1.5
	1		3
	2		4
4		5	
5		7	
5		7	
5		7	
10		9	
12		10	
	14		11
20		12	
	42		13
	43		14
49		15	
		4 $R_1 = \overline{72}$	**5** $R_2 = \overline{48}$

6 Computational Check:

$$R_1 + R_2 = \frac{n(n + 1)}{2}$$

$$72 + 48 = \frac{15(15 + 1)}{2}$$

$$120 = 120$$

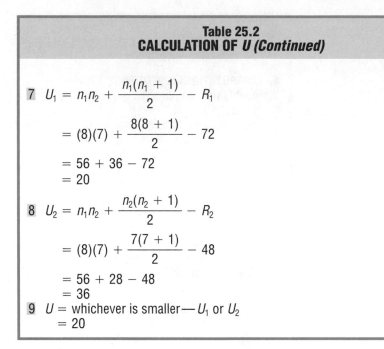

Table 25.2
CALCULATION OF *U* (Continued)

7 $U_1 = n_1 n_2 + \dfrac{n_1(n_1 + 1)}{2} - R_1$

$= (8)(7) + \dfrac{8(8 + 1)}{2} - 72$

$= 56 + 36 - 72$
$= 20$

8 $U_2 = n_1 n_2 + \dfrac{n_2(n_2 + 1)}{2} - R_2$

$= (8)(7) + \dfrac{7(7 + 1)}{2} - 48$

$= 56 + 28 - 48$
$= 36$

9 U = whichever is smaller — U_1 or U_2
$= 20$

Table 25.2. Finally, calculate values for both U_1 and U_2, and set the smaller of these two values equal to U, that is,

MANN-WHITNEY *U* TEST (TWO INDEPENDENT SAMPLES)

$$U_1 = n_1 n_2 + \frac{n_1(n_1 + 1)}{2} - R_1$$

$$U_2 = n_1 n_2 + \frac{n_2(n_2 + 1)}{2} - R_2 \qquad (25.1)$$

$$U = \text{the smaller of } U_1 \text{ or } U_2$$

in which n_1 and n_2 represent the sample sizes of groups 1 and 2, and R_1 and R_2 represent the sum of ranks for groups 1 and 2. The value of U equals 20 for the present study.

25.6 *U* TABLES

Critical values of U are supplied for values of n_1 and n_2 (no larger than 20 each) in Table E of Appendix D.[1] Notice that there are two sets of tables, one for nondirectional tests and one for directional tests. Both tables supply critical values of U for hypothesis tests at the .05 level (light numbers) and the .01 level (dark numbers).

··
[1]In the unlikely event that you'll be using this test with sample sizes larger than 20 each, use the large sample approximation discussed in more advanced statistics books, such as W. Conover, *Practical Nonparametric Statistics* (New York: Wiley, 1999).

To find the correct critical U, identify the entry in the cell intersected by n_1 and n_2, the sample sizes of groups 1 and 2. For the present study, given a nondirectional test at the .05 level of significance with an n_1 of 8 and an n_2 of 7, the value of the critical U equals 10.

25.7 DECISION RULE

An unusual feature of hypothesis tests involving U (and also T, described later in the chapter) is that *the null hypothesis will be rejected only if the observed U is **less** than or equal to the critical U*. Otherwise, if the observed U exceeds the critical U, the null hypothesis will be retained.

HYPOTHESIS TEST SUMMARY: MANN-WHITNEY *U* TEST FOR TWO INDEPENDENT SAMPLES (ESTIMATES OF TV VIEWING)

Research Problem:
Are high school students' estimates of their weekly TV-viewing time influenced by depicting TV viewing as (1) a favorable or (2) an unfavorable activity?

Statistical Hypotheses:
H_0: population distribution 1 = population distribution 2
H_1: population distribution 1 $\neq$ population distribution 2

Decision Rule:
Reject H_0 at the .05 level of significance if U equals or is less positive than 10 (from Table E, Appendix D, given $n_1 = 8$ and $n_2 = 7$).

Calculations:
$U = 20$ (See Table 25.2 on pages 445–446 for computations.)

Decision:
Retain H_0 at .05 level of significant because $U = 20$ is *more* positive than 10.

Interpretation:
There is no evidence that depicting TV viewing as a favorable or an unfavorable activity influences high school students' estimates of their weekly TV-viewing times.

Explanation of Topsy-Turvy Rule

To appreciate this topsy-turvy decision rule for the U test, let's look more closely at U. Although not apparent in Formula 25.1, *U represents the number of times that individual ranks in the lower ranking group exceed individual ranks in the higher ranking group.* When a maximum difference separates two groups—because no rank in the lower ranking group exceeds any rank in the higher ranking group—U equals 0. At the other extreme, when a minimum difference separates two groups—because, as often as not, individual ranks in the lower ranking group exceed individual ranks in the higher ranking group—U equals a large number given by the expression

$$\frac{n_1 n_2}{2}$$

which is 28 for the present study.

Ordinarily, the difference in ranks between groups is neither maximum nor minimum, and U equals some intermediate value that, to be interpreted, must be compared with the appropriate critical U value. In the present study, as the observed U of 20 exceeds the critical U of 10, only a moderate difference separates the two groups, and the null hypothesis is retained.

*Exercise 25.2 Does it matter whether encounter group leaders adopt either an aggressive or a supportive role to facilitate growth among group members? One randomly selected set of six graduate trainees is taught to be aggressive, and the other set of six trainees is taught to be supportive. Subsequently, each trainee is randomly assigned to lead a small encounter group. Without being aware of the nature of the experiment, a panel of experienced group leaders ranks each encounter group from least (1) to most (12) growth promoting, on the basis of anonymous diaries submitted by all members of each group. The results are as follows:

GROWTH-PROMOTING RANKS OF ENCOUNTER GROUPS	
AGGRESSIVE LEADER	SUPPORTIVE LEADER
1	9
2	6
4.5	12
11	10
3	7
4.5	8

(a) Use U to test the null hypothesis at the .05 level of significance.

(b) Specify the approximate p-value for this test result.

Answers on page 528

25.8 DIRECTIONAL TESTS

The assumption that population distributions have roughly similar variabilities and shapes is required whenever, because of a concern only about population differences in a particular direction, a directional test is desired for the Mann-Whitney U test (or for the Wilcoxon T test discussed next). Without this assumption, a false H_0 could reflect a complex pattern of inequalities; rather than a simple difference in location, between population distributions, and therefore only a less precise, nondirectional test would be appropriate.

Judging from the estimates for groups 1 and 2 in Table 25.1, both population distributions could have roughly similar variabilities and shapes (positively skewed). Accordingly, if there had been a concern only that population distribution 1 exceeded population distribution 2, a directional test would have been possible in this study.

Caution

Before conducting a directional test, always verify that the observed differences are in the direction of concern. For instance, if the preceding directional test had been used in the previous study, you should have verified that the mean rank for group 1 exceeded that for group 2 (as it actually does in this study). Otherwise, if the observed difference between mean ranks had been in the direction of no concern, the hypothesis test should have been halted and H_0 retained.

WILCOXON *T* TEST (TWO MATCHED SAMPLES)

The previous experiment failed to support the investigator's hunch that estimates of TV-viewing time could be influenced by depicting it as a favorable or an unfavorable activity. Noting the large differences among the estimates of students *within* the same group, the investigator might attempt to reduce this variability—and improve the precision of the analysis—by matching students with the aid of some relevant variable (see Section 20.8). For instance, some of the variability among estimates might be due to differences in home environment. The investigator could match for home environment by using pairs of students who are siblings. One member of each pair is assigned randomly to one group, and the other sibling is assigned automatically to the second group. As in the previous experiment, the questionnaires depict TV viewing as either a favorable or an unfavorable activity. The results for the eight pairs of students are listed in the middle portion of **Table 25.3.**

25.9 WHY NOT A *t* TEST?

It might appear that the eight difference scores in Table 25.3 could be tested with the *t* test for two matched samples. But once again, there's a complication. The set of difference scores appears to be symmetrical, but somewhat non-normal, with no obvious cluster of scores in the middle range. When sample sizes are small, as

Table 25.3
CALCULATION OF *T*

A. COMPUTATIONAL SEQUENCE

For each pair of observations, subtract the second observation from the first observation to obtain a difference score **1**.

Ignore difference scores of zero, and, without regard to sign, list the remaining difference scores from smallest to largest **2**.

Assign numerical ranks to the ordered difference scores (still without regard to sign) **3**.

List the ranks for positive difference scores in the plus ranks column **4** and list the ranks for negative difference scores in the minus ranks column **5**.

Sum the ranks for positive differences, R_+ **6**, and sum the ranks for negative differences, R_- **7**.

Determine n, the number of nonzero difference scores **8**.

Substitute numbers in formula **9** to verify that ranks have been assigned and added correctly.

Set *T* equal to whichever is smaller: R_+ or R_- **10**.

B. DATA AND COMPUTATIONS

| | OBSERVATIONS | | | | RANKS | | |
Pairs of Students	(1) TV Favorable	(2) TV Unfavorable	**1** Difference Scores	**2** Ordered Scores	**3** Ranks	**4** Plus Ranks	**5** Minus Ranks
A	2	0	2	2	1.5	1.5	
B	11	5	6	−2	1.5		1.5
C	10	12	−2	3	3	3	
D	6	6	0	5	4	4	
E	7	2	5	6	5	5	
F	43	33	10	8	6	6	
G	33	25	8	10	7	7	
H	5	2	3				
						6 $R_+ = \overline{26.5}$	**7** $R_- = \overline{1.5}$

8 $n = 7$

9 Computational check:

$$R_+ + R_- = \frac{n(n+1)}{2}$$

$$26.5 + 1.5 = \frac{7(7+1)}{2}$$

$$28 = 28$$

10 $T =$ whichever is smaller: R_+ or R_-
$T = 1.5$

in the present experiment, violations of the normality assumption can seriously impair the accuracy of the *t* test for two matched samples. One remedy is to rank all difference scores and to analyze the resulting ranked data with the Wilcoxon *T* test.

25.10 STATISTICAL HYPOTHESES FOR *T*

For the present study, the statistical hypotheses take the form

H_0: population distribution 1 = population distribution 2
H_1: population distribution 1 $\neq$ population distribution 2

where TV viewing is depicted favorably in population 1 and unfavorably in population 2.

As with the null hypothesis for U, that for T equates two entire population distributions. Strictly speaking, the rejection of H_0 signifies only that the two populations differ because of some unspecified inequality, or combination of inequalities, between the original population distributions. More precise conclusions about central tendencies are possible only if it can be assumed that both population distributions have roughly similar variabilities and shapes.

25.11 CALCULATION OF *T*

Table 25.3 shows how to calculate T. When ordering difference scores from smallest to largest, ignore all difference scores of zero, and *temporarily treat all negative difference scores as though they were positive.* Beginning with the smallest difference score, assign the consecutive numerical ranks, 1, 2, 3, and so forth, until all nonzero difference scores have been ranked. When two or more difference scores are the same (regardless of sign), assign them the mean of the numerical ranks that would have been assigned if the scores had been different. For example, each of the two difference scores 2 and -2 receives a rank of 1.5, the mean of ranks 1 and 2.

Once numerical ranks have been assigned, those ranks associated with positive difference scores should be listed in the plus ranks column, and those associated with negative difference scores should be listed in the minus ranks column. Next find the sum of all ranks for positive difference scores, R_+, and the sum of all ranks for negative difference scores, R_-. (Notice that the more one group of difference scores outranks the other, the larger the discrepancy between the two sums of ranks, R_+ and R_-, and the more suspect the null hypothesis will be.) To verify that the ranks have been assigned and added correctly, perform the computational check in Table 25.3. Finally, the value of T equals the smaller value, either R_+ or R_-, that is,

---

Wilcoxon T *test*

A test for ranked data when there are two matched groups.

> **WILCOXON *T* TEST (TWO MATCHED SAMPLES)**
> $$T = \text{the smaller of } R_+ \text{ or } R_- \qquad (25.2)$$

where R_+ and R_- represent the sum of the ranks for positive and negative difference scores. The value of T equals 1.5 for the present study.

25.12 *T* TABLES

Critical values of *T* are supplied for values of *n* up to 50 in Table F of Appendix D. There are two sets of tables, one for nondirectional tests and one for directional tests. Both tables supply critical values of *T* for hypothesis tests at the .05 and .01 levels of significance.

To find the correct critical *T* value, locate the cell intersected by *n*, the number of nonzero difference scores, and the desired level of significance, given either a nondirectional or a directional test. In the present example, in which *n* equals 7, the critical *T* equals 2 for a nondirectional test at the .05 level of significance.

HYPOTHESIS TEST SUMMARY: WILCOXON *T* TEST FOR TWO MATCHED SAMPLES (ESTIMATE OF TV VIEWING)

Research Problem:

If high school students are matched for home environment, will depicting TV viewing as (1) a favorable or (2) an unfavorable activity influence their estimates of weekly TV-viewing time?

Statistical Hypotheses:

H_0: population distribution 1 = population distribution 2
H_1: population distribution 1 $\neq$ population distribution 2

Decision Rule:

Reject H_0 at the .05 level if *T* equals or is *less* positive than 2 (from Table F, Appendix D, given $n = 7$).

Calculation:

$T = 1.5$ (See Table 25.3 on page 450 for computations.)

Decision:

Reject H_0 at .05 level of significance because $T = 1.5$ is *less* positive than 2.

Interpretation:

If high school students are matched for home environment, depicting TV viewing as a favorable or unfavorable activity will influence their estimates of TV-viewing time. Estimates tend to be larger when TV viewing is depicted favorably rather than unfavorably.

....

25.13 DECISION RULE

As with U, *the null hypothesis will be rejected only if the observed T is **less** than or equal to the critical T.* Otherwise, if the observed T exceeds the critical T, the null hypothesis will be retained. The properties of T are similar to those of U. The greater the discrepancy is in ranks between positive and negative difference scores, the smaller the value of T will be. In effect, T represents the sum of the ranks for the lower ranking set of difference scores. For example, when the lower ranking set of difference scores fails to appear in the rankings, because all difference scores have the same sign, the value of T equals zero, and the null hypothesis is suspect. In the present study, as the calculated T of 1.5 is less positive than the critical T of 2, the null hypothesis is rejected.

Published Report

Judging from the estimates for groups 1 and 2 in Table 25.3, both population distributions could have similar variabilities and shapes. Therefore, the rejection of the null hypothesis permits a more precise interpretation in terms of central tendencies that, in a published report, might read as follows:

> **Given that students have been matched for home environment, estimates are larger, on the average, when TV viewing is depicted favorably ($T = 1.5$, $p < .05$).**

*Exercise 25.3 Does a quit-smoking workshop cause a decline in cigarette smoking? The daily consumption of cigarettes is estimated for a random sample of nine smokers during each month before (1) and after (2) their attendance at a quit-smoking workshop, consisting of several hours of films, lectures, and testimonials. The results are as follows:

DAILY CIGARETTE CONSUMPTION		
SMOKER	BEFORE (1)	AFTER (2)
A	22	14
B	15	7
C	10	0
D	21	22
E	14	10
F	3	3
G	11	10
H	8	7
I	15	12

(a) Why might the Wilcoxon T test be preferred to the customary t test for these data?

(b) Use *T* to test the null hypothesis at the .05 level of significance.

(c) Specify the approximate *p*-value for this test result.
Answers on pages 528–529

KRUSKAL-WALLIS *H* TEST (THREE OR MORE INDEPENDENT SAMPLES)

Now let's consider a test for ranked data when there are more than two independent groups. Some parents are concerned about the amount of violence in children's TV cartoons. During five consecutive Saturday mornings, 10-minute cartoon sequences were randomly selected and videotaped from the offerings of each of three major TV networks, coded as A, B, and C. A child psychologist, who cannot identify the network source of each cartoon, ranked the 15 videotapes from least violent (1) to most violent (15). On the basis of these ranks, as shown in **Table 25.4,** can it be concluded that the underlying populations of cartoons for the three networks rank differently in terms of violence?

25.14 WHY NOT AN *F* TEST?

An inspection of the numerical ranks in Table 25.4 might suggest an *F* test for three independent samples within the context of a one-way ANOVA. However, when original observations are numerical ranks, as in the present example, there is no basis for speculating about whether the underlying populations are normally distributed with equal variances, as assumed in ANOVA. It is advisable to use a test, such as the Kruskal-Wallis *H* test, which retains its accuracy, even though these assumptions might be violated.

25.15 STATISTICAL HYPOTHESES FOR *H*

For the TV-cartoon study, the statistical hypotheses take the form

H_0: population A = population B = population C
H_1: H_0 is false

where A, B, and C represent the three major TV networks.

This null hypothesis equates three entire population distributions. Unless the population distributions can be assumed to have roughly similar variabilities and shapes, the rejection of H_0 will signify only that two or more of the populations differ in some unspecified manner, because of differences in central

Table 25.4
CALCULATION OF *H*

A. COMPUTATIONAL SEQUENCE

Find the sum of ranks for each group **1**.

Identify the sizes of group 1, n_1, group 2, n_2, group 3, n_3, and the combined groups, n **2**.

Substitute numbers in formula **3** and verify that ranks have been added correctly.

Substitute numbers in formula **4** and solve for *H*.

B. DATA AND COMPUTATIONS

Ranks		
(1)	**(2)**	**(3)**
A	**B**	**C**
8	4.5	10
4.5	14	15
2	12	6
13	7	1
10	3	10

1 $R_1 = 37.5$ $R_2 = 40.5$ $R_3 = 42$

2 $n_1 = 5$ $n_2 = 5$ $n_3 = 5$ $n = 5 + 5 + 5 = 15$

3   Computational check:

$$R_1 + R_2 + R_3 = \frac{n(n + 1)}{2}$$

$$37.5 + 40.5 + 42 = \frac{15(15 + 1)}{2}$$

$$120 = 120$$

4 $H = \dfrac{12}{n(n + 1)} \left[\dfrac{R_1^2}{n_1} + \dfrac{R_2^2}{n_2} + \dfrac{R_3^2}{n_3} \right] - 3(n + 1)$

$$= \frac{12}{15(15 + 1)} \left[\frac{(37.5)^2}{5} + \frac{(40.5)^2}{5} + \frac{(42)^2}{5} \right] - 3(15 + 1)$$

$$= \frac{12}{240} \left[\frac{4810.5}{5} \right] - 48$$

$$= .05[962.1] - 48$$

$$= 48.11 - 48 = 0.11$$

tendency, variability, shape, or some combination of these factors. When the original observations consist of numerical ranks, as in the present example, there's no obvious basis for speculating that the population distributions have similar shapes. Therefore, if H_0 is rejected, it will be impossible to pinpoint the precise differences among populations.

25.16 CALCULATION OF H

Table 25.4 shows how to calculate H. (If the original data had been quantitative rather than ranked, then the first step would have been to assign numerical ranks—beginning with 1 for the smallest, and so forth—for the *combined* three groups. In other words, the same ranking procedure is followed for H as for U in Section 25.5.) When ties occur between ranks, assign a mean rank. In Table 25.4, two cartoons are assigned ranks of 4.5, the mean of ranks 4 and 5.

Find the sums of ranks for groups 1, 2, and 3, that is, R_1, R_2, and R_3. (Notice that when the sample sizes are equal, the larger the differences are between these three sums, the more the three groups differ from each other, and the more suspect is the null hypothesis. Otherwise, to gain a preliminary impression when the sample sizes are unequal, compare the mean ranks of the various groups.) Use the computational check in Table 25.4 to verify that the ranks have been added correctly. Finally, the value of H can be determined from the following formula:

Kruskal-Wallis H test

A test for ranked data when there are more than two independent groups.

> ### KRUSKAL-WALLIS H TEST (THREE OR MORE INDEPENDENT SAMPLES)
>
> $$H = \frac{12}{n(n+1)} \left[\sum \frac{R_i^2}{n_i} \right] - 3(n+1) \qquad (25.3)$$

where n equals the combined sample size of all groups; R_i represents the sum of ranks of the ith group; and n_i represents the sample size of the ith group. Each sum of ranks, R_i, is squared and divided by its sample size. The value of H equals 0.11 for the present study.

25.17 χ^2 TABLES

When the sample sizes are very small, the critical values of H must be obtained from special tables. When each sample size consists of at least four observations, as is ordinarily the case, relatively accurate critical values can be obtained from the χ^2 distribution (Table D, Appendix D). As usual, the value of the critical χ^2 appears in the cell intersected by the desired level of significance and the number of degrees of freedom. The number of degrees of freedom, *df*, can be determined from

DEGREES OF FREEDOM *H* TEST

$$df = \text{number of groups} - 1 \qquad (25.4)$$

25.18 DECISION RULE

In contrast with the decision rules for *U* and *T*, *the null hypothesis will be rejected only if the observed H is equal to or more positive than the critical χ^2.* The larger the differences are in ranks among groups, the larger the value of *H* and the more suspect the null hypothesis will be. In the present study, since the observed *H* of 0.11 is less positive than the critical χ^2 of 5.99, the null hypothesis is retained.

HYPOTHESIS TEST SUMMARY: KRUSKAL-WALLIS *H* TEST FOR THREE INDEPENDENT SAMPLES (VIOLENCE IN TV CARTOONS)

Research Problem:
Does violence in cartoon programming, as judged by a child psychologist, differ for three major TV networks, coded as A, B, and C?

Statistical Hypotheses:
H_0: population A = population B = population C
H_1: H_0 is false.

Decision Rule:
Reject H_0 at the .05 level of significance if *H* equals or is more positive than 5.99 (from Table D, Appendix D, given *df* = 2).

Calculations:
H = 0.11 (See Table 25.4 on page 455 for calculations.)

Decision:
Retain H_0 at .05 level of significance because *H* = 0.11 is less positive than 5.99.

Interpretation:
There is no evidence that violence in cartoon programming differs for TV networks A, B, and C.

25.19 *H* TEST IS NONDIRECTIONAL

Because the sum of ranks for the ith group, R_i, is squared in Formula 25.3, the H test—like the F and χ^2 tests—is nondirectional.

*Exercise 25.4 A consumers' group wishes to determine whether motion picture ratings are, in any sense, associated with the number of violent or sexually explicit scenes in films. Five films are randomly selected from among each of the five ratings (NC-17—Seventeen and Over; R—Restricted; PG-13—Parental Guidance, Thirteen and Over; PG—Parental Guidance; and G—General), and a trained observer counts the number of violent or sexually explicit incidents in each film to obtain the following results:

NUMBER OF VIOLENT OR SEXUALLY EXPLICIT INCIDENTS				
NC-17	**R**	**PG-13**	**PG**	**G**
15	8	12	7	6
20	16	7	11	0
19	14	13	6	4
17	10	8	4	0
23	7	9	9	2

(a) Why might the *H* test be preferred to the *F* test for these data?

(b) Use *H* to test the null hypothesis at the .05 level of significance.

(c) Specify the approximate *p*-value for this test result.

Answers on page 529

25.20 GENERAL COMMENT: TIES

In addition to the customary assumption about random sampling, all tests in this chapter assume that no two observations are exactly the same. In other words, there shouldn't be any ties in ranks. Refer to more advanced statistics books for a corrected version of the test if (1) the observed value of U, T, or H is in the vicinity of its critical value but *not* statistically significant and (2) there are more than a few ties.[2]

Summary

This chapter described three different tests of the null hypothesis, using ranked data for two independent samples (Mann-Whitney U test), two matched samples (Wilcoxon T test), and three or more independent samples (Kruskal-Wallis H test). Being relatively free of assumptions, these tests can replace the t and F tests whenever populations cannot be assumed to be normally distributed with equal variances.

Once observations have been expressed as ranks, each test prescribes its own special measure of the difference in ranks between groups, as well as tables of critical values for evaluating significance.

---

[2]W. Conover, *Practical Nonparametric Statistics* (New York: Wiley, 1999).

Strictly speaking, U, T, and H test the null hypothesis that entire population distributions are equal. The rejection of H_0 signifies merely that populations differ in some unspecified manner. If populations are assumed to have roughly similar variabilities and shapes, then the rejection of H_0 will signify that the populations differ in their central tendencies.

Given a concern only about population differences in a particular direction, the U and T tests can be directional—if it can be assumed that populations have roughly similar variabilities and shapes.

Although the U, T, and H tests assume that there are no ties in ranks, the occurrence of ties can be ignored except in those cases where a test just fails to reach statistical significance and more than a few ties occur.

Important Terms

Nonparametric tests

Distribution-free tests

Mann-Whitney *U* test

Wilcoxon *T* test

Kruskal-Wallis *H* test

REVIEW EXERCISES

25.5. A group of "high risk" automobile drivers (three moving violations in one year) are required, according to random assignment, either to attend a traffic school or to perform supervised volunteer work. During the subsequent five-year period, these same drivers were cited as follows:

NUMBER OF MOVING VIOLATIONS	
TRAFFIC SCHOOL	VOLUNTEER WORK
0	26
0	7
15	4
9	1
7	1
0	14
2	6
23	10
7	
8	

(a) Why might the Mann-Whitney *U* test be preferred to the *t* test for these data?

(b) Use *U* to test the null hypothesis at the .05 level of significance.

(c) Specify the approximate *p*-value for this test result.

25.6 A social psychologist wishes to test the assertion that our attitude toward other people tends to reflect our perception of their attitude toward us. A randomly selected member of each of twelve couples who live together is told (in private) that his or her partner has rated that person at the high end of a 0 to 100 scale of trustworthiness, and the other member is told (also in private) that his or her partner has rated that person at the low end of the trustworthiness scale. Each person is then asked to estimate, in turn, the trustworthiness of his or her partner, yielding the following results. (According to the original assertion, the people in the trustworthy condition should give higher ratings than should their partners in the untrustworthy condition.)

	TRUSTWORTHINESS RATINGS	
COUPLE	TRUSTWORTHY (1)	UNTRUSTWORTHY (2)
A	75	60
B	35	30
C	50	55
D	93	20
E	74	12
F	47	34
G	95	22
H	63	63
I	44	43
J	88	79
K	56	33
L	86	72

(a) Use T to test the null hypothesis at the .01 level.

(b) Specify the approximate p-value for this test result.

25.7 Does background music influence the scores of college students on a reading comprehension test? Sets of ten randomly selected students take a reading comprehension test with rock, country-and-western, or classical music in the background. The results are as follows (higher scores reflect better comprehension):

READING COMPREHENSION SCORES		
ROCK (1)	COUNTRY-WESTERN (2)	CLASSICAL (3)
90	99	52
11	94	75
82	95	91
67	23	94
98	72	97
93	81	31
73	79	83
90	28	85
87	94	100
84	77	69

(a) Why might the H test be preferred to the F test for these data?

(b) Use H to test the null hypothesis at the .05 level of significance.

25.8 Use U rather than t to test the results in Exercise 19.5 on page 322.

25.9 Use T rather than t to test the effects of meditation described in Exercise 20.6 on page 342.

25.10 Use H rather than F to test the weight change data recorded in Exercise 22.13 on page 389.

25.11 Noting that the calculations for the H test tend to be much easier than those for the F test, one person always uses the H test. Any objection?

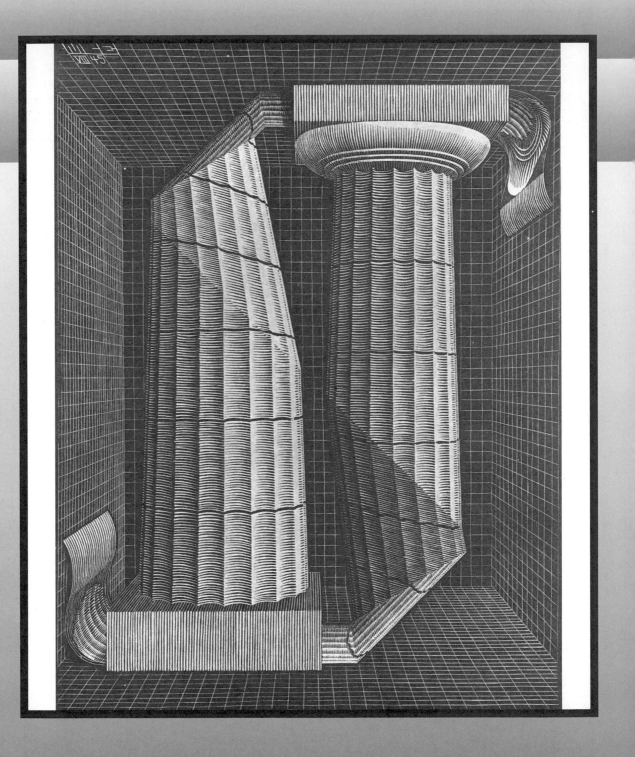

CHAPTER 26

Postscript: Which Test?

26.1   DESCRIPTIVE OR INFERENTIAL STATISTICS?
26.2   HYPOTHESIS TESTS OR CONFIDENCE INTERVALS?
26.3   QUANTITATIVE OR QUALITATIVE DATA?
26.4   DISTINGUISHING BETWEEN THE TWO TYPES OF DATA
26.5   ONE, TWO, OR MORE GROUPS?
26.6   CONCLUDING COMMENTS

Review Exercises

Although by no means exhaustive, the statistical tests in this book represent those most frequently encountered in more straightforward investigations, including many reported in the research literature. If you, yourself, initiate a test, there's a good chance that it also will be selected from among these tests. It's worthwhile, therefore, to review briefly the main themes of this book, particularly from the standpoint of selecting the appropriate statistical test for a given problem.

26.1 DESCRIPTIVE OR INFERENTIAL STATISTICS?

Descriptive Statistics

Is your intent descriptive because you wish merely to *summarize* existing data, or is your intent inferential because you wish to *generalize* beyond existing data? For instance, a data-oriented marriage counselor, who works with clients in groups, might suspect that some of the marital stress of current clients is attributable to the ages, in years, of their marital relationships. Accordingly, during an orientation session for the group, he describes the mean age—or if there are outliers, the median age—of their marital relationships. Wishing merely to summarize this information for current clients, the counselor's intent is descriptive, and it's appropriate to use any tools—tables, graphs, means, standard deviations, correlations—that enhance communication.

Inferential Statistics

On the other hand, assuming that the current group is representative of a much broader spectrum of clients, the counselor might use the same mean to estimate, possibly with a confidence interval, the mean age of marital relationships among *all* couples who seek professional help. Wishing to generalize beyond current clients, the counselor's intent is inferential, and it's appropriate to use confidence intervals and hypothesis tests as aids to generalizations.

26.2 HYPOTHESIS TESTS OR CONFIDENCE INTERVALS?

Traditionally, in the behavioral sciences hypothesis tests have been preferred to confidence intervals, and that preference probably would be expressed by the counselor if he chooses to conduct an investigation rather than a survey. For example, suspecting that the early years of marriage are both more stressful and more likely to produce clients, the counselor might use a *t* test to determine whether the mean age of marital relationships for a randomly selected group of clients is significantly less than that for a randomly selected group of nonclients.

The present review, as summarized in **Figure 26.1,** also reflects this preference for hypothesis tests, even though, if the null hypothesis is rejected, you should consider estimating the possible size of an effect or true difference with a confidence interval (or with some version of the squared correlation coefficient).

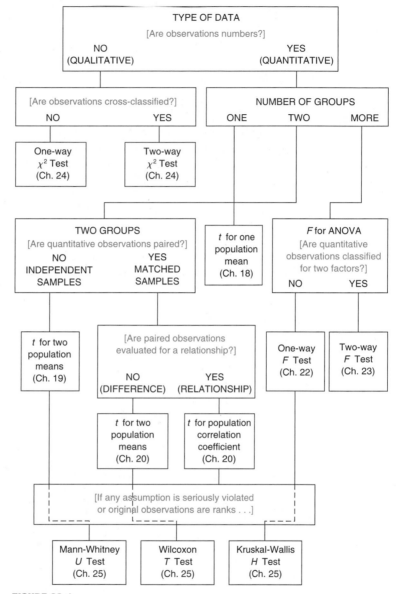

FIGURE 26.1
Guidelines for selecting the appropriate hypothesis test.

26.3 QUANTITATIVE OR QUALITATIVE DATA?

When attempting to identify the appropriate hypothesis test for a given situation, first decide whether observations are quantitative or qualitative. In other words, first decide whether the observations are quantitative because they are numbers that reflect an amount or a count, or qualitative because they are words (or codes) that reflect classes or categories.

Quantitative Data

Being numbers that reflect a count (of years), ages of marital relationships are quantitative observations. *When observations are quantitative, the appropriate statistical test should be selected from the various* t *or* F *tests or from their nonparametric counterparts*, as described below.

Qualitative Data

To illustrate the other possibility, when observations are qualitative, assume that the counselor wishes to test the prevalent notion that females are more likely than males to seek professional help. Now, clients are merely designated as either female or male, and because these observations reflect classes, they are qualitative. *When observations are qualitative, the appropriate test is either a one- or two-way χ^2 test*, as suggested in Figure 26.1.

One- or Two-way χ^2 Test?

If qualitative observations are categorized in terms of only one variable, as in the present case, the one-way χ^2 test is appropriate. If, however, qualitative observations are cross-classified in terms of two variables, the two-way χ^2 test is appropriate. The latter test is appropriate if, for example, clients are cross-classified in terms of both gender (female or male) and the marital status of their parents (married or separated) in order to test for any relationship between clients' gender and the marital status of their parents.

26.4 DISTINGUISHING BETWEEN THE TWO TYPES OF DATA

The distinction between quantitative and qualitative observations is crucial, and it's usually fairly easy, as suggested earlier. *Always first make the distinction between quantitative and qualitative data.* In those cases where you feel uncomfortable about this distinction, consider the following guidelines:

Focusing on a Single Observation

When you have access to the original observations, focus on any *single* observation. If an observation represents an amount or count, expressed numerically, it's quantitative; if it represents a class or category, described by a word or a code, it's qualitative.

Focusing on Numerical Summaries

When you don't have access to the original observations, focus on any numerical summaries specified for the data. If means and standard deviations are specified, the data are quantitative; if only frequencies are specified, the data are qualitative.

Focusing on Key Words

When, as in the case of the exercises at the end of this chapter, you have neither access to the original observations nor numerical summaries of data, read the word problem very carefully, attending to key words, such as "scores" or "means," which, if present, typify quantitative data or, if absent, typify qualitative data.

If All Else Fails

If all else fails, try *visualizing* the value of a single observation, whether a meaningful number (quantitative) or a word or numerical code (qualitative), on the basis of any information given. With just a little practice, you'll find that a careful reading, combined with an occasional speculation, usually reveals whether data are quantitative or qualitative.

26.5 ONE, TWO, OR MORE GROUPS?

Given that observations such as the ages of marital relationships are quantitative, either t or F tests are appropriate, assuming that no assumption is seriously violated. Now, the number of groups or samples is the key issue. More specifically, you must determine whether one, two, or more groups are involved—almost always a very straightforward task.

One Group

If only one group is involved—because, for example, the counselor wishes to determine whether, among the population of clients, the mean age of their marital relationships differs from a specific number, such as seven years (possibly to evaluate the popularly acclaimed "seven-year itch" as a source of marital stress)—then a t test for a single population mean is appropriate.

Two Groups

If, as suggested previously, the counselor wishes to determine whether the mean age of marital relationships for clients is significantly less than that for a group of nonclients, two groups are involved and a t test is appropriate. In the absence of any pairing, the two samples are independent, and the appropriate t test is for two population means (with independent samples).

Mean Difference or Correlation?

If observations are paired, the appropriate t test depends on the intent of the investigator—whether there is a concern about a mean difference or a correlation. If the paired observations are evaluated for a significant mean difference, the appropriate t test is for two population means (with matched samples). This would be the case if, for example, each client is paired or matched with a particular nonclient, possibly on the basis of age and income, and then a t test is based on the mean difference in marital ages between clients and nonclients.

If, on the other hand, the paired observations are being evaluated for a significant correlation, the appropriate t test is for a population correlation coefficient. This would be the case if, for example, the correlation between courtship ages and marital ages of clients' relationships is evaluated for significance, possibly to determine whether short courtships are associated with early marital difficulties.

More Than Two Groups

If the counselor wishes to determine whether significant differences exist among the mean ages of marital relationships for three randomly selected groups of clients with different ethnic backgrounds—African American, Asian American, and Hispanic—three population means are involved, and the F test for ANOVA is appropriate.

One- or Two-way *F* Test?

If quantitative observations are grouped according to the levels of only one factor, as in the present case, the F test for one-way ANOVA is appropriate. If, however, quantitative observations are grouped according to the levels of two factors—for instance, according to both ethnic background and gender—then the F test for two-way ANOVA is appropriate.

26.6 CONCLUDING COMMENTS

Nonparametric Tests

Figure 26.1 also includes the various nonparametric counterparts for selected t and F tests. Because these nonparametric tests are less likely to detect any effect, they are to be used only in those rare instances when some assumption is seriously violated (or when the original observations are ranked).

Use Figure 26.1

This chapter concludes with a series of exercises that require you to identify the appropriate statistical test from among those discussed in this book. Figure 26.1 should serve as a helpful guide when you're doing these exercises. For ease of reference, Figure 26.1 also appears inside the back book cover.

REVIEW EXERCISES

Note: In the following exercises, unless mentioned otherwise, no assumption has been seriously violated, and, therefore, the appropriate test should be selected from among t, F, or χ^2 tests. In your answer, specify the precise form of the test. For example, specify that the t test is for two population means with matched samples, or that the χ^2 test is a two-way test. Finally, remember always to decide first whether data are quantitative or qualitative.

*26.1 A political scientist wishes to determine whether males and females differ with respect to their attitudes toward defense spending by the federal government. She randomly selects equal numbers of males and females and asks each person if he or she thought the current level of defense spending should be increased, remain the same, or be decreased.

*26.2 Another political scientist also wishes to determine whether males and females differ with respect to their attitudes toward defense spending by the federal government. He randomly selects equal numbers of males and females. After being informed about the current budget for defense spending, each person is asked to estimate, to the nearest billion dollars, an appropriate level of defense spending.

*26.3 To determine whether speed reading influences reading comprehension, a researcher obtains two reading comprehension scores for each student in a group of high school students, once before and once after training in speed reading.

Answers on pages 529–530

26.4 Another investigator criticizes the design of the previous study, saying that high school students should have been randomly assigned to either the special training condition or a control condition and tested just once at the end of the study. Subsequently, she conducts this study.

*26.5 An educator wishes to determine whether chance can reasonably account for the fact that 40 of the top 100 students come from the northern district (rather than the eastern, southern, or western districts) of a large metropolitan school district.

*26.6 To determine whether a new sleeping pill has an effect that varies with dosage level, a researcher randomly assigns adult insomniacs, in equal numbers, to receive either 0, 4, 8, or 12 grams of the sleeping pill. The amount of sleeping time is measured for each insomniac during an 8-hour period after the administration of the dosage.

Answers on page 530

26.7 An investigator wishes to test whether creative artists are equally likely to be born under any of the 12 astrological signs.

26.8 To determine whether there is a relationship between the sexual codes of primitive tribes and their behavior toward neighboring tribes, an anthropologist consults available records, classifying each tribe on the basis of their sexual codes (permissive or repressive) and their behavior toward neighboring tribes (friendly or hostile).

*26.9 In a study of group problem solving, a researcher randomly assigns college students either to unstructured groups of 2, 3, or 4 students (without a leader) or to structured groups of 2, 3, or 4 students (with an arbitrarily designated leader), and measures the amount of time required to solve a complex puzzle.

*26.10 A school psychologist compares the reading comprehension scores of migrant children who, as a result of random assignment, are enrolled in either a special bilingual reading program or a traditional reading program.

*26.11 Another school psychologist wishes to determine whether reading comprehension scores are associated with the number of months of formal education, as reported on school transcripts, for a group of 12-year-old migrant children.

Answers on page 530

26.12 A century ago, the British surgeon Lister investigated the relationship between the operating room environment (presence or absence of disinfectant) and the fate of about 100 emergency amputees (survived or failed to survive).

26.13 A comparative psychologist suspects that specific chemicals in the urine of male rats trigger an increase in the activity of other rats. To check this hunch, she randomly assigns rats, in equal numbers, to either a sterile cage, a cage sprayed with a trace of the specific chemicals, or a cage sprayed thoroughly with the specific chemicals. Furthermore, to check out the possibility that reactions might be sex linked, equal numbers of female and male rats are assigned to the three cage conditions. An activity score is recorded for each rat during a 5-minute observation period in the specified cage.

26.14 A psychologist wishes to evaluate the effectiveness of a new desensitization counseling program on the subsequent performance of college students in a public speaking class. After being matched on the basis of the quality of their initial speeches, students are randomly assigned either to undergo desensitization training or to serve in a control group. Evaluation is based on scores awarded to students for their speeches at the end of the class.

26.15 An investigator wishes to determine whether, for a random sample of drug addicts, the mean score on the depression scale of a personality test differs from that which, according to the test documentation, represents the mean score for the general population.

26.16 Another investigator wishes to determine whether, for a random sample of drug addicts, the mean score on the depression scale of a personality test differs from the corresponding mean score for a random sample of nonaddicted people.

26.17 To determine whether cramming can increase Graduate Record Exam (GRE) scores, a researcher randomly assigns college students to either a specialized GRE test-taking workshop, a general test-taking workshop, or a control (non-test-taking) workshop. Furthermore, to check the effect of scheduling, students are randomly assigned, in equal numbers, to experience their workshop either during one long marathon weekend or during weekly sessions.

26.18 A criminologist suspects that there is a relationship between the degree of structure provided for paroled ex-convicts (supervised or unsupervised half-way house) and whether there is a violation of parole during the first six months of freedom.

26.19 A psychologist uses chimpanzees to test the notion that more crowded living conditions cause aggressive behavior. Chimps are randomly assigned to live in cages containing either one, several, or many other chimps. Subsequently, during periods spent in an observation cage, each chimp is assigned a score on the basis of its aggressive behavior toward a chimplike stuffed doll.

26.20 Assuming that, in the previous study, an inspection of the aggression scores reveals a serious violation of the normality assumption, specify an alternate test.

26.21 In an extrasensory perception (ESP) experiment, each of 30 subjects attempts to predict the one correct pattern from among five possible patterns during each of one hundred trials, and the mean number of correct predictions for all 30 subjects is compared with 20, the number of correct predictions per 100 trials on the assumption that subjects lack extrasensory perception.

26.22 A social scientist wishes to determine whether there is a relationship between the attractiveness scores (on a 100-point scale) assigned to college students by a panel of peers and their scores on a paper-and-pencil test of anxiety.

APPENDIX

A

Math Review

A.1   PRETEST

A.2   COMMON SYMBOLS

A.3   ORDER OF OPERATIONS

A.4   POSITIVE AND NEGATIVE NUMBERS

A.5   FRACTIONS

A.6   SQUARE ROOT RADICALS ($\sqrt{}$)

A.7   ROUNDING NUMBERS

A.8   POSTTEST

A.9   ANSWERS (WITH RELEVANT REVIEW SECTIONS)

This appendix summarizes many of the basic math symbols and operations used in this book. Little, if any, of this material will be entirely new, but—possibly because of years of little or no use—much may seem only slightly familiar. In any event, it's important that you master this material.

First, take the pretest in Section A.1, comparing your answers with those in Section A.9. Whenever errors occur, study the review section indicated for that set of answers. Then, after browsing through all review sections, take the posttest in Section A.8, again checking your answers with those in Section A.9. If you're still making lots of errors, repeat the entire process spending even more time studying the various review sections. If errors persist, consult your instructor for additional guidance.

A.1 PRETEST

Questions 1–6 Are the following statements true or false?

1. $(5)(4) = 20$ **2.** $4 > 6$ **3.** $7 \le 10$ **4.** $|-5| = 5$

5. $(8)^2 = 56$ **6.** $\sqrt{9} = 3$

Questions 7–30 Find the answers.

7. $\dfrac{5 - 3}{2 - 1} =$ **8.** $\sqrt{5 + 4 + 7} =$ **9.** $3(4 + 3) =$

10. $16 - \dfrac{10}{\sqrt{25}} =$ **11.** $(3)^2(10) - 4 =$ **12.** $[3^2 + 2^2]^2 =$

13. $\sqrt{\dfrac{2(3) - 2^2}{5 - 3}} =$ **14.** $\sqrt{\dfrac{(8 - 6)^2 + (5 - 3)^2}{2}} =$

15. $2 + 4 + (-1) =$ **16.** $5 - (3) =$ **17.** $2 + 7 + (-8) + (-3) =$

18. $5 - (-1) =$ **19.** $(-4)(-3) =$ **20.** $(-5)(6) =$

21. $\dfrac{-10}{2} =$ **22.** $\dfrac{4}{5} - \dfrac{1}{5} =$ **23.** $\dfrac{1}{4} + \dfrac{2}{5} =$

24. $\dfrac{2^2}{4} + \dfrac{3^2}{3} - \dfrac{2^2}{8} =$ **25.** $\left(\dfrac{2}{3}\right)\left(\dfrac{6}{7}\right) =$ **26.** $\sqrt{16 + 9} =$

27. $\sqrt{(4)(9)} =$ **28.** $\sqrt{4}\,\sqrt{9} =$ **29.** $\dfrac{\sqrt{25}}{\sqrt{100}} =$ **30.** $\sqrt{\dfrac{25}{100}} =$

Questions 31–35 Round to the nearest hundredth.

31. 98.769   **32.** 3.274   **33.** 23.765   **34.** 5476.375003

35. 54.1499

A.2 COMMON SYMBOLS

SYMBOL	MEANING	EXAMPLE
=	equals	$4 = 4$
$\neq$	doesn't equal	$4 \neq 2$
+	plus (addition)	$2 + 3 = 5$
−	minus (subtraction)	$3 - 2 = 1$
$\pm$	plus and minus	$4 \pm 2 = 4 + 2$ and $4 - 2$
( )( )	times (multiplication)*	$(3)(2) = 3(2) = 6$
$/, \dfrac{(\ )}{(\ )}$	divided by (division)	$6/2 = 3, \dfrac{(8)}{(2)} = 4$
>	is greater than	$4 > 3$
<	is less than	$5 < 8$
$\geq$	equals or is greater than	$z \geq 2$
$\leq$	equals or is less than	$t \leq 4$
$\sqrt{}$	the square root of†	$\sqrt{9} = 3$
$(\ )^2$	the square of	$(4)^2 = (4)(4) = 16$
| |	the absolute (positive) value of	$\|4\| = 4, \|-4\| = 4$
…	continuing the pattern	$1, 2, 3, \ldots, 8$ translates as: $1, 2, 3, 4, 5, 6, 7, 8$

When multiplication involves symbols, parentheses can be dropped. For instance,
$$(X)(Y) = X(Y) = XY$$
†*The square root of a number is that number which, when multiplied by itself, yields the original number.*

A.3 ORDER OF OPERATIONS

Expressions should be treated as single numbers when they appear in parentheses, square root signs, or in the top (or bottom) of fractions.

EXAMPLES

$$2(4 - 1) = 2(3) = 6$$

$$\sqrt{12 - 8} = \sqrt{4} = 2$$

$$\frac{8 - 4}{2 + 2} = \frac{4}{4} = 1$$

If all expressions contain single numbers, the order for performing operations is as follows:

1. square or square root
2. multiplication or division
3. addition or subtraction

EXAMPLES

$$10 + \frac{6}{\sqrt{4}} = 10 + \frac{6}{2} = 10 + 3 = 13$$

$$(3)(2)^2 - 1 = (3)(4) - 1 = 12 - 1 = 11$$

When expressions are nested, one within the other, work outward from the inside.

EXAMPLES

$$\sqrt{\frac{(6-3)^2 + (5-2)^2}{2}} = \sqrt{\frac{(3)^2 + (3)^2}{2}}$$

$$= \sqrt{\frac{9+9}{2}} = \sqrt{\frac{18}{2}} = \sqrt{9} = 3$$

$$\sqrt{\frac{3(4) - (2)^2}{4-2}} = \sqrt{\frac{12-4}{4}} = \sqrt{\frac{8}{2}} = \sqrt{4} = 2$$

A.4 POSITIVE AND NEGATIVE NUMBERS

In the absence of any sign, a number is understood to be positive.

EXAMPLE

$8 = +8$

To *add* numbers with unlike signs,

1. find two separate sums, one for all positive numbers and the other for all negative numbers
2. find the difference between these two sums
3. attach the sign of the larger sum

EXAMPLE

$2 + 3 + (-4) + (-3) = 5 + (-7) = -2$

To *subtract* one number from another,

1. change the sign of the number to be subtracted
2. proceed as in addition

EXAMPLES

$4 - (3) = 4 + (-3) = 1$

$4 - (-3) = 4 + 3 = 7$

To *multiply* (or *divide*) two signed numbers,

1. obtain the numerical result
2. attach a positive sign if the two original numbers have like signs or a negative sign if the two original numbers have unlike signs

EXAMPLES

$(-4)(-2) = 8; (4)(-2) = -8$

$\dfrac{4}{2} = 2; \dfrac{-4}{2} = -2$

A.5 FRACTIONS

A fraction consists of an upper part, the numerator, and a lower part, the denominator.

To *add* (or *subtract*) fractions, their denominators must be the same.

1. If denominators are the same, merely add (or subtract) numbers in the numerators and leave the number in the denominator unchanged.

EXAMPLES

$\dfrac{3}{5} + \dfrac{1}{5} = \dfrac{3 + 1}{5} = \dfrac{4}{5}$

$\dfrac{7}{10} - \dfrac{3}{10} = \dfrac{7 + (-3)}{10} = \dfrac{4}{10}$

2. If denominators are different, first find a common denominator. To obtain a common denominator, multiply both parts of each fraction by the denominators of all remaining fractions. Then proceed as in part (a):

EXAMPLES

$$\frac{2}{3} + \frac{1}{4} = \frac{(4)2}{(4)3} + \frac{(3)1}{(3)4} = \frac{8}{12} + \frac{3}{12} = \frac{11}{12}$$

$$\frac{4}{6} + \frac{2}{5} = \frac{(5)4}{(5)6} + \frac{(6)2}{(6)5} = \frac{20}{30} + \frac{12}{30} = \frac{32}{30}$$

To *add* (or *subtract*) fractions, sometimes it's more efficient to follow a different procedure. First, express each fraction as a decimal number—by dividing denominator into numerator—and then merely add (or subtract) the resulting decimal numbers.

EXAMPLES

$$\frac{3}{4} - \frac{1}{4} = .75 - .25 = .50$$

$$\frac{3}{10} + \frac{2}{6} + \frac{1}{5} = .30 + .33 + .20 = .83$$

To multiply fractions, multiply all numerators to obtain the new numerator, and multiply all denominators to obtain the new denominator.

EXAMPLES

$$\left(\frac{2}{3}\right)\left(\frac{3}{5}\right) = \frac{6}{15}$$

$$\left(\frac{2}{4}\right)\left(\frac{3}{4}\right) = \frac{6}{16}$$

A.6 SQUARE ROOT RADICALS ($\sqrt{}$)

The square root of a sum *doesn't* equal the sum of the square roots.

> **EXAMPLES**
>
> $$\sqrt{16 + 9} \neq \sqrt{16} + \sqrt{9}$$
> $$5 \neq 4 + 3$$

The square root of a product equals the product of the square roots.

> **EXAMPLE**
>
> $$\sqrt{(4)(9)} = (\sqrt{4})(\sqrt{9}) = (2)(3) = 6$$

The square root of a fraction equals the square root of the numerator divided by the square root of the denominator.

> **EXAMPLE**
>
> $$\sqrt{\frac{4}{16}} = \frac{\sqrt{4}}{\sqrt{16}} = \frac{2}{4}$$

A.7 ROUNDING NUMBERS

When the first term of the number to be dropped is 5 or more, increase the remaining number by one unit. Otherwise, leave the remaining number unchanged. In this book, for purposes of standardization, numbers usually are rounded to the nearest hundredth.

> **EXAMPLES**
> When rounding to the nearest hundredth.
> 21.86$\underline{6}$ rounds to 21.87
> 37.36$\underline{4}$ rounds to 37.36
> 102.64$\underline{5}$332 rounds to 102.65
> 87.98$\underline{4}$97 rounds to 87.98
> 52.10$\underline{5}$000 rounds to 52.11

A.8 POSTTEST

Questions 101–112 Find the answers.

101. $\sqrt{36} =$ **102.** $|24| =$ **103.** $(7)^2 =$ **104.** $5 \pm 3 =$

105. $3\sqrt{8 - (2)^2} =$ **106.** $\dfrac{1^2 + 4^2 + 5^2}{4^2 - 3^2} =$ **107.** $18 - (-3) =$

108. $(-10)(-8) =$ **109.** $\dfrac{3}{5} + \dfrac{2}{8} =$ **110.** $\dfrac{(2-3)^2}{2} + \dfrac{(6-4)^2}{3} =$

111. $\sqrt{9 + 9 + 9 + 9} =$ **112.** $\sqrt{25}\,\sqrt{4} =$

Questions 113–114 Round to the nearest tenth.

113. 107.45   **114.** 3.2499

···

A.9 ANSWERS (WITH RELEVANT REVIEW SECTIONS)

Pretest

1. True
2. False
3. True
4. True ⎫ Review Section A.2
5. False
6. True

7. 2
8. $\sqrt{16} = 4$
9. 21
10. 14
11. 86 ⎬ Review Section A.3
12. $(13)^2 = 169$
13. $\sqrt{1} = 1$
14. $\sqrt{4} = 2$

15. 5
16. 2
17. -2
18. 6 ⎬ Review Section A.4
19. 12
20. -30
21. -5

22. $\dfrac{3}{5}$ or .60

23. $\dfrac{13}{20}$ or .65

24. $\dfrac{84}{24}$ or 3.5 ⎬ Review Section A.5

25. $\dfrac{12}{21}$

26. 5

27. 6

28. 6   Review Section A.6

29. $\frac{1}{2}$

30. $\frac{1}{2}$

31. 98.77

32. 3.27

33. 23.77   Review Section A.7

34. 5476.38

35. 54.15

Posttest

101. 6

102. 24

103. 49   Review Section A.2

104. 8 and 2

105. 6

106. 6   Review Section A.3

107. 21

108. 80   Review Section A.4

109. $\frac{34}{40}$ or .85

           Review Section A.5

110. $\frac{11}{6}$ or 1.83

111. 6

112. 10   Review Section A.6

113. 107.5

114. 3.2   Review Section A.7

APPENDIX B

Levels of Measurement

B.1   **FOUR LEVELS OF MEASUREMENT**

B.2   **NOMINAL MEASUREMENT**

B.3   **ORDINAL MEASUREMENT**

B.4   **INTERVAL MEASUREMENT**

B.5   **RATIO MEASUREMENT**

B.6   **THE PROBLEM OF MEASUREMENT**

B.7   **NUMERICAL MEASUREMENT OF NONPHYSICAL CHARACTERISTICS**

B.8   *APPROXIMATING* **INTERVAL MEASUREMENT**

        SUMMARY

        REVIEW EXERCISES

It's tempting to think of observations, whether quantitative or qualitative, as bits of information that appear spontaneously, much like aliens from another galaxy. In fact, when making observations, we actively process these bits of information, using different sets of sorting or measuring rules known as levels of measurement. You can perform a statistical analysis without being familiar with levels of measurement—just as you can drive a car without knowing what's under the hood. Prudent drivers, however, do learn to deal with obvious malfunctions, such as a broken fan belt. Likewise, students of statistics study levels of measurement to aid their interpretation of different types of data.

B.1 FOUR LEVELS OF MEASUREMENT

There are four levels of measurement: nominal, ordinal, interval, and ratio. As shown in **Table B.1**, the four levels range in complexity from nominal, which possesses only one property, to ratio, which possesses four properties. Progressively more complex levels contain all properties of lesser levels, plus one new property (shown in Table B.1 with heavy type). For example, ratio measure-

		Table B.1 LEVELS OF MEASUREMENT		
LEVEL	**PROPERTIES**	**OBSERVATIONS REFLECT . . .**	**EXAMPLES**	**TYPE OF DATA**
Ratio	**true zero** equal intervals order classification	measurable differences in *total* amount	weight income reaction time family size	quantitative
Interval	**equal intervals** order classification	measurable differences in amount	Fahrenheit temperature IQ score* grade point average* verbal aptitude score*	quantitative
Ordinal	**order** classification	differences in degree	graded attitude toward abortion developmental stages academic letter grade movie ratings	qualitative
Nominal	**classification**	differences in kind	sex gender ethnic background political affiliation major in college	qualitative

Approximates interval measurement.

ment possesses all three properties of interval measurement, plus one new property, a true zero.

Shifts to more complex levels of measurement are accompanied by sets of observations that, because they contain more information, permit a wider variety of interpretations.

At the simplest level, interpretations (and statistical tools) are restricted because observations only reflect differences in kind. At more complex levels, interpretations (and statistical tools) are more varied because observations reflect measurable differences in amount. Let's look, in turn, at each level of measurement.

B.2 NOMINAL MEASUREMENT

If people are classified as either male or female, measurement is nominal. *The single property of **nominal** measurement is classification*—that is, the sorting of observations into different classes or categories. Reflecting only differences in kind, not differences in degree or amount, a nominal scale represents the most primitive form of measurement. Indeed, if you associate measurement only with stopwatches, yardsticks, and speedometers, it takes an extra effort to think of mere classification as measurement.

Behavioral and social scientists often deal with nominal data. For instance, mentally retarded children might be classified as educable or noneducable; psychotherapists as psychoanalysts, behaviorists, or humanists; communication networks as open or closed. The statistical analysis of nominal data requires the use of special tools—those appropriate for an analysis of qualitative data, as described elsewhere in this book.

B.3 ORDINAL MEASUREMENT

If replies to the question "Should all abortions be legal?" are classified as either *strongly agree, agree, neutral, disagree, or strongly disagree*, measurement is ordinal. *The distinctive property of **ordinal** measurement is order*. In the previous example, different replies reflect not merely differences in kind, as with nominal measurement, but also differences in degree—that is, differences based on more or less. *Strongly agree* represents more agreement than *agree*; by the same token, *agree* represents more agreement than *neutral,* and so forth. Notice that no claim can be made about the amount of difference between adjacent categories. Although *strongly agree* represents more agreement than *agree,* we don't know how much more.

Other examples of ordinal data include the designation of skilled workers as master craftsmen, journeymen, or apprentices; socioeconomic status as low, middle, or high.

In general, the same type of statistical analysis is performed for both ordinal and nominal data. The distinction between the two types of measurement is worth maintaining, nonetheless, because sometimes the analysis of ordinal data permits the use of a few extra statistical tools — those appropriate for the analysis of ordered qualitative data (or even those appropriate for the analysis of ranked data), as described elsewhere in this book.

B.4 INTERVAL MEASUREMENT

On a recent day, Anchorage, Boston, and St. Louis reported high temperatures of 40°F, 60°F, and 80°F, respectively. When Fahrenheit temperatures are reported, as stated here, measurement is interval. *The distinctive property of **interval** measurement is equal intervals.* This means that, anywhere along the Fahrenheit scale, equal intervals or equal differences in temperature readings always signify equal increases in the amount of heat or molecular motion (as revealed by the expansion of mercury in a glass tube). The Fahrenheit scale possesses equal intervals; it's appropriate to interpret the difference between 40°F and 60°F as representing exactly the same amount of heat as does the difference between 60°F and 80°F. In other words, interval measurement allows us to make precise claims about measurable differences in amount between observations.

The Fahrenheit temperature scale serves as one of the few clear-cut examples of interval measurement. Most measurement scales — such as body weight, age, speed — that clearly attain interval measurement (because of equal intervals) also qualify as ratio measurement and, therefore, are referred to as ratio data.

B.5 RATIO MEASUREMENT

Ratio measurement is the most advanced level of measurement. Often the product of familiar measuring devices — rulers, clocks, meters — it permits you to interpret one observation as exceeding another not only by a certain amount, as in interval measurement, but also by a *certain ratio*, such as "twice as much" or "four times more."

If people are weighed on a bathroom scale, measurement is ratio. *The distinctive property of **ratio** measurement is a true zero.* In the case of the bathroom scale, the achievement of a true zero is ridiculously easy. Simply verify that the scale registers 0 when not in use, that is, when weight is completely absent. Because the bathroom scale possesses a true zero, numerical readings reflect the *total amount* of a person's weight, and it's appropriate to describe one person's weight as a certain ratio of another's. It can be claimed that the weight of a 300-pound person is twice that of a 150-pound person.

In the absence of a true zero, numbers — much like the exposed tips of icebergs — fail to reflect the total amount being measured. For example, a reading of 0 on the Fahrenheit temperature scale does not reflect the complete absence of heat. In fact, true zero equals −459.4°F on this scale. It would be inappropriate, therefore, to claim that 80°F is twice as hot as 40°F. An appropriate claim could be salvaged by adding 459.4°F to each of these numbers: 80° becomes 539.4° and 40° becomes 499.4°. Clearly, 539.4° is not twice as hot as 499.4°.

Ratio data appear in the behavioral and social sciences as, for example, bar-press rates of rats in Skinner boxes and the amount of eye contact between pairs of human subjects. The analysis of both ratio and interval data employs a wide range of statistical tools—those appropriate for the analysis of quantitative data, as described elsewhere in this book.

B.6 THE PROBLEM OF MEASUREMENT

When physical characteristics, such as weight (or family size or annual income), are measured numerically, the attainment of ratio measurement is deceptively simple. To fully appreciate the achievement of a measurement scale with a true zero and equal intervals, remember that numbers are being used to represent the amount or magnitude of some characteristic. There's no automatic guarantee that a zero reading on a measurement scale really represents the complete absence of the characteristic, nor that the differences between consecutive numbers on a measurement scale really represent equal amounts of the characteristic.

B.7 NUMERICAL MEASUREMENT OF NONPHYSICAL CHARACTERISTICS

When nonphysical characteristics, such as intellectual aptitude, psychopathic tendency, or academic achievement, are measured numerically, the attainment of ratio or interval measurement is much more difficult. Let's look more closely at this problem, using the measurement of intellectual aptitude with IQ scores as the example. In the case of IQ scores, there is no instrument (such as the unoccupied bathroom scale) that registers 0 when intellectual aptitude is completely absent (true zero). There also is no external standard (such as the expansion of mercury in the Fahrenheit thermometer) to demonstrate that the addition of a fixed amount of intellectual aptitude always produces an equal increase in IQ scores (equal intervals).

In the absence of a true zero, it would be inappropriate to claim that an IQ score of 140 represents twice the amount of intellectual aptitude as a score of 70. By the same token, in the absence of equal intervals, it would be inappropriate to claim that the difference between scores of 120 and 140 represents the same amount of intellectual aptitude as the difference between scores of 100 and 120.

B.8 *APPROXIMATING* INTERVAL MEASUREMENT

Other interpretations are possible. One possibility is to treat IQ scores as attaining only ordinal measurement—that is, a score of 140 represents more intellectual aptitude than a score of 120 and so forth. A more common interpretation, adopted in this book, assumes that, although lacking a true zero, IQ scores provide a crude measure of corresponding differences in intellectual aptitude. Thus, the difference between scores of 120 and 140 represents a *roughly similar* amount of intellectual aptitude as the difference between scores of 100 and 120.

Insofar as numerical measures of nonphysical characteristics—intellectual aptitude, psychopathic tendency, academic achievement—tend to approximate interval measurement, the resulting data receive the same statistical treatment as do regular interval and ratio data. In other words, the analysis of these data supports the wide spectrum of statistical tools described elsewhere as appropriate for the analysis of quantitative data.

At this point, you might wish that a person could be injected with ten points of intellectual aptitude (or psychopathic tendency or academic achievement) as a first step toward an IQ scale with equal intervals and a true zero. Lacking this alternative, however, train yourself to look at numbers as products of measurement and to temper your numerical claims accordingly—particularly when, as often happens in the behavioral and social sciences, numerical data only seem to approximate interval measurement.

Summary

When making observations, we use four different sets of sorting or measuring rules known as levels of measurement. Distinctive properties of the four levels of measurement are classification (nominal), order (ordinal), equal intervals (interval), and true zero (ratio).

Shifts to more complex levels of measurement are accompanied by more informative observations that, in turn, permit a wider variety of interpretations and statistical analyses. Observations reflect differences in kind when measurement is nominal, differences in degree when measurement is ordinal, measurable differences in amount when measurement is interval, and, finally, measurable differences in total amount when measurement is ratio.

Even though the numerical measurement of various nonphysical characteristics fails to attain an interval or ratio level, the resulting data often are treated as approximating interval measurement. The limitations of these data shouldn't, however, be ignored completely when making numerical claims.

REVIEW EXERCISES

B.1   Indicate the level of measurement attained by the following sets of observations or data. Whenever appropriate, indicate that interval measurement is only approximated.

       Note: *Always assign the highest permissible level of measurement to a given set of observations.* For example, a list of annual incomes should be designated as ratio data because $0 signifies the complete absence of income. It would be wrong to describe annual income as interval data even though a $1,000 difference always signifies the same amount of income (equal intervals), or as ordinal data even though different incomes always can be ranked as more or less (order), or as nominal data even though different incomes always reflect different classes (classification).

   *(a) height

   *(b) religious affiliation

   *(c) math aptitude score

*(d) years of education

*(e) military rank

Answers on page 530

(f) favorite TV program

(g) place of birth

(h) diastolic blood pressure

(i) vocational goal

(j) grade point average

(k) daily intake of calories

(l) marital status

(m) highest academic degree

(n) blood type

(o) attitude toward total nuclear disarmament (favor, neutral, oppose)

(p) degree of test anxiety (high, medium, low)

(q) days absent from work

(r) academic letter grade

(s) taxable income

(t) astrological sign

(u) score for psychopathic tendency

(v) hair color

(w) reaction time

(x) mechanical aptitude score

(y) nationality of mother

(z) degree of satisfaction with life (100-point rating scale)

Answers To Selected Exercises

ANSWERS TO SELECTED EXERCISES

Chapter 1

1.1 **(a)** descriptive statistics   **(c)** descriptive statistics
   (b) inferential statistics   **(d)** inferential statistics

1.2 **(a)** qualitative   **(f)** quantitative
   (b) quantitative   **(g)** quantitative
   (c) quantitative   **(h)** qualitative
   (d) qualitative   **(i)** qualitative
   (e) qualitative   **(j)** quantitative

1.3 **(a)** correlation study
   (b) experiment; prescribed hours of sleep deprivation
   (c) correlation study
   (d) correlation study (An experiment requires that subjects be *assigned*, rather than choose, to participate in a given program.)
   (e) correlation study
   (f) experiment; different rehabilitation programs
   (g) experiment; on-campus or off-campus housing

Chapter 2

2.1

RATING	TALLY*	FREQUENCY
10	/	1
9	//	2
8	///	3
7	/N/	5
6	//	2
5	//	2
4	/	1
3	/N/ /	6
2	//	2
1	/	1
	Total	25

*Tally column usually is omitted from the finished table.

2.2 Calculating the class width,

$$\frac{123 - 69}{10} = \frac{54}{10} = 5.4$$

Round off to a convenient number, such as 5.

IQ	TALLY*	FREQUENCY
120–124	/	1
115–119		0
110–114	//	2
105–109	///	3
100–104	////	4
95–99	##/	6
90–94	## //	7
85–89	////	4
80–84	///	3
75–79	///	3
70–74	/	1
65–69	/	1
	Total	35

Tally column usually is omitted from the finished table.

2.3 Not all observations can be assigned to one and only one class (because of gap between 20–22 and 25–30 and overlap between 25–30 and 30–34). All classes aren't equal in width (25–30 versus 30–34). All classes don't have both boundaries (35–above).

2.4 Outliers are a summer income of $25,700; an age of 61; and a family size of 18. No outliers for GPA.

2.5

	RELATIVE FREQUENCY	
GRE	PROPORTION	PERCENT (%)
725–749	.01	1
700–724	.02	2
675–699	.07	7
650–674	.15	15
625–649	.17	17
600–624	.21	21
575–599	.15	15
550–574	.14	14
525–549	.07*	7
500–524	.02	2
475–499	.01	1
Totals	1.02	102%

From 13/200 = .065, which rounds to .07.

2.6

	(a)	(b) CUMULATIVE RELATIVE FREQUENCY	
GRE	CUMULATIVE FREQUENCY	PROPORTION	PERCENT (%)
725–749	200	1.00	100
700–724	199	1.00	100
675–699	196	.98	98
650–674	182	.91	91
625–649	152	.76	76
600–624	118	.59	59
575–599	76	.38	38
550–574	46	.23	23
525–549	19	.10	10
500–524	6	.03	3
475–499	2	.01	1

2.7   **(a)** The exact percentile rank for 1 romantic affair is 45.
        (b) The approximate percentile rank for weights between 200 and 209 pounds is 92 (because 92 is the cumulative percent for this interval).

2.8

MOVIE RATINGS	(a) FREQUENCY	(b) RELATIVE FREQUENCY (%)	(c) CUMULATIVE FREQUENCY
NC-17	2	10	20
R	4	20	18
PG-13	3	15	14
PG	8	40	11
G	3	15	3
Totals	20	100%	

(d) Percentile rank of 55 (from $\frac{11}{20}$ multiplied by 100).

2.9

BLOOD TYPE	(a) FREQUENCY	(b) RELATIVE FREQUENCY	
		PROPORTION	PERCENT (%)
O	14	.47	47
A	13	.43	43
B	2	.07	7
AB	1	.03	3
Totals	30	1.00	100%

(c) Not appropriate, because blood types can't be ordered from least to most.

2.14 **(a)**

| | SMALL TOWN | U.S. POPULATION |
AGE	RELATIVE FREQUENCY (%)	RELATIVE FREQUENCY (%)
65–above	21	13
60–64	11	4
55–59	9	4
50–54	8	6
45–49	9	7
40–44	8	8
35–39	6	9
30–34	5	8
25–29	5	7
20–24	4	7
15–19	4	7
10–14	4	7
5–9	3	7
0–4	3	7
Totals	100%	101%

(b) Among small town residents, there are relatively more older people and relatively fewer younger people.

2.16 **(a)** Hispanics (by 14.7 million)
(b) Asian Americans (by 119%)
(c) Whites increased by 8% while the general population increased by 18%
(d) Asian Americans and Hispanics are growing most rapidly. (Or some variation on this conclusion, such as nonwhites are growing more rapidly than whites.)

Chapter 3

3.1

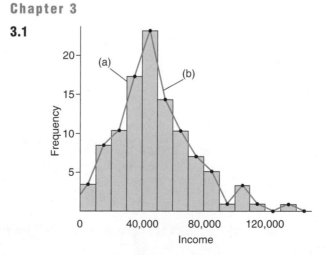

Note: Ordinarily, either (a), a histogram, or (b), a frequency polygon, only would be shown. When closing the left flank of (b), imagine extending a line to the midpoint of the first unoccupied class ($-10,000$ to -1) on the left, but stop the line at the vertical axis, as shown.

(c) Lopsided.

3.2

7	8				
8	5	8			
9	8	9	6		
10	8	2	9	6	4
11	8	7	1	3	
12	0	6	3	4	
13	2	7			
14	1	3			

Note: The *order* of the leaves within each stem depends on whether you entered IQ scores column by column (as above) or row by row.

3.3   **(a)** Positively skewed   **(d)** Bimodal
       (b) Normal   **(e)** Negatively skewed
       (c) Positively skewed

3.4

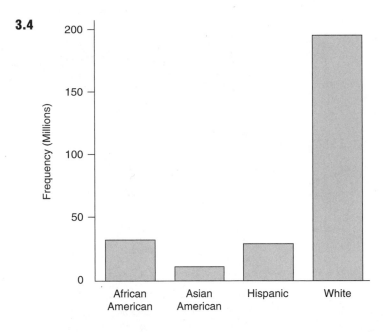

3.5   **(a)** Widths of two rightmost bars aren't the same as those of two left-most bars.
       (b) Histogram is more appropriate for quantitative data.
       (c) Height of vertical axis is too small relative to width of horizontal axis, causing the histogram to be squashed.

(d) Poorly selected frequency scale along the vertical axis, causing the histogram to be squashed.

(e) Bars have unequal widths. No wiggly lines along vertical axis indicating break between 0 and 50.

(f) Height of vertical axis is too large relative to the horizontal axis, causing differences between bars to be exaggerated.

3.6 **(a)** 5

(b) Negatively skewed

Chapter 4

4.1 mode = 63

4.2 mode = 21.4

4.3 median = 63

4.4 median = 21.15 (halfway between 20.9 and 21.4)

4.5 mean = 672/11 = 61.09

4.6 mean = 127.3/6 = 21.22

4.7 **(a)** median score exceeds mean score.

(b) mean age exceeds median age.

(c) mean amount exceeds median amount.

(d) median crowd size exceeds mean crowd size

4.8 mode = Fort Lauderdale (FL)

Impossible to find median because these qualitative data can't be ordered.

4.9 mode = 0, 2, 3, and 4 (multimodal)

median = 3

mean = 3.33

4.12 Two different averages are being used to describe the central tendency in a skewed distribution of pilots' salaries. Management is probably using the mean (because of its concern about total expenditures), while the pilots' association is probably using the median (because of its concern about the actual salaries of typical, middle-ranked pilots).

Chapter 5

5.1 **(a)** $35,000 to $45,000

(b) $30,000

(c) $50,000

(d) $38,000 to $42,000; $36,000; $44,000

5.2 $S = \sqrt{1.5} = 1.22$

5.3 **(a)** $S = \sqrt{\dfrac{8(137) - 729}{(8)^2}} = \sqrt{5.73} = 2.39$

(b) $S = \sqrt{\dfrac{9(325) - 1849}{(9)^2}} = \sqrt{13.28} = 3.64$

5.4 **(a)** range = 25; IQR = 65 − 60 = 5
　　　(b) range = 11; IQR = 4 − 1 = 3

5.6 **(a)** a_1 larger than a_2. Graduating high school seniors with very low SAT scores tend not to become college freshmen.
　　　(b) b_1 larger than b_2
　　　(c) c_2 larger than c_1
　　　(d) about the same
　　　(e) e_1 larger than e_2
　　　(f) f_2 larger than f_1

5.10 **(a)** A $70 per month raise would increase the original mean (by $70) but not change the original standard deviation. Raising everyone's pay by a constant amount has no effect on variability.
　　　(b) A 5% per month increase would increase both the original mean and standard deviation (by 5%). Raising everyone's pay by 5% generates a larger raise in actual dollars for higher paid employees and thus increases variability among monthly wages.

Chapter 6

Note: When answers are obtained directly from the standard normal table, the complete tabular entry (for instance, .0571) is listed along with the customary answer, rounded two digits to the right of the decimal point.

6.1 **(a)** 2.33　　**(d)** 0.00
　　　(b) −0.30　　**(e)** −1.50
　　　(c) −1.60

6.2 **(a)** .0359 = .04　　**(e)** .2743 = .27
　　　(b) .1664 = .17　　**(f)** .0040 = .00
　　　(c) .0013 = .00　　**(g)** .4750 = .48
　　　(d) .4505 = .45

6.3 **(a_1)** 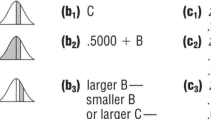　　**(b_1)** C′　　**(c_1)** $z = -1.00$
　　　　　　　　　　　　　　　　　　answer = .1587 = .16
　　　(a_2)　　　　**(b_2)** C　　**(c_2)** $z = 1.50$
　　　　　　　　　　　　　　　　　　answer = .0668 = .07

Chapter 7

7.1
　　　(a_1)　　**(b_1)** C　　　　　　**(c_1)** $z = 0.70$
　　　　　　　　　　　　　　　　　　　　　　.2420 = .24
　　　(a_2)　　**(b_2)** .5000 + B　　**(c_2)** $z = 0.15$
　　　　　　　　　　　　　　　　　　　　　　.5000 + .0596 =
　　　　　　　　　　　　　　　　　　　　　　.5596 = .56
　　　(a_3)　　**(b_3)** larger B—　　**(c_3)** $z = 0.20; z = 0.40$
　　　　　　　　　　smaller B　　　　　　　.1554 − .0793 =
　　　　　　　　　　or larger C—　　　　　.0761 = .08
　　　　　　　　　　smaller C　　　　　　　or .4207 − .3446
　　　　　　　　　　　　　　　　　　　　　　= .0761 = .08

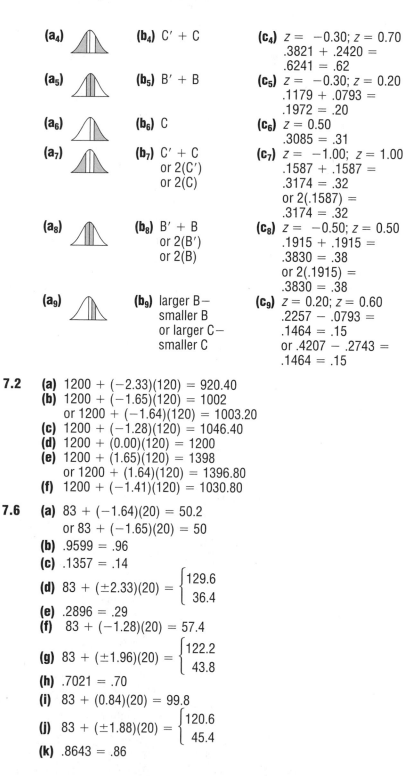

(**a₄**)

(**b₄**) C' + C

(**c₄**) $z = -0.30; z = 0.70$
$.3821 + .2420 =$
$.6241 = .62$

(**a₅**)

(**b₅**) B' + B

(**c₅**) $z = -0.30; z = 0.20$
$.1179 + .0793 =$
$.1972 = .20$

(**a₆**)

(**b₆**) C

(**c₆**) $z = 0.50$
$.3085 = .31$

(**a₇**)

(**b₇**) C' + C
or 2(C')
or 2(C)

(**c₇**) $z = -1.00; z = 1.00$
$.1587 + .1587 =$
$.3174 = .32$
or 2(.1587) =
$.3174 = .32$

(**a₈**)

(**b₈**) B' + B
or 2(B')
or 2(B)

(**c₈**) $z = -0.50; z = 0.50$
$.1915 + .1915 =$
$.3830 = .38$
or 2(.1915) =
$.3830 = .38$

(**a₉**)

(**b₉**) larger B−
smaller B
or larger C−
smaller C

(**c₉**) $z = 0.20; z = 0.60$
$.2257 - .0793 =$
$.1464 = .15$
or $.4207 - .2743 =$
$.1464 = .15$

7.2 **(a)** $1200 + (-2.33)(120) = 920.40$
(b) $1200 + (-1.65)(120) = 1002$
or $1200 + (-1.64)(120) = 1003.20$
(c) $1200 + (-1.28)(120) = 1046.40$
(d) $1200 + (0.00)(120) = 1200$
(e) $1200 + (1.65)(120) = 1398$
or $1200 + (1.64)(120) = 1396.80$
(f) $1200 + (-1.41)(120) = 1030.80$

7.6 **(a)** $83 + (-1.64)(20) = 50.2$
or $83 + (-1.65)(20) = 50$
(b) $.9599 = .96$
(c) $.1357 = .14$
(d) $83 + (\pm2.33)(20) = \begin{cases} 129.6 \\ 36.4 \end{cases}$
(e) $.2896 = .29$
(f) $83 + (-1.28)(20) = 57.4$
(g) $83 + (\pm1.96)(20) = \begin{cases} 122.2 \\ 43.8 \end{cases}$
(h) $.7021 = .70$
(i) $83 + (0.84)(20) = 99.8$
(j) $83 + (\pm1.88)(20) = \begin{cases} 120.6 \\ 45.4 \end{cases}$
(k) $.8643 = .86$

Chapter 8

8.1 (a) 0.33   (c) 0.75
      (b) −0.20   (d) 0.40

8.2 (a) A test score of 45 from distribution c because it converts to the largest z score (0.75).
      (b) Distribution b, because it yields a larger z score (2.40) than any other distribution.

8.3

	$\mu = 0$; $\sigma = 1$	$\mu = 50$; $\sigma = 10$	$\mu = 100$; $\sigma = 15$	$\mu = 500$; $\sigma = 100$
(a)	0.80	58	112	580
(b)	−1.67	33.3	74.95	333
(c)	1.80	68	127	680
(d)	−1.85	31.5	72.25	315

8.4 (a) 91.92   (c) 6.68
      (b) $100 + (1.48)(15) = 122.2$ (d) $500 + (.25)(100) = 525$

8.7 (a) $z = (.406 − .278)/.035 = 3.66$
         Williams's equivalent average for 1990 = $.260 + (3.66)(.027) = .359$
      (b) Brett's equivalent average for 1941 = $.278 + (3.75)(.035) = .409$. Not only does Brett's equivalent average for 1941 exceed .400, but it exceeds Williams's .406. Thus, it can be concluded that Brett's 1980 average of .390 (or a z of 3.75) is better, *relatively speaking,* than Williams's 1941 average of .406 (or a z of 3.66).
      (c) Hornsby's equivalent average for 1990 = $.269 + (3.83)(.028) = .376$

Chapter 9

9.1 (a) Positive. Students with low IQs tend to have low GPAs, or, conversely, students with high IQs tend to have high GPAs.
      (b) Negative. High rates of unemployment have been associated with low rates of inflation (and periods of economic depression).
      (c) Positive. The crime rate is higher, square mile for square mile, in the heart of a city than in a suburb.
      (d) Negative. As TV viewing increases, performance on academic achievement tests tends to decline.
      (e) Positive. Increases in car weight are accompanied by increases in cost due to gas consumption.
      (f) Negative. Increases in car weight are accompanied by decreases in miles per gallon.
      (g) Positive. Increases in educational level—grade school, high school, college—tend to be associated with increases in income.
      (h) Positive. Highly anxious people willingly spend more time performing a simple repetitive task than do less anxious people.

9.2 (a) I, D, F   (d) E, H
      (b) B, H, E   (e) True
      (c) E   (f) False. The relationship is positive.

9.3 **(a)** Cars with more total miles tend to have lower resale values.
(b) Students with more absences from school tend to score lower on math achievement tests.
(c) Little or no relationship between height and IQ.
(d) Taxpayers with higher gross annual incomes tend to claim higher amounts of tax deductions.
(e) Little or no relationship between anxiety level and college GPA.
(f) Older schoolchildren tend to have better reading comprehension.
(g) Longer pregnancies tend to be associated with heavier birthweights. Notice that a statement such as "Longer pregnancies *produce* heavier birthweights" goes beyond the observed correlation and implies a cause-effect relationship.

9.4 **(a)** simple cause-effect   **(d)** simple cause-effect
(b) complex   **(e)** complex
(c) complex   **(f)** complex

9.5 **(a)** 2   **(b)** above   **(c)** $1\frac{1}{2}$ **(d)** below   **(e)** $\frac{1}{3}$ **(f)** above

9.6 **(a)** 2   **(b)** below   **(c)** $1\frac{1}{2}$ **(d)** above   **(e)** $\frac{1}{3}$ **(f)** below

9.7 $r = \dfrac{(6)(32) - (12)(14)}{\sqrt{[(6)(28) - 144][(6)(42) - 196]}} = .65$

9.8 **(a)** .77854
(b)

1.00000	−.91413	−.81459
−.91413	1.00000	.77854
−.81459	.77854	1.00000

(c)

1.00000	−.91413	−.81459
−.91413	1.00000	.77854
−.81459	.77854	1.00000

(d) Either triangle is OK.

1.00000	−.91413	−.81459
−.91413	1.00000	.77854
−.81459	.77854	1.00000

(e) The relationship is negative between X and Y and between X and Z, and it is positive between Y and Z.
(f) 9
(g) Both of the negative correlations, but not the positive correlation since its *p*-value of .0134 isn't less than .0100.

9.12 **(a)** False. This statement would be true only if a perfect negative relationship (−1.00) described the relationship between TV viewing time and test scores.
(b) False. Correlation doesn't necessarily signify cause-effect.
(c) True
(d) True
(e) False. See (b).

(f) False. Although correlation doesn't necessarily signify cause-effect, it opens the *possibility* of cause-effect.

9.15 **(a)** No. The new correlation would have the same numerical value but the opposite sign, that is, it would equal $-.2981$. The change from positive to negative reflects the original tendency for males, the group with the larger code of 1, to have lower high school GPAs.

(b) Yes. The new negative correlation still reflects the original tendency of females, now coded as 0, to have higher high school GPAs than males, now coded as 1.

(c) Yes. The actual numerical value of the correlation, .2981, reflects only the pattern of predictability across pairs of z scores which, in turn, show no traces of the arbitrary codes assigned to females and males. The positive value of r reflects only the relatively higher coding of females (20) than males (10).

Chapter 10

10.1 **(a)** approximately 14 percent
(b) approximately 7–8 percent

10.2 **(a)** $b = (4/2)(.30) = .60; a = 8 - (.60)(13) = 0.20$
$Y' = .60(X) + .20$
(b) $Y' = (.60)(15) + .20 = 9.20$
(c) $Y' = (.60)(11) + .20 = 6.80$

10.3 **(a)** $S_{Y|X} = (4)\sqrt{1 - (.30)^2}$
$= 3.80$ (or 3.82 if you carried calculations more than two places to the right of the decimal point).

(b) Roughly indicates the average amount by which the prediction is in error.

10.4 **(a)** That $(.30)(.30) = .09$ or 9 percent of the variance in weekly reading time is predictable from educational level.
(b) That $1.00 - .09 = .91$ or 91 percent of the variance in weekly reading time *isn't* predictable from educational level.
(c) The 9 percent refers to the variance in weekly reading time for *all* people involved in the original correlation study.

10.5 **(a)** No. The value of r can't be interpreted as a proportion of a perfect relationship.
(b) The r^2 for foster parents and foster children, that is, $(.27)(.27) = .0729$, is only between one-fourth and one-third as strong as the r^2 for parents and children, that is, $(.50)(.50) = .25$.

Chapter 11

11.1 **(a)** Yes
(b) Yes
(c) Yes
(d) Yes
(e) No. Citizens of Wyoming aren't a subset of citizens of New York.
(f) Yes

(g) Yes

(h) No. All U.S. presidents aren't a subset of all registered Republicans.

(i) Yes

11.2 **(g)** All lab rats, similar to those used, that could undergo the same experiment.

(i) All possible tosses of a coin.

11.3 **(a)** False. Sometimes, just by chance, a random sample of 10 cards fails to represent the important features of the whole deck. More about this problem in Chapter 16.

(b) True

(c) False. Although unlikely, 10 hearts could appear in a random sample of 10 cards.

(d) True

11.4 **(a)** There are many ways. For instance, on the assumption that the classroom contains a maximum of 9 rows of seats and that each row contains a maximum of 9 seats, you could proceed as follows: Consult the tables of random numbers, using the first digit of each five-digit random number to identify the row (previously labeled 1, 2, 3, and so on), and the second digit of the same random number to locate a particular student's seat within that row. Repeat this process until five students have been identified. (If the classroom is larger, use additional digits so that every student can be sampled.)

(b) Once again, there are many ways. For instance, use the initial four digits of each random number (between 0001 and 3041) to identify the page number of the telephone directory and the next three digits (between 001 and 480) to identify the particular line on that page. Repeat this process, using seven-digit numbers, until 40 telephone numbers have been identified.

11.5 **(a)** For instance, if the first digit is odd (1, 3, 5, 7, or 9), the first subject is assigned to group A, and if the first digit is even (0, 2, 4, 6, or 8), the first subject is assigned to group B. To ensure equal groups, the second subject is assigned automatically to the group opposite that for the first subject. Repeat this procedure for the remaining five pairs of subjects.

There are other acceptable rules, all involving pairs of subjects (to ensure equal group sizes). For instance, if the first digit equals 0, 1, 2, 3, or 4, the first subject is assigned to group A; otherwise, the first subject is assigned to group B, and so on.

(b) Answer shows two possible assignment rules. In practice only one assignment rule actually would be used.

SUBJECT #	RANDOM NUMBER (TOP ROW, TABLE G)	ASSIGNMENT RULE 1*	OR	ASSIGNMENT RULE 2**
1	1	A		A
2	—	automatically B		automatically B
3	0	B		A
4	—	automatically A		automatically B
5	0	B		A

SUBJECT #	RANDOM NUMBER (TOP ROW, TABLE G)	ASSIGNMENT RULE 1*	OR	ASSIGNMENT RULE 2**
6	—	automatically A		automatically B
7	9	A		B
8	—	automatically B		automatically A
9	7	A		B
10	—	automatically B		automatically A
11	3	A		A
12	—	automatically B		automatically B

odd digits = group A; even digits = group B
**digits 0, 1, 2, 3, 4 = group A; digits 5, 6, 7, 8, 9 = group B*

11.6 **(a)** For instance, focus on the first two digits of each random number. Assign a subject to group 1 if the first two digits are between 00 and 24; to group 2 if the first two digits are between 25 and 49; to group 3 if the first two digits are between 50 and 74; and to group 4 if the first two digits are between 75 and 99.

(b) Random assignment could involve sets of 4 subjects. The first subject can be *randomly* assigned to any of the four groups; the second subject can be *randomly* assigned to any of the three remaining groups; the third subject can be *randomly* assigned to either of the two remaining groups; and the fourth subject is *automatically* assigned to the one remaining group. Repeat this process for each set of four subjects.

Chapter 12

12.1 **(a)** $\frac{1}{12}$ **(b)** $\frac{11}{12}$ **(c)** $\frac{2}{12}$

12.2 **(a)** $(\frac{1}{12})(\frac{1}{12}) = \frac{1}{144}$

(b) $(\frac{1}{12})(\frac{1}{12}) = \frac{1}{144}$

(c) $(\frac{2}{12})(\frac{2}{12}) = \frac{4}{144}$

12.3 **(a)** True

(b) larger than

(c) larger than

(d) False. You would have to know the actual value of the conditional probability of a person being born in December, given that the person's spouse was, in fact, born in December.

12.4 **(a)** .0250

(b) .0250

(c) .0250 + .0250 = .0500 = .05

(d) .4750 + .4750 = .9500 = .95

(e) .0049 + .0049 = .0098 = .01

(f) .0005 + .0005 = .0010 = .001

12.5 **(a)** $(\frac{1}{2})(\frac{1}{2}) = \frac{1}{4}$

(b) $(\frac{1}{2})(\frac{1}{2}) = \frac{1}{4}$

(c) $(\frac{1}{4}) + (\frac{1}{4}) = \frac{2}{4}$

12.8 **(a)** .98
(b) (.98)(.98)(.98)(.98)(.98)(.98) = .89
(c) 1 − .89 = .11
(d) (.02)(.02) = .0004
(e) 1.00 (According to Hoadley, the *Challenger* catastrophe has led to several improvements, including the addition of a third set of truly independent O-rings.)

12.10 **(a)** False
(b) Being a major league ballplayer and being left-handed are *dependent* outcomes because .10, the unconditional probability of being left-handed in the general population doesn't equal .18, the conditional probability of being left-handed, given that a person is a major league baseball player.

Chapter 13

13.1 **(a)**

(1) 2,2	(6) 4,2	(11) 6,2	(16) 8,2	(21) 10,2
(2) 2,4	(7) 4,4	(12) 6,4	(17) 8,4	(22) 10,4
(3) 2,6	(8) 4,6	(13) 6,6	(18) 8,6	(23) 10,6
(4) 2,8	(9) 4,8	(14) 6,8	(19) 8,8	(24) 10,8
(5) 2,10	(10) 4,10	(15) 6,10	(20) 8,10	(25) 10,10

$\overline{X}$	PROBABILITY
10	1/25
9	2/25
8	3/25
7	4/25
6	5/25
5	4/25
4	3/25
3	2/25
2	1/25

13.2 **(a)** μ **(b)** $\mu_{\overline{X}}$ **(c)** $\overline{X}$ **(d)** $\sigma_{\overline{X}}$ **(e)** S **(f)** σ

13.3 **(a)** False. It always equals the value of the population mean.
(b) True
(c) False. Because of chance, most sample means tend to be either larger or smaller than the mean of all sample means.
(d) True
(e) True
(f) True

13.4 **(a)** True
(b) False. It measures variability among sample means.
(c) False. It decreases in value with larger sample sizes.
(d) False. It equals the population standard deviation divided by the square root of the sample size.

(e) True
(f) True

13.5 **(a)** False. The shape of the population remains the same regardless of sample size.
(b) False. It requires that the sample size be sufficiently large—usually between 25 and 100.
(c) True
(d) False. It requires merely that sample size be sufficiently large, regardless of the shape of the population (which usually isn't known).
(e) False. It ensures that the shape of the sampling distribution approximates a normal curve, regardless of the shape of the population (which remains intact).
(f) True

Chapter 14

14.1 **(a)** $z = \dfrac{566 - 560}{30/\sqrt{36}} = \dfrac{6}{5} = 1.20$

(b) $z = \dfrac{24 - 25}{4/\sqrt{64}} = \dfrac{-1}{.5} = -2.00$

(c) $z = \dfrac{82 - 75}{14/\sqrt{49}} = \dfrac{7}{2} = 3.50$

(d) $z = \dfrac{136 - 146}{15/\sqrt{25}} = \dfrac{-10}{3} = -3.33$

14.2 **(a)** Different numbers appear in H_0 and H_1.
(b) Sample means (rather than population means) appear in H_0 and H_1.

14.3 **(a)** Sixth-grade boys in his school district average 6.2 pushups.
$H_0: \mu = 6.2$
(b) On the average, weights of packages of ground beef sold by a large supermarket chain equal 16 ounces.
$H_0: \mu = 16$
(c) The marriage counselor's clients average 11 interruptions per session.
$H_0: \mu = 11$

14.4 **(a)** Retain H_0 at .05 level of significance because $z = 1.74$ is less positive than 1.96.
(b) Retain H_0 at .05 level of significance because $z = 0.13$ is less positive than 1.96.
(c) Reject H_0 at .05 level of significance because $z = -2.51$ is more negative than -1.96.

14.5 **(a)** The observed difference between $50,300 and $51,500 can't be interpreted at face value, as it could have happened just by chance. A hypothesis test permits us to evaluate the effect of chance by measuring the observed difference relative to the standard error of the mean.
(b) All female members of the APA with a Ph.D. degree and a full-time teaching appointment.

(c) H_0: $\mu = 51{,}500$
(d) H_1: $\mu \neq 51{,}500$
(e) Reject H_0 if z equals or is more positive than 1.96 or if z equals or is more negative than -1.96.
(f) $z = \dfrac{50{,}300 - 51{,}500}{3000/\sqrt{100}} = \dfrac{-1{,}200}{300} = -4.00$
(g) Reject H_0 at .05 level of significance because $z = -4.00$ is more negative than -1.96.
(h) The average salary of all female APA members (with a Ph.D. and a full-time teaching appointment) is less than $51{,}500.

14.6 *Research Problem:*
Does the mean IQ of all students in the district differ from 100?
Statistical Hypotheses:
H_0: $\mu = 100$
H_1: $\mu \neq 100$
Decision Rule:
Reject H_0 at the .05 level of significance if z equals or is more positive than 1.96 or if z equals or is more negative than -1.96.
Calculations:
Given that $\overline{X} = 105$; $\sigma_{\overline{X}} = \dfrac{15}{\sqrt{25}} = \dfrac{15}{5} = 3$

$z = \dfrac{105 - 100}{3} = \dfrac{5}{3} = 1.67$

Decision:
Retain H_0 at .05 level of significance because $z = 1.67$ is less positive than 1.96.
Interpretation:
No evidence that the mean IQ of all students differs from 100.

Chapter 15

15.1 **(a)** H_0: $\mu \leq 0.54$
H_1: $\mu > 0.54$
Justification: to increase rainfall
(b) H_0: $\mu \geq 23$
H_1: $\mu < 23$
Justification: weight-reduction program
(c) H_0: $\mu \leq 134$
H_1: $\mu > 134$
Justification: life-prolonging drug

15.2 **(a)** Reject H_0 at .01 level of significance because $z = -2.34$ is more negative than -2.33.
(b) Reject H_0 at .01 level of significance because $z = -5.13$ is more negative than -2.33.
(c) Retain H_0 at .01 level of significance because $z = 4.04$ is *less negative* than -2.33.*
(d) Reject H_0 at .05 level of significance because $z = 2.00$ is more positive than 1.65.

(e) Retain H_0 at .05 level of significance because $z = -1.80$ is *less positive* than 1.65.*

(f) Retain H_0 at .05 level of significance because $z = 1.61$ is less positive than 1.65.

*Value of observed z is in direction of no concern.

15.3 **(a)** Reject H_0 at the .05 level of significance if z equals or is more positive than 1.96 or if z equals or is more negative than -1.96.

(b) Reject H_0 at the .01 level of significance if z equals or is more positive than 2.33.

(c) Reject H_0 at the .05 level of significance if z equals or is more negative than -1.65.

(d) Reject H_0 at .01 level of significance if z equals or is more positive than 2.58 or if z equals or is more negative than -2.58.

Chapter 16

16.1 **(a)** Correct decision (True H_0 is retained)
Type I error
Correct decision (False H_0 is rejected)
Type II error

(b)

DECISION	STATUS OF H_0	
	TRUE H_0(INNOCENT)	**FALSE H_0(GUILTY)**
Retain H_0 (Release)	**Correct Decision:** Innocent defendant is released.	**Type II Error:** Guilty defendant is released (Miss).
Reject H_0 (Sentence)	**Type I Error:** Innocent defendant is sentenced (False Alarm).	**Correct Decision:** Guilty defendant is sentenced.

16.2 A false H_0 will never be rejected.

16.3 **(a)** True
(b) True
(c) False. The one observed sample mean originates from the true sampling distribution.
(d) False. If the one observed sample mean has a value of 103, an incorrect decision would be made because the false H_0 would be retained.

16.4 **(a)** True
(b) False. The critical value of z (1.65) is based on the hypothesized sampling distribution.
(c) False. Since the true sampling distribution goes below 100, a sample mean less than or equal to 100 is possible, although not highly likely.
(d) True

16.5 **(a)** Because of small sample size, only very large effects will be detected.
  (b) Because of large sample size, even small, unimportant effects will be detected.

16.6 **(a)** The detection rate for the 12-point effect is larger than .90 because the true sampling distribution is shifted further into the rejection region for the false H_0.
  (b) The detection rate for the 5-point effect is smaller than .90 because the true sampling distribution is shifted further into the retention region for the false H_0.

16.7 **(a)** $H_0: \mu \leq 0.54$, that is, cloud seeding has no effect on rainfall.

	STATUS OF H_0	
DECISION	**TRUE H_0**	**FALSE H_0**
Retain H_0	**Correct Decision:** Conclude that there is no evidence that cloud seeding increases rainfall when in fact it doesn't.	**Type II Error:** Conclude that there is no evidence that cloud seeding increases rainfall when in fact it does.
Reject H_0	**Type I Error:** Conclude that cloud seeding increases rainfall when in fact it doesn't.	**Correct Decision:** Conclude that cloud seeding increases rainfall when in fact it does.

Chapter 17

17.1 $62,600

17.2 **(a)** $3.82 \pm 1.96 \left(\dfrac{.4}{\sqrt{64}} \right) = \begin{cases} 3.92 \\ 3.72 \end{cases}$
  (b) We can claim, with 95 percent confidence, that the interval between 3.72 and 3.92 includes the true *population mean* reading score for the fourth graders. All of these values suggest that, on the average, the fourth graders are underachieving.

17.3 **(a)** False. The interval from 507 and 527 describes possible values of the population mean for students who receive special training.
  (b) False. The interval from 507 to 527 refers to the possible values of the population mean.
  (c) False. We can be 95 percent confident that the mean for all subjects will be between 507 and 527.
  (d) True
  (e) False. We can be reasonably confident—but not absolutely confident—that the true population mean lies between 507 and 527.

(f) False. This particular interval either describes the one true population mean or fails to describe the one true population mean.
(g) True
(h) True
(i) True

17.4 **(a)** Switch to an interval having a lesser degree of confidence, such as 90 percent or 75 percent.
(b) Increase the sample size.

17.5 **(a)** False. The interval from 76 to 84 percent refers to possible values of the population proportion.
(b) True
(c) False. We can be reasonably confident—but not absolutely confident—that the true population proportion is between 76 and 84 percent.
(d) True
(e) True
(f) True

17.6 **(a)** $50,300 \pm (2.57) \left(\dfrac{3,000}{\sqrt{100}} \right) = \begin{cases} 51,071 \\ 49,529 \end{cases}$

$or\ 50,300 \pm (2.58) \left(\dfrac{3,000}{\sqrt{100}} \right) = \begin{cases} 51,074 \\ 49,526 \end{cases}$

(b) We can claim, with 99 percent confidence, that the interval between $49,529 and $51,071 (or $49,526 and $51,074) includes the *true population mean* salary for all female members of the American Psychological Association. All of these values suggest that, on the average, females' salaries are less than males' salaries.

17.9 **(a)** 3   **(d)** 3
(b) 3   **(e)** 5
(c) 4   **(f)** 4

Chapter 18

18.1 **(a)** ± 2.179 **(c)** 1.697
(b) -2.539 **(d)** ± 2.704

18.2 *Research Problem:*
Does the mean weight for all packages of ground beef drop below the specified weight of 16 ounces?
Statistical Hypothesis:
$H_0: \mu \geq 16$
$H_1: \mu < 16$
Decision Rule:
Reject H_0 at the .05 level of significance if t equals or is more negative than -1.833, given $df = 10 - 1 = 9$.
Calculations:
$$t = \frac{14.7 - 16}{.26} = -5.00$$

Decision:
> Reject H_0.

Interpretation:
> The mean weight for all packages drops below the specified weight of 16 ounces.

18.3 **(a)** $14.7 \pm (2.26)(.26) = \begin{cases} 15.29 \\ 14.11 \end{cases}$

(b) We can be 95 percent confident that the interval between 14.11 and 15.29 ounces includes the true population mean weight for all packages.

18.4 **(a)** $\overline{X} = \dfrac{147}{10} = 14.7$

(b) $s = \sqrt{\dfrac{10(2167) - (147)^2}{10(10 - 1)}} = 0.82$

$s_{\overline{x}} = \dfrac{.82}{\sqrt{10}} = 0.26$

18.5 **(a)** 18 hours
(b) 23 hours
(c) $df = 1$ in (a) and $df = 3$ in (b)
(d) When all observations are expressed as deviations from their mean, the sum of all deviations must equal zero.

18.6 *Research Problem:*
> On the average, do library patrons borrow books for longer or shorter periods than the currently specified loan period of 21 days?

Statistical Hypotheses:
> H_0: $\mu = 21$
> H_1: $\mu \neq 21$

Decision Rule:
> Reject H_0 at the .05 level of significance if t equals or is more positive than 2.365 or t equals or is more negative than -2.365, given $df = 8 - 1 = 7$.

Calculations:

$$\overline{X} = \frac{142}{8} = 17.75 \quad s = \sqrt{\frac{8(2652) - (142)^2}{8(8 - 1)}} = 4.33$$

$$s_{\overline{x}} = \frac{4.33}{\sqrt{8}} = 1.53 \quad t = \frac{17.75 - 21}{1.53} = -2.12$$

Decision:
> Retain H_0 at .05 level of significance because $t = -2.12$ is less negative than -2.365.

Interpretation:
> No evidence that, on the average, library patrons borrow books for longer or shorter periods than 21 days.

18.8 **(a)** *Research Problem:*

Is the temperature of earth getting warmer?

Statistical Hypotheses:

H_0: $\mu \leq 61.7$
H_1: $\mu > 61.7$

Decision Rule:

Reject H_0 at .01 level of significance if t equals or is more positive than 3.143, given $df = 7 - 1 = 6$.

Calculations:

$$s_{\bar{x}} = \frac{.22}{\sqrt{7}} = .08$$

$$t = \frac{62.14 - 61.70}{.08} = 5.50 \quad \left(\begin{array}{c} \text{or using deviation scores} \\ t = \dfrac{.44 - 0}{.08} = 5.50 \end{array} \right)$$

Decision:

Reject H_0 at .01 level of significance because $t = 5.50$ is more positive than 3.143.

Interpretation:

Temperature of earth is getting warmer.

(b) Because H_0 was rejected, a confidence interval is appropriate.

$$62.14 \pm (3.71)(.08) = \begin{cases} 62.44 \\ 61.84 \end{cases}$$

We can be 99 percent confident that the interval between 61.84 and 62.44 (°F) includes the true temperature of Earth during seven recent years.

Chapter 19

19.1 **(a)** H_0: $\mu_1 - \mu_2 = 0$ **(c)** H_0: $\mu_1 - \mu_2 \geq 0$
          H_1: $\mu_1 - \mu_2 \neq 0$ H_1: $\mu_1 - \mu_2 < 0$
      (b) H_0: $\mu_1 - \mu_2 \leq 0$ **(d)** H_0: $\mu_1 - \mu_2 = 0$
          H_1: $\mu_1 - \mu_2 > 0$ H_1: $\mu_1 - \mu_2 \neq 0$

19.2 **(a)** 2   **(d)** 2
      (b) 2   **(e)** 3
      (c) 1   **(f)** 1

19.3 **(a)** ± 2.080 **(c)** -2.423
      (b) 1.706   **(d)** ± 2.921

19.4 *Research Problem:*

Is there a difference, on the average, between the puzzle-solving times required by subjects who are told that the puzzle is difficult and by subjects who are told that the puzzle is easy?

Statistical Hypotheses:

H_0: $\mu_1 - \mu_2 = 0$
H_1: $\mu_1 - \mu_2 \neq 0$

Decision Rule:

Reject H_0 at .05 level of significance if t equals or is more positive than 2.101 or t equals or is more negative than -2.101, given $df = 10 + 10 - 2 = 18$.

Calculations:

$$\bar{X}_1 = \frac{158}{10} = 15.8 \quad \bar{X}_2 = \frac{90}{10} = 9.0$$

$$s_1^2 = \frac{10(3168) - (158)^2}{10(10 - 1)} = 74.62$$

$$s_2^2 = \frac{10(1036) - (90)^2}{10(10 - 1)} = 25.11$$

$$s_p^2 = \frac{(10 - 1)(74.62) + (10 - 1)(25.11)}{(10 - 1) + (10 - 1)} = 49.87$$

$$s_{\bar{X}_1 - \bar{X}_2} = \sqrt{\frac{49.87}{10} + \frac{49.87}{10}} = 3.16$$

$$t = \frac{(15.8 - 9.0) - 0}{3.16} = 2.15$$

Decision:

Reject H_0 at .05 level of significance because $t = 2.15$ is more positive than 2.101.

Interpretation:

Puzzle-solving times are longer, on the average, for subjects who are told that the puzzle is difficult than for those who are told that the puzzle is easy.

19.7 **(a)** *Research Problem:*

Is the mean performance of college students in an introductory biology course affected by grading policy?

Statistical Hypotheses:

$H_0: \mu_1 - \mu_2 = 0$

$H_1: \mu_1 - \mu_2 \neq 0$

Decision Rule:

Reject H_0 at .05 level of significance if t equals or is more positive than 2.042 or t equals or is more negative than -2.042, given $df = 20 + 20 - 2 = 38$ (read as 30 in Table B).

Calculations:

$$t = \frac{(86.2 - 81.6) - 0}{1.50} = 3.07$$

Decision:

Reject H_0 at .05 level of significance because $t = 3.07$ is more positive than 2.042.

Interpretation:

Introductory biology students have higher achievement scores, on the average, when awarded letter grades rather than a simple pass/fail.

(b) The calculated t ratio would have been equal to -3.07 rather than 3.07. Most important, however, the same interpretation would have been appropriate: Introductory biology students have higher achievement scores, on the average, when awarded letter grades rather than a simple pass/fail.

(c) $86.2 - 81.6 \pm (2.04)(1.50) = \begin{cases} 7.66 \\ 1.54 \end{cases}$

We can claim, with 95 percent confidence, that the interval between 1.54 and 7.66 (achievement score points) includes the true difference between population mean achievement scores. Being consistently positive, all of these values suggest that, on the average, higher achievement scores are attained by those students who received letter grades.

(d) Because of self-selection, groups might differ with respect to any one or several uncontrolled variables, such as motivation, aptitude, and so on, in addition to the difference in grading policy. Hence, any observed difference between the mean achievement scores for these two groups couldn't be attributed solely to the difference in grading policy.

19.8 *Research Problem:*
Does alcohol consumption cause an increase in mean performance errors on a driving simulator?
Statistical Hypotheses:
$H_0: \mu_1 - \mu_2 \leq 0$
$H_1: \mu_1 - \mu_2 > 0$
Decision Rule:
Reject H_0 at .05 level of significance if t equals or is more positive than 1.671, given $df = 60 + 60 - 2 = 118$ (read as 60 in Table B).
Calculations:
$$t = \frac{26.4 - 18.6}{2.4} = 3.25$$
Decision:
Reject H_0 at .05 level of significance because $t = 3.25$ is more positive than 1.671.
Interpretation:
Alcohol consumption causes an increase in mean performance errors on a driving simulator.

Chapter 20

20.1 **(a)** 2 independent samples   **(c)** 2 independent samples
(b) 2 matched samples   **(d)** 2 matched samples

20.2 **(a)** 1   **(d)** 2
(b) 3   **(e)** 2
(c) 2   **(f)** 1

20.3 **(a)** t test for two independent samples
(b) t test for two matched samples
(c) t test for two independent samples
(d) t test for one sample
(e) t test for two matched samples

20.4 *Research Problem:*

When schoolchildren are matched for home environment, does vitamin C reduce the mean frequency of common colds?

Statistical Hypotheses:

$H_0: \mu_D \geq 0$

$H_1: \mu_D < 0$

Decision Rule:

Reject H_0 at .05 level of significance if t equals or is more negative than -1.833, given $df = 10 - 1 = 9$.

Calculations:

$$\overline{D} = \frac{-15}{10} = -1.5 \qquad s_D = \sqrt{\frac{10(37) - (-15)^2}{10(10 - 1)}} = 1.27$$

$$s_{\overline{D}} = \frac{1.27}{\sqrt{10}} = 0.40 \qquad t = \frac{-1.5 - 0}{.40} = -3.75$$

Decision:

Reject H_0 at .05 level of significance because $t = -3.75$ is more negative than -1.833

Interpretation:

When schoolchildren are matched for home environment, vitamin C reduces the mean frequency of common colds.

20.5 *Research Problem:*

For the population of California taxpayers, is there a relationship between educational level and annual income?

Statistical Hypotheses:

$H_0: \rho = 0$

$H_1: \rho \neq 0$

Decision Rule:

Reject H_0 at .05 level of significance if t equals or is more positive than 2.060 or t equals or is more negative than -2.060, given $df = 27 - 2 = 25$.

Calculations:

$$t = \frac{.43 - 0}{\sqrt{\dfrac{1 - (.43)^2}{27 - 2}}} = \frac{.43}{.18} = 2.39$$

Decision:

Reject H_0 at .05 level of significance because $t = 2.39$ is more positive than 2.060.

Interpretation:

For the population of California taxpayers, there is a relationship (positive) between educational level and annual income.

20.6 *Research Problem:*

Does meditation cause an increase in mean GPAs for students—given that pairs of students are originally matched for their GPAs?

Statistical Hypotheses:

$H_0: \mu_D \leq 0$

$H_1: \mu_D > 0$

Decision Rule:

Reject H_0 at .01 level of significance if t equals or is more positive than 3.143, given $df = 7 - 1 = 6$.

Calculations:

$$\overline{D} = \frac{1.56}{7} = .22 \qquad s_D = \sqrt{\frac{7(.50) - (1.56)^2}{7(7-1)}} = .16$$

$$s_{\overline{D}} = \frac{.16}{\sqrt{7}} = .06 \qquad t = \frac{.22 - 0}{.06} = 3.67$$

Decision:

Reject H_0 at .01 level of significance because $t = 3.67$ is more positive than 3.143.

Interpretation:

Meditation causes an increase in mean GPAs when pairs of students are matched for their original GPAs.

20.9 **(a)** *Research Problem:*

Is there a decline in the mean running time of blood-doped athletes?

Statistical Hypotheses:

$H_0: \mu_D \leq 0$

$H_1: \mu_D > 0$

Decision Rule:

Reject H_0 at .05 level of significance if t equals or is more positive than 1.796, given $df = 12 - 1 = 11$.

Calculations:

$$\overline{D} = \frac{616}{12} = 51.33$$

$$s_D = \sqrt{\frac{(12)(80,022) - (616)^2}{12(12-1)}} = 66.33$$

$$s_{\overline{D}} = \frac{66.33}{\sqrt{12}} = 19.17 \qquad t = \frac{51.33 - 0}{19.17} = 2.68$$

Decision:

Reject H_0 at .05 level of significance because $t = 2.68$ is more positive than 1.796.

Interpretation:

Blood doping causes a decline in the mean running time (for athletes who serve as their own controls).

(b) Yes. Although the appearance of the test results would change (since now negative rather than positive difference scores would support the research hypothesis), H_0 still would have been rejected, and the interpretation would have been the same.

(c) $51.33 \pm (2.20)(19.17) = \begin{cases} 93.50 \\ 9.16 \end{cases}$

We can claim, with 95 percent confidence, that the interval between 9.16 and 93.50 seconds includes the true effect of blood doping. Being positive, all of these differences suggest that blood doping has the desired effect.

(d) Counterbalancing eliminates any possible bias due to the order of testing.

(e) The interval between tests would have been too short to eliminate the lingering effects of blood doping on those subjects who were tested first with real blood. Consequently, any effect due to blood doping would tend to be obscured.

Chapter 21

21.1 **(a)** $p < .001$ **(e)** $p > .05$
(b) $p < .05$ **(f)** $p > .05$
(c) $p < .01$ **(g)** $p < .05$
(d) $p > .05$

21.2 a_2, b_1, c_2, d_1, e_2

21.3 $a_2, b_1, b_2, c_1, c_2, e_1, e_2$

21.4 There is evidence that, on the average, solution time is longer when a puzzle is described as difficult rather than as easy [$t(18) = 2.15, p < .05$]. *Or* there is a statistically significant difference between the mean solution times, with the longer mean time occurring for the group told that the puzzle is difficult [$t(18) = 2.15, p < .05$].

21.5 **(a)** The t test results for the pooled variance estimate should be reported, because the p-value of .467 for the F value is much too large to question the assumption of equal population variances.
(b) $t = .66$; $df = 10$; $p = .522$
(c) Not significant, because the p-value of .522 for t is larger than .05.
(d) The observed t, plus any more deviant ts (in either direction) could have occurred with a probability of .522, given that the null hypothesis is true.

21.6 **(a)** Experiment A—because it is more likely to detect only a larger, more important effect.
(b) For experiment A,

$$r_{pb}^2 = \frac{(2.10)^2}{(2.10)^2 + 38} = .10 \text{ (between a medium and a large effect, according to Cohen's rule)}$$

For experiment B,

$$r_{pb}^2 = \frac{(2.10)^2}{(2.10)^2 + 198} = .02 \text{ (small effect, according to Cohen's rule)}$$

21.8 **(a)** $p < .01$
(b) Compared with a simple pass/fail, letter grades are associated with higher achievement scores among introductory biology students [$t(38) = 3.07, p < .01$].
(c) $r_{pb}^2 = \frac{(3.07)^2}{(3.07)^2 + 38} = .20$ (large effect, according to Cohen's rule)

Chapter 22

22.1

	TYPE OF VARIABILITY	
	BETWEEN GROUPS	**WITHIN GROUPS**
(a)	No	No
(b)	Yes	No
(c)	No	Yes
(d)	Yes	Yes

22.2 **(a)** random error
(b) treatment effect
(c) one

22.3 **(a)** 4.41   **(c)** 3.26
(b) 4.16   **(d)** 2.48

22.4 **(a)** *Research Problem:*
On the average, is aggressive behavior influenced by sleep deprivation?
Statistical Hypotheses:
$H_0: \mu_0 = \mu_{24} = \mu_{48} = \mu_{72}$
$H_1: H_0$ is false.
Decision Rule:
Reject H_0 at .05 level of significance if F equals or is more positive than 2.95, given $df_{between} = 3$ and $df_{within} = 28$.
Calculations:
$F = 10.84$. See (b) for more information.
Decision:
Reject H_0 at .05 level of significance because $F = 10.84$ is more positive than 2.95.
Interpretation:
Mean aggression score is influenced by sleep deprivation.

(b)

SOURCE	SS	df	MS	F
Between	154.12	3	51.37	10.84*
Within	132.75	28	4.74	
Total	286.87	31		

Significant at .05 level.

(c) $p < .01$
(d) There is evidence that the mean aggression score is influenced by sleep deprivation [$F(3,28) = 10.84, p < .01$].

22.5 $\eta^2 = \dfrac{154.12}{286.87} = .54$ (large effect, according to Cohen's rule)

22.6   Calculating Scheffé's critical value

$$(\bar{X}_1 - \bar{X}_2) = \pm\sqrt{(3)(2.95)(4.74)\left(\frac{1}{8} + \frac{1}{8}\right)}$$

$$= \pm\sqrt{10.49} = \pm 3.24$$

POPULATION COMPARISON	SCHEFFÉ'S TEST OBSERVED MEAN DIFFERENCE	SCHEFFÉ'S CRITICAL VALUE
$\mu_0 - \mu_{24}$	-1.50	± 3.24
$\mu_0 - \mu_{48}$	-3.37^*	± 3.24
$\mu_0 - \mu_{72}$	-5.87^*	± 3.24
$\mu_{24} - \mu_{48}$	-1.87	± 3.24
$\mu_{24} - \mu_{72}$	-4.37^*	± 3.24
$\mu_{48} - \mu_{72}$	-2.50	± 3.24

Significant at .05 level.

Subjects deprived of sleep for 48 or 72 hours have higher mean aggression scores than those deprived of sleep for 0 hours. Furthermore, subjects deprived of sleep for 72 hours have higher mean aggression scores than those deprived of sleep for 24 hours. No other differences are significant.

22.7   **(a)** 10.836
   (b) The observed F ratio (or any larger value of F) could have occurred with a probability close to zero (rounded off to .000 in the output), given that the null hypothesis is true.
   (c) Yes, because Sig. (p-value), rounded off to .000, is less than the specified .05 level of significance.
   (d) According to Scheffé's test, three pairs differ significantly at the .05 level: 0 and 48 hours; 0 and 72 hours; and 24 and 72 hours—in agreement with the findings for Exercise 22.6.

22.9   **(a)** *Research Problem:*
         On the average, are subjects' alarm reactions to potentially dangerous smoke affected by crowds of zero, two, and four confederates?
      Statistical Hypotheses:
         $H_0: \mu_0 = \mu_2 = \mu_4$
         $H_1: H_0$ is false.
      Decision Rule:
         Reject H_0 at .05 level if F equals or is more positive than 4.26, given $df_{between} = 2$ and $df_{within} = 9$.

Calculations:
$F = 13.70$. See (b) for more information.
Decision:
Reject H_0 at .05 level of significance because $F = 13.70$ is more positive than 4.26.
Interpretation:
Mean alarm reactions are affected by crowd size.

(b)

SOURCE	SS	df	MS	F
Between	106.05	2	53.03	13.70*
Within	34.87	9	3.87	
Total	140.92	11		

Significant at .05 level.

(c) Because H_0 has been rejected, Scheffé's test may be used.

POPULATION COMPARISON	SCHEFFÉ'S TEST OBSERVED MEAN DIFFERENCE	SCHEFFÉ'S CRITICAL VALUE
$\mu_0 - \mu_2$	−2.73	±4.18
$\mu_0 - \mu_4$	−6.90*	±3.85
$\mu_2 - \mu_4$	−4.17	±4.37

Significant at .05 level.

Mean alarm reaction of subjects in crowds of zero confederates is shorter than that of subjects in crowds of four confederates. There is no evidence, however, that the mean alarm reaction of subjects in crowds of two confederates differs either from that of those with zero or four confederates.

22.12 (a) 4
(b) 21
(c) A and B
(d) A, B, and C
(e) D. Because of the small number of subjects, only a larger treatment effect would have been detected.
(f) C. Because of the relatively large number of subjects, even a fairly small treatment effect would have been detected.
(g) For A, $p < .05$ For C, $p > .05$
For B, $p < .01$ For D, $p > .05$
(h) For outcome A, $\eta^2 = \dfrac{900}{8900} = .10$ (between a medium and large effect, according to Cohen's rule)

For outcome B, $\eta^2 = \dfrac{1500}{9500} = .16$ (large effect, according to Cohen's rule)

Chapter 23

23.1

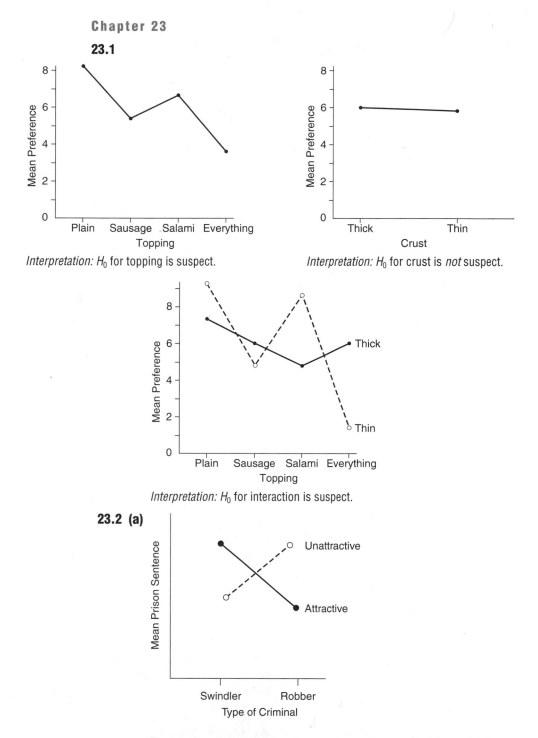

Interpretation: H_0 for topping is suspect.

Interpretation: H_0 for crust is *not* suspect.

Interpretation: H_0 for interaction is suspect.

23.2 (a)

The two lines should cross (in any manner). Note that the solid line represents the simple effect of type of criminal for attractive defendants, whereas the broken line represents the simple effect of type of criminal for unattractive defendants.

(b)

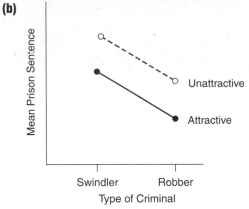

The two lines should be parallel and sloped (in any manner).

23.3 (a) *Research Problem:*

Do viewing time and type of program, as well as the interaction of these two factors, affect mean aggression scores?

Statistical Hypotheses:

H_0: no main effect for columns or viewing time (or $\mu_0 = \mu_1 = \mu_2 = \mu_3$).

H_0: no main effect for rows or type of program (or $\mu_{cartoon} = \mu_{real\ life}$).

H_0: no interaction effect.

H_1: H_0 is false.

Decision Rule:

Reject H_0 at the .05 level of significance if F_{column} equals or is more positive than 4.07, given 3 and 8 degrees of freedom; if F_{row} equals or is more positive than 5.32, given 1 and 8 degrees of freedom; and if $F_{interaction}$ equals or is more positive than 4.07, given 3 and 8 degrees of freedom.

Calculations:

$$\left.\begin{array}{l} F_{column} = 16.35 \\ F_{row} = \ \ 0.02 \\ F_{interaction} = \ \ 0.08 \end{array}\right\} \text{See (c) for more information.}$$

Decision:

Reject H_0 for column (viewing time) at .05 level of significance because $F = 16.35$ is more positive than 4.07.

Interpretation:

Viewing time affects mean aggression scores. There is no evidence, however, that mean aggression scores are affected either by the type of program or the interaction between viewing time and type of program.

(b)

SOURCE	SS	df	MS	F
Column (Viewing Time)	144.19	3	48.06	16.35*
Row (Type of Program)	0.07	1	0.07	0.02
Interaction	0.68	3	0.23	0.08
Error	23.50	8	2.94	
Total	168.44	15		

Significant of the .05 level.

(c) F_{column} = 16.53, $p < .01$
F_{row} = 0.02, $p > .05$
$F_{interaction}$ = 0.08, $p > .05$

(d) Viewing time affects mean aggression scores [$F(3,8) = 16.35$, $p < .01$]. There is no evidence that mean aggression scores are affected either by type of program [$F(1,8) = 0.02$, $p > .05$] or the interaction between viewing time and type of program [$F(3,8) = 0.08$, $p > .05$].
Note: Since it describes nonsignificant findings, the latter sentence would often be omitted from a published report.

23.4 $\eta^2_{column} = \dfrac{72}{352} = .20$ (large effect, according to Cohen's rule)

$\eta^2_{row} = \dfrac{192}{352} = .55$ (large effect, according to Cohen's rule)

$\eta^2_{interaction} = \dfrac{56}{352} = .16$ (large effect, according to Cohen's rule)

23.5 (a) 24
(b) 17.55
(c) 0.9561
(d) Reject H_0 at .05 level of significance for sauna and vitamin C, but not for interaction.

23.6 (a)

SOURCE	df
Food deprivation (F)	3
Reward magnitude (R)	2
F × R	6
Error	84
Total	95

(b) Effects

	FOOD DEPRIVATION	REWARD MAGNITUDE	INTERACTION
(1)	Yes	Yes	Yes
(2)	Yes	Yes	No
(3)	Yes	No	Yes
(4)	No	Yes	Yes
(5)	Yes	No	No
(6)	No	Yes	No
(7)	No	No	Yes
(8)	No	No	No

(c) If the interaction effects are viewed as particularly important, then all outcomes with interaction effects (1, 3, 4, and 7) would be preferred. If sheer number of effects is viewed as important, outcome 1 would be preferred.

(d) Outcome 8 probably would be least preferred, since it contains no effects.

23.8 (a)

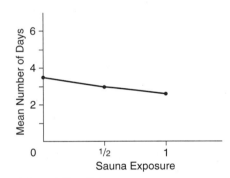

Interpretation: H_0 for vitamin C dosage is suspect.

Interpretation: H_0 for sauna exposure is suspect (although not as much as H_0 for vitamin C dosage).

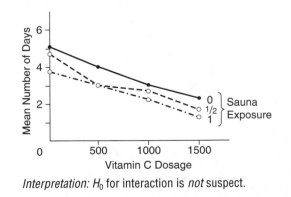

Interpretation: H_0 for interaction is *not* suspect.

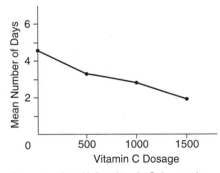

(b) *Research Problem:*

Do vitamin C dosage and sauna exposure, as well as the interaction of these two factors, affect the mean number of days of discomfort due to colds?

Statistical Hypotheses:

H_0: no main effect due to columns or vitamin C dosage (or $\mu_0 = \mu_{500} = \mu_{1000} = \mu_{1500}$)

H_0: no main effect due to rows or sauna exposure (or $\mu_0 = \mu_{1/2} = \mu_1$)

H_0: no interaction

H_1: H_0 is false.

Decision Rule:

Reject H_0 at the .05 level of significance if F_{column} equals or is more positive than 3.01, given 3 and 24 degrees of freedom; if F_{row} equals or is more positive than 3.40, given 2 and 24 degrees of freedom; and if $F_{interaction}$ equals or is more positive than 2.51, given 6 and 24 degrees of freedom.

Calculations:

$$\left.\begin{array}{r} F_{column} = 17.52 \\ F_{row} = 3.95 \\ F_{interaction} = 0.25 \end{array}\right\} \text{ See (c) for more information.}$$

Decision:

Reject H_0 for column (vitamin C) because $F = 17.52$ is more positive than 3.01 and reject H_0 for row (sauna) because $F = 3.95$ is more positive than 3.40.

Interpretation:

Both vitamin C dosage and sauna exposure affect the mean number of days of discomfort due to colds. There is no evidence, however, of an interaction between these two factors.

(c)

SOURCE	SS	df	MS	F
Column (Vitamin C)	33.64	3	11.21	17.52*
Row (Sauna)	5.05	2	2.53	3.95*
Interaction	0.95	6	0.16	0.25
Error	15.33	24	0.64	
Total	54.97	35		

Significant at .05 level.

(d) $F_{column} = 17.52, p < .01$
$F_{row} = 3.95, p < .05$
$F_{interaction} = 0.25, p > .05.$

(e) The mean number of days of discomfort due to common colds is affected by vitamin C dosage [$F(3, 24) = 17.52, p < .01$] and by sauna exposure [$F(2, 24) = 3.95, p < .05$].

Chapter 24

24.1 **(a)** H_0: $P_A = P_B = \frac{1}{2}$

(b) H_0: $P_{north} = P_{east} = P_{south} = P_{west} = \frac{1}{4}$

(c) H_0: $P_{Mon} = P_{Tue} = P_{Wed} = P_{Thu} = P_{Fri} = P_{Sat} = P_{Sun} = \frac{1}{7}$

(d) H_0: $P_{weekday} = \frac{5}{7}$; $P_{weekend} = \frac{2}{7}$

24.2 **(a)** *Research Problem:*
　　The attribute most desired by a population of college students is equally distributed among various possibilities.
　　Statistical Hypotheses:
　　　$H_0: P_{love} = P_{wealth} = P_{power} = P_{health} = P_{fame} = P_{family\ happiness} = \frac{1}{6}$
　　　$H_1: H_0$ is false.
　　Decision Rule:
　　Reject H_0 at .05 level of significance if χ^2 equals or is more positive than 11.07, given $df = 5$.
　　Calculations:

$$\chi^2 = \frac{(25-15)^2}{15} + \frac{(10-15)^2}{15} + \frac{(5-15)^2}{15}$$
$$+ \frac{(25-15)^2}{15} + \frac{(10-15)^2}{15} + \frac{(15-15)^2}{15}$$
$$= 23.33$$

　　Decision:
　　Reject H_0 at .05 level of significance because $\chi^2 = 23.33$ is more positive than 11.07.
　　Interpretation:
　　The attribute most desired by a population of college students is not equally distributed among various possibilities.

(b) $p < .001$

24.3 **(a)** Educational level and attitude toward right-to-abortion legislation are independent.
(b) Clients and nonclients are not distinguishable on the basis of—or are independent of—whether or not their parents are divorced.
(c) Employees' annual evaluations are independent of whether they work day or night shifts.

24.4 **(a)** *Research Problem:*
　　Is hair color related to susceptibility to poison oak?
　　Statistical Hypotheses:
　　　H_0: Hair color and susceptibility to poison oak are independent.
　　　$H_1: H_0$ is false.
　　Decision Rule:
　　Reject H_0 at .01 level if χ^2 equals or is more positive than 11.34 given $df = (2-1)(4-1) = 3$.
　　Calculations:

$$\chi^2 = \frac{(10-18)^2}{18} + \frac{(30-36)^2}{36} + \frac{(60-54)^2}{54}$$
$$+ \frac{(80-72)^2}{72} + \frac{(20-12)^2}{12} + \frac{(30-24)^2}{24}$$
$$+ \frac{(30-36)^2}{36} + \frac{(40-48)^2}{48} = 15.28$$

　　Decision:
　　Reject H_0 at .01 level of significance because $\chi^2 = 15.28$ is more positive than 11.34.

Interpretation:

There is a relationship between hair color and susceptibility to poison oak.

(b) $p < .01$

(c) There is a relationship between hair color and susceptibility to poison oak [$\chi^2(3, n = 300) = 15.28, p < .01$].

24.5 $\phi_c^2 = \dfrac{15.28}{300(2 - 1)} = .05$ (between a small and medium effect, according to Cohen's rule)

24.6 **(a)** 43   **(c)** 1.553
        (b) 34.37   **(d)** $p = .213$

24.8 **(a)** *Research Problem:*

Are people more likely to die after rather than before a major holiday?

Statistical Hypotheses:

$H_0: P_{before} = P_{after} = \frac{1}{2}$
$H_1: H_0$ is false.

Decision Rule:

Reject H_0 at .05 level of significance if χ^2 equals or is more positive than 3.84, given $df = 1$.

Calculations:

$$\chi^2 = \frac{(33 - 51.5)^2}{51.5} + \frac{(70 - 51.5)^2}{51.5} = 13.30$$

Decision:

Reject H_0 at .05 level of significance because $\chi^2 = 13.30$ is more positive than 3.84.

Interpretation:

People are more likely to die after rather than before a major holiday.

(b) $p < .001$

(c) More elderly California women of Chinese descent died of natural causes during a 1-week period after rather than before the Harvest Moon Festival [$\chi^2(1, n = 103) = 13.30, p < .001$].

24.11 **(a)** Degrees of Freedom (✓)

✓	✓	✓	✓	×
✓	✓	✓	✓	×
✓	✓	✓	✓	×
×	×	×	×	×

$df = (5 - 1)(4 - 1) = 12$

(b) Yes. All frequencies end in multiplies of ten, suggesting that the observed frequencies might be fictitious as is actually the case (both for this exercise and some others in this chapter) in order to simplify computations.

(c) *Research Problem:*

Is the religious preference of students related to their political affiliation?

Statistical Hypotheses:
H_0: Religious preference and political affiliation are independent.
H_1: H_0 is false.
Decision Rule:
Reject H_0 at .05 level if χ^2 equals or is more positive than 21.03, given $df - (5 - 1)(4 - 1) = 12$.
Calculations:
$$\chi^2 = \frac{(30 - 20)^2}{20} + \frac{(30 - 20)^2}{20} + \cdots + \frac{(100 - 40)^2}{40} = 220$$
Decision:
Reject H_0 at .05 level of significance because $\chi^2 = 220$ is more positive than 21.03.
Interpretation:
There is a relationship between the religious preference of students and their political affiliation.

Chapter 25

25.1 **(a)** 1, 2, 3, 4.5, 4.5, 6, 7, 9, 9, 9, 11
   (b) 1, 2.5, 2.5, 4.5, 4.5, 6, 7, 8, 9, 10
   (c) 1, 3.5, 3.5, 3.5, 3.5, 6, 7, 8.5, 8.5, 10

25.2 **(a)** *Research Problem:*
Do encounter groups with (1) aggressive leaders produce more or less growth (in members) than encounter groups with (2) supportive leaders?
Statistical Hypotheses:
H_0: Population distribution 1 = Population distribution 2
H_1: Population distribution 1 $\neq$ Population distribution 2
Decision Rule:
Reject H_0 at .05 level if U equals or is less than 5, given $n_1 = 6$ and $n_2 = 6$.
Calculations:
$U_1 = 31$
$U_2 = 5$
$U = 5$
Decision:
Reject H_0 at .05 level of significance because $U = 5$ equals 5.
Interpretation:
Encounter groups with aggressive leaders produce less growth than those with supportive leaders.
   (b) $p < .05$

25.3 **(a)** Each distribution of difference scores tends to be non-normal, with "heavy" tails and "light" middle.
   (b) *Research Problem:*
Does an antismoking workshop cause a decline in cigarette smoking?

Statistical Hypotheses:

H_0: Population distribution 1 $\leq$ Population distribution 2

H_1: Population distribution 1 $>$ Population distribution 2

Note: The directional H_1 assumes both population distributions have roughly similar shapes.

Decision Rule:

Reject H_0 at .05 level (directional test) if T equals or is less than 5, given $n = 8$.

Calculations:

$R_+ = 34$

$R_- = 2$

$T = 2$

Decision:

Reject H_0 at .05 level of significance because $T = 2$ is *less* positive than 5.

Interpretation:

Antismoking workshop causes a decline in smoking.

(c) $p < .05$

25.4 **(a)** Observed scores tend not to be normally distributed—there is no obvious cluster of scores in the middle range for each of the five groups.

(b) *Research Problem:*

Are motion picture ratings associated with the number of violent or sexually explicit scenes in films?

Statistical Hypotheses:

H_0: Population dist. NC-17 = Population dist. R = Population dist. PG-13 = Population dist. PG = Population dist. G

H_1: H_0 is false.

Decision Rule:

Reject H_0 at .05 level if H equals or is more positive than 9.49, given $df = 4$.

Calculations:

$$H = \frac{12}{25(25 + 1)} \left[\frac{(114)^2}{5} + \frac{(75.5)^2}{5} + \frac{(69)^2}{5} + \frac{(49.5)^2}{5} \right. $$
$$\left. + \frac{(17)^2}{5} \right] - 3(25 + 1)$$

$= 18.73$

Decision:

Reject H_0 at .05 level of significance because $H = 18.73$ is more positive than 9.49.

Interpretation:

Motion picture ratings are associated with the number of violent or sexually explicit scenes in films.

(c) $p < .001$

Chapter 26

26.1 Two-way χ^2

26.2 *t* for two population means (independent samples)

26.3 *t* for two population means (matched samples)

26.5 One-way χ^2

26.6 One-way *F*

26.9 Two-way *F*

26.10 *t* for two population means (independent samples)

26.11 *t* for population correlation coefficient

Appendix B

1. **(a)** ratio   **(d)** ratio
    (b) nominal   **(e)** ordinal
    (c) interval*

....

*Approximates interval measurement.

Tables

A   **PROPORTIONS (OF AREA) UNDER STANDARD
     NORMAL CURVE FOR VALUES OF z**

B   **CRITICAL VALUES OF t**

C   **CRITICAL VALUES OF F**

D   **CRITICAL VALUES OF χ^2**

E   **CRITICAL VALUES OF MANN-WHITNEY U**

F   **CRITICAL VALUES OF WILCOXON T**

G   **RANDOM NUMBERS**

Table A entries were computed by the second author.

Table B is taken from Table 12 of E. Pearson and H. Hartley. (Eds.), *Biometrika Tables for Statisticians*, Vol.1, 3d ed., University Press, Cambridge, 1966, with permission of the Biometrika Trustees.

Table C reprinted by permission from *Statistical Methods*, by George W. Snedecor and William G. Cochran, 8th ed., Iowa State University Press, Ames, Iowa, 1989.

Table D is taken from Table IV of R. A. Fisher and F. Yates, *Statistical Tables for Biological, Agricultural and Medical Research*, 6th ed., 1974, published by Longman Group, Ltd., London (previously published by Oliver and Boyd, Edinburgh), and by permission of the authors and publishers.

Table E is taken from the *Bulletin of the Institute of Educational Research* Vol. 1, No. 2, Indiana University, with permission of the publishers.

Table F is taken from F. Wilcoxon and R. A. Wilcox, *Some Rapid Approximate Statistical Procedures*, Lederle Laboratories, New York, 1964, with permission of the American Cyanamid Company.

Table G reprinted from Page 1 of *A Million Random Digits with 100,000 Normal Deviates*, Rand, 1994, RP-295, 200 pp. Used by permission.

Table A[a]
PROPORTIONS (OF AREA) UNDER STANDARD NORMAL CURVE FOR VALUES OF z

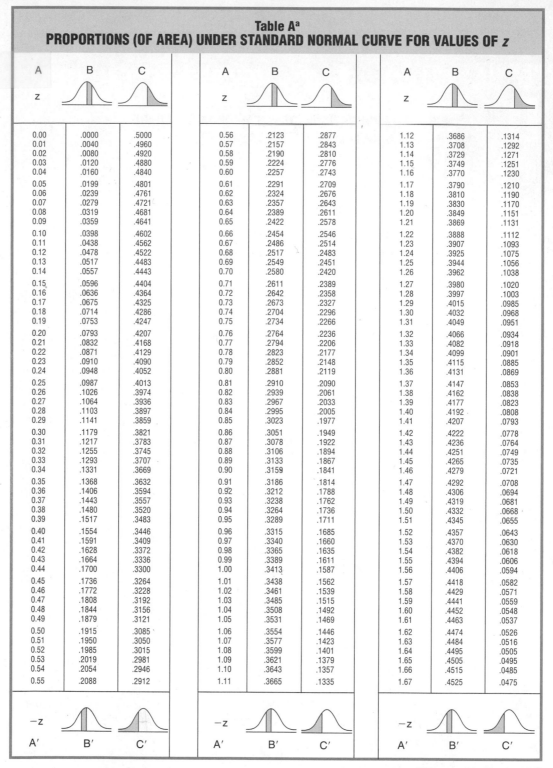

A z	B	C	A z	B	C	A z	B	C
0.00	.0000	.5000	0.56	.2123	.2877	1.12	.3686	.1314
0.01	.0040	.4960	0.57	.2157	.2843	1.13	.3708	.1292
0.02	.0080	.4920	0.58	.2190	.2810	1.14	.3729	.1271
0.03	.0120	.4880	0.59	.2224	.2776	1.15	.3749	.1251
0.04	.0160	.4840	0.60	.2257	.2743	1.16	.3770	.1230
0.05	.0199	.4801	0.61	.2291	.2709	1.17	.3790	.1210
0.06	.0239	.4761	0.62	.2324	.2676	1.18	.3810	.1190
0.07	.0279	.4721	0.63	.2357	.2643	1.19	.3830	.1170
0.08	.0319	.4681	0.64	.2389	.2611	1.20	.3849	.1151
0.09	.0359	.4641	0.65	.2422	.2578	1.21	.3869	.1131
0.10	.0398	.4602	0.66	.2454	.2546	1.22	.3888	.1112
0.11	.0438	.4562	0.67	.2486	.2514	1.23	.3907	.1093
0.12	.0478	.4522	0.68	.2517	.2483	1.24	.3925	.1075
0.13	.0517	.4483	0.69	.2549	.2451	1.25	.3944	.1056
0.14	.0557	.4443	0.70	.2580	.2420	1.26	.3962	.1038
0.15	.0596	.4404	0.71	.2611	.2389	1.27	.3980	.1020
0.16	.0636	.4364	0.72	.2642	.2358	1.28	.3997	.1003
0.17	.0675	.4325	0.73	.2673	.2327	1.29	.4015	.0985
0.18	.0714	.4286	0.74	.2704	.2296	1.30	.4032	.0968
0.19	.0753	.4247	0.75	.2734	.2266	1.31	.4049	.0951
0.20	.0793	.4207	0.76	.2764	.2236	1.32	.4066	.0934
0.21	.0832	.4168	0.77	.2794	.2206	1.33	.4082	.0918
0.22	.0871	.4129	0.78	.2823	.2177	1.34	.4099	.0901
0.23	.0910	.4090	0.79	.2852	.2148	1.35	.4115	.0885
0.24	.0948	.4052	0.80	.2881	.2119	1.36	.4131	.0869
0.25	.0987	.4013	0.81	.2910	.2090	1.37	.4147	.0853
0.26	.1026	.3974	0.82	.2939	.2061	1.38	.4162	.0838
0.27	.1064	.3936	0.83	.2967	.2033	1.39	.4177	.0823
0.28	.1103	.3897	0.84	.2995	.2005	1.40	.4192	.0808
0.29	.1141	.3859	0.85	.3023	.1977	1.41	.4207	.0793
0.30	.1179	.3821	0.86	.3051	.1949	1.42	.4222	.0778
0.31	.1217	.3783	0.87	.3078	.1922	1.43	.4236	.0764
0.32	.1255	.3745	0.88	.3106	.1894	1.44	.4251	.0749
0.33	.1293	.3707	0.89	.3133	.1867	1.45	.4265	.0735
0.34	.1331	.3669	0.90	.3159	.1841	1.46	.4279	.0721
0.35	.1368	.3632	0.91	.3186	.1814	1.47	.4292	.0708
0.36	.1406	.3594	0.92	.3212	.1788	1.48	.4306	.0694
0.37	.1443	.3557	0.93	.3238	.1762	1.49	.4319	.0681
0.38	.1480	.3520	0.94	.3264	.1736	1.50	.4332	.0668
0.39	.1517	.3483	0.95	.3289	.1711	1.51	.4345	.0655
0.40	.1554	.3446	0.96	.3315	.1685	1.52	.4357	.0643
0.41	.1591	.3409	0.97	.3340	.1660	1.53	.4370	.0630
0.42	.1628	.3372	0.98	.3365	.1635	1.54	.4382	.0618
0.43	.1664	.3336	0.99	.3389	.1611	1.55	.4394	.0606
0.44	.1700	.3300	1.00	.3413	.1587	1.56	.4406	.0594
0.45	.1736	.3264	1.01	.3438	.1562	1.57	.4418	.0582
0.46	.1772	.3228	1.02	.3461	.1539	1.58	.4429	.0571
0.47	.1808	.3192	1.03	.3485	.1515	1.59	.4441	.0559
0.48	.1844	.3156	1.04	.3508	.1492	1.60	.4452	.0548
0.49	.1879	.3121	1.05	.3531	.1469	1.61	.4463	.0537
0.50	.1915	.3085	1.06	.3554	.1446	1.62	.4474	.0526
0.51	.1950	.3050	1.07	.3577	.1423	1.63	.4484	.0516
0.52	.1985	.3015	1.08	.3599	.1401	1.64	.4495	.0505
0.53	.2019	.2981	1.09	.3621	.1379	1.65	.4505	.0495
0.54	.2054	.2946	1.10	.3643	.1357	1.66	.4515	.0485
0.55	.2088	.2912	1.11	.3665	.1335	1.67	.4525	.0475

| −z | | | −z | | | −z | | |
| A′ | B′ | C′ | A′ | B′ | C′ | A′ | B′ | C′ |

[a]*Discussed in Section 6.5.*

Table Aª (Continued)
PROPORTIONS OF AREA UNDER STANDARD NORMAL CURVE FOR VALUES OF z

A z	B	C	A z	B	C	A z	B	C
1.68	.4535	.0465	2.24	.4875	.0125	2.80	.4974	.0026
1.69	.4545	.0455	2.25	.4878	.0122	2.81	.4975	.0025
1.70	.4554	.0446	2.26	.4881	.0119	2.82	.4976	.0024
1.71	.4564	.0436	2.27	.4884	.0116	2.83	.4977	.0023
1.72	.4573	.0427	2.28	.4887	.0113	2.84	.4977	.0023
1.73	.4582	.0418	2.29	.4890	.0110	2.85	.4978	.0022
1.74	.4591	.0409	2.30	.4893	.0107	2.86	.4979	.0021
1.75	.4599	.0401	2.31	.4896	.0104	2.87	.4979	.0021
1.76	.4608	.0392	2.32	.4898	.0102	2.88	.4980	.0020
1.77	.4616	.0384	2.33	.4901	.0099	2.89	.4981	.0019
1.78	.4625	.0375	2.34	.4904	.0096	2.90	.4981	.0019
1.79	.4633	.0367	2.35	.4906	.0094	2.91	.4982	.0018
1.80	.4641	.0359	2.36	.4909	.0091	2.92	.4982	.0018
1.81	.4649	.0351	2.37	.4911	.0089	2.93	.4983	.0017
1.82	.4656	.0344	2.38	.4913	.0087	2.94	.4984	.0016
1.83	.4664	.0336	2.39	.4916	.0084	2.95	.4984	.0016
1.84	.4671	.0329	2.40	.4918	.0082	2.96	.4985	.0015
1.85	.4678	.0322	2.41	.4920	.0080	2.97	.4985	.0015
1.86	.4686	.0314	2.42	.4922	.0078	2.98	.4986	.0014
1.87	.4693	.0307	2.43	.4925	.0075	2.99	.4986	.0014
1.88	.4699	.0301	2.44	.4727	.0073	3.00	.4987	.0013
1.89	.4706	.0294	2.45	.4929	.0071	3.01	.4987	.0013
1.90	.4713	.0287	2.46	.4931	.0069	3.02	.4987	.0013
1.91	.4719	.0281	2.47	.4932	.0068	3.03	.4988	.0012
1.92	.4726	.0274	2.48	.4934	.0066	3.04	.4988	.0012
1.93	.4732	.0268	2.49	.4936	.0064	3.05	.4989	.0011
1.94	.4738	.0262	2.50	.4938	.0062	3.06	.4989	.0011
1.95	.4744	.0256	2.51	.4940	.0060	3.07	.4989	.0011
1.96	.4750	.0250	2.52	.4941	.0059	3.08	.4990	.0010
1.97	.4756	.0244	2.53	.4943	.0057	3.09	.4990	.0010
1.98	.4761	.0239	2.54	.4945	.0055	3.10	.4990	.0010
1.99	.4767	.0233	2.55	.4946	.0054	3.11	.4991	.0009
2.00	.4772	.0228	2.56	.4948	.0052	3.12	.4991	.0009
2.01	.4778	.0222	2.57	.4949	.0051	3.13	.4991	.0009
2.02	.4783	.0217	2.58	.4951	.0049	3.14	.4992	.0008
2.03	.4788	.0212	2.59	.4952	.0048	3.15	.4992	.0008
2.04	.4793	.0207	2.60	.4953	.0047	3.16	.4992	.0008
2.05	4798	.0202	2.61	.4955	.0045	3.17	.4992	.0008
2.06	.4803	.0197	2.62	.4956	.0044	3.18	.4993	.0007
2.07	.4808	.0192	2.63	.4957	.0043	3.19	.4993	.0007
2.08	.4812	.0188	2.64	.4959	.0041	3.20	.4993	.0007
2.09	.4817	.0183	2.65	.4960	.0040	3.21	.4993	.0007
2.10	.4821	.0179	2.66	.4961	.0039	3.22	.4994	.0006
2.11	.4826	.0174	2.67	.4962	.0038	3.23	.4994	.0006
2.12	.4830	.0170	2.68	.4963	.0037	3.24	.4994	.0006
2.13	.4834	.0166	2.69	.4964	.0036	3.25	.4994	.0006
2.14	.4838	.0162	2.70	.4965	.0035	3.30	.4995	.0005
2.15	.4842	.0158	2.71	.4966	.0034	3.35	.4996	.0004
2.16	.4846	.0154	2.72	.4967	.0033	3.40	.4997	.0003
2.17	.4850	.0150	2.73	.4968	.0032	3.45	.4997	.0003
2.18	.4854	.0146	2.74	.4969	.0031	3.50	.4998	.0002
2.19	.4857	.0143	2.75	.4970	.0030	3.60	.4998	.0002
2.20	.4861	.0139	2.76	.4971	.0029	3.70	.4999	.0001
2.21	.4864	.0136	2.77	.4972	.0028	3.80	.4999	.0001
2.22	.4868	.0132	2.78	.4973	.0027	3.90	.49995	.00005
2.23	.4871	.0129	2.79	.4974	.0026	4.00	.49997	.00003

−z			−z			−z		
A′	B′	C′	A′	B′	C′	A′	B′	C′

Table Bᵃ
CRITICAL VALUES OF *t*

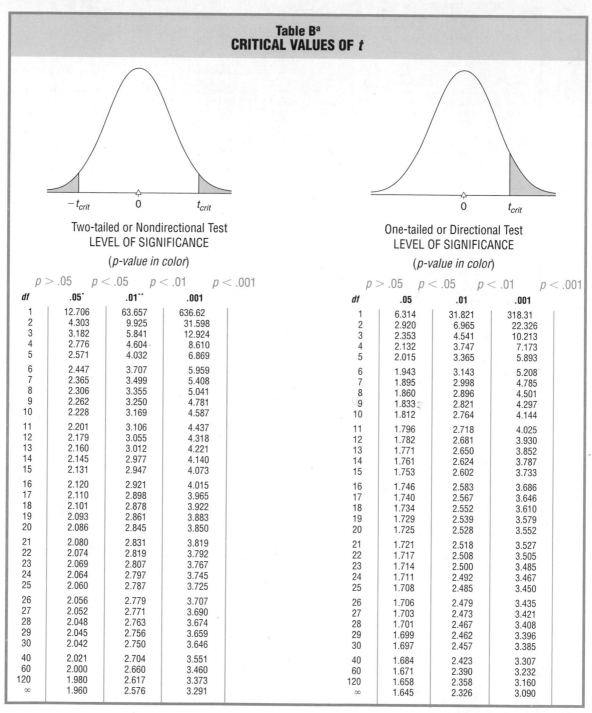

Two-tailed or Nondirectional Test LEVEL OF SIGNIFICANCE (*p-value in color*)				One-tailed or Directional Test LEVEL OF SIGNIFICANCE (*p-value in color*)			
$p > .05$	$p < .05$	$p < .01$	$p < .001$	$p > .05$	$p < .05$	$p < .01$	$p < .001$
df	.05*	.01**	.001	*df*	.05	.01	.001
1	12.706	63.657	636.62	1	6.314	31.821	318.31
2	4.303	9.925	31.598	2	2.920	6.965	22.326
3	3.182	5.841	12.924	3	2.353	4.541	10.213
4	2.776	4.604	8.610	4	2.132	3.747	7.173
5	2.571	4.032	6.869	5	2.015	3.365	5.893
6	2.447	3.707	5.959	6	1.943	3.143	5.208
7	2.365	3.499	5.408	7	1.895	2.998	4.785
8	2.306	3.355	5.041	8	1.860	2.896	4.501
9	2.262	3.250	4.781	9	1.833	2.821	4.297
10	2.228	3.169	4.587	10	1.812	2.764	4.144
11	2.201	3.106	4.437	11	1.796	2.718	4.025
12	2.179	3.055	4.318	12	1.782	2.681	3.930
13	2.160	3.012	4.221	13	1.771	2.650	3.852
14	2.145	2.977	4.140	14	1.761	2.624	3.787
15	2.131	2.947	4.073	15	1.753	2.602	3.733
16	2.120	2.921	4.015	16	1.746	2.583	3.686
17	2.110	2.898	3.965	17	1.740	2.567	3.646
18	2.101	2.878	3.922	18	1.734	2.552	3.610
19	2.093	2.861	3.883	19	1.729	2.539	3.579
20	2.086	2.845	3.850	20	1.725	2.528	3.552
21	2.080	2.831	3.819	21	1.721	2.518	3.527
22	2.074	2.819	3.792	22	1.717	2.508	3.505
23	2.069	2.807	3.767	23	1.714	2.500	3.485
24	2.064	2.797	3.745	24	1.711	2.492	3.467
25	2.060	2.787	3.725	25	1.708	2.485	3.450
26	2.056	2.779	3.707	26	1.706	2.479	3.435
27	2.052	2.771	3.690	27	1.703	2.473	3.421
28	2.048	2.763	3.674	28	1.701	2.467	3.408
29	2.045	2.756	3.659	29	1.699	2.462	3.396
30	2.042	2.750	3.646	30	1.697	2.457	3.385
40	2.021	2.704	3.551	40	1.684	2.423	3.307
60	2.000	2.660	3.460	60	1.671	2.390	3.232
120	1.980	2.617	3.373	120	1.658	2.358	3.160
∞	1.960	2.576	3.291	∞	1.645	2.326	3.090

ᵃ*Discussed in Section 18.3.*
*95% level of confidence.
**99% level of confidence.

Table C[a]
CRITICAL VALUES OF F

F_{crit}

.05 level of significance (light numbers)
.01 level of significance (**dark numbers**)

FINDING p-VALUE
If observed F is
… smaller than light number, p > .05
… between light and dark numbers, p < .05
… larger than dark number, p < .01

DEGREES OF FREEDOM IN NUMERATOR

Each cell shows light number (.05) / **dark number** (.01).

DEGREES OF FREEDOM IN DENOMINATOR	1	2	3	4	5	6	7	8	9	10	11	12	14	16	20	24	30	40	50	75	100	200	500	∞
1	161 / **4,052**	200 / **4,999**	216 / **5,403**	225 / **5,625**	230 / **5,764**	234 / **5,859**	237 / **5,928**	239 / **5,981**	241 / **6,022**	242 / **6,056**	243 / **6,082**	244 / **6,106**	245 / **6,142**	246 / **6,169**	248 / **6,208**	249 / **6,234**	250 / **6,258**	251 / **6,286**	252 / **6,302**	253 / **6,323**	253 / **6,334**	254 / **6,352**	254 / **6,361**	254 / **6,366**
2	18.51 / **98.49**	19.00 / **99.00**	19.16 / **99.17**	19.25 / **99.25**	19.30 / **99.30**	19.33 / **99.33**	19.36 / **99.34**	19.37 / **99.36**	19.38 / **99.38**	19.39 / **99.40**	19.40 / **99.41**	19.41 / **99.42**	19.42 / **99.43**	19.43 / **99.44**	19.44 / **99.45**	19.45 / **99.46**	19.46 / **99.47**	19.47 / **99.48**	19.47 / **99.48**	19.48 / **99.49**	19.49 / **99.49**	19.49 / **99.49**	19.50 / **99.50**	19.50 / **99.50**
3	10.13 / **34.12**	9.55 / **30.82**	9.28 / **29.46**	9.12 / **28.71**	9.01 / **28.24**	8.94 / **27.91**	8.88 / **27.67**	8.84 / **27.49**	8.81 / **27.34**	8.78 / **27.23**	8.76 / **27.13**	8.74 / **27.05**	8.71 / **26.92**	8.69 / **26.83**	8.66 / **26.69**	8.64 / **26.60**	8.62 / **26.50**	8.60 / **26.41**	8.58 / **26.35**	8.57 / **26.27**	8.56 / **26.23**	8.54 / **26.18**	8.54 / **26.14**	8.53 / **26.12**
4	7.71 / **21.20**	6.94 / **18.00**	6.59 / **16.69**	6.39 / **15.98**	6.26 / **15.52**	6.16 / **15.21**	6.09 / **14.98**	6.04 / **14.80**	6.00 / **14.66**	5.96 / **14.54**	5.93 / **14.45**	5.91 / **14.37**	5.87 / **14.24**	5.84 / **14.15**	5.80 / **14.02**	5.77 / **13.93**	5.74 / **13.83**	5.71 / **13.74**	5.70 / **13.69**	5.68 / **13.61**	5.66 / **13.57**	5.65 / **13.52**	5.64 / **13.48**	5.63 / **13.46**
5	6.61 / **16.26**	5.79 / **13.27**	5.41 / **12.06**	5.19 / **11.39**	5.05 / **10.97**	4.95 / **10.67**	4.88 / **10.45**	4.82 / **10.27**	4.78 / **10.15**	4.74 / **10.05**	4.70 / **9.96**	4.68 / **9.89**	4.64 / **9.77**	4.60 / **9.68**	4.56 / **9.55**	4.53 / **9.47**	4.50 / **9.38**	4.46 / **9.29**	4.44 / **9.24**	4.42 / **9.17**	4.40 / **9.13**	4.38 / **9.07**	4.37 / **9.04**	4.36 / **9.02**
6	5.99 / **13.74**	5.14 / **10.92**	4.76 / **9.78**	4.53 / **9.15**	4.39 / **8.75**	4.28 / **8.47**	4.21 / **8.26**	4.15 / **8.10**	4.10 / **7.98**	4.06 / **7.87**	4.03 / **7.79**	4.00 / **7.72**	3.96 / **7.60**	3.92 / **7.52**	3.87 / **7.39**	3.84 / **7.31**	3.81 / **7.23**	3.77 / **7.14**	3.75 / **7.09**	3.72 / **6.99**	3.71 / **6.99**	3.69 / **6.94**	3.68 / **6.90**	3.67 / **6.88**
7	5.59 / **12.25**	4.47 / **9.55**	4.35 / **8.45**	4.12 / **7.85**	3.97 / **7.46**	3.87 / **7.19**	3.79 / **7.00**	3.73 / **6.84**	3.68 / **6.71**	3.63 / **6.62**	3.60 / **6.54**	3.57 / **6.47**	3.52 / **6.35**	3.49 / **6.27**	3.44 / **6.15**	3.41 / **6.07**	3.38 / **5.98**	3.34 / **5.90**	3.32 / **5.85**	3.29 / **5.78**	3.28 / **5.75**	3.25 / **5.70**	3.24 / **5.67**	3.23 / **5.65**
8	5.32 / **11.26**	4.46 / **8.65**	4.07 / **7.59**	3.84 / **7.01**	3.69 / **6.63**	3.58 / **6.37**	3.50 / **6.19**	3.44 / **6.03**	3.39 / **5.91**	3.34 / **5.82**	3.31 / **5.74**	3.28 / **5.67**	3.23 / **5.56**	3.20 / **5.48**	3.15 / **5.36**	3.12 / **5.28**	3.08 / **5.20**	3.05 / **5.11**	3.03 / **5.06**	3.0 / **5.00**	2.98 / **4.96**	2.96 / **4.91**	2.94 / **4.88**	2.93 / **4.86**
9	5.12 / **10.56**	4.26 / **8.02**	3.86 / **6.99**	3.63 / **6.42**	3.48 / **6.06**	3.37 / **5.80**	3.29 / **5.62**	3.23 / **5.47**	3.18 / **5.35**	3.13 / **5.26**	3.10 / **5.18**	3.07 / **5.11**	3.02 / **5.00**	2.98 / **4.92**	2.93 / **4.80**	2.90 / **4.73**	2.86 / **4.64**	2.82 / **4.56**	2.80 / **4.51**	2.77 / **4.45**	2.76 / **4.41**	2.73 / **4.36**	2.72 / **4.33**	2.71 / **4.31**
10	4.96 / **10.04**	4.10 / **7.56**	3.71 / **6.55**	3.48 / **5.99**	3.33 / **5.64**	3.22 / **5.39**	3.14 / **5.21**	3.07 / **5.06**	3.02 / **4.95**	2.97 / **4.85**	2.94 / **4.78**	2.91 / **4.71**	2.86 / **4.60**	2.82 / **4.52**	2.77 / **4.41**	2.74 / **4.33**	2.70 / **4.25**	2.67 / **4.17**	2.64 / **4.12**	2.61 / **4.05**	2.59 / **4.01**	2.56 / **3.96**	2.55 / **3.93**	2.54 / **3.91**
11	4.84 / **9.65**	3.98 / **7.20**	3.59 / **6.22**	3.36 / **5.67**	3.20 / **5.32**	3.09 / **5.07**	3.01 / **4.88**	2.95 / **4.74**	2.90 / **4.63**	2.86 / **4.54**	2.82 / **4.46**	2.79 / **4.40**	2.74 / **4.29**	2.70 / **4.21**	2.65 / **4.10**	2.61 / **4.02**	2.57 / **3.94**	2.53 / **3.86**	2.50 / **3.80**	2.47 / **3.74**	2.45 / **3.70**	2.42 / **3.66**	2.41 / **3.62**	2.40 / **3.60**
12	4.75 / **9.33**	3.88 / **6.93**	3.49 / **5.95**	3.26 / **5.41**	3.11 / **5.06**	3.00 / **4.82**	2.92 / **4.65**	2.85 / **4.50**	2.80 / **4.39**	2.76 / **4.30**	2.72 / **4.22**	2.69 / **4.16**	2.64 / **4.05**	2.60 / **3.98**	2.54 / **3.86**	2.50 / **3.78**	2.46 / **3.70**	2.42 / **3.61**	2.40 / **3.56**	2.36 / **3.49**	2.35 / **3.46**	2.32 / **3.41**	2.31 / **3.38**	2.30 / **3.36**
13	4.67 / **9.07**	3.80 / **6.70**	3.41 / **5.74**	3.18 / **5.20**	3.02 / **4.86**	2.92 / **4.62**	2.84 / **4.44**	2.77 / **4.30**	2.72 / **4.19**	2.67 / **4.10**	2.63 / **4.02**	2.60 / **3.96**	2.55 / **3.85**	2.51 / **3.78**	2.46 / **3.67**	2.42 / **3.59**	2.38 / **3.51**	2.34 / **3.42**	2.32 / **3.37**	2.28 / **3.30**	2.26 / **3.27**	2.24 / **3.21**	2.22 / **3.18**	2.21 / **3.16**

[a]Discussed in Section 22.11.

Table Cª (Continued)
CRITICAL VALUES OF F

FINDING p-VALUE
If observed F is
... smaller than light number, $p > .05$
... between light and dark numbers, $p < .05$
... larger than dark number, $p < .01$

DEGREES OF FREEDOM IN NUMERATOR

DEGREES OF FREEDOM IN DENOMINATOR	1	2	3	4	5	6	7	8	9	10	11	12	14	16	20	24	30	40	50	75	100	200	500	∞
14	4.60 / 8.86	3.74 / 6.51	3.34 / 5.56	3.11 / 5.03	2.96 / 4.69	2.85 / 4.46	2.77 / 4.28	2.70 / 4.14	2.65 / 4.03	2.60 / 3.94	2.56 / 3.86	2.53 / 3.80	2.48 / 3.70	2.44 / 3.62	2.39 / 3.51	2.35 / 3.43	2.31 / 3.34	2.27 / 3.26	2.24 / 3.21	2.21 / 3.14	2.19 / 3.11	2.16 / 3.06	2.14 / 3.02	2.13 / 3.00
15	4.54 / 8.68	3.68 / 6.36	3.29 / 5.42	3.06 / 4.89	2.90 / 4.56	2.79 / 4.32	2.70 / 4.14	2.64 / 4.00	2.59 / 3.89	2.55 / 3.80	2.51 / 3.73	2.48 / 3.67	2.43 / 3.56	2.39 / 3.48	2.33 / 3.36	2.29 / 3.29	2.25 / 3.20	2.21 / 3.12	2.18 / 3.07	2.15 / 3.00	2.12 / 2.97	2.10 / 2.9	2.08 / 2.89	2.070 / 2.80
16	4.49 / 8.53	3.63 / 6.23	3.24 / 5.29	3.01 / 4.77	2.85 / 4.44	2.74 / 4.20	2.66 / 4.03	2.59 / 3.89	2.54 / 3.78	2.49 / 3.69	2.45 / 3.61	2.42 / 3.55	2.37 / 3.45	2.33 / 3.37	2.28 / 3.25	2.24 / 3.18	2.20 / 3.10	2.16 / 3.01	2.13 / 2.96	2.09 / 2.89	2.07 / 2.86	2.04 / 2.80	2.02 / 2.77	2.01 / 2.75
17	4.45 / 8.40	3.59 / 6.11	3.20 / 5.18	2.96 / 4.67	2.81 / 4.34	2.70 / 4.10	2.62 / 3.93	2.55 / 3.79	2.50 / 3.68	2.45 / 3.59	2.41 / 3.52	2.38 / 3.45	2.33 / 3.35	2.29 / 3.27	2.23 / 3.16	2.19 / 3.08	2.15 / 3.00	2.11 / 2.92	2.08 / 2.86	2.04 / 2.79	2.02 / 2.76	1.99 / 2.70	1.97 / 2.67	1.96 / 2.65
18	4.41 / 8.28	3.55 / 6.01	3.16 / 5.09	2.93 / 4.58	2.77 / 4.25	2.66 / 4.01	2.58 / 3.85	2.51 / 3.71	2.46 / 3.60	2.41 / 3.51	2.37 / 3.44	2.34 / 3.37	2.29 / 3.27	2.25 / 3.19	2.19 / 3.07	2.15 / 3.00	2.11 / 2.91	2.07 / 2.83	2.04 / 2.78	2.00 / 2.71	1.98 / 2.68	1.95 / 2.62	1.93 / 2.59	1.92 / 2.57
19	4.38 / 8.18	3.52 / 5.93	3.13 / 5.01	2.90 / 4.50	2.74 / 4.17	2.63 / 3.94	2.55 / 3.77	2.48 / 3.63	2.43 / 3.52	2.38 / 3.43	2.34 / 3.36	2.31 / 3.30	2.26 / 3.19	2.21 / 3.12	2.15 / 3.00	2.11 / 2.92	2.07 / 2.84	2.02 / 2.76	2.00 / 2.70	1.96 / 2.63	1.94 / 2.60	1.91 / 2.54	1.90 / 2.51	1.88 / 2.49
20	4.35 / 8.10	3.49 / 5.85	3.10 / 4.94	2.87 / 4.43	2.71 / 4.10	2.60 / 3.87	2.52 / 3.71	2.45 / 3.56	2.40 / 3.45	2.35 / 3.37	2.31 / 3.30	2.28 / 3.23	2.23 / 3.13	2.18 / 3.05	2.12 / 2.94	2.08 / 2.86	2.04 / 2.77	1.99 / 2.69	1.96 / 2.63	1.92 / 2.56	1.90 / 2.53	1.87 / 2.47	1.85 / 2.44	1.84 / 2.42
21	4.32 / 8.02	3.47 / 5.78	3.07 / 4.87	2.84 / 4.37	2.68 / 4.04	2.57 / 3.81	2.49 / 3.65	2.42 / 3.51	2.37 / 3.40	2.32 / 3.31	2.28 / 3.24	2.25 / 3.17	2.20 / 3.07	2.15 / 2.99	2.09 / 2.88	2.05 / 2.80	2.00 / 2.72	1.96 / 2.63	1.93 / 2.58	1.89 / 2.51	1.87 / 2.47	1.84 / 2.42	1.82 / 2.38	1.81 / 2.36
22	4.30 / 7.94	3.44 / 5.72	3.05 / 4.82	2.82 / 4.31	2.66 / 3.99	2.55 / 3.76	2.47 / 3.59	2.40 / 3.45	2.35 / 3.35	2.30 / 3.26	2.26 / 3.18	2.23 / 3.12	2.18 / 3.02	2.13 / 2.94	2.07 / 2.83	2.03 / 2.75	1.98 / 2.67	1.93 / 2.58	1.91 / 2.53	1.87 / 2.46	1.84 / 2.42	1.81 / 2.37	1.80 / 2.33	1.78 / 2.31
23	4.28 / 7.88	3.42 / 5.66	3.03 / 4.76	2.80 / 4.26	2.64 / 3.94	2.53 / 3.71	2.45 / 3.54	2.38 / 3.41	2.32 / 3.30	2.28 / 3.21	2.24 / 3.14	2.20 / 3.07	2.14 / 2.97	2.10 / 2.89	2.04 / 2.78	2.00 / 2.70	1.96 / 2.62	1.91 / 2.53	1.88 / 2.48	1.84 / 2.41	1.82 / 2.37	1.79 / 2.32	1.77 / 2.28	1.76 / 2.26
24	4.26 / 7.82	3.40 / 5.61	3.01 / 4.72	2.78 / 4.22	2.62 / 3.90	2.51 / 3.67	2.43 / 3.50	2.36 / 3.36	2.30 / 3.25	2.26 / 3.17	2.22 / 3.09	2.18 / 3.03	2.13 / 2.93	2.09 / 2.85	2.02 / 2.74	1.98 / 2.66	1.94 / 2.58	1.89 / 2.49	1.86 / 2.44	1.82 / 2.36	1.80 / 2.33	1.76 / 2.27	1.74 / 2.23	1.73 / 2.21
25	4.24 / 7.77	3.38 / 5.57	2.99 / 4.68	2.76 / 4.18	2.60 / 3.86	2.49 / 3.63	2.41 / 3.46	2.34 / 3.32	2.28 / 3.21	2.24 / 3.13	2.20 / 3.05	2.16 / 2.99	2.11 / 2.89	2.06 / 2.81	2.00 / 2.70	1.96 / 2.62	1.92 / 2.54	1.87 / 2.45	1.84 / 2.40	1.80 / 2.32	1.77 / 2.29	1.74 / 2.23	1.72 / 2.19	1.71 / 2.17
26	4.22 / 7.72	3.37 / 5.53	2.98 / 4.64	2.74 / 4.14	2.59 / 3.82	2.47 / 3.59	2.39 / 3.42	2.32 / 3.29	2.27 / 3.17	2.22 / 3.09	2.18 / 3.02	2.15 / 2.96	2.10 / 2.86	2.05 / 2.77	1.99 / 2.66	1.95 / 2.58	1.90 / 2.50	1.859 / 2.41	1.82 / 2.36	1.78 / 2.28	1.76 / 2.25	1.72 / 2.19	1.70 / 2.15	1.69 / 2.13

df																								
27	1.67 / 2.10	1.68 / 2.12	1.71 / 2.16	1.74 / 2.21	1.76 / 2.25	1.80 / 2.33	1.84 / 2.38	1.88 / 2.47	1.93 / 2.55	1.97 / 2.63	2.03 / 2.74	2.08 / 2.83	2.13 / 2.93	2.16 / 2.98	2.20 / 3.06	2.25 / 3.14	2.30 / 3.26	2.37 / 3.39	2.46 / 3.56	2.57 / 3.79	2.73 / 4.11	2.96 / 4.60	3.35 / 5.49	4.21 / 7.68
28	1.65 / 2.06	1.67 / 2.09	1.69 / 2.13	1.72 / 2.18	1.75 / 2.22	1.78 / 2.30	1.81 / 2.35	1.87 / 2.44	1.91 / 2.52	1.96 / 2.60	2.02 / 2.71	2.06 / 2.80	2.12 / 2.90	2.15 / 2.95	2.19 / 3.03	2.24 / 3.11	2.29 / 3.23	2.36 / 3.36	2.44 / 3.53	2.56 / 3.76	2.71 / 4.07	2.95 / 4.57	3.34 / 5.45	4.20 / 7.64
29	1.64 / 2.03	1.65 / 2.06	1.68 / 2.10	1.71 / 2.15	1.73 / 2.19	1.77 / 2.27	1.80 / 2.32	1.85 / 2.41	1.90 / 2.49	1.94 / 2.57	2.00 / 2.68	2.05 / 2.77	2.10 / 2.87	2.14 / 2.92	2.18 / 3.00	2.22 / 3.08	2.28 / 3.20	2.35 / 3.33	2.43 / 3.50	2.54 / 3.73	2.70 / 4.04	2.93 / 4.54	3.33 / 5.42	4.18 / 7.60
30	1.62 / 2.01	1.64 / 2.03	1.66 / 2.07	1.69 / 2.13	1.72 / 2.16	1.76 / 2.24	1.79 / 2.29	1.84 / 2.38	1.89 / 2.47	1.93 / 2.55	1.99 / 2.66	2.04 / 2.74	2.09 / 2.84	2.12 / 2.90	2.16 / 2.98	2.21 / 3.06	2.27 / 3.17	2.34 / 3.30	2.42 / 3.47	2.53 / 3.70	2.69 / 4.02	2.92 / 4.51	3.32 / 5.39	4.17 / 7.56
32	1.59 / 1.96	1.61 / 1.98	1.64 / 2.02	1.67 / 2.08	1.69 / 2.12	1.74 / 2.20	1.76 / 2.25	1.82 / 2.34	1.86 / 2.42	1.91 / 2.51	1.97 / 2.62	2.02 / 2.70	2.07 / 2.80	2.10 / 2.86	2.14 / 2.94	2.19 / 3.01	2.25 / 3.12	2.32 / 3.25	2.40 / 3.42	2.51 / 3.66	2.67 / 3.97	2.90 / 4.46	3.30 / 5.34	4.15 / 7.50
34	1.57 / 1.91	1.59 / 1.94	1.61 / 1.98	1.64 / 2.04	1.67 / 2.08	1.71 / 2.15	1.74 / 2.21	1.80 / 2.30	1.84 / 2.38	1.89 / 2.47	1.95 / 2.58	2.00 / 2.66	2.05 / 2.76	2.08 / 2.82	2.12 / 2.89	2.17 / 2.97	2.23 / 3.08	2.30 / 3.21	2.38 / 3.38	2.49 / 3.61	2.65 / 3.93	2.88 / 4.42	3.28 / 5.29	4.13 / 7.44
36	1.55 / 1.87	1.56 / 1.90	1.59 / 1.94	1.62 / 2.00	1.65 / 2.04	1.69 / 2.12	1.72 / 2.17	1.78 / 2.26	1.82 / 2.35	1.87 / 2.43	1.93 / 2.54	1.98 / 2.62	2.03 / 2.72	2.06 / 2.78	2.10 / 2.86	2.15 / 2.94	2.21 / 3.04	2.28 / 3.18	2.36 / 3.35	2.48 / 3.58	2.63 / 3.89	2.86 / 4.38	3.26 / 5.25	4.11 / 7.39
38	1.53 / 1.84	1.54 / 1.86	1.57 / 1.90	1.60 / 1.97	1.63 / 2.00	1.67 / 2.08	1.71 / 2.14	1.76 / 2.22	1.80 / 2.32	1.85 / 2.40	1.92 / 2.51	1.96 / 2.59	2.02 / 2.69	2.05 / 2.75	2.09 / 2.82	2.14 / 2.91	2.19 / 3.02	2.26 / 3.15	2.35 / 3.32	2.46 / 3.54	2.62 / 3.86	2.85 / 4.34	3.25 / 5.21	4.10 / 7.35
40	1.51 / 1.81	1.53 / 1.84	1.55 / 1.88	1.59 / 1.94	1.61 / 1.97	1.66 / 2.05	1.69 / 2.11	1.74 / 2.20	1.79 / 2.29	1.84 / 2.37	1.90 / 2.49	1.95 / 2.56	2.00 / 2.66	2.04 / 2.73	2.07 / 2.80	2.12 / 2.88	2.18 / 2.99	2.25 / 3.12	2.34 / 3.29	2.45 / 3.51	2.61 / 3.83	2.84 / 4.31	3.23 / 5.18	4.08 / 7.31
42	1.49 / 1.78	1.51 / 1.80	1.54 / 1.85	1.57 / 1.91	1.60 / 1.94	1.64 / 2.02	1.68 / 2.08	1.73 / 2.17	1.78 / 2.26	1.82 / 2.35	1.89 / 2.46	1.94 / 2.54	1.99 / 2.64	2.02 / 2.70	2.06 / 2.77	2.11 / 2.86	2.17 / 2.96	2.24 / 3.10	2.32 / 3.26	2.44 / 3.49	2.59 / 3.80	2.83 / 4.29	3.22 / 5.15	4.07 / 7.27
44	1.48 / 1.75	1.50 / 1.78	1.52 / 1.82	1.56 / 1.88	1.58 / 1.92	1.63 / 2.00	1.66 / 2.06	1.72 / 2.15	1.76 / 2.24	1.81 / 2.32	1.88 / 2.44	1.92 / 2.52	1.98 / 2.62	2.01 / 2.68	2.05 / 2.75	2.10 / 2.84	2.16 / 2.94	2.23 / 3.07	2.31 / 3.24	2.43 / 3.46	2.58 / 3.78	2.82 / 4.26	3.21 / 5.12	4.06 / 7.24
46	1.46 / 1.72	1.48 / 1.76	1.51 / 1.80	1.54 / 1.86	1.57 / 1.90	1.62 / 1.98	1.65 / 2.04	1.71 / 2.13	1.75 / 2.22	1.80 / 2.30	1.87 / 2.42	1.91 / 2.50	1.97 / 2.60	2.00 / 2.66	2.04 / 2.73	2.09 / 2.82	2.14 / 2.92	2.22 / 3.05	2.30 / 3.22	2.42 / 3.44	2.57 / 3.76	2.81 / 4.24	3.20 / 5.10	4.05 / 7.21
48	1.45 / 1.70	1.47 / 1.73	1.50 / 1.78	1.53 / 1.84	1.56 / 1.88	1.61 / 1.96	1.64 / 2.02	1.70 / 2.11	1.74 / 2.20	1.79 / 2.28	1.86 / 2.40	1.90 / 2.48	1.96 / 2.58	1.99 / 2.64	2.03 / 2.71	2.08 / 2.80	2.14 / 2.90	2.20 / 3.04	2.30 / 3.20	2.41 / 3.42	2.56 / 3.74	2.80 / 4.22	3.19 / 5.08	4.04 / 7.19
50	1.44 / 1.68	1.46 / 1.71	1.48 / 1.76	1.52 / 1.82	1.55 / 1.86	1.60 / 1.94	1.63 / 2.00	1.69 / 2.10	1.74 / 2.18	1.78 / 2.26	1.85 / 2.39	1.90 / 2.46	1.95 / 2.56	1.98 / 2.62	2.02 / 2.70	2.07 / 2.78	2.13 / 2.88	2.20 / 3.02	2.29 / 3.18	2.40 / 3.41	2.56 / 3.72	2.79 / 4.20	3.18 / 5.06	4.03 / 7.17
55	1.41 / 1.64	1.43 / 1.66	1.46 / 1.71	1.50 / 1.78	1.52 / 1.82	1.58 / 1.90	1.61 / 1.96	1.67 / 2.06	1.72 / 2.15	1.76 / 2.23	1.83 / 2.35	1.88 / 2.43	1.93 / 2.53	1.97 / 2.59	2.00 / 2.66	2.05 / 2.75	2.11 / 2.85	2.18 / 2.98	2.27 / 3.15	2.38 / 3.37	2.54 / 3.68	2.78 / 4.16	3.17 / 5.01	4.02 / 7.12
60	1.39 / 1.60	1.41 / 1.63	1.44 / 1.68	1.48 / 1.74	1.50 / 1.79	1.56 / 1.87	1.59 / 1.93	1.65 / 2.03	1.70 / 2.12	1.75 / 2.20	1.81 / 2.32	1.86 / 2.40	1.92 / 2.50	1.95 / 2.56	1.99 / 2.63	2.04 / 2.72	2.10 / 2.82	2.17 / 2.95	2.25 / 3.12	2.37 / 3.34	2.52 / 3.65	2.76 / 4.13	3.15 / 4.98	4.00 / 7.08
65	1.37 / 1.56	1.39 / 1.60	1.42 / 1.64	1.46 / 1.71	1.49 / 1.76	1.54 / 1.84	1.57 / 1.90	1.63 / 2.00	1.68 / 2.09	1.73 / 2.18	1.80 / 2.30	1.85 / 2.37	1.90 / 2.47	1.94 / 2.54	1.98 / 2.61	2.02 / 2.70	2.08 / 2.79	2.15 / 2.93	2.24 / 3.09	2.36 / 3.31	2.51 / 3.62	2.75 / 4.10	3.14 / 4.95	3.99 / 7.04

Table C[a] (Continued)
CRITICAL VALUES OF *F*

FINDING *p*-VALUE
If observed *F* is
... smaller than light number, $p > .05$
... between light and dark numbers, $p < .05$
... larger than dark number, $p < .01$

DEGREES OF FREEDOM IN NUMERATOR

DEGREES OF FREEDOM IN DENOMINATOR	1	2	3	4	5	6	7	8	9	10	11	12	14	16	20	24	30	40	50	75	100	200	500	∞
70	3.98 **7.01**	3.13 **4.92**	2.74 **4.08**	2.50 **3.60**	2.35 **3.29**	2.23 **3.07**	2.14 **2.91**	2.07 **2.77**	2.01 **2.67**	1.97 **2.59**	1.93 **2.51**	1.89 **2.45**	1.84 **2.35**	1.79 **2.28**	1.72 **2.15**	1.67 **2.07**	1.62 **1.98**	1.56 **1.88**	1.53 **1.82**	1.47 **1.74**	1.45 **1.69**	1.40 **1.62**	1.37 **1.56**	1.35 **1.53**
80	3.96 **6.96**	3.11 **4.88**	2.72 **4.04**	2.48 **3.56**	2.33 **3.25**	2.21 **3.04**	2.12 **2.87**	2.05 **2.74**	1.99 **2.64**	1.95 **2.55**	1.91 **2.48**	1.88 **2.41**	1.82 **2.32**	1.77 **2.24**	1.70 **2.11**	1.65 **2.03**	1.60 **1.94**	1.54 **1.84**	1.51 **1.78**	1.45 **1.70**	1.42 **1.65**	1.38 **1.57**	1.35 **1.52**	1.32 **1.49**
100	3.94 **6.90**	3.09 **4.82**	2.70 **3.98**	2.46 **3.51**	2.30 **3.20**	2.19 **2.99**	2.10 **2.82**	2.03 **2.69**	1.97 **2.59**	1.92 **2.51**	1.88 **2.43**	1.85 **2.36**	1.79 **2.26**	1.75 **2.19**	1.68 **2.06**	1.63 **1.98**	1.57 **1.89**	1.51 **1.79**	1.48 **1.73**	1.42 **1.64**	1.39 **1.59**	1.34 **1.51**	1.30 **1.46**	1.28 **1.43**
125	3.92 **6.84**	3.07 **4.78**	2.68 **3.94**	2.44 **3.47**	2.29 **3.17**	2.17 **2.95**	2.08 **2.79**	2.01 **2.65**	1.95 **2.56**	1.90 **2.47**	1.86 **2.40**	1.83 **2.33**	1.77 **2.23**	1.72 **2.15**	1.65 **2.03**	1.60 **1.94**	1.55 **1.85**	1.49 **1.75**	1.45 **1.68**	1.39 **1.59**	1.36 **1.54**	1.31 **1.46**	1.27 **1.40**	1.25 **1.37**
150	3.91 **6.81**	3.06 **4.75**	2.67 **3.91**	2.43 **3.44**	2.27 **3.14**	2.16 **2.92**	2.07 **2.76**	2.00 **2.62**	1.94 **2.53**	1.89 **2.44**	1.85 **2.37**	1.82 **2.30**	1.76 **2.20**	1.71 **2.12**	1.64 **2.00**	1.59 **1.91**	1.54 **1.83**	1.47 **1.72**	1.44 **1.66**	1.37 **1.56**	1.34 **1.51**	1.29 **1.43**	1.25 **1.37**	1.22 **1.33**
200	3.89 **6.76**	3.04 **4.71**	2.65 **3.88**	2.41 **3.41**	2.26 **3.11**	2.14 **2.90**	2.05 **2.73**	1.98 **2.60**	1.92 **2.50**	1.87 **2.41**	1.83 **2.34**	1.80 **2.28**	1.74 **2.17**	1.69 **2.09**	1.62 **1.97**	1.57 **1.88**	1.52 **1.79**	1.45 **1.69**	1.42 **1.62**	1.35 **1.53**	1.32 **1.48**	1.26 **1.39**	1.22 **1.33**	1.19 **1.28**
400	3.86 **6.70**	3.02 **4.66**	2.62 **3.83**	2.39 **3.36**	2.23 **3.06**	2.12 **2.85**	2.03 **2.69**	1.96 **2.55**	1.90 **2.46**	1.85 **2.37**	1.81 **2.29**	1.78 **2.23**	1.72 **2.12**	1.67 **2.04**	1.60 **1.92**	1.54 **1.84**	1.49 **1.74**	1.42 **1.64**	1.38 **1.57**	1.32 **1.47**	1.28 **1.42**	1.22 **1.32**	1.16 **1.24**	1.13 **1.19**
1000	3.85 **6.66**	3.00 **4.62**	2.61 **3.80**	2.38 **3.34**	2.22 **3.04**	2.10 **2.82**	2.02 **2.66**	1.95 **2.53**	1.89 **2.43**	1.84 **2.34**	1.80 **2.26**	1.76 **2.20**	1.70 **2.09**	1.65 **2.01**	1.58 **1.89**	1.53 **1.81**	1.47 **1.71**	1.41 **1.61**	1.36 **1.54**	1.30 **1.44**	1.26 **1.38**	1.19 **1.28**	1.13 **1.19**	1.08 **1.11**
∞	3.84 **6.64**	2.99 **4.60**	2.60 **3.78**	2.37 **3.32**	2.21 **3.02**	2.09 **2.80**	2.01 **2.64**	1.94 **2.51**	1.86 **2.41**	1.83 **2.32**	1.79 **2.24**	1.75 **2.18**	1.69 **2.07**	1.64 **1.99**	1.57 **1.87**	1.52 **1.79**	1.46 **1.69**	1.40 **1.59**	1.35 **1.52**	1.28 **1.41**	1.24 **1.36**	1.17 **1.25**	1.11 **1.15**	1.00 **1.00**

Table Dᵃ
CRITICAL VALUES OF χ^2

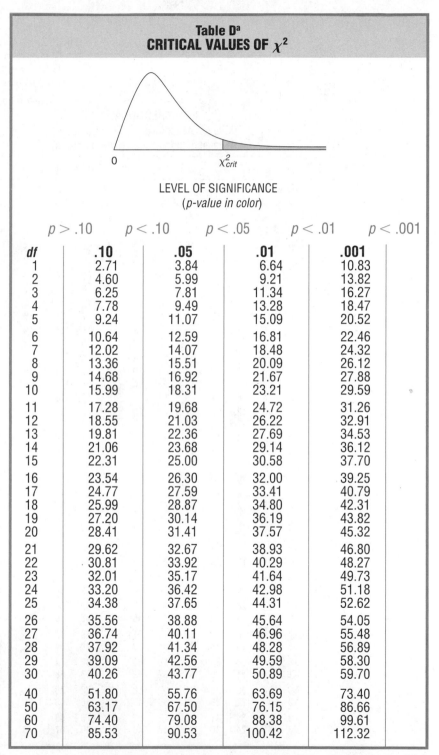

LEVEL OF SIGNIFICANCE
(*p-value in color*)

	$p > .10$	$p < .10$	$p < .05$	$p < .01$	$p < .001$
df	**.10**	**.05**	**.01**	**.001**	
1	2.71	3.84	6.64	10.83	
2	4.60	5.99	9.21	13.82	
3	6.25	7.81	11.34	16.27	
4	7.78	9.49	13.28	18.47	
5	9.24	11.07	15.09	20.52	
6	10.64	12.59	16.81	22.46	
7	12.02	14.07	18.48	24.32	
8	13.36	15.51	20.09	26.12	
9	14.68	16.92	21.67	27.88	
10	15.99	18.31	23.21	29.59	
11	17.28	19.68	24.72	31.26	
12	18.55	21.03	26.22	32.91	
13	19.81	22.36	27.69	34.53	
14	21.06	23.68	29.14	36.12	
15	22.31	25.00	30.58	37.70	
16	23.54	26.30	32.00	39.25	
17	24.77	27.59	33.41	40.79	
18	25.99	28.87	34.80	42.31	
19	27.20	30.14	36.19	43.82	
20	28.41	31.41	37.57	45.32	
21	29.62	32.67	38.93	46.80	
22	30.81	33.92	40.29	48.27	
23	32.01	35.17	41.64	49.73	
24	33.20	36.42	42.98	51.18	
25	34.38	37.65	44.31	52.62	
26	35.56	38.88	45.64	54.05	
27	36.74	40.11	46.96	55.48	
28	37.92	41.34	48.28	56.89	
29	39.09	42.56	49.59	58.30	
30	40.26	43.77	50.89	59.70	
40	51.80	55.76	63.69	73.40	
50	63.17	67.50	76.15	86.66	
60	74.40	79.08	88.38	99.61	
70	85.53	90.53	100.42	112.32	

ᵃ*Discussed in Section 24.5.*

Table E[a]
CRITICAL VALUES OF MANN-WHITNEY U

FINDING p-VALUE
If observed U is
... larger than light number, $p > .05$
... between light and dark numbers, $p < .05$
... smaller than dark number, $p < .01$

NONDIRECTIONAL TEST
.05 level of significance (light numbers)
.01 level of significance (dark numbers)

n_2 \ n_1	1	2	3	4	5	6	7	8	9	10	11	12	13	14	15	16	17	18	19	20
1	–	–	–	–	–	–	–	–	–	–	–	–	–	–	–	–	–	–	–	–
	–	**–**	**–**	**–**	**–**	**–**	**–**	**–**	**–**	**–**	**–**	**–**	**–**	**–**	**–**	**–**	**–**	**–**	**–**	**–**
2	–	–	–	–	–	–	–	0	0	0	0	1	1	1	1	1	2	2	2	2
	–	**–**	**–**	**–**	**–**	**–**	**–**	**–**	**–**	**–**	**–**	**–**	**–**	**–**	**–**	**–**	**–**	**–**	**0**	**0**
3	–	–	–	–	0	1	1	2	2	3	3	4	4	5	5	6	6	7	7	8
	–	**–**	**–**	**–**	**–**	**–**	**–**	**–**	**0**	**0**	**0**	**1**	**1**	**1**	**2**	**2**	**2**	**2**	**3**	**3**
4	–	–	–	0	1	2	3	4	4	5	6	7	8	9	10	11	11	12	13	13
	–	**–**	**–**	**–**	**–**	**0**	**0**	**1**	**1**	**2**	**2**	**3**	**3**	**4**	**5**	**5**	**6**	**6**	**7**	**8**
5	–	–	0	1	2	3	5	6	7	8	9	11	12	13	14	15	17	18	19	20
	–	**–**	**–**	**–**	**0**	**1**	**1**	**2**	**3**	**4**	**5**	**6**	**7**	**7**	**8**	**9**	**10**	**11**	**12**	**13**
6	–	–	1	2	3	5	6	8	10	11	13	14	16	17	19	21	22	24	25	27
	–	**–**	**–**	**0**	**1**	**2**	**3**	**4**	**5**	**6**	**7**	**9**	**10**	**11**	**12**	**13**	**15**	**16**	**17**	**18**
7	–	–	1	3	5	6	8	10	12	14	16	18	20	22	24	26	28	30	32	34
	–	**–**	**–**	**0**	**1**	**3**	**4**	**6**	**7**	**9**	**10**	**12**	**13**	**15**	**16**	**18**	**19**	**21**	**22**	**24**
8	–	0	2	4	6	8	10	13	15	17	19	22	24	26	29	31	34	36	38	41
	–	**–**	**–**	**1**	**2**	**4**	**6**	**7**	**9**	**11**	**13**	**15**	**17**	**18**	**20**	**22**	**24**	**26**	**28**	**30**
9	–	0	2	4	7	10	12	15	17	20	23	26	28	31	34	37	39	42	45	48
	–	**–**	**0**	**1**	**3**	**5**	**7**	**9**	**11**	**13**	**16**	**18**	**20**	**22**	**24**	**27**	**29**	**31**	**33**	**36**
10	–	0	3	5	8	11	14	17	20	23	26	29	33	36	39	42	45	48	52	55
	–	**–**	**0**	**2**	**4**	**6**	**9**	**11**	**13**	**16**	**18**	**21**	**24**	**26**	**29**	**31**	**34**	**37**	**39**	**42**
11	–	0	3	6	9	13	16	19	23	26	30	33	37	40	44	47	51	55	58	62
	–	**–**	**0**	**2**	**5**	**7**	**10**	**13**	**16**	**18**	**21**	**24**	**27**	**30**	**33**	**36**	**39**	**42**	**45**	**48**
12	–	1	4	7	11	14	18	22	26	29	33	37	41	45	49	53	57	61	65	69
	–	**–**	**1**	**3**	**6**	**9**	**12**	**15**	**18**	**21**	**24**	**27**	**31**	**34**	**37**	**41**	**44**	**47**	**51**	**54**
13	–	1	4	8	12	16	20	24	28	33	37	41	45	50	54	59	63	67	72	76
	–	**–**	**1**	**3**	**7**	**10**	**13**	**17**	**20**	**24**	**27**	**31**	**34**	**38**	**42**	**45**	**49**	**53**	**56**	**60**
14	–	1	5	9	13	17	22	26	31	36	40	45	50	55	59	64	67	74	78	83
	–	**–**	**1**	**4**	**7**	**11**	**15**	**18**	**22**	**26**	**30**	**34**	**38**	**42**	**46**	**50**	**54**	**58**	**63**	**67**
15	–	1	5	10	14	19	24	29	34	39	44	49	54	59	64	70	75	80	85	90
	–	**–**	**2**	**5**	**8**	**12**	**16**	**20**	**24**	**29**	**33**	**37**	**42**	**46**	**51**	**55**	**60**	**64**	**69**	**73**
16	–	1	6	11	15	21	26	31	37	42	47	53	59	64	70	75	81	86	92	98
	–	**–**	**2**	**5**	**9**	**13**	**18**	**22**	**27**	**31**	**36**	**41**	**45**	**50**	**55**	**60**	**65**	**70**	**74**	**79**
17	–	2	6	11	17	22	28	34	39	45	51	57	63	67	75	81	87	93	99	105
	–	**–**	**2**	**6**	**10**	**15**	**19**	**24**	**29**	**34**	**39**	**44**	**49**	**54**	**60**	**65**	**70**	**75**	**81**	**86**
18	–	2	7	12	18	24	30	36	42	48	55	61	67	74	80	86	93	99	106	112
	–	**–**	**2**	**6**	**11**	**16**	**21**	**26**	**31**	**37**	**42**	**47**	**53**	**58**	**64**	**70**	**75**	**81**	**87**	**92**
19	–	2	7	13	19	25	32	38	45	52	58	65	72	78	85	92	99	106	113	119
	–	**0**	**3**	**7**	**12**	**17**	**22**	**28**	**33**	**39**	**45**	**51**	**56**	**63**	**69**	**74**	**81**	**87**	**93**	**99**
20	–	2	8	13	20	27	34	41	48	55	62	69	76	83	90	98	105	112	119	127
	–	**0**	**3**	**8**	**13**	**18**	**24**	**30**	**36**	**42**	**48**	**54**	**60**	**67**	**73**	**79**	**86**	**92**	**99**	**105**

Table E[a] (Continued)
CRITICAL VALUES OF MANN-WHITNEY U

DIRECTIONAL TEST
.05 level of significance (light numbers)
.01 level of significance (**dark numbers**)

n_2	sig	1	2	3	4	5	6	7	8	9	10	11	12	13	14	15	16	17	18	19	20
1	.05	–	–	–	–	–	–	–	–	–	–	–	–	–	–	–	–	–	–	0	0
	.01	–	–	–	–	–	–	–	–	–	–	–	–	–	–	–	–	–	–	–	–
2	.05	–	–	–	–	0	0	0	1	1	1	1	2	2	2	3	3	3	4	4	4
	.01	–	–	–	–	–	–	–	–	–	–	–	–	**0**	**0**	**0**	**0**	**0**	**0**	**1**	**1**
3	.05	–	–	0	0	1	2	2	3	3	4	5	5	6	7	7	8	9	9	10	11
	.01	–	–	–	–	–	**0**	**0**	**1**	**1**	**1**	**2**	**2**	**2**	**3**	**3**	**4**	**4**	**4**	**4**	**5**
4	.05	–	–	0	1	2	3	4	5	6	7	8	9	10	11	12	14	15	16	17	18
	.01	–	–	–	–	**0**	**1**	**1**	**2**	**3**	**3**	**4**	**5**	**5**	**6**	**7**	**7**	**8**	**9**	**9**	**10**
5	.05	–	0	1	2	4	5	6	8	9	11	12	13	15	16	18	19	20	22	23	25
	.01	–	–	–	**0**	**1**	**2**	**3**	**4**	**5**	**6**	**7**	**8**	**9**	**10**	**11**	**12**	**13**	**14**	**15**	**16**
6	.05	–	0	2	3	5	7	8	10	12	14	16	17	19	21	23	25	26	28	30	32
	.01	–	–	–	**1**	**2**	**3**	**4**	**6**	**7**	**8**	**9**	**11**	**12**	**13**	**15**	**16**	**18**	**19**	**20**	**22**
7	.05	–	0	2	4	6	8	11	13	15	17	19	21	24	26	28	30	33	35	37	39
	.01	–	–	**0**	**1**	**3**	**4**	**6**	**7**	**9**	**11**	**12**	**14**	**16**	**17**	**19**	**21**	**23**	**24**	**26**	**28**
8	.05	–	1	3	5	8	10	13	15	18	20	23	26	28	31	33	36	39	41	44	47
	.01	–	–	**0**	**2**	**4**	**6**	**7**	**9**	**11**	**13**	**15**	**17**	**20**	**22**	**24**	**26**	**28**	**30**	**32**	**34**
9	.05	–	1	3	6	9	12	15	18	21	24	27	30	33	36	39	42	45	48	51	54
	.01	–	–	**1**	**3**	**5**	**7**	**9**	**11**	**14**	**16**	**18**	**21**	**23**	**26**	**28**	**31**	**33**	**36**	**38**	**40**
10	.05	–	1	4	7	11	14	17	20	24	27	31	34	37	41	44	48	51	55	58	62
	.01	–	–	**1**	**3**	**6**	**8**	**11**	**13**	**16**	**19**	**22**	**24**	**27**	**30**	**33**	**36**	**38**	**41**	**44**	**47**
11	.05	–	1	5	8	12	16	19	23	27	31	34	38	42	46	50	54	57	61	65	69
	.01	–	–	**1**	**4**	**7**	**9**	**12**	**15**	**18**	**22**	**25**	**28**	**31**	**34**	**37**	**41**	**44**	**47**	**50**	**53**
12	.05	–	2	5	9	13	17	21	26	30	34	38	42	47	51	55	60	64	68	72	77
	.01	–	–	**2**	**5**	**8**	**11**	**14**	**17**	**21**	**24**	**28**	**31**	**35**	**38**	**42**	**46**	**49**	**53**	**56**	**60**
13	.05	–	2	6	10	15	19	24	28	33	37	42	47	51	56	61	65	70	75	80	84
	.01	–	**0**	**2**	**5**	**9**	**12**	**16**	**20**	**23**	**27**	**31**	**35**	**39**	**43**	**47**	**51**	**55**	**59**	**63**	**67**
14	.05	–	2	7	11	16	21	26	31	36	41	46	51	56	61	66	71	77	82	87	92
	.01	–	**0**	**2**	**6**	**10**	**13**	**17**	**22**	**26**	**30**	**34**	**38**	**43**	**47**	**51**	**56**	**60**	**65**	**69**	**73**
15	.05	–	3	7	12	18	23	28	33	39	44	50	55	61	66	72	77	83	88	94	100
	.01	–	**0**	**3**	**7**	**11**	**15**	**19**	**24**	**28**	**33**	**37**	**42**	**47**	**51**	**56**	**61**	**66**	**70**	**75**	**80**
16	.05	–	3	8	14	19	25	30	36	42	48	54	60	65	71	77	83	89	95	101	107
	.01	–	**0**	**3**	**7**	**12**	**16**	**21**	**26**	**31**	**36**	**41**	**46**	**51**	**56**	**61**	**66**	**71**	**76**	**82**	**87**
17	.05	–	3	9	15	20	26	33	39	45	51	57	64	70	77	83	89	96	102	109	115
	.01	–	**0**	**4**	**8**	**13**	**18**	**23**	**28**	**33**	**38**	**44**	**49**	**55**	**60**	**66**	**71**	**77**	**82**	**88**	**93**
18	.05	–	4	9	16	22	28	35	41	48	55	61	68	75	82	88	95	102	109	116	123
	.01	–	**0**	**4**	**9**	**14**	**19**	**24**	**30**	**36**	**41**	**47**	**53**	**59**	**65**	**70**	**76**	**82**	**88**	**94**	**100**
19	.05	0	4	10	17	23	30	37	44	51	58	65	72	80	87	94	101	109	116	123	130
	.01	–	**1**	**4**	**9**	**15**	**20**	**26**	**32**	**38**	**44**	**50**	**56**	**63**	**69**	**75**	**82**	**88**	**94**	**101**	**107**
20	.05	0	4	11	18	25	32	39	47	54	62	69	77	84	92	100	107	115	123	130	138
	.01	–	**1**	**5**	**10**	**16**	**22**	**28**	**34**	**40**	**47**	**53**	**60**	**67**	**73**	**80**	**87**	**93**	**100**	**107**	**114**

[a]*Discussed in Section 25.6. To be significant, the observed* U *must equal or be* less than *the value shown in the table. Dashes in the table indicate that no decision is possible at the specified level of significance.*

Table Fª
CRITICAL VALUES OF WILCOXON *T*

FINDING *p*-VALUE
If observed *T* is
. . . larger than .05 number, $p > .05$
. . . between .05 and .01 numbers, $p < .05$
. . . smaller than .01 number, $p < .01$

LEVEL OF SIGNIFICANCE

n	NONDIRECTIONAL TEST .05	.01	n	.05	.01	n	DIRECTIONAL TEST .05	.01	n	.05	.01
5	–	–	28	116	91	5	0	–	28	130	101
6	0	–	29	126	100	6	2	–	29	140	110
7	2	–	30	137	109	7	3	0	30	151	120
8	3	0	31	147	118	8	5	1	31	163	130
9	5	1	32	159	128	9	8	3	32	175	140
10	8	3	33	170	138	10	10	5	33	187	151
11	10	5	34	182	148	11	13	7	34	200	162
12	13	7	35	195	159	12	17	9	35	213	173
13	17	9	36	208	171	13	21	12	36	227	185
14	21	12	37	221	182	14	25	15	37	241	198
15	25	15	38	235	194	15	30	19	38	256	211
16	29	19	39	249	207	16	35	23	39	271	224
17	34	23	40	264	220	17	41	27	40	286	238
18	40	27	41	279	233	18	47	32	41	302	252
19	46	32	42	294	247	19	53	37	42	319	266
20	52	37	43	310	261	20	60	43	43	336	281
21	58	42	44	327	276	21	67	49	44	353	296
22	65	48	45	343	291	22	75	55	45	371	312
23	73	54	46	361	307	23	83	62	46	389	328
24	81	61	47	378	322	24	91	69	47	407	345
25	89	68	48	396	339	25	100	76	48	426	362
26	98	75	49	415	355	26	110	84	49	446	379
27	107	83	50	434	373	27	119	92	50	466	397

ªDiscussed in Section 25.12. *To be significant, the observed* T *must equal or be* less than *the value shown in the table.
Dashes in the table indicate that no decision is possible at the specified level of significance.*

Table G[a]
RANDOM NUMBERS

ROW NUMBER										
00000	10097	32533	76520	13586	34673	54876	80959	09117	39292	74945
00001	37542	04805	64894	74296	24805	24037	20636	10402	00822	91665
00002	08422	68953	19645	09303	23209	02560	15953	34764	35080	33606
00003	99019	02529	09376	70715	38311	31165	88676	74397	04436	27659
00004	12807	99970	80157	36147	64032	36653	98951	16877	12171	76833
00005	66065	74717	34072	76850	36697	36170	65813	39885	11199	29170
00006	31060	10805	45571	82406	35303	42614	86799	07439	23403	09732
00007	85269	77602	02051	65692	68665	74818	73053	85247	18623	88579
00008	63573	32135	05325	47048	90553	57548	28468	28709	83491	25624
00009	73796	45753	03529	64778	35808	34282	60935	20344	35273	88435
00010	98520	17767	14905	68607	22109	40558	60970	93433	50500	73998
00011	11805	05431	39808	27732	50725	68248	29405	24201	52775	67851
00012	83452	99634	06288	98033	13746	70078	18475	40610	68711	77817
00013	88685	40200	86507	58401	36766	67951	90364	76493	29609	11062
00014	99594	67348	87517	64969	91826	08928	93785	61368	23478	34113
00015	65481	17674	17468	50950	58047	76974	73039	57186	40218	16544
00016	80124	35635	17727	08015	45318	22374	21115	78253	14385	53763
00017	74350	99817	77402	77214	43236	00210	45521	64237	96286	02655
00018	69916	26803	66252	29148	36936	87203	76621	13990	94400	56418
00019	09893	20505	14225	68514	46427	56788	96297	78822	54382	14598
00020	91499	14523	68479	27686	46162	83554	94750	89923	37089	20048
00021	80336	94598	26940	36858	70297	34135	53140	33340	42050	82341
00022	44104	81949	85157	47954	32979	26575	57600	40881	22222	06413
00023	12550	73742	11100	02040	12860	74697	96644	89439	28707	25815
00024	63606	49329	16505	34484	40219	52563	43651	77082	07207	31790
00025	61196	90446	26457	47774	51924	33729	65394	59593	42582	60527
00026	15474	45266	95270	79953	59367	83848	82396	10118	33211	59466
00027	94557	28573	67897	54387	54622	44431	91190	42592	92927	45973
00028	42481	16213	97344	08721	16868	48767	03071	12059	25701	46670
00029	23523	78317	73208	89837	68935	91416	26252	29663	05522	82562
00030	04493	52494	75246	33824	45862	51025	61962	79335	65337	12472
00031	00549	97654	64051	88159	96119	63896	54692	82391	23287	29529
00032	35963	15307	26898	09354	33351	35462	77974	50024	90130	39333
00033	59808	08391	45427	26842	83609	49700	13021	24892	78565	20106
00034	46058	85236	01390	92286	77281	44077	93910	83647	70617	42941
00035	32179	00597	87379	25241	05567	07007	86743	17157	85394	11838
00036	69234	61406	20117	45204	15956	60000	18743	92423	97188	96338
00037	19565	41430	01758	75379	40419	21585	66674	36806	84962	85207
00038	45155	14938	19476	07246	43667	94543	59047	90033	20826	69541
00039	94864	31994	36168	10851	34888	81553	01540	35456	05014	51176
00040	98086	24826	45240	28404	44999	08896	39094	73407	35441	31880
00041	33185	16232	41941	50949	89435	48581	88695	41944	37548	73043
00042	80951	00406	96382	70774	20151	23387	25016	25298	94624	61171
00043	79752	49140	71961	28296	69861	02591	74852	20539	00387	59579
00044	18633	32537	98145	06571	31010	24674	05455	61427	77938	91936
00045	74029	43902	77557	32270	97790	17119	52527	58021	80814	51748
00046	54178	45611	80993	37143	05335	12969	56127	19255	36040	90324
00047	11664	49883	52079	84827	59381	71539	09973	33440	88461	23356
00048	48324	77928	31249	64710	02295	36870	32307	57546	15020	09994
00049	69074	94138	87637	91976	35584	04401	10518	21615	01848	76938
00050	09188	20097	32825	39527	04220	86304	83389	87374	64278	58044
00051	90045	85497	51981	50654	94938	81997	91870	76150	68476	64659
00052	73189	50207	47677	26269	62290	64464	27124	67018	41361	82760
00053	75768	76490	20971	87749	90429	12272	95375	05871	93823	43178
00054	54016	44056	66281	31003	00682	27398	20714	53295	07706	17813
00055	08358	69910	78542	42785	13661	58873	04618	97553	31223	08420
00056	28306	03264	81333	10591	40510	07893	32604	60475	94119	01840
00057	53840	86233	81594	13628	51215	90290	28466	68795	77762	20791
00058	91757	53471	61613	62669	50263	90212	55781	76514	83483	47055
00059	89415	92694	00397	58391	12607	17646	48949	72306	94541	37408

[a]Discussed in Section 11.5.

APPENDIX

E

Glossary

Glossary

Numbers in parentheses indicate section in which term is introduced and defined.

Addition rule: Add together the separate probabilities of several mutually exclusive outcomes to find the probability that any one of these outcomes will occur. (12.2)

Alpha (α): The probability of a type I error, that is, the probability of rejecting a true null hypothesis. (16.3) Also see *Level of significance*. (14.7)

Alternative hypothesis (H_1): The opposite of the null hypothesis. Often identified with the research hypothesis. (14.6)

Analysis of variance (ANOVA): An overall test of the null hypothesis for more than two population means. (22.1)

Average: Usually refers to the mean. (4.7)

Bar graph: Bar-type graph for qualitative data. Gaps between adjacent bars. (3.5)

Beta (β): The probability of a type II error, that is, the probability of retaining a false null hypothesis. (16.4)

Bimodal: Describes any distribution with two obvious peaks. (4.1)

Central limit theorem: A statement that the shape of the sampling distribution of the mean will approximate a normal curve if the sample size is sufficiently large. (13.6)

Chi-square test (χ^2): A test of the null hypothesis for qualitative data, expressed as frequencies. (24.1)

Class interval width: The distance between the two tabled boundaries, after each boundary has been expanded by one-half of one unit of measurement. (2.5)

Conditional probability: The probability of one outcome, *given* the occurrence of another outcome. (12.3)

Confidence interval: A range of values that, with a known degree of certainty, includes an unknown population characteristic, such as a population mean. (17.3)

Confidence interval for $\mu_1 - \mu_2$ (or μ_D): A range of values that, in the long run, includes the unknown difference between population means a certain percent of the time. (19.10)

Correlation coefficient (r): A number between -1 and 1 that describes the relationship between pairs of variables. (9.3)

Correlation matrix: A table showing correlations for all possible pairs of variables. (9.11)

Correlation study: A study with two dependent variables. (1.9)

Counterbalancing: Reversing the order of conditions for equal numbers of subjects. (20.9)

Critical z score: A z score that separates common from rare outcomes and hence dictates whether the null hypothesis should be retained or rejected. (14.7)

Cumulative frequency distribution: A frequency distribution showing the total number of observations in each class and all lower-ranked classes. (2.10)

Curvilinear relationship: A relationship that can be described with a curved line. (9.2)

Data: A collection of observations from a survey or an experiment. (1.5)

Decision rule: Specifies precisely when the null hypothesis should be rejected (because the observed value qualifies as a rare outcome). (14.7)

Degrees of freedom (df): The number of values free to vary, given one or more mathematical restrictions. (18.11) Also, in analysis of variance, the number of deviations free to vary in any sum of squares term. (22.9)

Dependent variable: A variable that is measured, counted, or recorded by the investigator. (1.9)

Descriptive statistics: The area of statistics concerned with organizing and summarizing information about a collection of actual observations. (1.2)

Difference score (*D*): The arithmetic difference between each pair of scores in two matched samples. (20.3)

Directional test: See *One-tailed test*.

Distribution-free tests: Tests, such as U, T, and H, that make no assumptions about the form of the population distribution. (25.2)

Effect: Any difference between a true and a hypothesized population mean. (16.4) Also, any difference between two (or more) population means. (19.3) (21.8) See *Treatment effect*.

Estimated standard error of the mean ($s_{\bar{x}}$): The version of the standard error of the mean that is used whenever the unknown population standard deviation must be estimated. (18.9)

Expected frequency (*f_e*): The hypothesized frequency for each category, given that the null hypothesis is true. Used with the chi-square test. (24.3)

Experiment: A study with an independent and dependent variable. (1.9)

Frequency distribution: A collection of observations, produced by sorting observations into classes that show their frequencies of occurrence. (2.1)

Frequency distribution for grouped data: A frequency distribution produced whenever observations are sorted into classes of *more than one* value. (2.2)

Frequency distribution for ungrouped data: A frequency distribution produced whenever observations are sorted into classes of *single* values. (2.1)

Frequency polygon: A line graph for quantitative data. (3.2)

***F* ratio:** Ratio of the between-group mean square (for subjects treated differently) to the within-group mean square (for subjects treated similarly). (22.10)

Grouped data: Observations organized into classes of more than one value. (2.2)

Histogram: A bar-type graph for quantitative data, with no gaps between adjacent bars. (3.1)

Hypothesis test: Indicates whether an effect is present. (17.9)

Hypothesized sampling distribution: Centered about the population mean, this distribution is used to generate the decision rule. (16.4)

Independent outcomes: The occurrence of one outcome has no effect on the probability that the other outcome will occur. (12.3)

Independent variable: A variable that is manipulated by the investigator. (1.9)

Inferential statistics: The area of statistics concerned about generalizing beyond actual observations. (1.3)

Interaction: The product of inconsistent simple effects. (23.4)

Interquartile range (IQR): The range for the middle 50 percent of all observations. (5.11)

Interval measurement: Locates observations along a scale having equal intervals. (B.4)

Kruskal-Wallis *H* test: A test for ranked data when there are more than two independent groups. (25.16)

Least significant difference (LSD) test: A multiple comparison test. Also known as a *protected t* test. (22.17)

Least squares prediction equation: The equation that minimizes the total of all squared predictive errors for known Y scores in the original correlation analysis. (10.4)

Level of confidence: The percent of time that a series of confidence intervals includes the unknown population characteristic, such as the population mean. (17.7)

Level of significance (*α*): The degree of rarity among random outcomes required to reject the null hypothesis. (14.7)

Linear relationship: A relationship that can be described with a straight line. (9.2)

Main effect: The effect of a single factor when any other factor is ignored. (23.2)

Mann-Whitney *U* test: A test for ranked data when there are two independent groups. (25.5)

Margin of error: That which is added to and subtracted from some sample value, such as the sample proportion or sample mean, to obtain the limits of a confidence interval. (17.10)

Mean: The balance point for a frequency distribution, found by dividing the total of all observations by the number of observations. (4.3)

Mean square (*MS*): A variance estimate obtained by dividing a sum of squares by its degrees of freedom. (22.7)

Measures of central tendency: A general term for the various averages. (4.1)

Measures of variability: A general term for various measures of the amount of variation or differences among observations in a distribution. (5.1)

Median: The middle value when observations are ordered from least to most. (4.2)

Mode: The value of the most frequent observation. (4.1)

Multimodal: A distribution with more than two peaks. (4.1)

Multiple comparisons: The series of possible comparisons whenever, as in analysis of variance, more than two population means are involved. (22.15)

Multiplication rule: Multiply together the separate probabilities of several independent outcomes to find the probability that these outcomes will occur together. (12.3)

Mutually exclusive outcomes: Outcomes that can't occur together. (12.2)

Negative relationship: Occurs insofar as pairs of observations tend to occupy dissimilar and opposite relative positions in their respective distributions. (9.1)

Negatively skewed distribution: A distribution that includes a few extreme observations with relatively small values in the negative direction. (3.4)

Nominal measurement: Sorts observations into different classes or categories. (B.2)

Nondirectional test: See *Two-tailed test*.

Nonparametric tests: Tests, such as U, T, and H, that evaluate entire population distributions rather than specific population characteristics. (25.2)

Normal curve: A theoretical curve noted for its symmetrical bell-shaped form. (6.1)

Null hypothesis (H_0): A statistical hypothesis that usually asserts that nothing special is happening with respect to some characteristic of the underlying population. (14.5)

Observed frequency (f_0): The obtained frequency for each category. Used with the chi-square test. (24.3)

One-tailed (or directional) test: Rejection region is located in just one tail of the sampling distribution. (15.4)

One-way ANOVA: The simplest type of analysis of variance where population means differ only with respect to one dimension or factor. (22.1)

One-way χ^2 test: Evaluates whether observed frequencies for a single qualitative variable are adequately described by hypothesized or expected frequencies. (24.1)

Ordinal measurement: Arranges observations in terms of order. (B.3)

Outlier: A very extreme observation. (2.7)

Percentile rank of an observation: Percentage of observations in the entire distribution with similar or smaller values than that observation. (2.11)

Pearson correlation coefficient (r): A number between -1 and 1 that describes the linear relationship between pairs of quantitative variables. (9.3)

Point estimate: A single value that represents some unknown population characteristic, such as the population mean. (17.2)

Pooled variance estimate (s_p^2): The most accurate estimate of the variance (assumed to be the same for both populations) based on a combination of two sample variances. (19.12)

Population: Any complete set of observations. (4.3)(11.2)

Population mean (μ): The balance point for a population, found by dividing the total value of all observations in the population by the population size. (4.3)

Population size (*N*): The total number of observations in the population. (4.3)

Positively skewed distribution: A distribution that includes a few extreme observations with relatively large values in the positive direction. (3.4)

Positive relationship: Occurs insofar as pairs of observations tend to occupy similar relative positions in their respective distributions. (9.1)

Power (1 − *β*): The probability of detecting a particular effect. (16.8)

Power curves: Cross-reference the likelihood of detecting any possible effect with different sample sizes. (16.8)

Probability: The proportion or fraction of times that a particular outcome is likely to occur. (12.1)

***p*-value:** The degree of rarity of a test result, given that the null hypothesis is true. (21.2) Also referred to as Sig. (Significance) by some statistical software programs, such as SPSS. (21.7)

Qualitative data: A set of observations where any single observation is a word or code that represents a class or category. (1.6)

Quantitative data: A set of observations where any single observation is a number that represents an amount or count. (1.5)

Random error: The combined effects (on the scores of individual subjects) of all uncontrolled factors. (22.2)

Random sample: A sample produced when all potential observations in the population have equal chances of being selected. (11.4)

Range: The difference between the largest and smallest observations. (5.2)

Ratio measurement: Locates observations along a scale having a true zero. (B.5)

Relative frequency distribution: A frequency distribution showing the frequency of each class as a part or fraction of the total frequency for the entire distribution. (2.8)

Repeated measures: Whenever the same subject is measured more than once. (20.9)

Research hypothesis: Usually identified with the alternative hypothesis, this is the informal hypothesis or hunch that inspires the entire investigation. (14.6) See *Alternative hypothesis.*

Sample: Any subset of observations from a population. (4.3)(11.3)

Sample mean ($\overline{X}$): The balance point for the sample, found by dividing the total value of all observations in the sample by the sample size. (4.3)

Sample size (*n*): The total number of observations in the sample. (4.3)

Sample standard deviation (*s*): The version of the sample standard deviation, with $n - 1$ in its denominator that is used to estimate the unknown population standard deviation. (18.8)

Sampling distribution of mean: The probability distribution of means for all possible random samples of a given size from some population. (13.1)

Sampling distribution of *t*: The distribution of t values that would be obtained if a value of t were calculated for each sample mean among all possible random samples of a given size from some population. (18.2)

Sampling distribution of *z*: The distribution of z values that would be obtained if a value of z were calculated for each sample mean among all possible random samples of a given size from some population. (14.2)

Sampling distribution of $\overline{X}_1 - \overline{X}_2$: Differences between sample means based on all possible pairs of random samples. (19.5)

Scatterplot: A special graph containing a cluster of dots that represents all pairs of observations. (9.2)

Scheffé's test: A multiple comparison test that, regardless of the number of comparisons, never permits the cumulative probability of at least one type I error to exceed the specified level of significance. (22.16)

Sig: See *p-value.*

Simple effect: The effect of one factor at a single level of another factor. (23.4)

Squared correlation coefficient (r^2): The proportion of total variance in one variable that is predictable from its relationship with the other variable. (10.8)

Squared Cramèr's phi coefficient (ϕ_c^2): A very rough estimate of the proportion of predictability between two qualitative variables. (24.14)

Squared curvilinear correlation (η^2): The proportion of variance in the dependent variable that can be explained by the independent variable. (22.14)

Squared point biserial correlation (r_{pb}^2): The proportion of variance in the dependent variable that can be explained by the independent variable. (21.9)

Standard deviation: A rough measure of the average amount by which observations deviate from their mean. (5.5)

Standard error of the mean ($\sigma_{\bar{X}}$): Being the standard deviation of the sampling distribution of the mean, it's a rough measure of the average amount by which sample means deviate from the population mean. (13.5)

Standard error of $\bar{X}_1 - \bar{X}_2$: A rough measure of the average amount by which any difference between sample means deviates from the difference between population means. (19.7)

Standard error of prediction ($S_{y|x}$): A rough measure of the average amount of predictive error. (10.6)

Standard normal curve: The one tabled normal curve with a mean of 0 and a standard deviation of 1. (6.4)

Standard score: Any score expressed relative to a known mean and a known standard deviation. (8.2)

Statistical significance: Not an indication of importance, but merely that the null hypothesis is probably false. (21.8)

Stem and leaf display: A device for sorting quantitative data on the basis of leading and trailing digits. (3.3)

Sum of squares (SS): The sum of squared deviations of some set of scores about their mean. (22.8)

t ratio: A replacement for the z ratio whenever the unknown population standard deviation must be estimated. (18.4)

t test for a population mean: A test to determine whether the sample mean qualifies as a common or rare outcome under the null hypothesis. This test requires the estimation of the population standard deviation from the sample. (18.10)

t test for two population means (independent samples): A test to determine whether the difference between sample means qualifies as a common or rare outcome, given that the two samples are independent. (19.13) See *Two independent samples.*

t test for two population means (matched samples): A test to determine whether sample mean difference qualifies as a common or a rare outcome under the null hypothesis, given that the two samples are matched. (20.13) See *Two matched samples.*

Transformed standard score (z'): A standard score that, unlike a z score, usually lacks negative signs and decimal points. (8.3)

Treatment effect: The existence of at least one difference between population means categorized by the independent variable. (22.2)

True sampling distribution: Centered about the true population mean, this distribution produces the one observed mean (or z). (16.4)

Two independent samples: Observations in one sample aren't paired, on a one-to-one basis, with observations in the other sample. (19.2)

Two matched samples: Each observation in one sample is paired, on a one-to-one basis, with a single observation in the other sample. (20.2)

Two-tailed (or nondirectional) test: Rejection regions are located in both tails of the sampling distribution. (15.4)

Two-way ANOVA: A more complex type of analysis of variance where population means differ with respect to two dimensions or factors. (23.1)

Two-way χ^2 test: Evaluates whether observed frequencies reflect the independence of two qualitative variables. (24.8)

Type I error: Rejecting a true null hypothesis. (16.2)

Type II error: Retaining a false null hypothesis. (16.2)

Ungrouped data: Observations organized into classes of single values. (2.1)

Unit of measurement: The smallest possible difference between scores within a particular set of data. (2.5)

Variable: A characteristic or property that can take on different values. (1.8)

Variability between groups (in ANOVA): Variability among scores of subjects who, being in different groups, receive different experimental treatments. (22.2)

Variability within groups (in ANOVA): Variability among scores of subjects who, being in the same group, receive the same experimental treatment. (22.2)

Variance: The mean of all squared deviations from the mean. (5.3)

Variance estimate (in ANOVA): See *Mean square*.

Variance interpretation of r^2: The proportion of variance predictable from the existing correlation. (10.8)

Wilcoxon T test: A test for ranked data when there are two matched groups. (25.11)

X-axis: Abscissa, or horizontal axis of any graph. (3.1)

Y-axis: Ordinate, or vertical axis of any graph. (3.1)

z score: A score that indicates how many standard deviations an observation is from the mean of the distribution. (6.3)

z test for a population mean: A hypothesis test that evaluates how far the observed sample mean deviates, in standard error units, from the hypothesized population mean. (14.2)

Formulas

Formulas are listed in their order of appearance in the book. Page numbers specify where formulas are introduced and defined.

DESCRIPTIVE STATISTICS   PAGE

SAMPLE MEAN	$\bar{X} = \dfrac{\Sigma X}{n}$	66
POPULATION MEAN	$\mu = \dfrac{\Sigma X}{N}$	67
SAMPLE STANDARD DEVIATION (DEFINITION FORMULA)	$S = \sqrt{\dfrac{\Sigma(X - \bar{X})^2}{n}}$	84
POPULATION STANDARD DEVIATION (DEFINITION FORMULA)	$\sigma = \sqrt{\dfrac{\Sigma(X - \mu)^2}{N}}$	85
SAMPLE STANDARD DEVIATION (COMPUTATION FORMULA)	$S = \sqrt{\dfrac{n\Sigma X^2 - (\Sigma X)^2}{n^2}}$	85
z SCORE	$z = \dfrac{X - \mu}{\sigma}$	98
CONVERTING z SCORE TO ORIGINAL SCORE	$X = \mu + (z)(\sigma)$	115
TRANSFORMED STANDARD SCORE	$z' =$ desired mean $+\ (z)($desired stand. dev.$)$	126

CORRELATION

CORRELATION COEFFICIENT (z SCORE FORMULA)	$r = \dfrac{\Sigma z_X z_Y}{n}$	144
CORRELATION (COMPUTATION FORMULA)	$r = \dfrac{n\Sigma XY - (\Sigma X)(\Sigma Y)}{[\sqrt{n\Sigma X^2 - (\Sigma X)^2}][\sqrt{n\Sigma Y^2 - (\Sigma Y)^2}]}$	148

PREDICTION

LEAST SQUARES PREDICTION EQUATION	$Y' = bX + a$	164
SOLVING FOR b	$b = \dfrac{S_Y}{S_X}(r)$	165

PAGE

SOLVING FOR *a*	$a = \overline{Y} - b\overline{X}$	165
STANDARD ERROR OF PREDICTION	$S_{Y|X} = S_Y\sqrt{1 - (r)^2}$	168

INFERENTIAL STATISTICS

ADDITIONAL RULE FOR MUTUALLY EXCLUSIVE OUTCOMES	$Pr(A \text{ or } B) = Pr(A) + Pr(B)$	199
MULTIPLICATION RULE FOR INDEPENDENT OUTCOMES	$Pr(A \text{ and } B) = [Pr(A)][Pr(B)]$	201
MEAN OF SAMPLING DISTRIBUTION	$\mu_{\overline{X}} = \mu$	217
STANDARD ERROR OF MEAN	$\sigma_{\overline{X}} = \dfrac{\sigma}{\sqrt{n}}$	219
z RATIO FOR SINGLE POPULATION MEAN	$z = \dfrac{\overline{X} - \mu_{\text{hyp}}}{\sigma_{\overline{X}}}$	232
CONFIDENCE INTERVAL FOR μ (BASED ON *z*)	$\overline{X} \pm (z_{\text{conf}})(\sigma_{\overline{X}})$	280

t TESTS

t RATIO FOR SINGLE POPULATION MEAN	$t = \dfrac{\overline{X} - \mu_{\text{hyp}}}{s_{\overline{X}}}$	294
CONFIDENCE INTERVAL FOR μ (BASED ON) *t*	$\overline{X} \pm (t_{\text{conf}})(s_{\overline{X}})$	295
SAMPLE STANDARD DEVIATION FOR INFERENTIAL STATISTICS (DEFINITION FORMULA)	$s = \sqrt{\dfrac{\Sigma(X - \overline{X})^2}{n - 1}}$	298
SAMPLE STANDARD DEVIATION FOR INFERENTIAL STATISTICS (COMPUTATION FORMULA)	$s = \sqrt{\dfrac{n\Sigma X^2 - (\Sigma X)^2}{n(n - 1)}}$	298

PAGE

ESTIMATED STANDARD ERROR	$s_{\overline{X}} = \dfrac{s}{\sqrt{n}}$	299
t RATIO FOR TWO POPULATION MEANS (TWO INDEPENDENT SAMPLES)	$t = \dfrac{(\overline{X}_1 - \overline{X}_2) - (\mu_1 - \mu_2)_{\text{hyp}}}{s_{\overline{X}_1 - \overline{X}_2}}$	312
CONFIDENCE INTERVAL FOR $\mu_1 - \mu_2$ (TWO INDEPENDENT SAMPLES)	$\overline{X}_1 - \overline{X}_2 \pm (t_{\text{conf}})(s_{\overline{X}_1 - \overline{X}_2})$	314
ESTIMATED STANDARD ERROR (TWO INDEPENDENT SAMPLES)	$s_{\overline{X}_1 - \overline{X}_2} = \sqrt{\dfrac{s_p^2}{n_1} + \dfrac{s_p^2}{n_2}}$	317
POOLED VARIANCE ESTIMATE	$s_p^2 = \dfrac{(n_1 - 1)s_1^2 + (n_2 - 1)s_2^2}{(n_1 - 1) + (n_2 - 1)}$	317
DIFFERENCE SCORE	$D = X_1 - X_2$	327
t RATIO FOR TWO POPULATION MEANS (TWO MATCHED SAMPLES)	$t = \dfrac{\overline{D} - \mu_{D\text{hyp}}}{s_{\overline{D}}}$	328
CONFIDENCE INTERVAL FOR μ_D (TWO MATCHED SAMPLES)	$\overline{D} \pm (t_{\text{conf}})(s_{\overline{D}})$	330
ESTIMATED STANDARD ERROR (TWO MATCHED SAMPLES)	$s_{\overline{D}} = \dfrac{s_D}{\sqrt{n}}$	336
SAMPLE STANDARD DEVIATION (DIFFERENCE SCORES)	$s_D = \sqrt{\dfrac{n\Sigma D^2 - (\Sigma D)^2}{n(n-1)}}$	336
t RATIO FOR SINGLE POPULATION CORRELATION COEFFICIENT	$t = \dfrac{r - \rho_{\text{hyp}}}{\sqrt{\dfrac{1 - r^2}{n - 2}}}$	339
PROPORTION OF EXPLAINED VARIANCE (TWO SAMPLES)	$r_{\text{pb}}^2 = \dfrac{t^2}{t^2 + df}$	356

ANOVA

F RATIO	$F = \dfrac{\text{variability between groups}}{\text{variability within groups}}$	366

MEAN SQUARE: GENERAL EXPRESSION	$MS = \dfrac{SS}{df}$	370

| SUMS OF SQUARES (ONE WAY) | $SS_{total} = SS_{between} + SS_{within}$ | 372 |

| MEAN SQUARE BETWEEN GROUPS | $MS_{between} = \dfrac{SS_{between}}{df_{between}}$ | 374 |

| MEAN SQUARE WITHIN GROUPS | $MS_{within} = \dfrac{SS_{within}}{df_{within}}$ | 375 |

| F RATIO (ONE WAY) | $F = \dfrac{MS_{between}}{MS_{within}}$ | 375 |

| PROPORTION OF EXPLAINED VARIANCE (ONE-WAY ANOVA) | $\eta^2 = \dfrac{SS_{between}}{SS_{total}}$ | 380 |

| SCHEFFÉ'S CRITICAL VALUE | $(\overline{X}_i - \overline{X}_j)_{crit} =$ $\pm \sqrt{(df_{between})(F_{crit})(MS_{within})\left(\dfrac{1}{n_i} + \dfrac{1}{n_j}\right)}$ | 383 |

| SUMS OF SQUARES (TWO WAY) | $SS_{total} = SS_{column} + SS_{row} + SS_{interaction} + SS_{within}$ | 401 |

χ^2 TEST

EXPECTED FREQUENCY (ONE WAY)	$f_e = $ (expected proportion) (total sample size)	418

| χ^2 RATIO | $\chi^2 = \Sigma \dfrac{(f_o - f_e)^2}{f_e}$ | 419 |

| EXPECTED FREQUENCY (TWO WAY) | $f_e = \dfrac{\text{(column total)(row total)}}{\text{overall total}}$ | 427 |

| PROPORTION OF PREDICTABILITY (TWO-WAY χ^2) | $\phi_c^2 = \dfrac{\chi^2}{N(k-1)}$ | 432 |

RANKED DATA

<table>
<tr>
<td>MANN-WHITNEY U TEST
(TWO INDEPENDENT
SAMPLES)</td>
<td>$$U_1 = n_1 n_2 + \frac{n_1(n_1 + 1)}{2} - R_1$$

$$U_2 \doteq n_1 n_2 + \frac{n_2(n_2 + 1)}{2} - R_2$$

U = the smaller of U_1 or U_2</td>
<td>446</td>
</tr>
<tr>
<td>WILCOXON T TEST
(TWO MATCHED SAMPLES)</td>
<td>T = the smaller of R_+ or R_-</td>
<td>451</td>
</tr>
<tr>
<td>KRUSKAL-WALLIS H TEST
(THREE OR MORE
INDEPENDENT SAMPLES)</td>
<td>$$H = \frac{12}{n(n + 1)} \left[\sum \frac{R_i^2}{n_i} \right] - 3(n + 1)$$</td>
<td>456</td>
</tr>
</table>

Index

A

Addition rule. *See Probability*
Alpha (α) error. *See Type I error.*
Alternative hypothesis, 235, 309–310, 327–328. *See also Hypothesis.*
Analysis of variance,
   alternative hypothesis, 368, 397
   ANOVA tables, 377, 405
   assumptions, 369, 400
   degrees of freedom, 372–374, 404–405
   effect size, 379–380, 406–407
   F ratio, 366, 376, 395–396
   F test, 366–367
   interaction, 394, 396–400
   interpretation with graphs, 393–394
   meaning of, 366–368
   mean squares, 370, 371–374, 405
   multiple comparisons, 381, 383–384
   null hypothesis, 362, 397
   one-way test, 362
   other types, 369, 401
   published reports, 377–378
   sample size selection, 379, 401
   Scheffe's test, 381–383, 407
   squared curvilinear correlation, 379–380, 406–407
   sum of squares, 371–372, 401–404
   tables, 376–377, 406
   two-way test, 392–397
   variability between groups, 364–365, 396
   variability within groups, 364–365, 396
   variance estimates, 369–370, 401
ANOVA, 362
Arithmetic mean. *See Mean.*
Average(s),
   and skewed distributions, 68–69
   common usage, 71–72
   for qualitative data, 70–71
   for quantitative data, 62–67
   mean, 65–67
   median, 63–64
   mode, 62
   which?, 68

B

Bar graph, 51
Beta (β) error. *See Type II error.*
Bimodal distribution, 49–50, 62

C

Central limit theorem, 220–222
Central tendency, measures of, 62
Chi square,
   alternative hypothesis, 416, 425
   degrees of freedom, 419–421, 427
   expected frequencies, 418, 425–427
   formula, 419
   null hypothesis, 416–417, 425
   one-way test, 416, 420, 466
   p-values, 430–431
   precautions, 431–432
   published reports, 430–431
   sample size selection, 432
   squared Cramer's coefficient, 432–433
   tables, 420, 427
   two-way test, 424–425, 427, 466
Class intervals, 21–24
Cohen, J., 270, 357, 380, 433
Cohen's rule of thumb,
   for analysis of variance, 380, 407
   for chi square, 433
   for *t* test, 357
Common outcome, 117, 203–204, 229
Complex prediction equations, 178
Computer printouts, 55–57
   Minitab, 56–57, 384, 435
   SAS, 152–153, 353, 408–409, 434
   SPSS, 152–153, 353–354, 385, 407–408
Conditional probability, 202–203
Confidence, level of, 281
Confidence interval,
   and effect of sample size, 282
   compared to hypothesis test, 283–284, 315
   defined, 276
   false, 278

Confidence interval (*cont.*)
  for difference between population means,
    matched samples, 329–331
    independent samples, 314–316
  for single population mean, 295–297
  for population percent, 284–285
  interpretation of, 281
  other types of, 286
  true, 278
Conover, W., 446, 458
Correlation,
  and cause-effect, 9, 143–144
  and outliers, 149–151
  and r^2, 170–175
  coefficient, Cramer's phi, 151, 432–434
    Pearson r, 140–142
    point biserial, 151
    squared, 355–357
    population, 338–341
    Spearman rho, 151
  hypothesis test for, 338–341
  matrix, 151–154
  other types of, 151
  meaning of, 141–142, 147, 176
  formulas, 144, 148
  scatterplots for, 136–139
  study, 9
Counterbalancing, 333
Critical z scores, 236
  table of, 253
Cumulative frequency distribution,
  for qualitative data, 31–32
  for quantitative data, 28–29
Curvilinear relationship, 139

D

Data,
  defined, 4
  grouped, 19–20
  overview, 4–6
  qualitative, 5
  quantitative, 5
  ungrouped, 18
Decision rule, 236–237
Degrees of freedom,
  defined, 299–301
  in analysis of variance, 372–374, 404–405
  in chi square, 419–421, 427
  in correlation, 339

  in one sample, 299–302
  in two samples, 318, 328, 339
Dependent outcomes, 201–202
Dependent variable, 7–9
Descriptive statistics, 15–181
  compared to inferential statistics,
  defined, 3
Deviations from mean, 67, 78
Difference,
  between population means, 309
  between sample means, 311–312
Difference score, 326–327
Directional and nondirectional tests, 247–250
Distribution,
  bimodal, 49
  multimodal,
  normal, 47, 95–97
  sampling, 212–213
  shape of, 49–51
  skewed, negatively, 50
    positively, 50
Distribution-free tests, 442

E

Effect, 263. *See also Main effect.*
Effect size, 263–270, 354–355, 406–407
Error,
  prediction, 168–170
  random, 365–367
  statistical and nonstatistical, 285
  type I, 260
  type II, 260
Estimate,
  interval. *See Confidence interval.*
  point, 276
Expected frequency, 418, 425–427
Explained variance. *See Variance interpretation of r^2.*
Experiment, 8, 143–144, 192–193

F

F test
  for means, 366–367, 395, 396. *See also Analysis of variance.*
  for variances, 353
False alarm, 260. *See also Type I error.*
Fisher's r-to-z transformation, 341
Frequency distribution, 18
  cumulative, 28–29
  for qualitative data, 31–32,

for quantitative data, 18–24
  grouped, 19–24
  ungrouped, 18
gaps between boundaries, 21–22
guidelines for constructing, 20
interpreting, 33
relative, 27–28
Frequency polygon, 43–45

G

Gosset, W., 291
Graphs,
  constructing, 53–55
  for qualitative data, 51–52
  for quantitative data, 42–48
  misleading, 52–53
Greek letters, significance of, 215, 217

H

H (Kruskall-Wallis) Test,
  and ties, 458
  as replacement for F, 454
  calculation, 455–456
  decision rule, 457
  degrees of freedom, 457
  tables, 456–457
Histogram, 42–43
Homogeneity of variance, assumption of, 317,
    316–317, 353, 380, 400
Homoscedasticity, assumption of, 170
Howell, D., 294, 316–317, 353, 400
Huff, D., 52
Hypothesis,
  alternative, 235
    choice of, 247, 250, 309–310, 327
    directional, 247–250
    nondirectional, 247–248
  null,
    defined, 234–235
    secondary status of, 247
  research, 235, 246–247
Hypothesis tests,
  and four possible outcomes, 259–260
  and p-values, 351
  and step-by-step procedure, 233–238
  compared to confidence intervals, 283–284, 315
  for qualitative data. *See Chi square.*
  for quantitative data. *See F, t, and z tests.*
  for ranked data. *See H, T, and U tests.*

  guidelines for selecting, 465
  less structured approach to, 348, 350–351
  meaning of, 244–246
  published reports of, 352, 377–378, 430–431

I

Importance, checking, 355, 379–380, 432–433
Independent,
  observations, 431–432
  outcomes, 201
  samples, 308
  variable, 7–8
Inferential statistics, 182–461
  compared to descriptive statistics, 464
  defined, 3
Interaction, 394, 396–400
Internet demonstrations,
  confidence intervals, 283
  experiments, 335
  explained variance, 176
  guessing correlations, 141
  histogram, 54
  hypothesis test, 253
  mean and median, 69
  one-way chi-square, 423
  outliers, 151
  power, 270
  prediction, 164
  sampling distributions, 223
  two-way chi-square, 433
Internet sites,
  Gallup Poll, 188
  Minitab, 56
  SAS, 56
  SPSS, 56
  U.S. Census Bureau, 187
Interquartile range, 87–88
Interval measurement, 486
  approximating, 487–488
Interval estimate. *See Confidence interval.*

K

Keppel, G., 400, 407
Kruskall-Wallis test. *See H test.*

L

Least squares prediction, 163–167
Levels of measurement. *See Measurement.*

Level of significance,
  and *p*-values, 351
  as Type I error, 262
  choice of, 252–253
  defined, 237
Levene's test, 353–354
Linear relationship, 139
Lopsided distribution. *See Skewed distribution.*
LSD Test, 384

M

Main effect, 393
Mann-Whitney test. *See U test.*
Margin of error, 284–285
Matched samples, 326
Matching subjects, 326, 331–332
Math background, 10
Mean,
  and skewed distributions, 68–69
  as balance point, 67, 301
  as measure of position, 81–82
  for qualitative data, 71
  for quantitative data, 65–67
  of difference scores, 328
  of population, 67
  of sample, 66
  of sampling distribution of mean, 217–218
  sampling distribution of, 212–215
  special status of, 70
  standard error of, 218–220
Mean absolute deviation, 78–79
Mean squares, 370, 374–375, 405
Measurement
  and type of data, 484
  approximating interval, 487–488
  levels of,
    interval, 486
    nominal, 485
    ordinal, 485
    ratio, 486–487
  problem of, 487
Median,
  for qualitative data, 70–71
  for quantitative data, 63–64
Minitab printouts, 56–67, 385, 435
Minium, E., 270
Miss, 260. *See also Type II error.*
Mode,
  and bimodal distributions, 62
  and multimodal distributions, 62

  for qualitative data, 70
  for quantitative data, 62
Multimodal distribution, 62
Multiple comparisons, 81
  with Scheffé's test, 381–383, 407
  other tests of, 383–384
Multiple regression equations, 178
Multiplication rule. *See Probability.*
Mutually exclusive outcomes, 199

N

Negatively skewed distribution, 50, 68–69
Negative relationship, 134, 146
Nominal measurement, 485
Nondirectional test and directional test, 247–250
Nonparametric tests, 442, 468
Normal bivariate population, 340–341
Normal distribution,
  and central limit theorem, 220–222
  and percentile ranks, 126, 128
  and *z* scores, 97–99
  compared to *t* distribution, 292, 297
  general properties, 96–97
  problems, finding proportions, 106–112
    finding scores, 112–117
    guidelines for solving, 118
  standard, 98–99, 292
  tables, 99–101
Null hypothesis, 234–235, 250–251, 309, 327, 362, 416–417, 425, 443, 451, 454–436
Numerical codes, 6

O

Observed frequency, 425
Omega squared, 357, 380
One-tailed and two-tailed tests, 247–251
Ordinal measurement, 485
Outcomes, common and rare, 149–150, 203–205, 229–230
Outlier, 26, 149–151
Overview,
  data, 4–6, 465–467
  hypothesis tests, 244–246, 295
    guidelines to, 465
  hypothesis tests or confidence intervals, 283–284, 464
  surveys or experiments, 194
  three *t* tests, 333–335, 468
  one-and two-way ANOVA, 395–396, 468

P

p-value,
   and level of significance, 351
   approximate, 349
   defined, 348
   exact, 350
   finding, 349–350
   merits of, 350–351
   published reports of, 350, 352
   reported by others, 350
Parameter, 442
Parametric test, compared to nonparametric test, 442
Pearson, K., 140
Pearson r, 140–142. *See also Correlation.*
Percentile ranks, 30–31, 128–129
Percents or proportions, 28
Placebo effect, 258
Point biserial correlation, 151
   squared, 355–356
Point estimate, 276
Pooled variance estimate, 317–318
Population,
   correlation coefficient, 338
   defined, 186
   hypothetical, 187, 308–309
   mean, 67
   mean of difference scores, 326–327
   percent, 284
   real, 186–187
   standard deviation, 85
Positive relationship, 137–138, 146
Positively skewed distribution, 50, 68–69
Power, 269
Power curves, 269–270, 357
Prediction,
   and more complex equations, 178
   and r^2, 170–175
   assumptions, 170
   equation, 163–167
   error, 168–170
   least squares, 164–167
   standard error of, 168–170
Probability,
   and addition rule, 198–200
   and multiplication rule, 200–201
   and statistics, 203–205
   as area under curve, 204
   conditional, 202–203
   defined, 198

Percent,
   population, 284
   sample, 284
Protected *t* test, 38

Q

Qualitative data,
   averages for, 70–71
   defined, 5
   frequency distributions for, 31–32
   graph for, 51
   hypothesis test for, 416–438
   measures of variability for, 83
   ordered, 31, 70–71, 83, 486
Quantitative data,
   averages for, 62–68
   defined, 5
   frequency distributions for, 18–24
   graphs for, 42–48
   hypothesis tests for, 227–393
   measures of variability for, 76–82

R

Random assignment of subjects, 192–193
Random error, 365–367
Random numbers, tables of, 190–191
Random sample,
   and hypothetical populations, 192
   defined, 188–189
   fish-bowl method for, 189
   tables of random numbers for, 190–191
Range, 77
Ranked data, 442–459
Ranks,
   assigning, 444, 451, 456
   critical ties in, 458
Rare outcome, 117, 204–205, 229–230
Ratio measurement, 486–487
Regression. *See Prediction.*
Rejecting null hypothesis, 246
Relationship between variables,
   curvilinear, 139
   linear, 139
   negative, 136, 137–138, 146
   perfect, 139
   positive, 135, 137–138, 146
   strength of, 139
Relative frequency distribution, 27–28

Repeated measures, 332–333
Reports, published, 352, 377, 430–431, 453
Research hypothesis, 235, 246–247
Research problem, 234
Retaining null hypothesis, 246

S

Sample(s),
    all possible, 212–213
    correlation coefficient, 338–339
    defined, 187
    mean, 66
    mean of difference scores, 328
    percent, 284
    standard deviation, 298
    random, 187–189
    variance, 317
Sample size,
    and probability of type II error, 266–268
    and standard error, 219, 267–268
    equality of, 316–317, 369, 401
    selection of,
        for confidence interval, 283–284
        for one sample, 268–269
        for two samples, 357
Sampling distribution,
    constructed from scratch, 213–215
    of difference between means,
        matched samples, 328
        independent samples, 311–312
    of F, 367–368
    of mean, defined, 212
        hypothesized, 228–230
        hypothesized and true, 263
        mean, 217–218
        shape, 220–222
        standard error, 218–220
    of t, 291–292
    of z, 230–232
    of χ^2, 421
    other types of, 223
Sampling variability of mean, 218–220
SAS printouts, 154, 353, 408–409, 434
Scatterplot, 136–139
Scheffé's test, 381–383, 407
Sig., 152. *Also see p-value.*
Significance, level of. *See Level of significance.*
Simple effect, 399
Skewed distribution, 50–51, 68–69

Spearman correlation coefficient, 151
SPSS printouts, 56, 152–153, 353–354, 385, 407–408
Square root, 79
Squared correlation coefficient. *See Variance*
    interpretation of r^2.
Squared curvilinear coefficient. *See Variance*
    interpretation of η^2.
Standard deviation,
    in descriptive statistics,
        and mean absolute deviation, 79
        and normal distribution, 96–97
        formulas, 83–85
        general properties, 79–81
        measure of distance, 81–82
    in inferential statistics, 87
        degrees of freedom, 299–302
        formulas, 298
Standard error,
    of difference between means,
        matched samples, 336
        independent samples, 317–318
    of mean, 218–220, 298–299
    of prediction, 168–170
Standard normal distribution, *See Normal distribution.*
Standard scores,
    general properties, 125–127
    T scores, 126–127
    transformed, 125–127
    z scores, 97–99, 124–125, 421
Statistics,
    descriptive, 3, 17–181
    inferential, 3, 183–461
Statistical significance, 354–355
Stem and leaf display, 46–48
"Student." *See Gosset, W.*
Sum of squares, 371–372, 401–404
Summation sign, 66

T

Test of hypothesis, *See Hypothesis test.*
Ties in ranks. *See Ranks.*
Treatment effect, 364, 395
t Test,
    and degrees of freedom, 299–302
    and F test, 381, 379
    and z test, 291, 297
    assumptions, 299, 311–312, 340–341
    expressed as ratio, 294, 312, 328
    for correlation coefficient, 338–341

for one population mean, 290–291
for two population means,
   matched samples, 329–337
   independent samples, 319–320
protected, 384
tables, 292–293
T Test (Wilcoxen),
   and ties, 458
   as replacement for t, 449–450
   calculation, 450–451
   decision rule, 453
   tables, 452
Two-tailed and one-tailed tests, 247–250
Type I error,
   and effect of multiple tests, 381
   defined, 260
   probability of, 261–262
Type II error,
   and difference between true and hypothesized
        population means, 263–266
   and sample size, 266–269
   defined, 260
   minimizing, 266–268
   probability of, 263–266

U

U Test (Mann-Whitney),
   and ties, 458
   as replacement for t, 443
   calculation, 444–446
   decision rule, 447–448
   tables, 447
Unexplained variance. *See Variance interpretation of r^2.*
Unit of measurement,
   and correlation, 147
   defined, 22

V

Variable
   defined, 7
   dependent, 8–9
   independent, 7–8
Variability,
   measures of, 76
      for qualitative data, 83

for quantitative data, 76–82
interquartile range, 87–88
mean absolute deviation, 78
range, 77
standard deviation,
   descriptive statistics, 79–87
   inferential statistics, 297–298
variance,
   descriptive statistics, 77–79
   inferential statistics, 317–318, 369–370, 397
Variance,
   defined, 78
   estimates of,
      in analysis of variance, 369–370, 397
      pooled, 317–318
   in descriptive statistics, 77–79
   weakness of, 78
Variance, homogeneity of. *See Homogeneity of
   variance.*
Variance interpretation
   and cause-effect, 177
   of r^2, 170–175, 355–356
   of η^2, 379–381, 406–407
   of ϕ_e^2, 432–433

W

Web site, for book, 10
Wilcoxen test. *See T test.*

Z

z Score,
   and correlation, 144–147
   and hypothesis tests, 231–232
   and normal distribution, 97–99
   and non–normal distributions, 124–125
   and other standard scores, 125–127
   critical, 236–237, 253
   defined, 97
z Test,
   compared to t test, 290, 297
   for population mean, 230–233
   for population percent, 284
   for two population means, 312
   tables of critical values, 253